The Correspondence of John Tyndall

VOLUME I

THE CORRESPONDENCE OF JOHN TYNDALL

VOLUME I

The Correspondence, May 1840–August 1843

Edited by
Geoffrey Cantor and Gowan Dawson

General Editors
James Elwick, Bernard Lightman, and Michael S. Reidy

UNIVERSITY OF PITTSBURGH PRESS

Published by the University of Pittsburgh Press, Pittsburgh, Pa., 15260

Manufactured in the United States of America
Printed on acid-free paper
10 9 8 7 6 5 4 3 2 1

Cataloguing-in-Publication data is available from the British Library

ISBN 13: 978-0-8229-4470-6
ISBN 10: 0-8229-4470-7

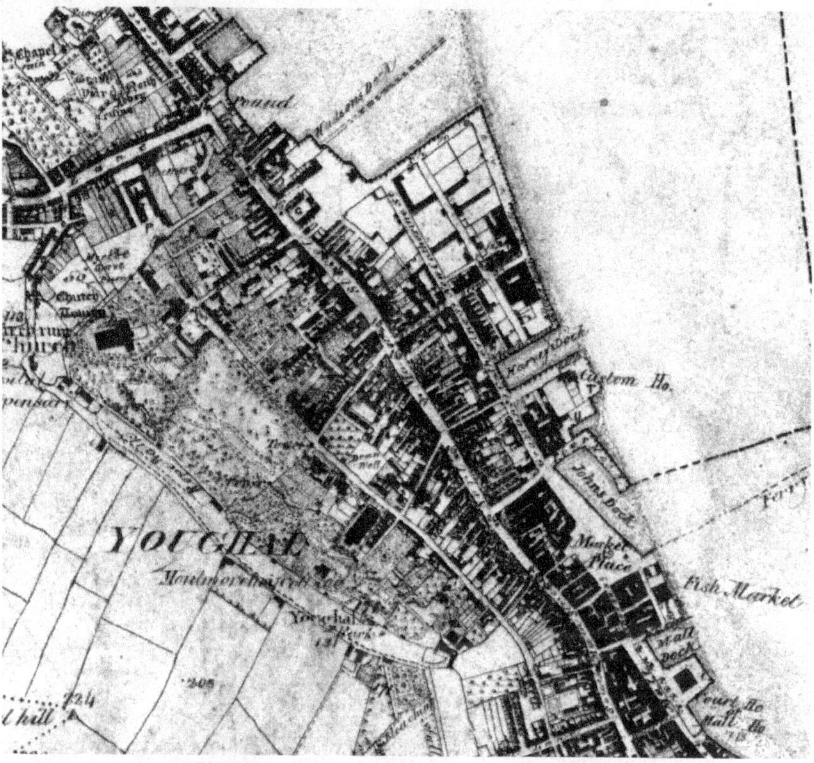

Frontispiece: Map of Youghal (detail) based on measurements provided by John Parker and drawn by Anthony Linedale, Phillip Evans, John Tyndall (who served in the drawing office at Youghal from May 1840 to June 1841) and other civil assistants. The manuscript includes the annotation: 'I certify that I have had this Plan examined, and now think it fit for publication. M A Waters. | Capn Royl Engineers | & Major | 30th Augt 1842', which suggests that it had undergone various revisions. This map thus appears to have been the final drawing for the 6 inch to the mile Irish Ordnance Survey map of the Youghal area, County Cork, which was published in 1842. Although reproduced here in black and white, in the original manuscript buildings are shown in red and some information, such as heights above sea level, is in either red or blue. Reproduced with the permission of the National Archives of Ireland (NAI OS/105C/853).

CONTENTS

CONTENTS

ACKNOWLEDGEMENTS

With such a large-scale, collaborative research project, the debt the general editors owe to researchers, funding agencies and institutions keeps growing. First, we are indebted to those who helped obtain Tyndall's letters. This includes the staff at the Royal Institution of Great Britain, especially Professor Frank James, Head of Collections, as well as Sir Roland Jackson, Visiting Fellow at the Royal Institution, and Archivist Jane Harrison. Approximately 1,400 further letters were obtained from the collections of the Alexander Turnbull Library; Alpine Club; American Philosophical Society; the Archiv der Berlin-Brandenburgischen Akademie der Wissenschaften; Berkshire Record Office; Bibliothèque National de France; British Library; Cambridge University; Columbia University; Deutsches Museum; Garrett Herman; Harvard University; Huntington Library; Imperial College London; Institution of Engineering and Technology; Lehigh University; London School of Economics; Martin Nevin; National Archives; National Library of Ireland; National Library of Scotland; National Portrait Gallery; National Research Council of Canada Archives; Natural History Museum, London; Oxford University; private collectors wishing to remain anonymous; Roland Jackson; Royal Botanic Gardens; Royal College of Physicians; Royal Dublin Society; Royal Geographical Society; Royal Greenwich Observatory; Royal Society; Royal Society of Chemistry; Roger Meyenberg; Science Museum, London; Sheffield Archives; Smithsonian Institution; Staatsbibliothek zu Berlin; St Bride Library; St John's College, Cambridge University; Trinity College Dublin; Universitatsbibliothek Braunschweig; University College London; University of Birmingham; University of Edinburgh; University of Glasgow; University of Heidelberg; University of Liverpool; University of London; University of Manchester; University of Newcastle; University of Reading; University of St Andrews; Uppsala University; Wellcome Library; West Sussex Record Office; Women's Library; Yale University; and York University.

Second, we thank all the scholars who have worked with graduate students who transcribed letters, or scholars who will be editing one of the volumes of the correspondence, or who will have done both, which include Melinda Baldwin, Ruth Barton, Geoffrey Belknap, William Brock, Janet Browne, Geoffrey Cantor,

Gowan Dawson, Miguel De Arce, Graeme Gooday, Piers Hale, John Lynch, Roy MacLeod, Norman McMillan, Iwan Rhys Morus, Elizabeth Neswald, Richard Noakes, Greg Radick, Simon Schaffer, Efram Sera-Shriar and Matthew Stanley. We also have been extremely fortunate to have the participation of hard working postdoctoral researchers who have been largely responsible for running the day-to-day operations of the project, including Melinda Baldwin and Joshua Howe.

The bulk of the transcriptions and translations of Tyndall's letters were undertaken by over 100 graduate and undergraduate students worldwide, sometimes moving between two or even three institutions during the life of the project. This included Michael Ainsworth, Marcella Agustina, Assunta Alegiani, Mark C. Ambrogio, Micah Anshan, Jenn Bazar, Poonam Bhalla, Steve Bunn, Céline Coppin, Joseph D'Addario, Dylan De Jong, Andrew Deneault, Justin Derry, Adrian Dimoff, James Finkelstein, Trinley Gongya, Zena Gopal, Juliana Goriounov, Chris Grafos, Alicia Grudzinskas, Victor Hainagiu, Avram Heisler, Matthew Hyland, Shannon Higson, Bernty Isopp, Alexandra Jocic, Kristina Karakolis, Matthew Kaufman, Matthew Kirshenblatt, Cynthia Klinowski, Rajeshkumar Macwan, Alan Mark, Rachel Mason Dentinger, Benjamin Mitchell, Daniela Monaldi, Sylvia Nickerson, Joseph Oryshak, Raymond Paglicauan, Martin Parrot, Sven Pinczewski, Lina Pinto Garcia, Evan Radford, Sheri Repucci, Kathryn Roberts, Nedra Rodrigo, Nicolas Sanchez Guerrero, Laurie Schell, Tatyana Shestakov, Kylie Stasila, Sean Steele, Fiona Stewart, Val Strawczynski, Christina Sulmona, Mary Sunderland, Amy Teitel, Allison Vandervalk, Benno Wilhelm, Noa Yaari, and David Zylberberg at York University; Michael Barton, Michael Conrad, Jake Rubow, Shannon Zera, Alexandre Manigault, Estella Terrazas, Jerry Jessee, Paige Madison, Dan Zizzamia, Clinton Colgrove and Craig Townsend at Montana State University; Louis Gerdelan, Jeremiah Rankin and Stephen Sequeira at the University of Auckland; Mike Finn, James Stark, Dominic Berry, Mark Steadman, Jonathan Averett, Adam Peardon, Thomas Brouwer at the University of Leeds; Sarah Swenson and Kate Sheppard at the University of Oklahoma; Geoff Belknap and Stephen Courtney at the University of Cambridge; James Lippard, Ashley Klott and Amy Shira Teitel at Arizona State University; Jeannie Warmington and Kathryn Gallagher at Aberystwyth University; James Lippard and Eric Adamson at NYU's Gallatin School of Individualized Study; Zach Meisel at Michigan State University; Kenneth Corbett, Monika Pries-Klassen, and Sarah Lea Harford at Brock University; Nazia Parveen at the University of Leicester; Ben Carver at the University of Exeter; and Julie Hipperson at Imperial College London. We want to make special mention of Efram Sera-Shriar, who worked on the project while he was a graduate student at York University and at Leeds University. Efram made a crucial contribution to the development of the transcribing system.

Funding from federal agencies in both Canada and the United States enabled the Tyndall Project to expand to four countries, incorporate the research expertise of over twenty scholars, and mentor and fund graduate students. We are extremely grateful for the funding we have received from the Mellon Foundation, a Partnership Development Grant and a Standard Research Grant through the Social Sciences and Humanities Research Council of Canada (SSHRC), and a Collaborative Research Grant, Division of Science and Technology Studies, at the National Science Foundation (award # 0924426). We are indebted to Janet Friskney, Research Officer in the Faculty of Liberal Arts and Professional Studies at York University, whose guidance in applying for grants was indispensable.

As the project gained momentum and visibility, scholars from around the world began to volunteer to help with different aspects of Tyndall's life and letters. This has included Roland Jackson at the Royal Institution; Ian Elliott, an astrophysicist at the Royal Dublin Society; Sarah Dry, a graduate of Cambridge University; Ciaran Toal, a PhD student at Queen's University Belfast; and Roger Meyenberg and Renato Arnold at the Kollegium Spiritus Sanctus (in Brig, Switzerland). Of special significance was the group of scholars in Ireland with a local interest in Tyndall, including Norman McMillan, Miguel De Arce and Martin Nevin.

We would like to express our appreciation to Pickering & Chatto Publishers, especially Mark Pollard and Stephina Clarke, for their unflagging support and for providing us with the benefit of their experience. They have been ideal partners for this ambitious correspondence project.

Though the project has been funded by federal agencies in both Canada and the United States, the project would not have been possible without additional support from our home institutions, particularly York University, Montana State University and Leeds University, which helped fund conferences and workshops focusing on Tyndall and the larger project. To all of our colleagues, who have patiently listened to us talk about John Tyndall for the past decade, we owe a debt of gratitude.

James Elwick, Bernard Lightman and Michael Reidy

FOREWORD

In 1867 John Tyndall inherited most of Michael Faraday's positions at the Royal Institution. This strengthened his already influential position in mid-Victorian science, culture and society. During the middle third of the nineteenth century, Faraday had developed and enhanced the Royal Institution's reputation, established by Humphry Davy early in the century, as the venue for excellent, well publicized and engaging scientific lectures and also as the location for fundamental research especially in natural and chemical philosophy.

Faraday held a high opinion of Tyndall and provided him with considerable personal and institutional support including helping with his appointment as Professor of Natural Philosophy at the Royal Institution in 1853. At first sight such support might seem surprising. Faraday and Tyndall held opposing views on a whole range of issues from religion, to whether force or matter was the primary feature of the universe, to the value of the royal family. Their social interactions were also markedly different: Tyndall, as his letters illustrate, was gregarious in line with the norms of the scientific community, whereas Faraday was notoriously unsociable; Faraday was firm but tactful in expressing his views, whilst Tyndall could make intemperate statements, for which Faraday would, on occasion, chide him. But they also held much in common. Both believed strongly in the educational value of science, which was why when opposing spiritualism in the 1850s they strategically rooted their criticisms in the educational system. In terms of constructing scientific knowledge both believed in the power and primacy of experiment. But here the generational gap between them is striking. Faraday was possibly the last major natural philosopher who could experimentally and theoretically construct, with his field theory of electro-magnetism, a powerful and original way of understanding the universe without the use of analytical mathematics. By Tyndall's time, a generation later, mathematics had become essential to understanding the world and Faraday's approach had come to be seen as old fashioned.

Nevertheless, Tyndall's lack of mathematics, compared with his rivals such as William Thomson (later Lord Kelvin) and James Clerk Maxwell, did not prevent him from making some major experimental discoveries and interpreting them in terms of molecular action. Starting with some very careful experiments on diamagnetism, which made his original scientific reputation, Tyndall's research covered a wide variety of different scientific disciplines. His discoveries included what we now call the greenhouse effect, understanding the motion of glaciers, falsifying the notion of spontaneous generation and explaining why the sky is blue. Furthermore, Tyndall published accessible books on topics such as the first biography of Faraday, as well as on heat, light, sound and glaciers, all of which went through innumerable editions and translations.

Much more significant in terms of both quality and quantity than anything that his friend Thomas Huxley did, such contributions to the construction and circulation of knowledge about the world should have ensured Tyndall a significant place in histories of nineteenth-century science. Yet aside from his Presidential Address to the British Association at its Belfast meeting in 1874 where he argued strongly for a pantheistic (which most contemporaries read as atheistic and materialistic) view of the world, Tyndall has, until fairly recently, been little studied, compared with many of his contemporaries. One of the reasons for this seems to be that he was not accorded a Victorian life and letters after his death. This was not for the want of trying by his widow, Louisa Tyndall, who more than twenty years his junior, lived until 1940. She devoted her forty-seven years of widowhood to assembling her husband's letters, papers, diaries, laboratory notebooks, newspaper cuttings and the like, having them typed up (alas then destroying some of the originals) in preparation for a major life and letters. This never happened and five years after her death the physicist Arthur Eve and the science teacher Clarence Creasey, together with the lawyer and civil servant Claud Schuster (who contributed the chapter on Tyndall's mountaineering exploits), published a rather superficial and unsatisfactory biography. Fortunately, all Tyndall's manuscripts and transcripts collected and made by Louisa Tyndall as well as her own papers were placed in the archives of the Royal Institution. The Tyndall collection, now kept in nearly one hundred boxes over about thirty shelf metres, forms the Royal Institution's third largest special collection after the papers of George Porter and Lawrence Bragg.

Tyndall wrote and received an enormous number of letters, many thousands of which have survived and have been located by this project. Their publication is a major addition to the documentation of nineteenth-century scientific practice which already exists in correspondences such as those of Faraday, Maxwell, Charles Darwin, Joseph Henry, André-Marie Ampère, Joseph Banks, Macedonio Melloni, Marc Pictet, Jöns Berzelius, Henry Fox Talbot and Ferdinand von Mueller, with Davy waiting in the wings, to name but the most obvious. The projects which publish scholarly editions of letters signify recognition of the historical importance of the specific individual involved. By including all correspondence it is possible to appreciate areas of significance that have previously been neglected. Faraday's and Tyndall's involvement with the British lighthouse service in its various institutional guises is a good example here. Furthermore, collections of all known letters permit the precise mapping of the social networks to which scientific practitioners belonged. When the letters are between practitioners themselves, this provides clues as to how various pieces of knowledge were constructed and also evidence of how careers were built. When letters were exchanged beyond the scientific community, then patterns of how knowledge and scientific ideology were circulated into wider society can be studied.

But whatever the status or content of a letter, they all share one feature and that is that the writer has to ensure that the recipient understands what he or she receives. Such meanings provide clues as to context and shared tacit knowledge (even if because of brevity modern editors cannot always appreciate that understanding) of the author and recipient. Finally, sources such as letters and diaries illustrate the sheer messiness of everyday life through which even the most distinguished figures lived and help us reconstruct the complex processes that are needed to produce some sort of ordered sense of the world. Tyndall's correspondence will add immeasurably to our understanding of that process and give him the recognition that his achievements, influence and place in Victorian society deserve.

FRANK A. J. L. JAMES

GENERAL INTRODUCTION

Though it would be natural for the introduction to John Tyndall's collected correspondence to begin with the exact date of his birth, we are unable to do so. We know that he was born at Leighlin Bridge, Ireland, on the 2nd of August, but the year is much more problematic. According to the older scholarship, and Louisa, the year is 1820. But recently there have been some doubts about this date, and 1822 seems more plausible. As Tyndall wrote to a close friend in 1857, 'It may be a year more or less for aught I know.'[1] That we didn't even realize that no one actually knew the date until the correspondence project began is suggestive of how much we will gain from a detailed analysis of his letters. While Tyndall's date of birth is elusive, the precise date of his death is not. He died on 4 December 1893, of an accidental overdose of chloral, a powerful narcotic he had been taking for insomnia. His wife administered it. Distraught by her fatal error, Louisa Tyndall spent the next forty-seven years collecting and transcribing much of her husband's correspondence with the intention of writing a definitive 'life and letters' of her husband. The task proved far too daunting, and by the time of her own death in 1940 she had published none of her husband's correspondence.

The basic facts of Tyndall's life seem to form the backbone of a proverbial rags-to-riches story. Born in Leighlin Bridge, County Carlow, Ireland, Tyndall's father John was a struggling shoemaker and leather dealer. In 1839 John Jr joined the Irish Ordinance Survey as a civil assistant, where he impressed his superiors with his drafting abilities and dexterity with numbers. In August 1842 he was transferred to the English Survey in Preston. When Tyndall protested about the inefficiency of the Survey's administration and its unfairness to the Irish assistants, he was dismissed in November 1843. The following July he found a position in a private surveyor's office in Preston. In August 1847 he accepted an appointment as teacher of mathematics at Queenwood College in Hampshire, but left after one year to earn his doctoral degree at Marburg University, Germany. He returned to England in 1851, taking up his old position at Queenwood while he looked for a scientific post. In 1853 he was elected Professor of Natural Philosophy of the Royal Institution, where he remained for the rest of his scientific career. Tyndall's climb up the social ladder from humble family circumstances was capped in spectacular fashion when, in 1867, he succeeded Michael Faraday as Superintendent of the Royal Institution.[2]

Of course, Tyndall's life cannot be summed up so simply. He lived several interconnected lives. The first was spent primarily in his laboratory at the Royal Institution; the second was either in front of fashionable audiences or behind the scenes, directing and shaping the sciences of the day; and the third he spent with mountain guides, suspended high on the sides of alpine cliffs.

In his first life, Tyndall became one of the most influential experimental physicists in the Victorian era. His delicate experiments on atmospheric phenomena in particular led to major discoveries. While nitrogen and oxygen, the main constituents of the atmosphere, were largely transparent to infrared radiation, he found that compound gases, such as carbon dioxide and water vapour, were relatively powerful absorbers of heat. The significance struck him immediately. In his Bakerian Lecture read to the Royal Society of London in 1861, he announced the first experimental confirmation of what is now known as the natural greenhouse effect.[3]

His ingenuity for setting up experiments was exemplified in all of his major research topics. He demonstrated how large particles in the atmosphere scattered light, experimentally showing why the sky is blue. He also showed how constant heating and cooling could rid liquids of microbes and their spores, leading to an improved method of sterilization now known as Tyndallization. What Tyndall called his 'Hours of Exercise in the Attic and the Laboratory' earned him the Rumford Medal of the Royal Society in 1869 and honourary membership in thirty scientific societies worldwide by the time of his death.

Tyndall's life as an experimental physicist overlapped with his second great life's work as one of the most outspoken champions of science in the nineteenth century. His public lectures at the Royal Institution introduced fashionable audiences to the latest discoveries in the fields of physics and chemistry. His accomplishments as a lecturer also supported his work behind the scenes, where he deftly defended science from its critics. He was a central figure of a larger cohort of influential scientists, which included Thomas Henry Huxley, Herbert Spencer and Joseph Dalton Hooker, who demanded that naturalistic explanations be used to explain the workings of nature. The closely knit group formed the X-Club in 1864 in an attempt to direct the course of British science and to lobby for its support. As part of his defence of science, Tyndall unflinchingly participated in the heady debates over evolution and a naturalistic worldview. Indeed, he is often remembered for two debates in particular. In 1872, he defended the need for an experimental verification of prayer, what became known as the 'Prayer-Gauge Debate'. Two years later, he gave the presidential address at the British Association for the Advancement of Science in Belfast, where he dramatically declared that men of science 'shall wrest from theology, the entire domain of cosmological theory'.[4]

Tyndall's third life was spent in the high altitudes of the Swiss Alps, where he travelled almost every summer from 1856 until his death in 1893. He was one of the figures responsible for establishing mountaineering as a sport during the 'golden age of mountaineering'. He climbed Mont Blanc three times and was the first to spend the night on its summit. He was also one of the first alpinists to study the approaches to the Matterhorn, and the first to turn the mountain into a pass, climbing the more difficult traverse up the Lion's Ridge from Italy and down the Hornli Ridge to Switzerland. He also made the first solo ascent of the Monte Rosa, Europe's second highest peak. Most 'amateurs' like Tyndall hired professional guides, so Tyndall's solo climbing was both bold and controversial. His crowning achievement in mountaineering took place on 19 August 1861, when he made the first successful ascent of the majestic Weisshorn, a solitary snow-covered peak in the Pennine Alps.

Yet these three lives of Tyndall's were inseparably mixed. They often fed off one another, with a discovery or event in one life sparking something in another. He gained prominence as an experimental scientist, a reputation he then used to defend science and shape its popular appeal; conversely, his popularizations of subjects often generated new research questions for him to revisit in the laboratory. As he climbed mountains and traversed glaciers he conducted experiments upon them, and then reciprocally used his fame as a scientist to popularize the burgeoning sport of mountaineering. Though it is analytically clarifying to speak of Tyndall's three lives, these three strands all wove together and enriched one another.

The Historical Importance of Correspondence

Tyndall's contemporaries believed that an individual's life could be narrated most authentically through the reproduction of large portions of his or her correspondence and other personal papers.[5] When substantial numbers of letters were incorporated into a biography the result was an important genre of writing known as the 'life and letters'. By the late Victorian era, many of the eminent scientists of the period had become the subject of a 'life and letters' written by a family member. Darwin's son Francis completed his *Life and Letters of Charles Darwin* in 1887. Later, in 1900, Huxley's son Leonard produced *The Life and Letters of Thomas Henry Huxley*. And it wasn't just male family members who undertook this task. Katharine Lyell, sister-in-law to the distinguished geologist Charles Lyell, edited *Life, Letters, and Journals of Sir Charles Lyell* (1881). Louisa Tyndall's wish to write the definitive 'life and letters' of her husband was not so unusual.

While valuable, such 'lives and letters' were often incomplete as historical records because family members were understandably concerned with protecting and consolidating the reputations of their loved ones. Such works still tended only to present the publicly constructed face of the eminent Victorian. For historians, the importance of a more extensive collection of correspondence resides in the insight it gives us into dimensions of scientific life and culture that cannot be found in other contemporary sources. It lets us get behind the scenes to observe historical actors when they have left the public stage. Through letters we can glimpse the sometimes hidden aspects of a scientist's personal life, whether it be their relationship to members of their family, their romantic entanglements, their closest friendships, or even their patrons. It is here that we can frequently observe their innermost feelings as they experience the ups and downs of a scientific life, or as they conduct unreported laboratory work that seemed to promise a new discovery but which ended up going nowhere. Letters often help the historian to reconstruct the intellectual and social forces that shape the thoughts and practices of the scientist. It is in correspondence that we find the scientist planning out his or her career strategies or working with allies to respond to an opponent. An examination of a scientist's correspondence undermines the ill-informed conception of scientists as isolated geniuses. Our interpretations of important scientific figures, including the significance of their achievements, are often revised in light of their correspondence.

While correspondence opens up the life of the scientist to the historians' gaze, it also reveals the structure of the scientific world in which the scientist lives. Letters can be seen as communication technologies that disclose the nature of the information network of which the scientist is a part.[6] The study of correspondence fits nicely into the current interest in the circulation of knowledge. Nineteenth-century scientists used correspondence as one of the main ways to collect, process and disseminate data. We have only to think about Darwin's extensive correspondence network, which he used both to gather scientific information from all corners of the globe and to subsequently publicize his discoveries. Writing and responding to letters was just as important for scientists as the time spent in the field or in the laboratory. Letters exchanged between scientists were not merely a daily fact of life, but, as Janet Browne argues, 'comprised a great deal of what it meant to be a man of science'.[7] Correspondence can shed light on hitherto unknown or obscure individuals, allowing us to better understand the role of 'invisible technicians' and 'subordinate laborers' in science at a particular time. Moreover it is becoming clear that letters were instrumental in the rise of the scientific journal and the introduction of the peer-review system.[8] In sum, correspondence is an invaluable resource into the personal life of the scientist, the community of practitioners and the nature of the scientific process.

Browne has noted that 'today is something of a golden age for the publication of major editions of correspondence'.[9] In the field of nineteenth-century British science numerous projects have recently been completed, including the correspondence of Joseph Banks, Michael Faraday and William Huggins.[10] More projects are currently underway, such as the correspondence of Charles Darwin and Alfred Russel Wallace.[11] In fact, we are just at the beginning of what will be a long process, as we have barely begun work on the correspondence of those figures widely seen as minor in the pantheon of scientific heroes. Even the letters of major figures, such as Darwin, have not been made available in their entirety. Scientific correspondence projects may be among the most vital forms of research for historians of science in the decades to come.

The Importance of Tyndall's Correspondence

Since little of Tyndall's correspondence has been available previously, it is a particularly crucial resource for scholars. Some of his correspondence was reproduced in Arthur Stewart Eve and Clarence Hamilton Creasey's *Life and Work of John Tyndall* (1945), but these letters were carefully chosen and had been heavily expunged of all things unpalatable. A catalogue produced in microfilm format of the Tyndall correspondence at the Royal Institution is part of James R. Friday, Roy M. MacLeod and Philippa Shepherd's *John Tyndall, Natural Philosopher* (1974). Subsequently, Microform Academic Publishers produced a microfilm edition of the Royal Institution letters, but over two thirds of those letters remain in their handwritten form and include no annotations. A few of Tyndall's letters appeared in various other published correspondences – Charles Darwin, T. H. Huxley and Michael Faraday, for instance – but the vast majority of Tyndall's epistolary bulk has been inaccessible to scholars in published form.

The lack of a standard *Life and Letters*, along with the inaccessibility of his correspondence, has obscured Tyndall's central place in Victorian science, culture and society. Humanities scholars working in areas such as cultural studies, gender studies, Irish studies and historians of mountaineering, sport and tourism will find much of interest. Tyndall communicated with a number of important cultural figures, including Thomas Carlyle, Alfred Lord Tennyson and John Ruskin. His letters while in Germany provide insight into the impact of German Romanticism and idealism on British thought. Letters pertaining to Tyndall's Irish origins, and his later opposition to Home Rule, offer interesting political commentary on the 'Irish question'. An opportunity exists for scholars of gender to explore the rhetoric of masculinity that runs through Tyndall's correspondence with his closest male friends, particularly those involved in alpinism.

Undoubtedly, however, the vast majority of the letters concern scientific issues. A list of Tyndall's correspondents reads like a 'who's who' of nineteenth-century science. Among his correspondents were George Biddell Airy, David Brewster, Robert Bunsen, George Busk, Rudolf Clausius, E. du Bois-Reymond, Charles Darwin, Michael Faraday, James David Forbes, Edward Frankland, Asa Gray, Hermann Helmholtz, John Herschel, Thomas Archer Hirst, J. D. Hooker, T. H. Huxley, James Joule, Charles Lyell, Julius Robert Mayer, Louis Pasteur, Lyon Playfair, Edward Sabine, Herbert Spencer, George Gabriel Stokes, J. J. Sylvester, and William Thomson. Scholars remain fascinated by these larger than life figures – Tyndall's friends and enemies, mentors and mentees. His correspondence will further illuminate their lives while highlighting Tyndall's seminal role as an intermediary in the broader developments taking place within Victorian science.

Tyndall's published correspondence therefore provides new insights into the development of such scientific disciplines as physics, glaciology, climatology and modern biology. It also illuminates the ways in which scientific expertise was used to achieve social status. Tyndall's time at the University of Marburg in Germany, where he studied under Robert Bunsen and obtained his PhD in 1850, is key to understanding how and why the laboratory became so important to Tyndall and his allies – and, thereafter, to Western science – furthering our approach to the laboratory/field border in the Victorian era. This formative period in Germany, his subsequent familiarity with the Continent, and his later tour of the United States generated an enormous network of friends and acquaintances among non-British and non-English-speaking scientists. The resulting letters show how Tyndall became an international broker of scientific work, and reveal patterns of how knowledge and scientific practices travelled (or sometimes failed to travel) between different countries, languages and cultures.

Because he spent most of his career at the Royal Institution, Tyndall's letters throw light upon one of the most influential scientific institutions of his period while adding to our understanding of popular lectures and audiences more generally. His appointment to a variety of government positions, including scientific advisor on lighthouses, offers the opportunity of examining the relationships between scientists and the state. And, of course, his spirited defence of naturalism and Darwinism throws further light on members of the X-Club, aiding our understanding of the debates and heated controversies surrounding science and religion. While we gain an intimate knowledge of Tyndall, we also gain insight into the broader scientific community during this significant period in the specialization, professionalization and popularization of science.

The History of the Tyndall Correspondence Project

In a very real sense, the Project began over one hundred years ago, when Louisa Tyndall began collecting her husband's letters. Without her work many of the letters would have remained dispersed and likely lost forever. It was Louisa who wrote to Tyndall's friends to retrieve them. Although Louisa never completed her 'life and letters', the largest collection of Tyndall letters (containing over 6,000 letters), now at the Royal Institution, is the result of her labour. Because Louisa's role was so crucial in preserving the core of the correspondence, we looked upon her as one of our colleagues and editors.

The current Project began in 2006. Bernard Lightman received funding from the Social Sciences and Humanities Research Council of Canada (SSHRC) in the spring of that year to support his research on a scholarly biography of Tyndall, which included the collection and transcription of some of his letters. It quickly became evident that the Tyndall correspondence represented a project in and of itself. Lightman hired a postdoctoral researcher, James Elwick, and several graduate students in August 2006 to begin the larger project of finding, collecting, digitizing and transcribing all of Tyndall's letters. The SSHRC grant paid for the production of electronic facsimiles of the original manuscripts housed at the Royal Institution.

Since the SSHRC grant was intended to fund the work of a Tyndall biography, only a portion of the grant could be devoted to the costs of digitization and transcription. Thus, Lightman sought and received further funding from the Mellon Foundation in June 2008. This funding enabled the York team to proceed with further digitization and transcription and to look beyond the Royal Institution collection for extant letters. More letters were found in the collections of the British Library, St Andrews University Library, University College London, the Royal Society, the Royal Botanic Gardens, Imperial College London, Cambridge University, the West Sussex Record Office, the Alpine Club, the Huntington Library, and at other locations throughout the world. Owing to the funding of SSHRC and the Mellon Foundation, the bulk of Tyndall's letters were located and digitized by the York team.

But the pace of transcribing was agonizingly slow. Michael Reidy proposed that we enlarge the transcription operation by applying for a Collaborative Research Grant through the National Science Foundation (NSF) in 2009. The funding enabled us to create an international, collaborative correspondence project. Scholars at eleven universities from around the world hired graduate students to undertake the preliminary transcription of letters. Along with York University and Montana State University, contributing institutions included Aberystwyth University (Iwan Morus), Arizona State University (John Lynch), University of Auckland (Ruth Barton), Brock University (Elizabeth Neswald), University of Cambridge (Simon Schaffer), University of Exeter (Richard Noakes), University of Leeds (Graeme Gooday), University of Leicester (Gowan Dawson), New York University (Matthew Stanley) and University of Oklahoma (Piers Hale).

We developed a multi-tiered process of transcription, whereby each letter underwent four progressively comprehensive stages by increasingly experienced graduate students. This process became the basis for the new, de-centralized, international system and it is the process we currently use today. Each letter begins as a collection of digital images, which may include the original letter, Louisa Tyndall's typed transcription, or both. After an initial straight word-for-word transcription, each successive stage adds further coherence and accuracy to the transcription. The most experienced scholars – known affectionately as the 'closers' – work on the final stage, and are the last to review the letters before they are put in a volume to be sent to co-editors for editing.

As the NSF grant drew to a close, Reidy organized a workshop for those who had indicated an interest in co-editing one of the early volumes. It took place in June 2012 at a dude ranch in Big Sky, Montana. Situated amidst mountains, this rustic setting was the perfect location for a meeting to work out, in a collaborative fashion, the editorial policy for the volumes, the publication schedule and to begin to assign specific volumes. We came out of that workshop with a definite plan on how to move from the transcription to the publication phase of the project. Each volume will have two co-editors, preferably with different areas of scholarly expertise, to better deal with the polymathic Tyndall. The first volume of letters was sent to the co-editors in February of 2013, to be published in January 2015. A new volume will be published every six months. In addition to the letters, each volume will contain an introduction by the co-editors, editorial notes, a timeline, a biographical register and an index by letter number.

The history of the project has been one of explosive growth and continual rejigging of its organization. We have been challenged by the scope of the project and by being located outside the countries where Tyndall lived and worked. However, these problems have been surmounted by the project being an international and collaborative one. Indeed it seems highly appropriate that such a system has emerged to publish the correspondence of a scientist who himself corresponded and collaborated so broadly across national borders.

Notes

1. Tyndall to Hooker, April 1857, RI MS JT/1/TYP/8/2543.
2. For short biographical accounts see: W. H. Brock, 'Tyndall, John (1820–1893)', in *Oxford Dictionary of National Biography*, ed. H. C. G. Matthew and B. Harrison (Oxford: Oxford University Press, 2004,) vol. 55, pp. 789–94; J. D. Burchfield, 'Tyndall, John (1820–93)', in *The Dictionary of Nineteenth-Century British Scientists*, ed. B. Lightman (Bristol: Thoemmes Continuum, 2004), vol. 4, pp. 2053–58.
3. J. Tyndall, 'On the Absorption and Radiation of Heat by Gases and Vapours, and on the Physical Connexion of Radiation, Absorption, and Conduction', *Philosophical Transactions of the Royal Society of London*, 151 (1861), pp. 1–36.
4. J. Tyndall, *Address delivered before the British Association assembled at Belfast. With Additions* (London: Longmans, Green, and Co., 1874), p. 197.
5. R. D. Altick, *Lives and Letters: A History of Literary Biography in England and America* (New York: Alfred A. Knopf, 1966), pp. 193–5.

6. J. Browne, 'Corresponding Naturalists', in *The Age of Scientific Naturalism*, ed. B. Lightman and M. Reidy (London: Pickering & Chatto, 2014), pp. 157–69.
7. Ibid., p. 159.
8. Ibid., p. 158.
9. Ibid., p. 169.
10. N. Chambers, *The Scientific Correspondence of Sir Joseph Banks, 1765–1820*, 6 vols (London: Pickering & Chatto, 2007); F. James, *The Correspondence of Michael Faraday*, 6 vols (London: Institution of Electrical Engineers, 1991–2011); B. Becker, *Selected Correspondence of William Huggins*, 2 vols (London: Pickering & Chatto, 2014).
11. See http://wallaceletters.info/content/homepage, the website of the Alfred Russel Wallace Correspondence Project.

SELECT BIBLIOGRAPHY OF MAJOR SCHOLARLY WORKS ON JOHN TYNDALL

Bartlett, W., 'Preaching Science: John Tyndall and the Rhetoric of Victorian Scientific Naturalism' (PhD dissertation, Rutgers University, 1995).

Barton, R., 'John Tyndall, Pantheist: A Rereading of the Belfast Address', *Osiris*, 3 (1987), pp. 111–34.

—, 'Tyndall, John (1820–93)', in W. J. Mander and A. P. F. Sells (eds), *The Dictionary of Nineteenth-Century British Philosophers*, 4 vols (Bristol: Thoemmes Continuum, 2004), vol. 2, pp. 1137–41.

Blinderman, C., 'John Tyndall and the Victorian New Philosophy', *Bucknell Review*, 9 (1961), pp. 281–90.

Brock, W. H., 'Tyndall, John (1820–1893)', in *Oxford Dictionary of National Biography* (Oxford: Oxford University Press, 2004), vol. 55, pp. 789–94.

Brock, W. H., N. D. McMillan and R. C. Mollan (eds), *John Tyndall: Essays on a Natural Philosopher*. Dublin: Royal Dublin Society, 1981.

Brown, M., 'Darwin at Church: John Tyndall's Belfast Address', in J. H. Murphy (ed.), *Evangelicals and Catholics in Nineteenth-Century Ireland* (Dublin: Four Courts Press, 2005), pp. 235–46.

Burchfield, J. D., 'John Tyndall at the Royal Institution', in F. A. J. L. James (ed.), *The Common Purposes of Life: Science and Society at the Royal Institution of Great Britain* (Burlington, VT: Ashgate, 2002), pp. 147–68.

—, 'Tyndall, John (1820–93)', in B. Lightman (ed.), *The Dictionary of Nineteenth-Century British Scientists*, 4 vols (Bristol: Thoemmes Continuum, 2004), vol. 4, pp. 2053–8.

Comtois, P., 'John Tyndall and the Floating Matter of the Air', *Aerobiologia*, 17 (2001), pp. 193–202.

Conant, J. (ed.), *Pasteur's and Tyndall's Study of Spontaneous Generation* (Cambridge, MA: Harvard University Press, 1953).

Cosslett, T., 'Science and Value: The Writings of John Tyndall', *Prose Studies 1800–1900: History, Theory, Criticism*, 2:1 (1978), pp. 41–57.

Daub, E. E., 'The Hidden Origins of the Tait-Tyndall Controversy: The Thomson-Tyndall Conflict', in *Proceedings, XIVth International Congress of the History of Science, Tokyo and Kyoto, Japan, 19–27 August, 1974* (Tokyo: Japan Science Council, 1974), pp. 241–4.

Dawson, G., and B. Lightman (eds), *Victorian Scientific Naturalism: Community, Identity, Continuity* (Chicago, IL: University of Chicago Press, 2014).

DeArce, M. 'Darwin's Irish Correspondence', *Biology and Environment: Proceedings of the Royal Irish Academy* section B, 108B (2008), pp. 43–56.

DeArce, M., et al., 'Further Notes on the Genealogy and Social History of the Carlow Family of John Tyndall (1820–1893)', *Carloviana: The Journal of the Carlow Historical and Archeological Society* (2013–14), pp. [1]–[7].

DeYoung, U., *A Vision of Modern Science: John Tyndall and the Role of the Scientist in Victorian Culture* (New York: Palgrave Macmillan, 2011).

Eaton, R. D., 'In the "World of Death and Beauty": Risk, Control, and John Tyndall as Alpinist', *Victorian Literature and Culture*, 41:1 (2013), pp. 55–73.

Eve, A. S., and C. H. Creasey, *Life and Work of John Tyndall* (London: Macmillan & Co., Ltd, 1945).

Farley, J., and G. Geison. 'Science, Politics and Spontaneous Generation in Nineteenth-century France: the Pasteur-Pouchet Debate', *Bulletin of the History of Medicine*, 48:2 (1974), pp. 161–98.

Fleming, J. R., 'John Tyndall, Svante Arrhenius, and Early Research on Carbon Dioxide and Climate', in J. R. Fleming (ed.), *Historical Perspectives on Climate Change* (Oxford: Oxford University Press, 1998), pp. 65–82.

Foster, J. W., 'Darwin in Ireland: John Tyndall and the Irish Churches', in J. W. Foster, *Recoveries: Neglected Episodes in Irish Cultural History, 1860–1912* (Dublin: University College Dublin Press, 2002).

Friday, J. R., R. M. MacLeod and P. Shepherd (eds), *John Tyndall: Natural Philosopher, 1820–1893. Catalogue of Correspondence, Journals and Collected Papers.* (London: Mansell, 1974).

Gentry, J. W., 'The Legacy of John Tyndall in Aerosol Science', *Journal of Aerosol Science*, 28:8 (1997), pp. 1365–72.

Gieryn, T. F., 'John Tyndall's Double Boundary-Work: Science, Religion, and Mechanics in Victorian England', in T. F. Gieryn, *Cultural Boundaries of Science: Credibility on the Line* (Chicago, IL: University of Chicago Press, 1999), pp. 37–64.

Gray, P. H., 'The Problem of Free Will in a Scientific Universe: Rene Descartes to John Tyndall', *Journal of General Psychology*, 80 (1969), pp. 57–72.

Haugrud, R. A., 'Tyndall's Interest in Emerson', *American Literature*, 41:4 (1970), pp. 507–17.

Higgins, K., and J. McEvoy, *Tyndall Mountain Club Kilkenny Ireland: A History* (Kilkenny: Kilkenny People Printing, 1991).

Howard, J., '"Physics and Fashion": John Tyndall and his Audiences in Mid-Victorian Britain', *Studies in the History and Philosophy of Science*, 35:4 (2004), pp. 729–58.

Hulme, M., 'On the Origin of "The Greenhouse Effect": John Tyndall's 1859 Interrogation of Nature', *Weather*, 64:5 (2009), pp. 121–3.

Jackson, R., 'John Tyndall and the Royal Medal that Was Never Struck', *Notes and Records of the Royal Society of London*, 68:2 (2014), pp. 151–64.

—, 'John Tyndall and the Early History of Diamagnetism', *Annals of Science* (2014), pp. 1–55.

Jenkins, A., 'Spatial Imagery in Nineteenth-Century Representations of Science: Faraday and Tyndall', in C. Smith and J. Agar (eds), *Making Space for Science: Territorial Themes in the Shaping of Knowledge* (New York: Palgrave MacMillan, 1998), pp. 181–91.

Kearney, R., 'John Tyndall and Irish Science', in R. Kearney, *Postnationalist Ireland: Politics, Culture, Philosophy* (London: Routledge, 1997), pp. 169–77.

Kim, S., *John Tyndall's Transcendental Materialism and the Conflict between Religion and Science in Victorian England* (Lewiston, Queenston, Lampeter: Mellen University Press, 1996).

Libera, S. M., 'John Tyndall and Tennyson's "Lucretius"', *Victorian Newsletter*, 45 (1974), pp. 19–22.

Lightman, B., *The Origins of Agnosticism: Victorian Unbelief and the Limits of Knowledge* (Baltimore, MD: Johns Hopkins University Press, 1987).

—, 'Scientists as Materialists in the Periodical Press: Tyndall's Belfast Address', in G. Cantor and S. Shuttleworth (eds), *Science Serialized: Representations of the Sciences in Nineteenth-Century Periodicals* (Cambridge, MA: MIT Press, 2004), pp. 199–237.

—, *Evolutionary Naturalism in Victorian Britain: The "Darwinians" and Their Critics* (Aldershot: Ashgate, 2009).

Lightman, B., and M. Reidy (eds), *The Age of Scientific Naturalism: Tyndall and His Contemporaries* (London: Pickering & Chatto, 2014).

McCabe, I. M., 'Second Best as a Researcher, Second to None as a Populariser?: The Atmospheric Science of John Tyndall FRS (1820–1893)' (PhD dissertation, University College London, 2012).

McMillan, N., and P. Hogan, *John Tyndall: "X"emplar of Scientific & Technological Education* (Dublin: N.C.E.A., 1980).

—, 'John Tyndall 1820–1893', in M. McCartney and A. Whitaker (eds), *Physicists of Ireland: Passion and Precision* (Bristol and Philadelphia: Institute of Physics Pub., 2003).

Mackowiak, J. R., 'The Poetics of Mid-Victorian Materialism in the Writings of John Tyndall, W. K. Clifford and Others' (PhD dissertation, University of Cambridge, 2008).

Meadows, J., 'John Tyndall: Then and Now', *Proceedings of the Royal Institution of Great Britain*, 66 (1995), pp. 239–49.

Molan, R. C., *It's Part of What we Are: Some Irish Contributors to the Development of the Chemical and Physical Sciences. Vol. 1, Richard Boyle (1566–1643) to John Tyndall (1820–1893)* (Dublin: Royal Dublin Society, 2007).

Mullin, R. B., 'Science, Miracles, and the Prayer-Gauge Debate', in R. L. Numbers and D.C. Lindberg (eds), *When Science and Christianity Meet* (Chicago, IL: University of Chicago Press, 2003), pp. 203–24.

Murphy, J., 'The Irish Catholics in Science Debate: John Tyndall, Cardinal Cullen and the Uses of Science at Castleknock College in the Nineteenth Century', in J. Adelman and E. Agnew (eds), *Science and Technology in Nineteenth-Century Ireland* (Dublin: Four Courts Press, 2011), pp. 127–35.

O'Flynn, B., *John Tyndall: A Biographical Note* (Kilcoole: Clo Ceo Draoidheachta, 1999).

O'Gorman, F., '"The Eagle and the Whale?": John Ruskin's Argument with John Tyndall', in M. Wheeler (ed.), *Time and Tide: Ruskin and Science* (London: Pilkington Press, 1996), pp. 45–64.

—, 'John Tyndall as Poet: Agnosticism and "A Morning on Alp Lusgen"', *Review of English Studies*, 48:191 (1997), pp. 353–8.

—, 'Some Ruskin Annotations of John Tyndall', *Notes and Queries*, 44:3 (1997), pp. 348–9.

—, '"The Mightiest Evangel of the Alpine Club": Masculinity and Agnosticism in the Alpine Writing of John Tyndall', in A. Bradstock et al. (eds), *Masculinity and Spirituality in Victorian Culture* (New York: Palgrave, 2000), pp. 134–48.

Phillipson, E. A., 'What Did Edward Jenner and John Tyndall Have in Common?', *Clinical & Investigative Medicine*, 25:4 (2002), pp. 137–8.

Reidy, M. S., 'John Tyndall's Vertical Physics: From Rock Quarries to Icy Peaks', *Physics in Perspective,* 12:2 (2010), pp. 122–45.

—, 'Coming Down, or How Mountaineering Changed Science', *Alpine Journal: A Record of Mountain Adventures and Scientific Observations Since 1863,* 116 (2013), pp. 53–62.

Sackmann, W., 'John Tyndall (1820–93) and his Relationship to the Alps and Switzerland', [in German] *Gesnerus*, 50 (1993), pp. 66–78.

Sarton, G., 'Faraday to Tyndall', *Isis*, 31 (1940), pp. 303–4.

Sawyer, P. L., 'Ruskin and Tyndall: The Poetry of Matter and The Poetry of Spirit', in J. Paradis and T. Postlewait (eds), *Victorian Science and Victorian Values: Literary Perspectives* (New York: New York Academy of Sciences, 1981), pp. 217–46.

Sopka, K., 'An Apostle of Science Visits America: John Tyndall's Journey of 1872–1873', *Physics Teacher*, 10 (1972), pp. 369–75.

Strick, J., *Sparks of Life: Darwinism and the Victorian Debates over Spontaneous Generation* (Cambridge MA: Harvard University Press, 2000).

Sugiyama, S., 'The Significance of the Particulate Conception of Matter in John Tyndall's Physical Researches', *Historia Scientiarum*, 2 (1992), pp. 119–38.

Taylor, C. A., 'John Tyndall's Demonstrations on Sound', *Proceedings of the Royal Institution of Great Britain, London*, 48 (1976), p. 37.

Thomas, G., 'Microbes in the Air: John Tyndall and the Spontaneous Generation Debate', *Microbiology Today* (November 2005), pp. 164–6.

Thompson, D. 'The Development of Scientific Education during the Second Half of the Nineteenth Century with Special Reference to the Influence of John Tyndall' (MA dissertation, University of Sheffield, 1955).

—, 'John Tyndall (1820–1893): A Vocational Enterprise', *Vocational Aspects of Secondary and Further Education*, 9:18 (1957), pp. 38–48.

—, 'John Tyndall and the Royal Institution', *Annals of Science*, 13 (1957), pp. 9–21.

Wainwright, M., 'Re-examination of Some of John Tyndall's Studies on Microbial Antagonism', *Transactions of the British Mycological Society*, 85:3 (1985), pp. 526–9.

Weed, L. A., 'John Tyndall and His Contribution to the Theory of Spontaneous Generation', *Annals of Medical History*, 4 (1942), pp. 55–62.

Williamson, D. E., 'The Double-Beam Infrared Gas Analyzer of John Tyndall', *American*

Scientist, 39:4 (1951), pp. 672–81.

Wiseman, E. J., 'John Tyndall: His Contributions to the Defeat of the Theory of Spontaneous Generation of Life', *School Science Review*, 159 (1965), pp. 362–7.

Yamalidou, M., 'John Tyndall, the Rhetorician of Molecularity. Part One. Crossing the Boundary Towards the Invisible', *Notes and Records of the Royal Society of London*, 53:2 (1999), pp. 231–42.

—, 'John Tyndall, the Rhetorician of Molecularity. Part Two. Questions Put to Nature', *Notes and Records of the Royal Society of London*, 53:3 (1999), pp. 319–31.

Young, H., *A Record of the Scientific Work of John Tyndall, D.C.L., LL.D., F.R.S. (1850–1888)* (London: Printed for Private Circulation at the Chiswick Press, 1935).

ACKNOWLEDGEMENTS

The editors of this volume would like to express their appreciation to the many people who generously assisted their editorial labours by providing information and advice on a wide range of issues, especially Bill Brock, Barbara Cantor, Aideen Ireland, Roland Jackson, Frank James, Chris Kenny, Shay Kinsella, Irena McCabe, Deidre McGing, Daniela Monaldi, Jon Topham and Ann Wilson. They would particularly like to thank Jane Harrison, who welcomed them to the archives room at the Royal Institution, Julie Hipperson, who assisted with research in London, and Martin Nevin, the editor of *Carloviana*, for providing a wealth of information about County Carlow and its nineteenth-century inhabitants. For his careful and dedicated research on the archives of the Ordnance Survey of Ireland (deposited in the National Archives of Ireland) the editors are indebted to Mark Tynan, whose 'Tyndall and the Irish Ordnance Survey' forms an appendix to this volume. Gowan Dawson would also like to thank the Leverhulme Trust for the award of a Research Fellowship which greatly assisted his work on the volume.

For permission to reproduce the map of Youghal we gratefully acknowledge the National Archives of Ireland. We are most grateful to Laura Vann for drawing the other three maps in this volume.

GEOFFREY CANTOR AND GOWAN DAWSON

LIST OF ILLUSTRATIONS

LIST OF ABBREVIATIONS

CS	*Carlow Sentinel*
Griffith's Valuation	*The General Valuation of Rateable Property in Ireland* (Dublin: General Valuation Office, 1847–64); available at http://www.askaboutireland.ie/griffith-valuation/
Irish Church Records	Records available at http://churchrecords.irishgenealogy.ie/churchrecords/
Journal	John Tyndall's Journal (RI MS JT/2/1–22)
LT	Louisa Tyndall, John Tyndall's widow
LT, 'Biography'	Louisa Tyndall's typescript biography of her husband (RI MS JT/5/9–11)
NA OS	National Archives, Kew, Ordnance Survey Records
NA WO	National Archives, Kew, War Office Records
NAI OS	National Archives of Ireland, Dublin, Ordnance Survey Records
OED	*Oxford English Dictionary*
RI MS JT	Royal Institution, London, John Tyndall Papers.
Slater's	I. Slater's *National and Commercial Directory of Ireland* (Manchester and London: I. Slater, 1846)

INTRODUCTION TO VOLUME 1

The 230 letters in this volume chart John Tyndall's emergence into early adulthood, spanning from his arrival in Youghal in May 1840 as a civil assistant with just a year's experience working on the Irish Ordnance Survey to his pseudonymous authorship of an open letter to the Prime Minister Robert Peel protesting the pay and conditions on the English Survey in August 1843. The letters, which include Tyndall's earliest extant correspondence, encompass some of the most significant events of the early 1840s, including the sectarian violence that in Ireland accompanied the 1841 General Election, and the Chartist strikes and riots that spread through north-west England in the summer of 1842. Tyndall's correspondents also discuss their experiences of, among other topics, British military expansion in India and economic migration to North America. The letters show the development of many of the traits and talents, both mathematical and literary, that would subsequently make Tyndall one of most prominent men of science in Victorian Britain, but they also afford broader insights into a period of almost unprecedented social upheaval and cultural and technological change. This introduction considers the key contexts and events pertaining to the period covered by these letters that shaped Tyndall's development into adulthood.

Early Life

Although John Tyndall's family probably came from the Tyne Valley in north-east England, his ancestors are said to have settled in Ireland late in the seventeenth century, first at Wexford and subsequently at Leighlin Bridge, County Carlow.[1] Owing to a family quarrel, Tyndall's father (John Tyndall, Snr) did not inherit property from his own father but instead earned his living as a constable in the Royal Irish Constabulary, a land agent and, latterly, as shoe maker and leather dealer. A Protestant and staunch Orangeman, he was self-educated but possessed a number of books and was keen for his children to gain a good education. The regular correspondence between father and son shows their close relationship, as well as a degree of petulance on both sides. Tyndall's mother Sarah (née McAssey) was descended from farmers with lands in Ballybrommell to the east of Leighlin Bridge. Tyndall and his older sister Emma were the only definitely known children of the marriage. While this volume contains a large number of letters both to and from Tyndall's father and a far smaller number to and from his sister, there is only one letter to his mother and not a single letter from her.[2]

Leighlin Bridge, where Tyndall was born, stands on the River Barrow with fertile lands to the east and a range of rising hills to the west. Local farmers cultivated wheat, barley, oats and, of course, potatoes. Sheep and cattle farming flourished and County Carlow was famous for the quality of its butter.[3] Yet, partly due to many maltsters and distillers being forced out of business by the high level of excise duty, Leighlin Bridge was not flourishing economically and one commentator even noted that although 'formally a place of some importance', by the early 1840s it showed signs of a slow decline.[4]

Little is known about Tyndall's early years, and even his date of birth – usually given as 2 August 1820 – is open to question.[5] During his childhood the family moved from Leighlin Bridge to Nurney (between Leighlin Bridge and Ballybrommell) and then to Castlebellingham, County Louth, before returning to Leighlin Bridge in 1836. Tyndall then entered the National School at Ballinabranna under the instruction of John Conwill, a Roman Catholic with an excellent reputation as a teacher, especially of grammar and mathematics. While his father encouraged learning, Conwill provided Tyndall with the tools for effective self-expression and also with a firm grounding in mathematics.

Irish Ordnance Survey

Tyndall's career prospects would have been very limited had not the Ordnance Survey of Ireland moved one of its divisions to Leighlin Bridge in April 1838. Moreover, although the Survey employed members of the Corps of Royal Engineers and Royal Sappers and Miners, large numbers of locally-recruited civil assistants were also required. On 1 April 1839 Tyndall joined the Survey and worked in the 5th Division of C District, which covered the central and central-southern counties of Ireland. Although he sometimes undertook fieldwork, Tyndall's principal activities were based in the Division's office where, with twenty or more other civil assistants, he engaged in 'printing and writing fair plans' and drawing.[6]

Following earlier surveys of parts of England and Wales undertaken by the Board of Ordnance, in 1824 the Lord Lieutenant of Ireland proposed an accurate and comprehensive survey and valuation of Ireland that was required for administrative purposes, including the calculation of taxes and determining whether a person's property was of sufficient value to be included on the voting register.[7] A select committee of the House of Commons considered this proposal and argued that the maps produced should show the townlands and their boundaries, from which the valuations of lands and properties could be ascertained. The committee also appointed the Ordnance Survey, under the direction of Major (later Colonel) Thomas Colby, to undertake this work because it was a military organization with extensive experience of surveying and possessed the best available instruments.[8] With a staff drawn from the Royal Engineers and Royal Sappers and Miners, Colby soon moved to Ireland in order to undertake the Survey and produce a 6 inch to 1 mile map of all the Irish counties. Although Colby made many innovations and was greatly admired for his leadership, the project proved far more difficult than expected and he had to hire an army of civil assistants; at its peak some 2,000 men were employed. Moreover, the project vastly exceeded its projected cost but Colby refused to compromise the high standard he set for the task, which resulted in the publication of over 2,000 sheet maps. Most of the work of the Irish Survey ended in 1842 (although it was not formally concluded until 1846) with many of the civil assistants – Tyndall included – being transferred to the Survey of England.

Thus Tyndall participated in the final years of the Irish Survey and, after leaving Leighlin Bridge at the beginning of May 1840, saw service in County Cork at Youghal, Kinsale and Cork. He was considered a 'superior Draughtsman and Surveyor' and was better paid than most other office staff.[9] He also undertook some fieldwork, including a survey of the Parish of Killeagh (near Youghal), whose inhabitants spoke only Gaelic, which he did not understand.[10] Although he worked diligently for the Irish Survey he was clearly frustrated by many aspects of it. For example, because of the pressure to complete the Survey, he had to work long hours and was refused permission to visit his parents during either Christmas 1840 or Christmas 1841. Also, while he respected some of the military men under whom he served (especially Corporals James Davy and James Mulligan and Lieutenant George Wynne), he was highly critical of some of the superior officers, especially Lieutenant Robert Fenwick who was a Divisional commander. Low morale was endemic among the civil assistants who complained of low pay and many resigned in order to take up better-paid jobs on the Tithe Survey in England. As several letters show, Tyndall seriously considered resigning from the Survey and instead taking employment in either the Bank of Ireland or the Inland Revenue. He even contemplated emigrating to America, as did his childhood friend William Wright and many other Irishmen and women.

Particularly after leaving Leighlin Bridge, he formed close friendships with other civil assistants, especially William Ginty, John Tidmarsh and Phillip Evans (known as Jim), and also with some of the military men, such as Private Thomas Foy. He often shared accommodation with these friends and they cooked for each other. They worked together, discussed issues of mutual interest and wrote to each other when dispersed to different locations. On Sundays they travelled together, including the occasion when Tyndall, Tidmarsh and other friends hired a rowing boat at Youghal and were in serious danger of drowning when the wind picked up and 'the sea gradually got into little mountains'.[11] Most importantly, they confided in each other about girlfriends and encouraged each other in their romantic escapades. The letters in this volume mention Charlotte Shine, Mary Edwards, Bessy, Sally, Ellen (or the 'lady of the raven plume'), Peri and 'that lovely creature Miss Stoul'.[12] Two of the letters are from an unidentified female admirer of Tyndall's.[13] Yet Tyndall and his friends also shared their antipathy towards some of the other men working for the Survey, such as the despised Anthony Linedale, who Tyndall considered to be dishonest and a liar. Thus, over the period covered by these letters, we see Tyndall develop from a youth immersed principally in his family and local community into a confident young man with extensive and diverse social experience and the ability to adapt to new and challenging circumstances. He was also engaged in a number of close and loyal friendships, many of which continued long after the period covered by these letters.

Sectarian Politics

Religion and politics provide a crucial context for many of the letters. Ireland was divided between Protestants, who included a very high proportion of the gentry, and Roman Catholics, many of whom spoke Gaelic as their first and possibly only language. As in many other fairly rural areas, County Carlow was in effect owned by such landed families as the Steuarts, the Kavanaghs, the Bruens, the Alexanders and the Fishbournes, who often spent much of their time in England. Land, wealth and power were in Protestant hands; the magistrates were predominantly Protestant, and Members of Parliament, whether Whig or Tory, were generally Protestant landowners.[14] Although Ireland had possessed its own Parliament until 1801, the country was subsequently ruled from Westminster and there was a continual struggle over the conditions that voters needed to fulfil, and what were seen as attempts to disenfranchise all but the wealthiest Catholics.

Since the seventeenth century a number of Irish nationalist movements had emerged (often accompanied by violence) in response to the depressed social and economic condition of the Catholic population. In the early 1840s Irish nationalism focused on Daniel O'Connell, who had earlier spearheaded the movement for Catholic Emancipation, and the Repeal Association, which sought to repeal the Union between Great Britain and Ireland and thus enable Ireland to free itself from Westminster's dominance. The Repealers held mass meetings and used the tactic of agitation to influence voters, especially those Catholic tenants who would be pressured by their landowners to vote Conservative. Catholic priests generally supported the Repealers and encouraged their flocks to join the Association.

Carlow was a major focus for Irish – and indeed British – politics during the hard-fought parliamentary elections of 1840 and 1841. The County was divided into two constituencies: Carlow Borough, which returned one Member of Parliament, and the remainder of Carlow County, which returned two. Elections for this latter constituency had a particularly contentious history. In the General Election of January 1835, the Conservative landowners Colonel Henry Bruen and Thomas Kavanagh were initially returned, but were unseated on petition and were replaced in June by the Liberals Nicholas Aylward Vigors and Alexander Raphael. A counter-petition to Parliament then resulted in Bruen and Kavanagh regaining the seats. The death of Kavanagh in February 1837 created a By-Election, which Vigors won with a majority of thirty-six over his Conservative rival. Then, in August 1837, a General Election resulted in the Liberal-Repealer Vigors being returned, together with his fellow Liberal John Ashton Yates, who was from a Liverpool dissenting family; they beat the Conservatives Henry Bruen and Thomas Bunbury by eighty-seven votes. An attempt to persuade Parliament to overthrow this result failed.

The death of Vigors on 26 October 1840 created a vacancy in Parliament. His old rival Henry Bruen now faced Frederick Ponsonby, the Liberal candidate who was resented by the Repealers whose views he did not share. Owing to this split among his opponents, Bruen was returned to Parliament, much to the relief of Tyndall and his father.[15] A mere seven months later a General Election was called at which the Liberals under Lord Melbourne ceded defeat to Robert Peel's Conservatives. The 1841 election in County Carlow attained national significance because O'Connell had travelled to Carlow to lead the agitation in favour of his son, Daniel, Jnr. Moreover, County Carlow had previously been represented by one Conservative (Bruen) and one Liberal (Yates, who now stood again). In this potentially combustible situation, troops were deployed to maintain order in Carlow. Both sides publicly claimed that the other was responsible for extensive intimidation of voters and fear gripped both Protestants and Catholics. One particular event in the run-up to the election impinged directly on the Tyndall family; Tyndall's uncle Caleb fired a shotgun and injured two people when an O'Connellite mob entered Leighlin Bridge. Although Caleb was tried and exonerated, he was hailed as a hero by fellow Protestants and condemned by Catholics as a Protestant thug. This incident and the trial that followed are reported in several letters.[16]

At the election in the Court House, voters openly expressed their preferences; their declarations being greeted by boos from the opposition. As voting was public, not private, a voter would fear for his life if he voted for the candidate from the opposing religious camp. Writing to his son on 19 July 1841, Tyndall's father was greatly relieved to report 'the glorious intelligence of Colonel Bruen's return for Carlow, also that of Mr Bumbury's after one of the most tremendous struggles that ever took place in any County'.[17] Although the Conservative candidates won, albeit by a mere eight votes, election-related violence continued and considerable antagonism remained between the two religious communities. Indeed, Tyndall's father complained of 'exclusive dealing' since Catholics no longer supported Protestant tradesmen and vice versa.[18]

During the 1841 election Tyndall had himself been attacked at Kinsale 'by a party of fishermen with green boughs in their hands', reporting that 'a dozen of them gripped me', and he remained throughout his life a bitter opponent of Irish nationalism including the Irish Home Rule movement that began in the early 1870s.[19] His disdain for Roman Catholicism, which seemingly underpinned his famous (or infamous) Belfast Address at the British Association meeting in 1874, likewise can be traced back to his father's stern anti-Catholic attitude and to his early experiences reported in these letters.[20] Yet in a public debate held in Youghal in April 1841 on the subject of Roman Catholicism versus Protestantism, Tyndall represented the Catholic side. John Tidmarsh later recalled that Tyndall 'had everything well prepared, studied, and noted'. He 'led off splendidly' and ended with 'a rare flow of most telling language, and the room rung with applause'. William Ginty, who spoke for Protestantism, was completely overwhelmed.[21] Not only does this incident demonstrate Tyndall's rhetorical skill, which was to serve him so well later in life, it also shows his ability to adopt the role of devil's advocate, revealing that even in his early twenties he was capable of intellectual sophistication and agility.

English Ordnance Survey

With the work of C District mostly complete by the summer of 1842, Tyndall left Ireland for the first time on 5 August, travelling overnight on the steamship PS *Ocean* from Cork to Liverpool. This 'modern Babylon', which was then exceeded only by London as the busiest port of the British Empire, immediately presented a range of experiences, including its 'eternal din', profoundly different from his previous life in Ireland.[22] Tyndall's letters home during his initial months in England record the striking modernity of a society undergoing exponential social, technological and cultural change, ranging from the railways (which Tyndall calls by the old-fashioned name 'steam coach'), Mechanics' Institutes, the Uniform Penny Post, illustrated newspapers and the penny press.[23] There are even discussions of a proposed 'aerial carriage' that would extend the steam technologies of the railway to flying machines, making travel between England and Ireland much quicker and cheaper.[24] Tyndall's first railway journey, two days after arriving in Liverpool, took him to Preston, where he joined the 1st Division of the English Ordnance Survey.[25] Only four days after arriving in this heavily industrialized Lancashire mill town, Tyndall witnessed, at first hand, a more threatening aspect of English modernity.

On 12 August 1842 a mass meeting of 3,000 workers called a strike in Preston's cotton factories in support of both improved wages and the 1838 People's Charter demanding democratic reform of the political system. The following day the strikers and their Chartist supporters were fired upon by soldiers, with four men being killed. Tyndall himself witnessed the bloody tumult 'in Lune Street, down which we could look from our office', and his approving response to the military intervention, reporting that the 'thing was checked in its bud, the people saw that decisive measures were about to be taken – this quelled the insurrectionary spirit', indicates that, at this stage, he still shared the hardline conservatism his father had instilled in him back in Ireland.[26] Tyndall, though, had already expressed an admiration for the 'bold Infidel who never crouches to servility' – probably the Welsh social reformer Robert Owen – to his teacher Conwill, and, in the very different circumstances of England in the early 1840s and with his 'memory of the Preston riots still vivid', it is hardly surprising that his political affinities soon began to shift markedly.[27]

Parts of England and Wales had already been surveyed during the period of the Irish Ordnance Survey, but only at the relatively small scales of 1 or 2 inches to the mile. The purpose of the English Ordnance Survey, which received government authorization in October 1840 as work in Ireland was starting to wind down, was to provide detailed maps at the larger scale of 6 inches to 1 mile that had been pioneered by the Irish Survey. In its initial stages, the English Survey's focus was on areas in the north of the country whose mining and manufacturing interests had recently given them enormous economic and strategic importance. With the exception of some of the Royal Engineers in charge of the five Divisions, almost everyone involved in the English Survey had previously been employed in Ireland, with the civil assistants, according to Tyndall, exclusively Irish. This prior experience ought to have ensured that the English Survey would run more smoothly than its Irish counterpart, which had required continual revisions, re-measurements, re-levellings and other expensive alterations.

The 6-inch scale survey of the six northern counties of Cumberland, Durham, Lancashire, Northumberland, Westmoreland and Yorkshire, however, had been approved by Melbourne's Whig government, which, as discussed above, was soon after defeated by Peel's Conservatives in the 1841 General Election. After years of public expenditure exceeding public income, which had enabled the Irish Survey to regularly overspend its annual grant, Peel immediately instituted a programme of retrenchment to reduce the deficit. As a result, the English Survey's grant was reduced dramatically, and its activities came under continual scrutiny from a government committed to squeezing public spending. While Tyndall's father had celebrated when the 'bloody whigs' were 'kicked out of office' and exulted that 'Peel at the head of the torys' was 'now holding the reigns [*sic*] of Government', his son soon had to face the uncomfortable practical realities of Peelite austerity.[28] Pay rises recommended on the Irish Survey were rescinded (or simply ignored) in England, and many civil assistants were left with an income that barely kept them from hunger or out of the dreaded work house. By March 1843 Ginty was warning 'Beware Tyndall – there is great talk of a reduction!', and in the following months numerous civil assistants and labourers, beginning with twenty-five from each Division, were summarily dismissed from the Survey.[29] At the time that Tyndall was transferred to England in August 1842 the Survey employed 1,595 men, but within three years this had fallen by almost half to 820.[30] To make matters worse, these stringent measures were implemented by English officers of the Royal Engineers, notably Captain Henry Tucker, commander of the 2nd and 3rd Divisions, and Lieutenant William George Hamley, commander of the 5th Division, who considered their civil assistants as 'a parcel of Irish pigs' and highhandedly dismissed their grievances.[31] For Tyndall at this time, England was a 'country whose sympathies often run counter to the name of Irishman'.[32]

In the late 1830s the higher pay received by civil assistants employed by private contractors on the English Tithe Survey had prompted many surveyors trained on the Irish Ordnance Survey, including Ginty, to resign and take jobs in England. In fact, the ensuing shortage of personnel was probably one of the reasons why Tyndall had found an opening on the Irish Survey in April 1839. By the early 1840s, the Tithe Survey, which produced maps of landholdings from which the new tithe rent charge could be calculated, still offered superior prospects to the Ordnance Survey, and while in Preston Tyndall planned, albeit unsuccessfully, to obtain a position at the Tithe Map Office in London. The Tithe Commissioners, advised by their assistant Robert Kearsley Dawson, required maps at a scale of 26⅔ inches to 1 mile. Colby, who now superintended both the Irish and English Surveys, proposed that the Ordnance Survey could make maps at the Tithe Commissioners' preferred scale for an additional 2*d.* per acre above the cost for a scale of 6 inches to 1 mile (which was 1*s.* 4*d.* per acre), adding £68,000 to the cost of the surveying of the six northern counties. This would involve the English Survey producing both Ordnance and Tithe maps at the larger scale, and, in addition to the government funds allocated for the Ordnance maps, receiving payment from individual parishes who were responsible for financing the production of Tithe maps. Twenty-three parishes and townships in Lancashire and Yorkshire were mapped by the Ordnance Survey at the 26⅔ inch-scale, with, significantly, the publicly-funded Royal Engineers effectively undercutting the private surveyors and civil engineers who had previously been employed by the parishes, and (allegedly) pocketing the parish payments themselves. When a Survey of London, at the even larger scale of 5 feet to 1 mile, was proposed in early 1842, the suggestion that the Exchequer, apparently with Peel's personal approval, should stump up another £104,000 for what Ginty called 'the mean R.E. monopolists' provoked their hard-pressed civil assistants into revolt.[33]

Having responded so unsympathetically to the Preston Strike in August 1842, Tyndall, who observed that he 'could pass very quietly thro' ten thousand Chartists and be taken for a brother', had, by the following summer, seemingly come to embrace many of the radical objectives of the strikers.[34] Tyndall's bitter experience of the iniquities of the English Survey, intensified by his reading of Thomas Carlyle's impassioned pleas on behalf of impoverished 'Lancashire weavers ... copied into Preston newspapers at this time', appears to have led to a profound reconsideration of his political beliefs, which he professed in a long letter to his parents back in Leighlin Bridge.[35] Unfortunately, this particular missive, as well as his father's reply, is missing, although their contents can be surmised from Tyndall's subsequent acknowledgement to his disconsolate father:

> I knew you would not relish it, and my mother poor woman, why I had more hope of radicalizing even you than her. I shall make just one remark in connexion with our two last letters and that is that the God of protestantism never intended that it [i.e. the Kingdom of God on earth] should be established by unjust means.[36]

Tyndall's own radicalization while in Preston, during which he came to believe that 'without truckling to the ape and tiger of the mob, a man might hold the views of a radical', prompted him to take the lead in a protest against the pay and condition of the Survey's civil assistants, which was combined with an exposé of the Royal Engineers' monopolistic misuse of public funds.[37] One of his collaborators in this radical agitation, Ginty, confessed that he was 'now within a trick of being a Repealer', and even urged that they should send a copy of their written protest to the 'illustrious Daniel O'Connell!!!' in order to 'kick up a shindy'.[38] Tyndall had not reneged on his erstwhile principles to this extent (although Ginty did ask 'Suppose John Tyndall and Jim Evans chanced to be born Roman Catholics – would they be repealers?'), but by the late summer of 1843 he was composing a respectful though nonetheless defiant letter of protest addressed to the Conservative Prime Minister that would eventually lead to his dismissal from the Ordnance Survey.[39]

Authorship

By the end of the period covered by these letters, Tyndall had taken the initial steps towards becoming an author, publishing, under the pseudonym Spectator, the first of a series of five letters to Robert Peel in the *Liverpool Mercury*.[40] These missives earnestly remonstrating against the Royal Engineers' mismanagement of both the English and Irish Surveys, according to the *Mercury*'s editor John Smith, warranted 'insertion on public grounds', even if the Master-General of the Ordnance, George Murray, considered them 'undeserving of any reply'.[41] Tyndall's open letter to the Prime Minister on 1 September 1843, however, was not his first experience of authorship, and he had in fact been contributing to other newspapers for almost two years, although, notably, in a genre that was generally considered more suitable for the expression of private sentiments.

In the spring of 1841 Tyndall began appending short poetic passages to his letters in imitation of his teacher John Conwill's practice of beginning each of his own epistles with a couplet or quatrain (which he called a 'motto'), usually extolling the virtues of his favourite student.[42] By the summer, teacher and pupil were swapping longer verses in their letters, inciting each other to compose 'poetical effusion[s]' prompted by the subjects, including friendship, personal autonomy and poverty, they discussed elsewhere in their correspondence.[43] In October, Tyndall, with growing confidence and self-assurance, reflected that the 'bit of a rhyme to stick at the commencement of my next letter to Mr Conwill ... would have a much better effect in the columns of the Sentinel', and accordingly sent his latest effusion to the *Carlow Sentinel*'s editor.[44] Verse was a regular component of many newspapers at this time, and Tyndall's poem, a satire in rhyming couplets on Irish nationalism entitled 'The Leighlin "Orators" – or, the Late Repeal Meeting', appeared in the *Sentinel* on 16 October 1841.[45] This was Tyndall's first appearance in print, even if he carefully concealed his authorship by signing the poem W. S., the initials of his droll alias amongst his fellow civil assistants, Walter Snooks. Two further poems, both once more combining romantic evocations of Irish history and myth with bitter invective against the perpetrators of the country's current travails, appeared in the *Sentinel* before mid-November, eliciting praise from Tyndall's family – who intuited his authorship despite the pseudonymous initials – as well as the *Sentinel*'s proprietor Henry Malcolmson, who, as Tyndall's father told him, declared 'that his paper should be at your command'.[46]

Yet by January 1842 W. S. had 'become dumb', and Malcolmson's patronage was seemingly snubbed, perhaps because of the increased demands of the Irish Survey as the work of the C District reached completion.[47] When, in February 1843, Tyndall once more began submitting poems to the *Carlow Sentinel* they were now considered 'not altogether suitable and ... of so trifling a nature' as to induce Thomas Harris Carroll, who had become proprietor as well as editor after Malcolmson fell ill, to request that he instead try his hand at prose 'sketches of public men'.[48] Far from abandoning his chosen genre, though, Tyndall continued to produce a steady flow of verse after his transfer to the English Survey. Indeed, only a month after Carroll's curt rejection, Ginty proclaimed: 'Tyndall is neither more nor less than one of the British Poetical Luminarys! A star in Literature!! An Author!!! ... The newspapers English and Irish teem with his astounding effusions'.[49] Ginty, who also suggested that Tyndall would soon be made Poet Laureate, was no doubt indulging in hyperbole, and the only poem that Tyndall is definitely known to have published in the English press at this time, the elegiac 'On Leaving Westmorland' in the *Preston Chronicle* for 28 January 1843, was actually presented as if it were Ginty's own.[50] This mischievous false signature implied that the poem was Ginty's lovelorn paean to a local girl (who was already engaged to another man), and prompted consternation – as well as some mirth – when it was realized that the widely-read *Chronicle* might be seen by the girl and her relations. Sharing informal and often comical verse in their letters was an important mode of sociability among the civil assistants on the Ordnance Survey, and the conspiratorial camaraderie with which Ginty and Tyndall swapped poems on their respective love interests (in a manner similar to the earlier letters with Conwill) also entered into Tyndall's public appearances as a poet in newspapers. This element of private intrigue was perhaps reflected in the only known public response to Tyndall's youthful poems, with the *Chronicle*'s editor Lawrence Dobson complaining that 'they have one capital fault ... sense being, in many instances, sacrificed to sound'.[51]

Ginty's and Tyndall's letters also reveal their shared passion for reading contemporary verse, particularly that of the Irish poet Thomas Moore and, most especially, the infamous Lord Byron. Like many other young men of his generation, Tyndall's receptiveness to the Byronic cult of brooding, sardonic aloofness helped shape his youthful experiences. During the excitement of first crossing the Irish Sea in August 1842, for instance, he 'could not help bursting forth in the sentiment and language of Byron'.[52] Ironically, it was the same poetic influence that impelled Tyndall's first publication in prose, with Archibald McLachlan, the embittered civil assistant who supplied him with damning details of the Ordnance Survey's mismanagement and its employees' pay and conditions, urging in August 1843: 'Come, let us begin, at once ... recollect, "Who would be free, themselves must strike the Blow"'.[53] The missive containing this stirring quotation from *Childe Harold's Pilgrimage* (1812–18), which cast Tyndall in the role of self-determining Byronic hero, was received just days before he completed his open letter to Peel and instructed Ginty to send it to the *Liverpool Mercury*. That audacious first letter of remonstrance, which neither McLachlan nor Ginty, nor any other civil assistant, had been willing to write themselves, marked Tyndall's emergence as an author of prose, and also vindicated the courageous motto he had earlier chosen for himself: 'I neither seek the combat nor shun it when it comes'.[54] It was nevertheless written in the same spirit of Byronic heroism that had seen his friends refashion his poetic pseudonym Walter Snooks as 'childe Walter', and which, long before his reading of Carlyle's *On Heroes and Hero-Worship* (1841) in July 1847, afforded a model of defiant exceptionalism and distrust of vested authority that would continue to inform Tyndall's authorial persona in the future.[55]

Natural Philosophy

While Tyndall developed his skill as an author and gained the self-confidence to mount an incisive criticism of the Ordnance Survey's leadership, there is little evidence in these early letters that he would later pursue a career in science and become one of its leading Victorian practitioners. Conwill's declaration 'that Tyndall possesses a mind which for its vividness and effusions is not inferior to that of a Newton', may seem insightful with hindsight but should not be understood as indicating his probable career path.[56] The letters in this volume show that Conwill provided Tyndall with a good grounding in Euclidean geometry and number theory. However, the only physics problems discussed in these letters concern the fall of bodies through a hole in the centre of the earth and the application of the inverse square law to the intensity of light.[57] Louisa Tyndall later claimed that as a child Tyndall lapped up the stories his father told him about Newton and that he possessed a 'desire to get to the meaning of things'. When 'quite a lad' he read 'with intense pleasure' the articles on aerostatics, electricity and phlogiston in the *Encyclopædia Britannica* and subscribed to the *Engineer's and Mechanic's Magazine*.[58] He later joined the Mechanics' Institution in Preston and, in the winter of 1842/3, attended its 'very interesting course of lectures'.[59] He later recalled that John Clay, a local clergyman, had lectured on mechanics, Moses Holden lectured on astronomy, the surgeon William Corless lectured on physiology and that lectures were also given on 'general physics, chemistry, [and] botany'.[60] This weekly series of diverse lectures at the Preston Mechanics' Institution appears to have been the first occasion when Tyndall encountered such a broad range of sciences. Despite the above exceptions, the sciences played a minor role in Tyndall's life during the period covered in this volume and his immersion in physics and chemistry only began in earnest with his appointment to Queenwood College in August 1847, especially his encounters with its head teacher George Edmondson and Edward Frankland who taught science at the school.

Although Tyndall clearly did not envisage a scientific career at this stage, four aspects of his work on the Ordnance Survey can be recognized as relevant to his later pursuit of science, especially his experimental research. Drawing on his mathematical studies with Conwill, Tyndall used practical mathematics, principally trigonometry, in surveying. As he later noted, 'I habitually made incursions into the domains of the calculator and computer, and thus learned all their art'.[61] Second, as a draughtsman working in the Survey's offices in Youghal, Cork and Preston, he honed his skills at drawing; for example, in February 1842 he informed Conwill that 'I'm improving myself greatly in my office business, my drawing of the city of Cork is thought a good deal about'.[62] His ability to make accurate drawings proved useful in the illustrations he deployed in his research notebooks, research papers and monographs. Third, Tyndall gained valuable practical experience working in the field with scientific equipment. In particular, he mastered the theodolite, enabling him to make measurements with a very high degree of accuracy.[63] He later demonstrated a similar manual dexterity in the design and utilization of scientific instruments. Finally, through his Ordnance Survey work in both field and office Tyndall came to appreciate the importance of accuracy in making measurements. He had repeatedly witnessed the waste of time and money as parts of the Survey were rendered worthless and had to be re-measured and re-surveyed because the personnel, civil assistants and senior military officers alike, were not competent to achieve the necessary accuracy.[64]

This bitter experience of waste and improvidence was an important lesson for Tyndall and one that he sought to implement throughout his later career in science. Indeed, he considered that accuracy in measurement underpinned the progress of science, urging that 'Science is an organic growth, and accurate measurements give coherence to the scientific organism'.[65] As in many other instances, the early experiences chronicled in the letters in this volume had a profound influence on Tyndall's mature personality and the development of his scientific interests and aptitudes.

Notes

1. Although it is now known as Leighlinbridge, in the early 1840s the town was called Leighlin Bridge (as it is throughout the letters in this volume). The historical usage has been retained in this introduction and in the notes to the letters.
2. See letter 0006.
3. D., 'On the Agriculture of the County of Carlow', *Quarterly Journal of Agriculture*, 5 (1834–5), pp. 178–97.
4. *Slater's*, p. 63.
5. In an unpublished biography of her husband, Louisa Tyndall stated that 'he was born on the 2nd of August in the year 1820', adding in a note that no official record had survived and although 'Family records point to 1820 ... Tyndall himself inclined to the previous year' (LT, 'Biography', vol. 1, pp. 1–2). Other evidence suggests that Tyndall may have been born in 1821 or even 1822.
6. NAI OS/1/16.
7. See J. H. Andrews, *A Paper Landscape: The Ordnance Survey in Nineteenth-Century Ireland* (Oxford: Clarendon Press, 1975), pp. 21–34.
8. *Report from the Select Committee on the Survey and Valuation of Ireland* (London: Her Majesty's Stationary Office, 1824).
9. Letter 0126.
10. See letter 0043; M. Tynan, 'Appendix: John Tyndall's Role in the Ordnance Survey of Ireland'.
11. J. Tidmarsh, 'Reminiscences of John Tyndall', *Southern Australian Register*, 24 January 1894, p. 6.
12. Letters 0151 and 0082.
13. See letters 0091 and 0099.
14. Only with the Catholic Emancipation Act of 1829 could Catholics sit in Parliament.
15. See letters 0023 and 0026.

16. See letters 0073, 0078, 0079 and 0082.
17. Letter 0078.
18. Letter 0073.
19. Letter 0075; green symbolised Irish nationalism.
20. See J. Tyndall, *Address Delivered Before the British Association Assembled at Belfast* (London: Longmans, Green, 1874).
21. Tidmarsh, 'Reminiscences', p. 6; letters 0057, 0058 and 0059.
22. Letter 0161.
23. Ibid.
24. Letter 0199.
25. The six northern counties of the English Ordnance Survey seem to have together comprised one District, in which there were five Divisions commanded from Divisional Offices in, respectively, Preston, Wakefield (which was responsible for both the 2nd and 3rd Divisions), Manchester and Liverpool (NA OS3/412). Much less is known about the organization of the English Survey than that of its Irish counterpart because most of the records for the 1840s were destroyed when the English Survey's headquarters in Southampton was bombed during the Second World War.
26. J. Tyndall, 'On Unveiling the Statue of Thomas Carlyle', in *New Fragments* (London: Longmans, Green, 1892), pp. 392–7, on p. 392; letter 0164.
27. Letter 0135; J. Tyndall, 'Personal Recollections of Thomas Carlyle', in *New Fragments*, pp. 347–91, on p. 349.
28. Letter 0090.
29. Letter 0194.
30. R. Oliver, 'The Ordnance Survey in Great Britain 1835–1870' (PhD thesis, University of Sussex, 1985), p. 113.
31. Letter 0180.
32. Letter 0228.
33. Letter 0185. On the London Survey, see Oliver, 'Ordnance Survey in Great Britain', pp. 116, 136.
34. Letter 0164.
35. Tyndall, 'Statue of Thomas Carlyle', p. 393.
36. Letter 0213.
37. Tyndall, 'Personal Recollections', p. 349.
38. Letters 0224 and 0226.
39. Letter 0224.
40. Spectator, 'Ordnance Survey: Royal Engineers', *Liverpool Mercury*, 1 September 1843, p. 296.
41. Letter 0228; quoted in Oliver, 'Ordnance Survey in Great Britain', p. 114.
42. Letter 0028.
43. Letter 0069.
44. Letter 0109.
45. W. S., 'The Leighlin "Orators" – or, the Late Repeal Meeting', *Carlow Sentinel*, 16 October 1841, p. [3].
46. Letter 0108. The poems were W. S., 'Carlow', *Carlow Sentinel*, 30 October 1841, p. [3]; and W. S., 'The Testimonial', *Carlow Sentinel*, 13 November 1841, p. [3].
47. Letter 0119.
48. Letter 0204.
49. W. Ginty to T. C. Higginson, 24 March 1843, RI MS JT 1/TYP/11/3598.

50. 'On Leaving Westmorland. By W. Ginty', *Preston Chronicle*, 28 January 1843, p. [4]. Tyndall probably also published, under the pseudonym Wat Ripton, 'An Hibernian's Song to ____ ____', *Preston Chronicle*, 20 May 1843, p. [4].
51. 'To Correspondents', *Preston Chronicle*, 13 January 1844, p. [2].
52. Letter 0158.
53. Letter 0223.
54. Letter 0055.
55. Letter 0205. On Tyndall's reading of Carlyle's *On Heroes and Hero-Worship*, see the entry for 18 July 1847 in Journal, vol. 13a, p. 231.
56. Letter 0048.
57. See letters 0166 and 0167,
58. LT, 'Biography', vol. 1, pp. 7, 21. Louisa probably meant the *Practical Mechanic and Engineer's Magazine*, a monthly that commenced publication in October 1841.
59. Letter 0187.
60. J. Tyndall, 'Address Delivered at the Birkbeck Institution on October 22, 1884', in *New Fragments*, pp. 224–47, on p. 226; 'Preston Institution for the Diffusion of Knowledge', *Preston Chronicle*, 7 October 1843, p. [4].
61. Tyndall, 'Address', p. 227; letter 0124.
62. Letter 0124.
63. Tyndall, 'Address', pp. 227–8.
64. See, for instance, letter 0229.
65. J. Tyndall, 'The Rainbow and its Congeners', in *New Fragments*, pp. 199–223, on p. 201.

EDITORIAL PRINCIPLES

Our aim is to include all letters to and from John Tyndall that are currently extant. Some of the letters are from published sources, such as letters written by Tyndall to newspapers or journals, but the majority of them are either the originals or typescripts produced by Louisa Tyndall from originals. One central editorial principle has been to transcribe, wherever possible, from original letters rather than Louisa's typescripts. This was because of Louisa's conscious and unconscious editorial interventions, a point discussed in more detail below. We have also aimed to reproduce, as accurately as possible, the text of the letters as they were written. Spelling mistakes that have appeared in the handwritten letters have therefore been preserved; mistakes appearing where only the typescript letter has survived are silently corrected as in our judgement they are almost always typographical errors. But, in general, non-standard spellings (e.g. the use of dialect) has been retained even when in a typescript only letter (though there is the possibility that the spelling is Louisa's, not her husband's). We have included illustrations that are part of the letter.

The letters are presented in chronological order of writing. To indicate the beginning of a new year, the date of that year has been inserted in a large font before the first letter from that year.

Both the transcription of non-English language letters and the English translations are presented. The transcription of the non-English letter will appear first, followed by the English translation in a smaller font. Some foreign letters are in an original hand while others exist only as transcriptions by Louisa or others who often did not have a full command of the language. In the latter case these letters contain frequent spelling, grammatical and interpretation errors that do not seem to have been committed by the original author. In these typescript-only cases, and only in these cases, spelling and grammatical errors have been corrected, as have poor interpretations where a better word choice was obvious from the context. If the transcription and translation are based on a typescript transcription, this will be indicated by 'LT Transcript Only'.

There is a standard format for each letter. The first line lists the author, if it is to Tyndall, or the recipient, if it is from Tyndall; the date; and the letter number. The second line lists the place the letter was written (if known), and any information supplied by the writer in their salutation, such as the date or time of day when the letter was written. Information from stationery – for instance from institutional letterhead – is also supplied in the second line if extant. The opening salutation, the text of the letter, and closing then follow. Then come the source and editorial notes, the latter indicated in the text by Arabic numerals.

Editorial notes are intended to provide the contextualization necessary to understand the contents of each letter. There are basically six types of editorial notes. Many of them are informative notes on persons, real or mythical, mentioned in the letters. If a person referred to in a letter is not described in a note, they have been mentioned more than twice in the volume and they will have an entry in the Biographical Register at the back of the volume. For the rest – those mentioned once or twice – biographical information will appear in the editorial note appended to the letter in which he or she is first mentioned. When there is a second reference to that person a note will refer back to the letter in which he or she was first mentioned. Biographical notes on well-known figures such as Faraday or Huxley stress what that person was doing at the time when the letter was written rather than providing a full overview of the person's life.

A second type of editorial note identifies allusions and quotations. In the case of quotations, where possible, the first edition of the work is cited unless a different edition is mentioned in the letter. If a quote is not in English a translation is given. Glosses on obscure places, abstruse words, and words used in an unfamiliar sense comprise the third type of editorial note.

The fourth type provides information about the context of the letter, drawing on such sources as contemporary publications, scholarly sources, Tyndall's personal journals and field notes, and other letters in the Tyndall Correspondence. The fifth type of editorial note is cross-references to other letters in the Correspondence, referring to the letter number. The last type is information on significant textual changes that Tyndall made to the original that is thought to be relevant to the meaning of the letter.

There are a series of conventions for indicating when editors have given additional information about the text of a letter:

When Louisa Tyndall has made an annotation or marginal insertion that the editors have deemed important enough to retain, we have included her annotation in an editorial footnote.

When editors other than LT have made insertions they are indicated by unitalicized square brackets, e.g. [word].

When certain words are ambiguous and we have had to conjecture their meaning they are indicated by italicized square brackets, e.g. *[*words*]*; when we have considered such words illegible the number of illegible words are indicated in full italicization and square brackets, e.g. *[3 words illeg]*.

If a word or series of words are not present or have been destroyed– by inkblots, holes, or mould, for example– they are indicated as missing in italicized angular brackets, e.g. *<3 words missing>*.

In extremely rare cases where it is clear that Louisa herself has intentionally destroyed a few words, they are indicated as excised in italicized angular brackets, e.g. *<3 words excised>*.

Finally, a brief word about Louisa. A Tyndall correspondence project would be very difficult, perhaps even impossible, without the work done by Louisa, much of which seems to have been done in order to write a biography of her late husband. This biography was not completed. We have therefore treated Louisa as a *de facto* member of the editorial team. Some of her handwritten comments on the typewritten letters are extremely helpful and thus have been retained as footnotes to the relevant passage. Conversely, many of these handwritten annotations seem to have been inserted to help write her biography – for instance, summaries of passages or interesting quotes. When the editors have deemed these notes of Louisa's add no new information, they have not been included for publication.

It is also important to note that Louisa was not always reliable as a transcriber. She wasn't always accurate and may have gone back and edited portions of Tyndall's writing to make his letters seem more consistent or grammatically correct. She also left out details that she may have considered too private or too salacious for public consumption. In cases where there is an original, handwritten letter, therefore, we have considered that as taking precedence over a typed transcription.

NOTE ON MONEY

Money features prominently in these letters, which include discussions of the pay received by civil assistants and the overall cost of the Ordnance Surveys. In the 1840s the currency used in both England and Ireland was the pound sterling, usually abbreviated to pound (£). A pound consisted of 20 shillings and each shilling contained 12 pennies. In the notes to this volume the editors have used the following format: £3 4*s*. 5*d*. to express three pounds, four shillings and five pence. The letter L (for livres) has sometimes been used in place of the £ sign and the *d*. (derived from denarius, a small coin (Latin)) indicates pennies.

Several forms of representation were used in these letters. Sometimes the numbers for pounds, shillings and pence were separated by dots, forward slashes or spaces; thus £3 4*s*. 5*d*. could be written as £3.4.5, £3/4/5 or £3 4 5. A zero could be represented by an n-dash; thus 7 shillings was often written as 7/- or just 7/.

A farthing was a quarter of a penny (¼*d*.).

A crown was 5 shillings and half a crown equalled two shillings and sixpence, often written as 2/6.

A guinea was 21 shillings, that is £1 1*s*.

TIMELINE OF JOHN TYNDALL'S LIFE

c. 1822	John Tyndall born at Leighlin Bridge, County Carlow, Ireland
c. 1836	Attends John Conwill's National School in Ballinabranna
1839	Begins employment as civil assistant with Ordnance Survey of Ireland in Carlow
1840	Becomes civil assistant in the Ordnance Survey Office at Youghal, Cork County
1841	First appearance in print (poem in *Carlow Sentinel* under pseudonym 'W[alter] S[nooks]')
1842	Joins English Ordnance Survey in Preston
1843	Leads written criticism of Ordnance Survey in *Liverpool Mercury*; dismissed November
1844	Home in Ireland, mostly unemployed, until obtaining a position with the firm of Nevins and Lawton, Surveyors, of Manchester
1845–7	Works for Richard Carter, Surveyor, of Halifax
1845	Meets Thomas Archer Hirst in Halifax
1847	Begins teaching mathematics and surveying at Queenwood College, Hampshire
1848–50	At University of Marburg, working for PhD with Robert Bunsen and Hermann Knoblauch
1850	Returns to England: meets Michael Faraday and attends his first British Association meeting, before returning to Marburg
1851	Moves from Marburg to Berlin (April–June); then returns to Queenwood
1851	Meets Thomas Henry Huxley
1852	Elected Fellow of the Royal Society
1853	First lecture at Royal Institution (RI)
1853	Meets Herbert Spencer
1853	Appointed Professor of Natural Philosophy at RI
1853	Awarded Royal Medal for magnetic work; declines when controversy arises over award
1857	Climbs Mont Blanc for first time with Hirst
1859	Demonstrates existence of greenhouse gases and climatic implications
1859	Joins Government School of Mines as Professor of Natural Philosophy
1860	*The Glaciers of the Alps* published
1860	Meets John Lubbock and his wife, Ellen Lubbock
1861	First ascent of the Weisshorn
1861	Delivers his first Christmas Lectures at RI
1862	*Mountaineering in 1861* published
1862	Begins to champion Mayer's priority in discovery of conservation of energy

1863	*Heat Considered as a Mode of Motion* published
1864	Awarded Royal Society's Rumford Medal
1864	First meeting of X Club
1865	*On Radiation* published
1865	Engages in public controversy over the efficacy of prayer
1866	Succeeds Michael Faraday as scientific adviser to Trinity House
1867	Michael Faraday dies; Tyndall becomes Superintendent of the RI
1867	*Sound: A Course of Eight Lectures* published
1868	Successfully climbs Matterhorn
1868	Discovers the cause of light scattering, to be known as 'Tyndall Effect'
1868	*Faraday as a Discoverer* published
1868	Lecture on 'Scientific Materialism' at the British Association
1868	Resigns from Royal School of Mines
1869	Joins Metaphysical Society
1870	'On the Scientific Use of the Imagination' lecture at British Association
1870	*Researches on Diamagnetism* published
1870	*Three Scientific Addresses* published
1870	*Researches on Diamagnetism and Magne-Crystallic Action* published
1871	Meets Louis Pasteur for first time while in Paris
1871	*Hours of Exercise in the Alps* published
1871	*Light and Electricity* published
1871	*Fragments of Science* published
1872	Debates over how to measure the efficacy of prayer, aka the 'Prayer-Gauge Debate'
1872	*Contributions to Molecular Physics in the Domain of Radiant Heat* published
1872	*The Forms of Water in Clouds and Rivers, Ice and Glaciers* published
1872–3	USA Lecture Tour
1873	*Six Lectures on Light* published
1874	As President of the BAAS, delivers the 'Belfast Address'
1876	Marries Louisa Charlotte Hamilton
1877	Develops 'Tyndallization' (lengthy sterilization to destroy heat-resistant spores)
1877	Tyndall and Louisa build summer cottage 'Alp Lusgen' at Belalp, northern side of Valais, above Brieg
1877	*Fermentation and its Bearings on Phenomena of Disease* published
1879	*Fragments of Science*, which had gradually expanded, first published in two volumes
1881	*Essays on the Floating Matter of the Air* published
1885	Tyndall and Louisa build Hindhead retreat, Surrey Downs
1887	Resigns from RI
1890	Fight with Gladstone in *The Times* over Irish Home Rule
1892	Hirst dies
1892	*New Fragments* published
1893	Dies Haslemere, Surrey, accidentally poisoned

TIMELINE OF EVENTS IN JOHN TYNDALL'S LIFE SPECIFIC TO VOLUME 1

Year	Month	Tyndall biographical details	Notable social, political and cultural events
c. 1822		John Tyndall born Leighlin Bridge, County Carlow, Ireland (2 August *c.* 1822) to John Tyndall (1792–1847) and Sarah McAssey (d. 1867).	George IV becomes King
1824			Ordnance Survey (O.S.) of Ireland begins
1828–33		John Tyndall, Snr joins Irish Constabulary; family moves to Nurney, then Castlebellingham in Louth County	
1829			Catholic Emancipation
1830			Manchester & Liverpool Railway begins first regular commercial rail service; William IV becomes King; Irish land valuation exercise begins
1831			Michael Faraday discovers electromagnetic induction; British Association for the Advancement of Science founded; Charles Darwin begins voyage on HMS *Beagle*
1832			First Reform Act
1833			First of the eight Bridgewater Treatises published
1834			Poor Law Amendment Act
1835			Henry Fox Talbot exposes the first known photographic negative

Year	Month	Tyndall biographical details	Notable social, political and cultural events
1836		Tyndall's family returns to Leighlin Bridge; Tyndall attends National School, taught by John Conwill	Tithe Commutation Act creates the Tithe Survey
1837			Victoria ascends throne; William Whewell's *History of the Inductive Sciences* published
1838			Irish Poor Law introduced; Charles Dickens's *Oliver Twist* published; railway boom begins
1839		Begins employment as civil assistant with O.S. of Ireland in Carlow	First Anglo-Afghan War begins; First Anglo-Chinese War begins
1840	January		Uniform Penny Post introduced
	February		
	March		
	April		
	May	Transferred to County Cork O.S. office	
	June		
	July		English O.S. approved
	August		
	September		Louis Agassiz proposes 'ice ages' in his *Études sur les glaciers*
	October	County Carlow by-election	
	November		
	December		
1841	January		
	February		
	March	Friend William Wright and wife Sally emigrate to USA	
	April		
	May		New Zealand becomes a British colony
	June	Transferred to Kinsale O.S. office	
	July	General Election; Conservatives win County Carlow. Uncle Caleb shoots at O'Connellite mob, acquitted; Tyndall roughly handled by Catholic fishermen	Robert Peel forms Conservative Government; *Punch* founded
	August		
	September		

Year	Month	Tyndall biographical details	Notable social, political and cultural events
	October	First appearance in print – 'The Leighlin "Orators"' in *Carlow Sentinel*, pseudonym W[alter] S[nooks].	Fire at the Tower of London
	November		
	December	Applies for leave to go to Carlow to see family for Christmas – denied	
1842	January	Transferred to Cork O.S. office	Retreat from Kabul leads to end of First Anglo-Afghan War
	February	John Tyndall Sr tells Tyndall of the difficulty of getting any situation for him	
	March		Robert Peel's budget speech announces reintroduction of income tax (except in Ireland) and retrenchment in public spending; J. R. Mayer calculates mechanical equivalent of heat
	April		
	May	On leave from Survey, in Carlow to see family	*Illustrated London News* founded
	June	Receives letter from William Wright, now in Cincinnati, inviting JT to join them there	
	July		
	August	Joins English O.S. in Preston. Witnesses strike of 3,000 cotton workers; soldiers fire upon them the next day, to Tyndall's approval	End of First Anglo-Chinese War
	September		
	October	Tyndall takes French lessons, to read 'many clever Engineers'. Goal is to work in map office of Tithe Commissioners in London	
	November		
	December	Joins Mechanics' Institution, Preston; attends science lectures	
1843	January		Daniel O'Connell announces 'Year of Repeal' campaign

Year	Month	Tyndall biographical details	Notable social, political and cultural events
	February	William Ginty starts urging Tyndall to protest treatment of the Irish surveyors in England	
	March		First of the 'monster meetings' demanding Repeal of the Act of Union, at Trim, County Meath; annexation of the Sindh province in northern India
	April	Layoffs of surveyors begins	Thomas Carlyle's *Past and Present* published
	May		
	June		
	July	Tyndall actively considers emigrating to USA. Preparations to publish open letter to Peel in various newspapers begin	SS *Great Britain* launched
	August	Letter protesting treatment of surveyors sent to editor of the *Liverpool Mercury*	James Joule presents his calculations of mechanical equivalent of heat at BAAS
	September	First letter appears in the *Liverpool Mercury*	

MAPS

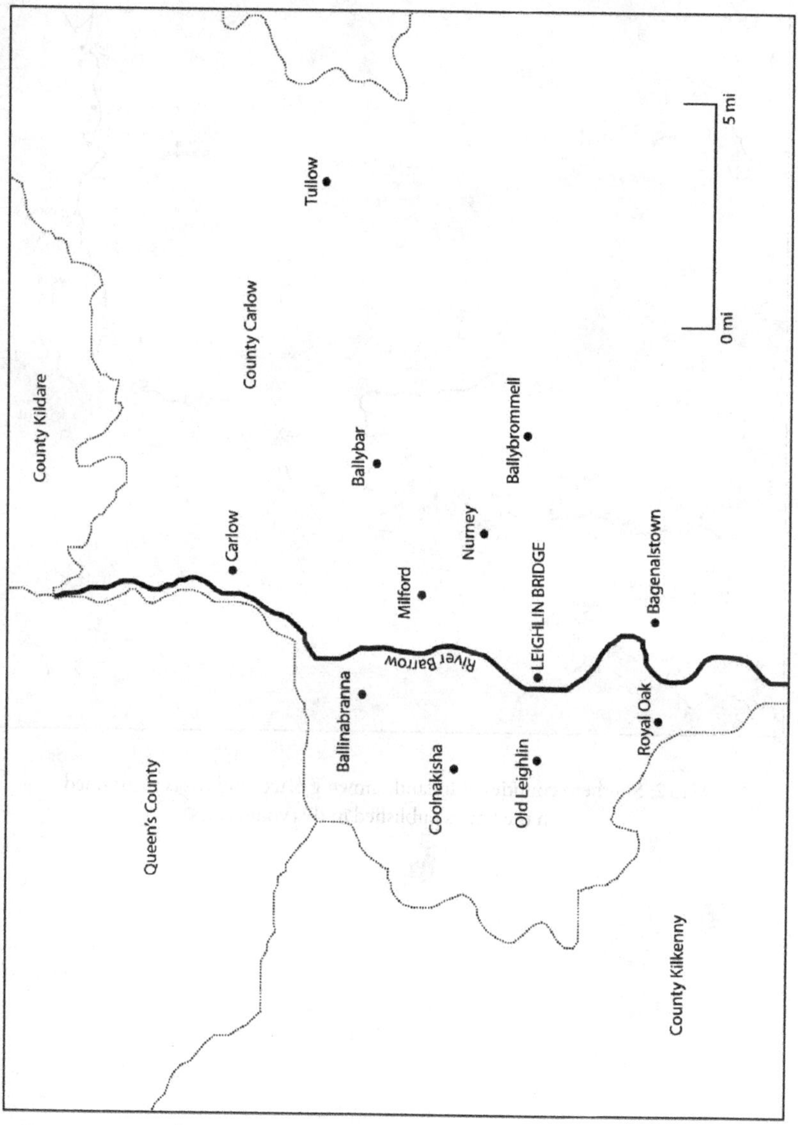

Map 1: County Carlow, showing the River Barrow and places mentioned in the letters published in this volume.

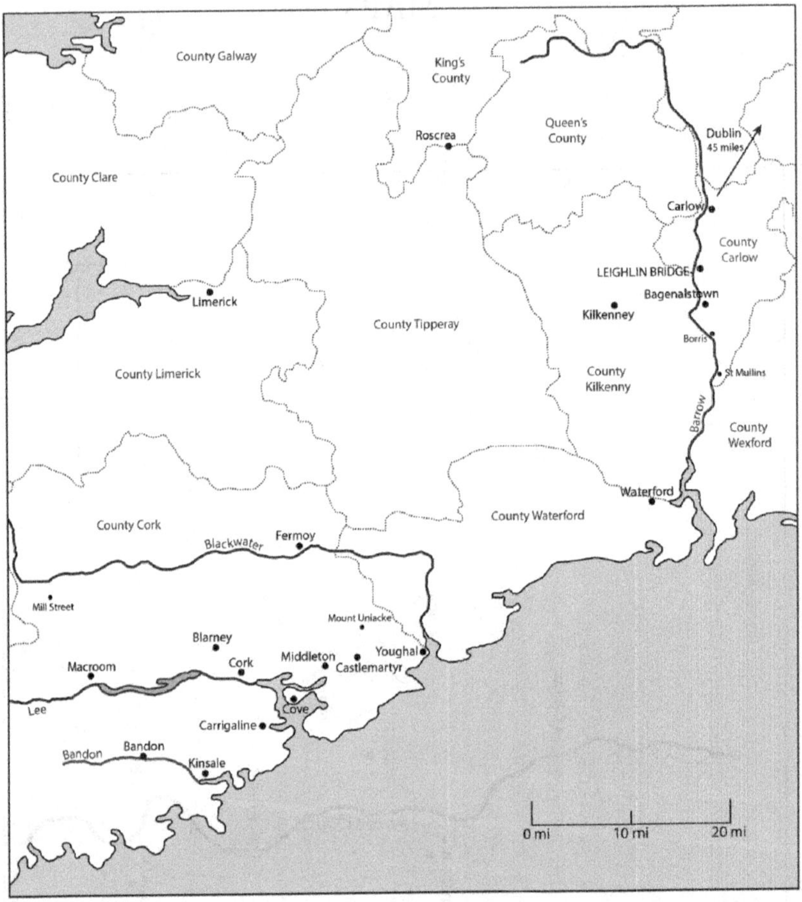

Map 2: Southern counties of Ireland, showing places and rivers mentioned in the letters published in this volume.

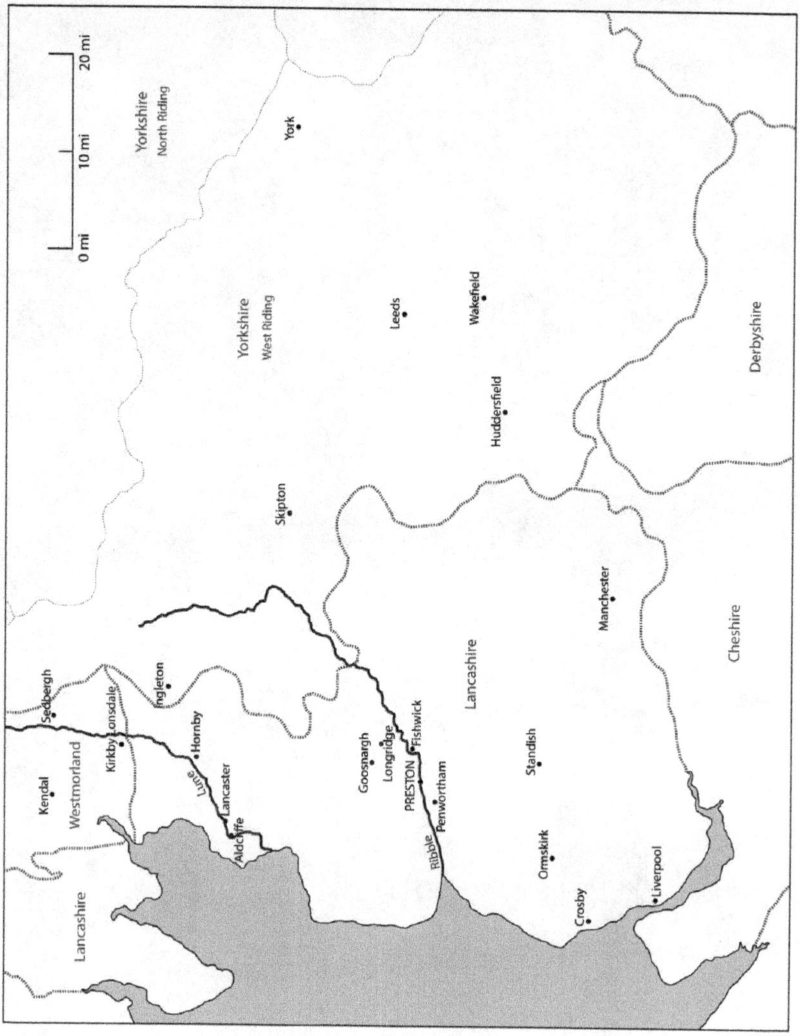

Map 3: North-west England, showing places and rivers mentioned in the letters published in this volume.

1840

To John Tyndall, Snr, 6 May 1840 0001

Mr John Tyndall | Leighlin Bridge | Co. Carlow | Youghal[1] | May 6th, 1840

My dear Father

I hasten with pleasure to relieve the anxiety which I know must fill your and my mother's mind on my account. I would have written to you immediately on my arrival[2] but I thought it better to delay a little until I could send you word of my being settled; this did not take place until this morning. So now I hope I have left you no room for impeaching me with ingratitude or unkindness – now for a description of the town and its inhabitants:– The town consists of one large long street with several minor ones branching from it at both sides, very good shops and extensive business carried on in the main street, a very good Quay – ships of considerable burden come in here. The structure of the town is remarkable to a Carlow eye who seldom views anything which modern improvement has not graced.[3] The houses are generally 3 and 4 stories high and very narrow which calls to the beholders mind their antiquity. The Cork 'you see' sounded rather harsh[4] to me on my arrival but I am now becoming accustomed to it. My lodging is in one of the small streets before alluded to; it leads from the main street to the quay. I and Evans stop together we are quite comfortable, we pay 10/- a week for our board.[5] We intend to change our mode of living and board ourselves when we get a little more acquainted with the manners of the people. Tell my mother to have the house clean and neat for me against my arrival at Christmas as if she does not I'll have wigs on the green.[6] Tell her not to fret on my account as her reason must tell her that I could never rise to any degree of respectability if I were to spend my days among my friends in Leighlin. Remember me to my uncle Caleb[7] and his family, to Mrs Payne, to John Mooney and Billy Henry. Tell John that if he was here listening to old Ocean's roar, he would forget his awl[8] and spend his days in making enigmas on its sublimity. Tell Mr Conwill if you see him that I will write to him shortly, remember me to both the Mr's Kellys old and young.[9] I cannot forget Jem Walsh as I have a parade of his handywork[10] at present before me – I am after eating a great mess of fish. Farewell

Your affectionate son | John Tyndall

When you write direct letters to me – | Ordnance Survey Office, Youghal.

Tell Jerry Murray that John is very well. Remember me to Mr and Mrs Murray I have seen William, his work lies 3 or 4 miles from town

RI MS JT/1/10/3182
LT Transcript Only

---

1. *Youghal*: Tyndall had arrived in Youghal, County Cork, 'a considerable market town and sea-port' with a population of about 10,000 (*Slater's*, p. 336).
2. *my arrival*: Tyndall probably arrived in Youghal on 3 May 1840, having left Leighlin Bridge on 29 April.
3. *a Carlow eye ... not graced*: Carlow's modern public buildings included 'the new cathedral for Roman Catholic worship [which] claims pre-eminence, for the beauty of its style and architecture; a new courthouse; a new jail; a lunatic asylum, also a new building' for 120 patients (*Dublin Penny Journal*, 26 July 1834; reprinted in *Carloviana*, 1 (pt 16, 1967), pp. 21–2).
4. *The Cork ... rather harsh*: Tyndall, from County Carlow, notes the difference in accent of the inhabitants of County Cork.
5. *we pay ... our board*: As Tyndall and Evans were each paid 2*s*. 6*d*. a day (15*s*. per week), they were each left with 10*s*. a week after paying for board and lodging.
6. *wigs on the green*: an altercation.
7. *my uncle Caleb*: Caleb Tyndall.
8. *awl*: a small tool with which holes may be pierced (*OED*); John Mooney was either a shoemaker or carpenter.
9. *Mr's Kellys old and young*: not identified.
10. *his handywork*: Tyndall's boots; see letter 0043 in which Tyndall complains that the 'old strong boots that Jem Walsh made me' had now worn out.

From John Tyndall, Snr, 8 May 1840 0002

To John Tyndall | Ordnance Survey Office | Youghal | Leighlin Bridge | May 8th. 1840.

Dear John

I received your letter on the 7th instant[1] and am happy to hear that you are well and settled in your lodging. I was very uneasy to hear from you as the weather was so very hot during your time on the road I was afraid you had caught a severe cold. Another thing that caused me to fret much was the unguarded manner you acted the evening you left this[2] by walking from the back to the front of the coach and it going at the time. For God's sake for mine and your mother's be more watchfull. And as the sea is so convenient to you do not make too free with it or swim beyond your depth. Your mother and Emma are frettin for your absence, but I think it all foolishness. The Cork 'you see'[3] I believe are a curious sort of people but I hope not half so bad as the South Boors.[4] I was told that you and seven or eight of the men were together in one house, I am glad to find that such is not the case. Mr Stewart told me he saw you on the coach towards Kilkenney,[5] and that if anything came round for your benefit during your absence him and others would not forget you. John Mooney says he wont go to Youghal to hear old Ocean's roar but that he is very well pleased to hear from you. And to prevent wigs on the green[6] the house must be neat and clean for you against Christmas. I hope Corporal Davey is in better health than when he left Leighlin and let him know that he has my best wishes both for his temporal as well as for his eternal welfare.[7] Remember me to Phillip Evans and all the boys. I myself your mother, Emma and all your friends send their love to you

I remain your affectionate father | John Tyndall

P.S. | Dont neglect writing to me often, and to mention Emma in your next.

RI MS JT/1/10/3183
LT Transcript Only

---

1. *your letter on the 7th instant*: letter 0001, which is dated 6 May.
2. *left this*: presumably left this place.
3. *'you see'*: compare letter 0001, n. 4.
4. *the South Boors*: not identified.
5. *the coach towards Kilkenney*: The Royal Mail coach, which left Leighlin Bridge at 2.30 each afternoon passed through Kilkenny (about 12 miles south-west of Leighlin Bridge) on its way to Cork.
6. *wigs on the green*: an altercation.
7. *I hope Corporal ... eternal welfare*: Tyndall's father had known James Davy when he had been stationed in Leighlin Bridge. Davy experienced poor health and in September 1840 a doctor treated him for throat and neck problems and recommended that he be transferred to a 'milder climate' (NAI OS/2/16).

To John Tyndall, Snr, [13] May 1840 0003

Youghal | Wednesday evening | May 18th, 1840.[1]

My dear Father

Again do I raise my pen to show that I do not forget you believe me no other motive induces me to write at present for to tell you the truth I have got very little to say.

You told me in your last[2] that my mother and Emma were fretting on my account. It has given me much concern to find that I am the cause of any uneasiness to my mother tell her that she wont know me at Christmas. I'm getting so fat and rearing such a fine whisker.

You bid me mention Emma believe me I would have done so in my last did I not feel quite assured that she was as well as she could be. Tell her that while she enjoys such a good appetite as that which she possessed when I left her there can be no fear of her.

You may depend on it that I will give good heed to the advice you gave me in your last. Good people are scarce and I will take care of myself. Remember me to John Mooney

farewell | Your son | John Tyndall

The moment I have any news I will transmit it all to you | Write soon | J.T.

RI MS JT/1/10/3184
LT Transcript Only

---

1. *Wednesday evening | May 18th, 1840*: The date may be incorrect as 18 May 1840 was a Monday; probably 13 May.
2. *your last*: letter 0002.

To John Tyndall, Snr, 27 May 1840 0004

Youghal | May 27th, 1840.

My dear Father

The receipt of your letter[1] has been a source of much gratification to me I was daily expecting to hear from you The reason you gave for not writing before 'tis my opinion is quite unfounded, for the intelligence which your letter conveyed respecting the defeat of the Radicals was quite new to me, and sounded melodious to my ear. You spoke of their political <u>death</u> and it is my sincere wish that they may never enjoy the benefit of a <u>resurrection</u>. I heard some vague rumours respecting the triumph of Conservatism[2] but nothing decisive until I received your letter.

My mode of living has been altered since I last wrote to you. I have changed my lodgings[3] and am at present boarding myself. Evans and a lad from Cork[4] stop here along with me. We pay 5/- a week for two rooms.[5] The place is greatly detached – no bustle or noise about it, it being a good distance off the main street, it is much more comfortable than my formal lodgings. I think I will content myself and stop here during my stay in Youghal.

I am just after returning from the field after my day's work, I have been employed there these few days back. The Survey of this town has just commenced, I will be engaged on it. The practice which I will have both at chain and instrument[6] will be of great service to me. I was obliged to buy a straw hat last night being quite sun-burnt for the last three days. I have written to Mr Conwill I posted the letter for him[7] the evening before I received yours. I told him in a

P.S.[8] to desire you to write to me but the receipt of yours has rendered that needless.

I am extremely glad to hear that my mother and Emma are well My mother will be glad to hear that I bought the straw hat as she told me to do so more than once before my departure. Send me word in your next how Emma Tyndall[9] got over the small pox and whether she will be marked – as bad as I like her I would be very sorry for such an occurrence.

Tell John Mooney that if he was with me here I could give him a fine view of the moon or north star thro' the telescope of the instrument[10]

Farewell | Yours &c. | John Tyndall

Send me word how John M'Gee's affairs[11] stand. I had almost forgot to mention Corporal Davey I am sorry to state that his health is still very precarious.

RI MS JT/1/10/3185
LT Transcript Only

---

1. *your letter*: letter missing.
2. *the defeat ... triumph of Conservatism*: The 'defeat of the Radicals' probably refers to the rejection by the Commons on 20 May of a motion for committee of Lord Stanley's Irish Registration Bill, which would have restricted the franchise in Ireland. This increased the likelihood of the Bill becoming law and thus represented a defeat for Melbourne's Whig administration. It might also refer to the Conservative candidate wresting the Ludlow seat from a Radical on 22 May.
3. *my lodgings*: Tyndall now lodged at the house of Mr McGrath, Cork Lane, Youghal.
4. *a lad from Cork*: John Tidmarsh, who would have been 16 or 17 years old.
5. *5/- a week for two rooms*: Both Tyndall and Evans were earning 15*s*. a week and Tidmarsh's pay was probably 14*s*. per week, but possibly less.
6. *chain and instrument*: Previously Tyndall had worked mainly as a draughtsman in the Survey office and therefore had limited experience at fieldwork. The chain used for measuring distance is 66 feet in length and the instrument was probably a theodolite – comprising a small telescope on a tripod and used for measuring angles.
7. *the letter for him*: letter missing.
8. *a P.S.*: a postscript.
9. *Emma Tyndall*: Emma, the daughter of Caleb Tyndall, later Emma Young.
10. *the telescope of the instrument*: the telescope that formed part of the theodolite (or possibly the level) used in surveying.
11. *John M'Gee's affairs*: Like other maltsters in County Carlow – 'one of the largest malting districts in Ireland' (*Parliamentary Papers*, 31 (1833), pp. 98–9) – M'Gee had suffered financially under the excessive excise duty imposed on malt by Parliament. He was rendered bankrupt in July 1840 (*Freeman's Journal*, 28 November 1839, p. [3] and 23 July 1840, p. [3]).

From John Conwill [30 May] 1840 0005

'My dear Tyndall'[1]

> 'With the man thou lovest ne'er contend
> But bear a thousand frailties from thy friend.'[2]

I feel confident that I have no reason to offer an apology for not furnishing you with a solution to my triangle question;[3] as your personal knowledge of my principles precludes further ceremony on my part.

I had the consummate pleasure of reading your soul-thrilling epistle for your dearly beloved mother,[4] on whom your language had the effect of an electric shock; so powerful were its effects that in an ecstacy of joy she exclaimed, Oh! my darling son, I knew he had those innate qualifications born with him, though they are aptly denied by logicians. Nor were your sentiments heard with less enthusiasm by your father and sister and my mother. I almost forgot, your mother expects that you will not pamper your whiskers too much, because she has been informed that Munster[5] men are not fond of hairy faces.

I now stand far above the reach of my most inveterate opponents.[6] I am a first class teacher under the auspicies of the Board,[7] receiving twenty pounds per annum. Next August there will be an examination of teachers in Graigue Carlow National School,[8] at which I hope to make an able display, and hold up to public gaze many empirics who are impeding the progress of literature and debilitating the march of intellect through the base power of lordly influence.[9]

I fear you are greatly annoyed by those satellites or wandering stars by which you are encompassed: that you may hurl them from their orbits and make them run lawless through the sky, I shall now commence with the questions.[10]

To bisect an isosceles triangle by a line parallel to the base, by the first book of Euclid.

Let ABC be the [isosceles] triangle [Figure 1[11]]. Make the [quadrilateral] HGOT = [in area to the] ΔABC, having the ∠GOT = ∠BAC: if HGOT be a rhombus, make AF = GO; draw FE parallel to BC, and you have ABC bisected as required.

If HTOG be not a rhombus, produce GO if necessary, to P, bisect GP in N, erect NO perpendicular to GP, making LNG2=LACD, CD being perpendicular to AB; from Q as centre and inter- QG describe the ONMGKLP; produce 2N and TO to K and L; join GL and describe a rhombus on GL having ∠GLT = ∠BAC or ∠GOT; make AF=KT, and draw FE parallel to BC and the thing is done.[12]

Demonstration. Join GH, KP, LP, 2P, SP and LH. Produce K2 to N and draw L2N. ∠NG2 and GN2 - ∠ACD and ADC,.. ∠G2K-BAC, whence it is easy to show that the circle passes through P, that ∠GK2X ∠PK2 - ∠ABCX ∠ACB; that ∠G2NX ∠P2N-2 ∠GBP, that ∠G2MX ∠P2M-2 ∠GLP; and that ∠G2NX ∠P2N - ∠G2MX ∠P2M... ∠GKP-GLP: wherefore PLT is a straight line, and it may be shown that SGP- LGH, and of course [quadrilateral] GSTK=[quadrilateral] HGOT=ΔABC. Consequently ΔGST = ΔAFE = half ΔABC.[13]

Q. E. D.[14]

To bisect any triangle by a line parallel to the base without passing the first book of Euclid.[15]

Let ABC be the given triangle [but not isosceles]. It is easy to make the isosceles ΔABD [where D is on BC, such that AD = AB]; bisect the ΔABD by means of the parallel line TH according to the preceding method and produce FH to I; GI bisects ΔABC.

Demonstration. It is plain ΔAOD - ΔBOC, hence EO-OF, EF being parallel to AB. If not one is greater than the other; let EO be the greater then ΔEODX ΔEOA would be greater than ΔFOCX ΔFOB, that is AOD would be greater than BOC, an absurdity; wherefore ΔEODX _EOG- ΔFOCX ΔFOH, and whence the ΔAGE ΔBHF; from this FG-HI; if not, let FG be greater than HI; then ΔFGEX _FGA will be greater than HIFX ΔJHB, that is AGE is greater than ΔBHF, an absurdity; ergo FH-IG, and ΔFHD - ΔIHC- _CAB.[16]

Q. E. D.

Given the hypothenuse of a right angled triangle to construct it, so that the rectangle under the difference of the legs[17] and hypothenuse will be equal the area of the triangle.

Given the sum of the base and perpendicular and the perpendicular from the right angle on the hypothenuse, to construct the triangle by the first book of Euclid.

I cannot send you any more problems in this letter, when I write again, I will send you ten more, which will constitute the twelve apostles, in a geometrical point of view. With all due respect I remain your ever affectionate teacher

J. Conwill

RI MS JT/1/11/3510
LT Transcript Only

---

1. *'My dear Tyndall'*: In LT's typescript of this and other letters from John Conwill, the opening greeting 'My dear Tyndall' appears in quotation marks. These quotation marks, which occur only in letters from Conwill, have been retained since they may possess some meaning which the editors have not been able to ascertain. Regarding the date above, 30 May 1840, in this letter Conwill refers to Sarah Tyndall's receipt of 'your soul-thrilling epistle'. It is not clear whether he is referring to letter 0006 (if so, one of the two is wrongly dated) or to a letter that is missing.
2. *With the man ... thy friend*: 'Oh, never with the man thou lov'st contend! | But bear a thousand frailties from thy friend', Thomas Fitzgerald, 'Golden Verses of Pythagoras'.
3. *a solution to my triangle question*: not identified as Tyndall's preceding letter to Conwill is missing.
4. *soul-thrilling epistle ... beloved mother*: Tyndall appears to claim that letter 0006 – next

letter – was the first sent to his mother. If that is the case then this letter and letter 0006 are in the wrong order.

5. *Munster*: The province in the south west of Ireland comprising counties Clare, Cork (where Tyndall was stationed), Kerry, Limerick, Tipperary and Waterford.
6. *my most inveterate opponents*: probably the school inspectors whom Conwill despised.
7. *the Board*: The National Board of Education had been initiated by Parliament in 1831 with two members drawn from each of the three main religious communities – Roman Catholic, Church of Ireland and Presbyterian. National schools were required to demarcate between secular and religious subjects; thus, Conwill confined his religious teaching to Saturdays and 9–10 a.m. on weekdays, but parents could withdraw their children from those lessons.
8. *Graigue Carlow National School*: *Slater's* (p. 21) lists Michael Henessy as the master of the National School at Graigue, the suburb of Carlow on the west side of the River Barrow.
9. *many empirics ... lordly influence*: presumably a reference to those aristocrats and landowners who wished to limit school education to pragmatic subjects.
10. *the questions*: A number of errors occur in LT's transcription of the following paragraphs. The editors have tried to reconstruct the geometrical arguments to the best of their ability and have added a diagram; LT's typescript does not include the diagrams that would have appeared in the original letters.
11. *Figure 1*:

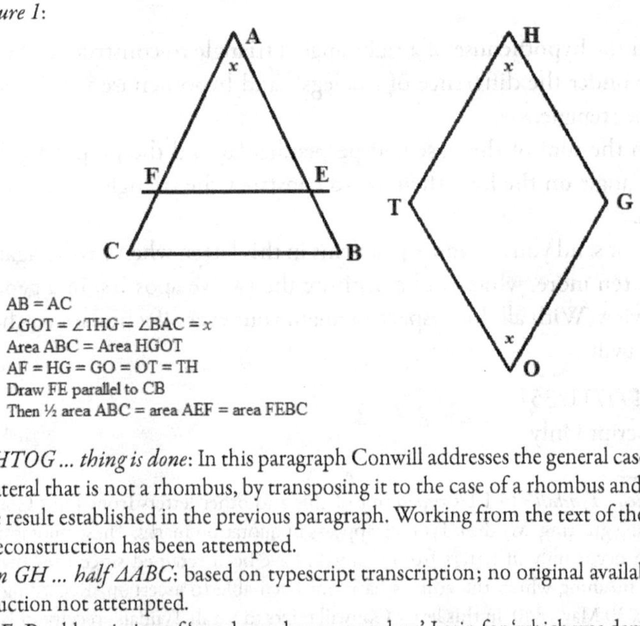

12. *If HTOG ... thing is done*: In this paragraph Conwill addresses the general case of a quadrilateral that is not a rhombus, by transposing it to the case of a rhombus and then using the result established in the previous paragraph. Working from the text of the typescript a reconstruction has been attempted.
13. *Join GH ... half ΔABC*: based on typescript transcription; no original available. Reconstruction not attempted.
14. *Q. E. D.*: abbreviation of 'quod erat demonstrandum', Latin for 'which was demonstrated'.
15. *To bisect ... book of Euclid*: Cf. method used by Miles Bland, *Geometrical Problems Deducible from the First Six Books of Euclid* (Cambridge: A. Smith, 1819), p. 172.
16. *It is plain ... CAB*: based on typescript transcription; no original available. Reconstruction not attempted.
17. *legs*: A leg is 'one of the sides of a triangle, viewed as standing upon a base' (*OED*).

To Sarah Tyndall 8 June 1840 0006

Youghal | June 8th, 1840.

My dear Mother

It is with inexpressible delight that I am now enabled to write to you, I am fully convinced that this will be received by you with as much pleasure as I feel in writing it, you will no doubt wonder at my using the word 'enabled'. I could no doubt have sent you a letter before this, but I did not like to send it empty but the moment I have scraped together the first pound I hasten with great satisfaction to make it and the letter fellow travellers. You will give John Mooney 5 shillings in payment for some books which I got from him at my departure. I am sorry that I am obliged to make the pound so much less, but I know you will not think me the less goodnatured on that account. Believe me it would give me five times as much pleasure were it a five pound note I was going to send you. Apologize for me to John Mooney I did not expect to keep him so long out of his money but being so unsettled when I first came here I was put to great expense which rendered me utterly unable to save a pound any sooner. Though I ask you to apologize to John I am fully convinced that he did not harbour the slightest doubt of my integrity. I am <u>very</u> comfortably situated here. Evans and I and a lad from Cork named Tidmarsh, a nephew of Tidmarsh of Kilkenny,[1] live together, we pay 5 shillings a week[2] for two very comfortable rooms and we call for what we please in the way of board. Mr Read had just come to stop with us for a month or so, I was informed by Wm[3] Heydon that my father was ill I'm astonished that you did not let me know it if such was the case, do not neglect it when you receive this. I hope Emma still preserves her appetite it would give me much concern to find that she was losing it. I hope you do not waste your time in thinking of me. I would not have you forget me totally, but it would be a cause of deep regret to me to find that I was the cause of one sleepless hour or melancholy thought to you.

I remain my dear Mother | Your affectionate Son | John.

Upon second thought I find that it is safer for me to send you the half note first and upon receipt of it for you to let me know and I will immediately transmit the other half.

RI MS JT/1/10/3186
LT Transcript Only

---

1. *Tidmarsh of Kilkenny*: *Slater's* lists James M. Tidmarsh & Co. – Woolen and Linen Drapers and Haberdashers – of High Street, Kilkenny.
2. *we pay 5 shillings a week*: At that time Tidmarsh was paid 1*s*. 4*d*. per day and Tyndall 2*s*. 6*d*., so their combined income was 23*s*. per 6-day week.
3. *Wm*: William.

From John Tyndall, Snr 4 July 1840 0007

Mr John Tyndall | Ordnance Survey | Youghal | Leighlin Bridge | July 4th. 1840.

My dear John

I received your letter of the 29th of June,[1] and I and your mother are very happy to hear from you, and that you are well. I would have written to you before I received your last, but knowing that letters had passed between you and your mother I thought it needless to trouble with more so soon. I hope you are pleased with the work you are at present employed at. I think it more wholesome and more instructive than any you could get. I believe it is very critical but with all that I know you will be able to surmount all its difficulties and walk triumphant over the demesne of that doting old lady who thought you were about to take possession of her property.[2] I was speaking to Mrs Steuart this day, she desires me to let you know that she is very glad to hear you are going on well, and that you have her best wishes for your welfare. She desires also that Mr Reid would write to her and let know how he is going on. John McGee's affairs[3] are in a very bad state, he has served notices on all his creditors and expects to come out under the benefit of the insolvent act[4] about the middle of this month. James Treacy has been liberated under the same act.[5] Their situations are deplorable. I am sorry you were not at home for the last six days back as I had an old friend and school-fellow by the name of Ryan with me during that period. He is from New York in America. This is the second time I have seen him these thirty three years. He is a most intelligent man and desires that you would make yourself as perfect an Engineer as possible, then you need not fear to make out a respectable living. He has a property worth fifteen hundred a year, and is a magistrate and a member of Congress for the district of New York. He has left a pamphlet[6] for you which I will send by the Miss Murphys[7] when they are going to Youghal. You might often hear me speak of Robinson Tyndall, he is worth forty thousand pounds, and his brother William also very wealthy. Mr Conwill will send you the twelve Apostles in the next letter,[8] and wonders you did not write to him since you received his. Send me word how Corporal Davey's health now is. My opinion is that it is much better, when he was able to convey the old lady's report from Mr Wynne to you. I would be glad to hear of his perfect recovery, Radical though he be.[9] I saw William Wright in Carlow on the 22nd of June, he has a sore knee and came home to get it cured. He told me that he expected to get a situation in Cork and if he did he would go to Youghal to see you. John Kennedy[10] son to the clerk of Clonmelch church requested I would ask you to exert all the interest you have with Corporal Davey to get a situation as writer in the office, and I would be glad that you would do so. John Maginneys[11] has died about a fortnight ago after a few days illness with fever on the brain. I, your mother and all your friends here are well, and conclude with sending our love to you

I am Your affectionate Father | John Tyndall

I wrote this letter very bad John, but the next may be better.

RI MS JT/1/10/3187
LT Transcript Only

---

1. *your letter of the 29th of June*: letter missing.
2. *demesne of that ... her property*: refers to an earlier incident, probably while Tyndall was surveying in the field near Leighlin Bridge. The issue of access to properties by surveyors working for the Survey was addressed in the Ordnance Survey Act of 1841; see letter 150, n. 6.
3. *John McGee's affairs*: see letter 0004, n. 11.
4. *the insolvent act*: Bankrupts were 'discharged of all suits and imprisonment, upon delivering up all their estate and effects to their creditors upon oath, at the sessions or assizes' ('Insolvency, Act of' in A. Rees, *The Encyclopædia*, vol. 19 (London: Longman, 1819), n. p.). Under the Insolvent Debtors (Ireland) Act (1813) those imprisoned for debt could apply to the Court for the Relief of Insolvent Debtors to be released.
5. *James Treacy ... same act*: James Treacy, who is described as a maltster of Leighlin Bridge, had his petition heard at the Court for the Relief of Insolvent Debtors, 3 Lower Ormond-Quay, Dublin on 1 July 1840 (*Freeman's Journal*, 10 June 1840, p. [1]). Treacy's petition was evidently successful.
6. *a pamphlet*: not identified.
7. *the Miss Murphys*: not identified.
8. *the twelve Apostles in the next letter*: see letters 0005, 0009, 0017, 0035 and 0039.
9. *Radical though he be*: Although Davy and Tyndall, Snr, respected one another they clearly disagreed over politics. '[I]t is likely we shall never perfectly agree below. Yet my friend if we agree to love the Lord and serve him with a perfect heart we shall agree when we meet above ...' James Davy to John Tyndall, Snr, 18 August 1840 (RI MS JT/1/11/3555).
10. *John Kennedy*: There is no evidence that Kennedy was hired by the Ordnance Survey.
11. *John Maginneys*: The Parish of Carlow burial register (p. 117) records the burial of John McGinnis of Turlow Street, Carlow, on 9 June 1840 (Irish Church Records).

To John Tyndall, Snr 18 August 1840 0008

Youghal, Tuesday morning | Aug 18th, 1840.

My dear Father

I am just after returning from a week's sojourn in the country. I saw a letter of yours which Corpl. Davey received yesterday[1] in which it was stated that I had received three letters from you and never answered one, this surprised me greatly as I did not receive a scroll since I last wrote to you. So it appears that either yours or mine must have been mislaid I hope however that a like fate wont befall this. The time I spent in the country[2] was not the most agreeable. It was in a farmer's house I took up my abode – very unlike the Carlow farm houses. Evans was with me I could not procure a room for our use so we were obliged to sleep in the same room as the farmer and his wife. I'm sure I could enjoy myself in the County Carlow listening to the country folk breaking their jokes and telling their stories; not so here – the eternal jabber of Irish[3] kept up by them resembled more the clatter of geese on a pond than a language spoken by human beings. My work lay in one of the wildest parts of the County along the sea shore. You know that when the tide goes out it leaves in some places an extensive strand which is generally composed of either sand or mud. Unfortunately it was with the latter description that I had to deal you might sometimes see me wading thro' the mud with my shoes in my hands at a great distance from the shore while the howling of the wind, the distant roar of the Atlantic and the cry of the Gull descending from the regions above all added to the solemn dreariness of the scene. I have however closed accounts with the farmer and returned to Youghal. I'm now employed in this parish. I'm still in the field. You'd scarcely know me if you saw me at present I'm quite bronzed with the weather. I enjoy excellent health I'm nearly as strong as an ox; every nerve has its proper tone, every sinew is braced, (my feet?) have partaken a little of the nature of a horse's hoof. So I think to sum up the whole I may consider myself a smart hardy chap. If possible find out for me where Wm Wright lives, I'd like to know as it's a long time since I heard from him. Mr Wynne is about, on account of Mrs Wynne's health, he is going to some foreign station in about a month.[4] There are several opinions respecting the fate of the division when he goes, some think it will be removed to Cork under the eye of Major Waters. Some think to Fermoy[5] to join another division there, while others think that we will get another officer in his place, this last opinion is I think the most probable. Give my love to my mother and Emma. Remember me to my uncle Caleb and his family, to my uncle Ned,[6] to Mr Holden and John, to Mr Brophy,[7] send me word how Mary Anne[8] is getting on, to both the Mr Kellys, Mr Williams, Mr Henry, Jerry Murray, Mr Hugh and Mrs Murray, Mr Young[9] – tell him to be prepared for a raking in Position[10] when I meet him – Jemmy Kinsela[11] and Jem Walsh, oh what a string of friends! and last tho' not the least to Mr Mooney, tell him I have forgotten all my philosophy[12] among the Cork savages. I'm afraid that there are others which I forget to mention but they all have my best wishes

I remain your affectionate son | John Tyndall

Write soon.

RI MS JT/1/10/3188
LT Transcript Only

1. *letter ... received yesterday*: letter missing, but see James Davy to John Tyndall, Snr, 18 August 1840 (RI MS JT/1/11/3555).
2. *The time I spent in the country*: on a surveying expedition.

3. *the eternal jabber of Irish*: Tyndall understood little (if any) Gaelic, the predominant language in rural parts of County Cork.
4. *he is going ... about a month*: Wynne left Youghal on 18 September 1840, but it is not clear whether he was immediately posted abroad, although he later saw service in several countries. His wife, whom he married on 16 December, was named Anne (née Osborne).
5. *Fermoy*: Town on the River Blackwater (which meets the sea at Youghal), about 20 miles north-west of Youghal.
6. *my uncle Ned*: Edward Styles.
7. *Mr Brophy*: not identified.
8. *Mary Anne*: possibly Maryann Commins or Maryanne Kearney.
9. *both the Mr Kellys ... Mr Young*: most people on this list have not been identified.
10. *raking in Position*: may refer to the strategy of fighting ships to align themselves so that all their guns would be pointing at an enemy ship.
11. *Jemmy Kinsela*: not identified.
12. *Mr Mooney ... all my philosophy*: The perpetual curate at Leighlin Bridge Paris Church was named the Rev. Peter Mooney (*Slater's*); see also letter 0011. However, the Mr Mooney referred to here was the father of John Mooney who was probably a Catholic as he later tried to prevent Roman Catholics from dealing with Tyndall's father; see letter 0084.

From John Conwill 6 September 1840 0009

Leighlin Bridge, Sunday Morning

'My dear Tyndall'

I am constrained to become a Sabbath breaker[1] to respond to the calls of 'friendship', and notwithstanding your animadversions on the disparity of our situations, and the many inconveniences under which you labour, I possitively assure you that I have undergone many fatigues these three months endeavouring to reclaim the brutal conduct and crude, undigested minds of several of my pupils, who were first placed under the jurisdiction of some pharisaical[2] dolt, whose alter-scraping evolutions became the shield of his stultified capacity. I had, moreover, to be very punctual as to my attendance at Mr Cullen's[3] every evening, and my calls every Saturday have been numerous; so that that disparity so much complained of by the pilgarlic in the County Cork,[4] exists not between him and the poor waywarn Argyle[5] in the County Carlow.

To be brief, had I committed no other crime than that of writing to my pupils on the Sabbathday, I entertain not the slightest doubt but the recording angel would drop a tear upon such fault and blot it out forever.[6]

Your graphic sketch[7] of the romantic scenes of the County Cork, in which you have displayed so much of the imagery of fancy and the natural effusions of an enlightened genius, has filled my mind with thoughts elate: who can read your plain, unvarnished description of the curlew's wild whistle, the roaring of old ocean's waves, and the thundering of the hollow winds, without being struck with the sublimity of your gigantic mind? I cannot pass this subject over without introducing a few lines from the works of the inimitable Ovid: –[8]

> You, though from heaven remote, to heaven will move
> With strength of mind, and tread the abyss above;
> And penetrate, with an interior light,
> Those upper depths, which nature hid from sight.
> Pleas'd you will be to walk along the sphere
> Of shining stars, and travel with the year,
> To leave the heavy earth and scale the height
> Of Atlas, who supports the heavenly weight;
> To look from upper light, and thence survey
> Mistaken mortals wand'ring from the way.

I presume to say that you will not deem me guilty of anything bordering on flattery in the preceding lines, as you know I am far above such fulsome conduct. Why I did so is, the meed of your merit deserves more than my humble pen can confer.

When I read the contents of your luminous epistle for your dearly beloved mother, her cheek became tinged by means of the sentimental glow that pervaded her whole frame, her eyes sparkled with joy and pourtrayed the gladness with which her heart was replete; then, then!! during her enthusiastic emotions, she exclaimed oh! my son, oh! my darling John, you are another Milton;[9] and in conclusion she ordered me to solicit Mick Bergin[10] to investigate the 47.E.B.'s,[11] that we might form an idea of the County Cork folk, when they would be addressed for pen and ink: I acceded to her wishes, and what was Mick's answer? It was, that when the people of the County Cork became acquainted with the 'refined terms', pen and ink, he would illustrate the Pythagorean theorem.[12]

Mr Butler was with me on Friday and after he had gone through his usual etiquette of examination he wrote a beautiful panegyric on me and my pupils.[13] You shall hear again from me in the latter part of this month, or early in October. Inclosed I have transmitted to you another portion of my geometrical apostles.[14]

I am yours very truly | John Conwill

RI MS JT/1/11/3511
LT Transcript Only

---

1. *a Sabbath breaker*: 6 September 1840 was a Sunday.
2. *pharisaical*: legalistic, self-righteous, hypocritical (*OED*).
3. *my attendance at Mr Cullen's*: Probably the farmer named Cullen who lived in the townland of Craan (one side of the road is in Craan and the other side is in the townland of Rathornan) about a mile outside Leighlin Bridge on the road to Milford. Conwill lived on the opposite side of the road in the townland of Rathornan but in a field or two from

the road and almost on the banks of the River Barrow.

4. *pilgarlic in the County Cork*: a pilgarlic is a pitiable, lowly, or foolish person (*OED*); the recipient of this insult has not been identified but was probably mentioned in a missing letter from Tyndall.
5. *poor waywarn Argyle*: not identified.
6. *the recording angel ... out forever*: In Christian thought the recording angel is assigned by God with the reponsibility of recording an individual's actions, both good and bad. Conwill here expresses the hope that his desecration of the Sabbath by writing to Tyndall would not be recorded as a sinful act.
7. *Your graphic sketch*: Tyndall had sent Conwill examples of his compositions, some of which are deposited in RI/JT8/1/2.
8. *Ovid*: Lines from *Metamorphoses*, XV.81–4 and 213–18, by the Roman poet Ovid (43 BC – 17 AD), describing the heavens according to the Pythagorean system.
9. *Milton*: John Milton (1608–74) was the foremost English poet and man of letters during the Commonwealth, a period of considerable religious and political upheaval. His best-known work is the epic poem *Paradise Lost* (1667).
10. *Mick Bergin*: not identified.
11. *the 47.E.B.'s*: not identified, probably a transcription error by LT.
12. *Pythagorean theorem*: the theorem articulated by the Ionian mathematician Pythagoras (*c*. 570 BC–*c*. 495 BC) that relates to a right-angled triangle; the square of the hypotenuse being equal to the sum of the squares of the other two sides.
13. *a beautiful ... my pupils*: the school inspector's report.
14. *Inclosed ... geometrical apostles*: enclosure missing.

From William Wright 6 September 1840 0010

Roscrea[1] Sep. 6th 1840.

My dear John

Few moments for the last six months has given me greater pleasure than those I experienced on the receipt of your letter[2] of this morning week It has been a great cause of uneasiness to me of late that I should have so slighted your kindness. But all is now past never to be remembered. Oh, if I was near you now I would give your paw such a shake as that you would remember it for years to come.

I am glad to find that the fair sex has made some impression on you,[3] I hope you will not let it slip; if you do take my word for it that you will be sorry, although I have had very little experience about the women so that I am a bad judge. But this I know, that opportunity once lost can never be recalled, so keep tight to the work and do not for a moment relax your hold.

When I had the pleasure of seeing your father in Carlow I had an intention of accepting a situation in Cork which had been offered me by a brother-in-law of Thomas's,[4] but some engagements which my present master had engaged in prevented me, along with engagements of my own with one of the fair ones which I am determined to fall or rise by. In my next I will give you a more lengthened detail of my intentions. I am sorry that it is not in my power to give you a description of Roscrea, my pen, or scribble as I should call it, would fall short of giving you an idea of the ugliness of it and the surrounding country. In short it is the most wretched place that I have had the good luck to be in. You will oblige me by giving my respects to the ever to be remembered L Bridge[5] folk which are along with you in Youghal. Mind Jack keep up a constant fire upon me with your kind letters and I hope we will never fall in to our old error again. Farewell and may you be always as well as I wish. Do not forget but let me have one from you shortly

Yours truly | William Wright.

RI MS JT/1/11/3883
LT Transcript Only

1. *Roscrea*: market town in County Tipperary 38 miles west of Carlow.
2. *your letter*: letter missing.
3. *that the fair sex ... on you*: one example is Tyndall's romantic encounter in July 1840 with 'Miss Stedman' as related in RI MS JT/8/1/2, pp. 2–10.
4. *a brother-in-law of Thomas's*: not identified.
5. *L Bridge*: Leighlin Bridge.

To John Tyndall, Snr 26 September 1840 0011

Youghal Saturday morning | Sep. 26th, 1840.

My dear Father

I suppose you now think it high time for me to attempt something in the shape of a letter to you. My correspondence with Emma was the cause of my delaying so long, and my having nothing of any consequence to say was another strong preventative to my scribbling. But as I feel that anything no matter how trifling when it comes from you carries an interest with it which chains it to my heart. So now I begin to think that the feeling may be reciprocal and that you on your part feel grateful if you receive only a sheet ful of fudge from me.

I am now in the office tugging away at the old work. We have got a new officer his name is Whittingham. Great alterations are expected very shortly. Mr Wynne is leaving on account of Mrs Wynne's health; he is going to some foreign station in about a month.[1] I asked Mr Wynne for a testimonial as to my character and capabilities he refused me, alleging as a reason that if he gave me one every other man would expect the same he said there was no man he would feel happier in bearing testimony to and that if ever I would refer to him he would respond with pleasure to the call. I regretted his refusal very much as I know it would be of great service to me at a future day however I cannot say that I have been unfairly dealt with. Mr Roberts wrote to him for the same purpose he refused him also, so you see there was no partiality in the case.

Corpl. Davey is on the point of leaving, as his disease was never totally eradicated and the doctor finally asserted that his removal to a more favourable climate was the only chance he had of ultimate recovery, he is to be off on this day week.[2]

These are the only incidents connected with the office which have transpired since you last heard from me. Send me word how all at home are and how Emma likes the reading, tell my Mother I'll dirty her painted room for her at Christmas with the mud of Cork. Remember me to Mr Holden and Mr Mooney,[3] has John[4] forgotten his philosophy yet? Send me word how Mr M'Gees and Miss M'Gees affairs[5] stand

farewell | I remain your affectionate Son | John Tyndall

RI MS JT/1/10/3189
LT Transcript Only

1. *Mr Wynne ... a month.*: This sentence appears only in the transcription of this letter in LT, 'Biography', vol. 1, p. 19. Wynne's next posting was to the island of Zanthe (now called Zakynthos or Zante) in the Ionian Sea, a British protectorate where he served with the 97th Regiment of Foot.
2. *Corpl. Davey ... this day week*: Although he was due to be transferred earlier, Corporal James Davy left the Irish Survey on 10 October 1840 (NAI OS/1/17).
3. *Mr Mooney*: see letter 0008, n. 12.
4. *John*: John Mooney.
5. *Mr M'Gees and Miss M'Gees affairs*: see letter 0004, n. 11.

From William Wright 27 September 1840 0012

Roscrea Sep. 27th 1840

My dear John

It was after a ride of upwards of forty miles when I received your kind favour which you will please to receive as an apology for my not writing to you on the following day which was Sunday. If I neglect to write on Sundays I am obliged to defer it to the following Sunday, as I have not sufficient time on week days.

You seem to wonder that I should have come to the hasty conclusion that you were at last caught by sly cupid, but I assure you that the most inattentive reader of your letter[1] could not be but led astray by the beautiful description you gave me of Miss Fisher;[2] and your humble friend, especially knowing your naturally warm feelings, must have thought that you were really in love. I am sorry that my advice of keeping a tight hold goes for nothing. I am much obliged to you for your kind advice and shall attempt to follow it as far as in my power it lies John, I shall endeavour as well as I can to give you the particulars of my situation, hoping that you will keep it a secret, as you are the third person in the world who knows anything of the matter. As I have often seen the cup dashed from the lips before it could be tasted, I would not wish it were known and as I stand upon rather unfair ground as the fair one is sister to my employer[3] and he would not at all sanction it were he to hear it I would instantly get my dirty bit of paper[4] which might frustrate my hopes altogether. I have got a promise from her that no matter what may interfere, another shall never enjoy either hand or heart as long as I stay unmarried. But Jack a woman's promise in such a case as that is I believe very fragile and on that account I fear a separation. On the other hand, it is not a foolish promise of an eighteen year old heifer[5] as she is two years my senior[6] – a sensible steady girl as far as I can see – but you know John persons in my situation see with partial eyes. In person she is the picture of a comfortable homely woman, has nothing of that comliness of features which you describe in your letter and necessarily <*words missing*> appreciate notwithstanding she is <*words missing*> she is not so tall as myself and I <*words missing*> seldom see a squarer pair than we will make. I am afraid I have got tiresome to you and will take my farewell

Your well wisher till death | William Wright

I hope the Father and Mother are well, mine when I heard, were all well. We must not forget Leighlin my old boy, I hope those who are from it along with you are enjoying good health.

RI MS JT/1/11/3884
LT Transcript Only

---

1. *your letter*: letter missing.
2. *Miss Fisher*: not identified.
3. *sister to my employer*: Sarah Maria Neale, who may have been the sister of Henry Neale, listed in *Slater's* as a baker and grocer in Roscrea.
4. *my dirty bit of paper*: his dismissal notice.
5. heifer: derogatory term for a woman or a girl, likening her to a heifer (a young cow) (*OED*).
6. *she is two years my senior*: According to a gravestone in the Parker Cemetery, Arkansas City, Kansas, Sarah was born in 1815, which would make her at least 3 years older than William, who was born in 1819.

From John Tyndall, Snr 5 October 1840 0013

Mr John Tyndall | Ordnance Survey Office | Youghal
Leighlin.Bridge | Oct. 5th. 1840.

My dear John

It was with the most heartfelt pleasure I received your letter,[1] before I read its contents I knew that you were in good health from the very form of the letters on the superscription.[2] You mention that you asked Mr Wynne for a testimonial[3] as to your character and capabilites and was refused. Dont be the least dismayed in consequence of that, for you need not fear but that your own good conduct and abilities will be able to carry you triumphant over all the difficulties that may be placed in your way, and your own good sense will tell you that to expect to pass through this life without disappointment is more than falls to the lot of man to meet with. You have returned to the office, but if I had my will I would much rather have you in the field, because it would be a deal better for your health. I am sorry to hear the Corporal Davey is leaving the division[4] for I always thought him a good man, and let him know that he has my best wishes for his welfare. You speak of great alterations in the office very shortly. It is but of little consequence to you what alterations they may think proper to make as you have an home to come to and what is much better an hearty welcome, for your mother and I would feel the greatest pleasure in seeing you step in this moment with the mud of Cork on your shoes to dirty her painted room. I was speaking to Mr Conwill yesterday, he told me that as soon as an examination which is to take place in Carlow was over he would write to you and let you know every particular connected with it. You requested to know how John and Miss McGee's affairs[5] stand at present. Why worse they cannot be. John is still a prisoner in Dublin, and Celia is at present collecting subscriptions at a pound each to raffle for the piano, formerly the property of Mrs McGee. We are on the eve of a great contest in this county for the Office of Poor Law Guardians.[6] Mr Steuart, Mr James Thomas, Mr Wm. Fishbourne Jun.,[7] on the protestant side; and on the Roman Catholic, Edward Lyons of Finiscourt, Patt Foley of Old Leighlin, John Cummins of Coolnakisha[8] – these six out of which three will be returned – one for the Barony of Idrone West or the Ballynockan side of the bridge.[9] The same strife will take place through the whole county and the Priests are using every art to have their pets returned. May they be disappointed is my sincere wish. Only think of Curly Foley's son putting up for Guardian, whose father held five or six acres of land from your grandfather, but miracles will never cease.

Mr and Mrs Steuart are glad to hear that you are well, and wonders much that Mr Reid never wrote to them to let them know how he was going on. Dean Bernard often enquires for you. I will now conclude with letting you know that your friends here are in good health and joins me your mother and Emma in sending their love to you

I am your affectionate Father | John Tyndall.

Write to me as soon as possible. You wont be able to read this letter it is written so bad, no matter, you must do your best

RI MS JT/1/10/3190
LT Transcript Only

1. *your letter*: see letter 0011.
2. *the superscription*: writing on the outside, especially the appearance of the address.
3. *you asked ... a testimonial*: see letter 0011.
4. *Corporal Davey ... the division*: Davey had served in the 5th Division, C District, but was

due to leave the Irish Survey for England on 8 October 1841, although he did not depart until 10 October (NAI OS/1/17).

5. *John and Miss McGee's affairs*: John M'Gee had been imprisoned for debt; see letter 0004, n. 11.
6. *great contest ... Law Guardians*: Parliament had passed the Irish Poor Relief act in 1838, which created elected Boards of Guardians that were empowered to create workhouses that were funded from rates levied locally. For the contest in Idrone West, see *CS*, 3 October 1840, p. [3] and 17 October, p. [3]. The first three contestants listed stood as Conservatives, the other three as Liberals. Steuart, Cummins and Lyons were elected with, respectively, 665, 564 and 560 votes.
7. *James Thomas, Mr Wm. Fishbourne Jun.*: Both James Thomas (d. 1850), of the Friary, Leighlin Bridge, and William Fishbourne Jun. (1799–1875) of Fonthill (which is 4 miles north of Leighlin Bridge on the west side of the River Barrow), were wealthy gentry.
8. *Edward Lyons ... Coolnakisha*: Edward Lyons possessed an extensive farm at Finniscourt, about 3 miles south of Leighlin Bridge; Patt Foley of Old Leighlin was possibly a member of the Foley family that later included Patrick Foley (1850–1926), who became Bishop of Kildare and Leighlin; John Cummins (*c*.1787–1867) owned a substantial farm at Coolnakisha and also land in the townland of Ballybar which was used for horse racing.
9. *Barony of Idrone ... the bridge*: to the west of the River Barrow.

To John Tyndall, Snr 7 October 1840 0014

Youghal | Oct. 7th, 1840.

My dear Father

You may see by the date of this that I lose no time in answering your letter of the 5th.[1] You prognosticated right when you judged of the good state of my health from the superscription of my last letter, I was indeed both stout and strong – able to demolish nearly the whole of a threepenny loaf[2] every morning. I am not at all dismayed on account of Mr. Wynne's refusal I certainly would hold a character[3] from him a valuable document and I am sure a useful one but as he thought proper not to give it it gives me no uneasiness. Before I proceed further I must thank you for the cordial manner in which you told me that I had a home to return to in the event of dismissal no such thing I hope is going to occur the alterations I spoke of merely effected the work but had nothing to do with individuals.

I am extremely sorry to hear of the continued embarrassment of Mr M'Gees' affairs[4] it certainly grieves me to hear of the sad change which has taken place. Mrs M'Gee I fear will find all her philosophy necessary to support her under the transition.

I have read with feelings of disgust the intention of Foley (I had almost said Fooly) & co to offer themselves as candidates for the office of poorlaw guardians[5] in opposition to such men as Captain Steuart.

'Whose station poised above the vulgar herd

Looks down on jarring parties undeterred.'[6]

When I see such stumps of upstart dignity standing forward and with brazen impudence opposing men who are as much superior to them as the midnight moon to the yelping of the shephard's cur,[7] I already see in anticipation the tragedy of Robespierre acted over again when everything great and noble was made to bend before the withering Simoon of democratic tyranny.[8] I feel extremely grateful to Mr and Mrs Steuart and the Dean[9] for the interest they take in my welfare. It shall be my highest ambition to preserve that good opinion which I think I may flatter myself they have already conceived concerning me.

Corpl. Davey leaves us on tomorrow[10] his health has been much improved of late but he tells me that even were he quite recovered he would prefer leaving the Survey.

You know Ginty who left the office in Leighlin to go to England, he has returned to us.[11]

I am glad to hear that all my friends are well especially my Mother and Emma tell Emma she may expect to hear from me shortly

Farewell | I remain your affectionate son | John Tyndall

RI MS JT/1/10/3191
LT Transcript Only

1. *your letter of the 5th*: letter 0013, in which a number of the issues raised in the present letter were introduced.
2. *a threepenny loaf*: a substantial loaf of bread; see letters 0053, n. 3 and 0062, n. 2.
3. *character*: character reference; the testimonial mentioned in letter 0011.
4. *Mr M'Gees' affairs*: see letter 0004, n. 10.
5. *Foley ... poorlaw guardians*: see letter 0013. Foley was unsuccessful but the other two Liberals were appointed Poor Law Guardians (*CS*, 17 October, p. [3]).
6. *Whose station ... undeterred*: This couplet appears to have been composed by Tyndall.
7. *shephard's cur*: sheepdog.
8. *tragedy of Robespierre ... democratic tyranny*: Maximilien Robespierre (1758–94) was the main proponent of the Reign of Terror that began officially on 5 September 1793 and swept through France claiming numerous lives. It was implemented in order to bring social justice by means of the Revolution. However, in a speech before the Convention on 26 July 1794 Robespierre defended himself against the charge of dictatorship and accused members of the Convention of conspiracy. Unable to substantiate this claim, the Convention ordered his arrest and he was executed two days later. A simoom is a strong, dry, dust-laden wind.
9. *the Dean*: Dean Richard Barnard.
10. *Corpl. Davey leaves us on tomorrow*: Davey departed two days later on 10 October.
11. *Ginty ... returned to us*: William Ginty had left the Irish Survey in the spring of 1839 (to join the Tithe Survey in England, which attracted a number of Irish surveyors owing to better pay) but rejoined on 7 October 1840 with a salary of 2*s*. a day.

From Maria Payne 9 October 1840 0015

Bagnalstown Oct 9th 1840.

My dear John

I am sure you think me a sad girl for not having answered your letter before now but I assure you that I am not a willfull offender – there's not a day in the week that I am unemployed except Friday. You have touched a chord that always vibrates in my heart. There is something very strong in an appeal from an old friend and I am not one to turn a deaf ear to yours – but I really am at a loss for any information worth communicating. Things are going on here as usual, only that we are just now going to have a race in Ballybar – the 26th[1] of this month it takes place – I believe we all intend going <*words missing*> As the world and <*words missing*> are to be there, they have commenced building the stand house on a very large scale and in future they are to be continued. So much for the sporting intelligence. Now the melancholy part follows as a matter of course. Well then, by the way, we are kilt[2] crying for my uncle John and family[3] and Tom Wall.[4] Seriously speaking they are gone to Vandiems land[5] but I think we must endeavour to live without them and reconcile ourselves in the best manner we can to the separation. I am sure you have a variety of nice acquaintances to say nothing of the many ardent admirers in Youghal. Emma is anxious to know if you are as fond of going to evening prayer there as you were in Leighlin Bridge I am told you are turned methodist[6] and a most exemplary character, I suppose when you come home you will be making converts though I tell you before hand that you will never make one of me. Dear John you must not condemn George[7] he has had the small pox lately and was in a delicate state of health. He desires to be remembered to you also Ann McGhee and Emma and my mother. Maryann and Bridget Commins send their compliments

farewell dear John | I must now subscribe myself | your affectionate coz | Maria Payne

P.S. Write to me again. Emma and Ann McGhee want to know what's become of the young men or what you have done with them

RI MS JT/1/10/3137
LT Transcript Only

---

1. *26th*: The race meeting at Ballybar (Carlow) was held on 27 October (*Bell's Life in London and Sporting Chronicle*, 1 November 1840, p. 2).
2. kilt: probably killed (died), pronounced kilt in local Irish accent.
3. *my uncle John and family*: 'uncle John' has not been identified but may possibly be John Styles or John Payne.
4. *Tom Wall*: probably the Tom Wall, born 22 June 1812, to Dowling and Sarah Wall (Irish Church Records).

5. *Vandiems land*: Van Diemen's Land, nowadays called Tasmania.
6. *turned methodist*: In his *Fragments of Science for Unscientific People; a Series of Detached Essays, Lectures, and Reviews* (8th edn, 2 vols (London: Longmans, Green, and Co., 1892), vol. 1, p. 11) Tyndall later recalled that when he was 'a boy' he used to read a Methodist magazine, which can be identified as the *Wesleyan-Methodist Magazine*.
7. *George*: probably George Payne.

To Emma Tyndall 13 October 1840 0016

Youghal, Tuesday night | Oct. 13th, 1840.

My dear Emma[1]

You no doubt think it wrong for me to delay without writing you, circumstances however prevented me. My family consisted originally of four lads, namely, Evans, Tidmarsh and Higginson, you know the first the second is a lad from Cork and the third a young gentleman from Dublin. While we remained together we contrived to live calmly and quietly but we have lately received an accession to our numbers in the persons of Messrs Roberts and Wright a son of Capt. Wright the Chief Constable[2] whom you may have seen in Leighlin. So you see that my family at present is rather large and indeed sometimes my children are very unruly so that my spare time being taken up in keeping them in order I have scarcely a moment to devote to the writing of a letter. Necessity, however, is the parent of invention and my anxiety to hear how you are getting on at your history enabled me to find a method of writing to you. Do you not now feel a pleasure in perusing the account of events which belong to the olden times? I have no doubt but you felt it irksome at first to sit down and put your ideas under restraint, and I must feel grateful to you when I consider that it was to please me that you put them under that restraint; the result, however, will prove that I had not only my own pleasure but your interest at heart.

Words cannot convey an idea of the happiness I felt on hearing of my mother's good health. Emma be kind to her; she is now descending the vale of years,[3] and, depend upon it, that when she is gone you will find inexpressible pleasure in the reflection that you were the solace of her declining years. That event, however, is I trust, far distant.[4] I expect to spend some happy days at home with you and her yet; when, having returned from my ramblings, I may be able to lay some of the fruits of my labour as well earned trophies at her feet. Write to me soon and tell my father to write to me also. I would fain hear from you all every day. God be with you.

Your affectionate brother | John Tyndall

I'll hold you a wager of a shilling that you'll not imitate this fold.

RI MS JT/1/10/3192
LT Transcript Only

1. *Emma*: John Tyndall's sister.
2. *Wright a son ... the Chief Constable*: probably Gustavus Wright, a civil assistant on the Survey. Captain Robert Wright served as Chief Constable of Carlow and Ballickmoyler, Queen's County (now County Laois).
3. *the vale of years*: evokes the Christian phrase 'the vale of tears', which would refer to the tribulations of this life which are left behind when we die and enter heaven.
4. *That event ... far distant*: their mother died in 1867.

From John Conwill 19 October 1840 0017

Ballinabrauna M.N.S.[1] | October 19th 1840

'My dear Tyndall'

The examination of the teachers in Mr Butler's district[2] has been deferred until next November; therefore I have taken up my pen to communicate to you Mr Butler's report of my exertions, and also a few geometrical ticklers for any of those pedantic gabblers, who may have the audacity to come in contact with you on the doctrine of Old Euclid.[3] But ere I commence with these subjects, I have waved the liberty of making the following observations: I had the pleasure, yea, the heart-felt pleasure of drinking three large cups of tea with your much valued mother, on yesterday evening, and it was so strong, that I fancied that its perfumes smelt strongly at the family table of my farfamed pupil in Youghal; bravo! my boy, I knew there was something of magic in you; you have discovered a mode by which you can conjure up a set of boys well able to masticate a large salmon or a good stout cod-fish. Why your mather almost split her sides laughing; I forget, I meant that her mouth was extended beyond its natural width, at hearing that you were able to procure your sons ready cut and dry; she at last exclaimed you droll fellow, John, the ladies will not resign their old way of making children. As for Emma, she is the most assiduous girl whom I know with regard to her study of history. I must now transcribe Mr Butler's report, and leave you to judge: 'Visited this School October 12th 1840. Ninety boys present, their answering and the orderly appearance of the school, do great credit to Mr Conwill's exertions.'

Geometrical questions. [Given ΔABC,] from BC draw any two straight lines BD [D being where it cuts AC] and CE [E being where it cuts AB], intersecting F; draw AF, DE and bisect them [respectively] in O, N; join ON, CN & CO. Prove by the first B[ook] of E[uclid] that ΔCON is the fourth of [the area of] the trapezium ADFE.

Bisect CB in M. Demonstrate by the first book of Euclid, that M, N, O are in directum.[4]

Make a triangle equal [in area] to any trapezium, suppose ADFE, one side of the figure is to be one of the triangle. All to be done by the first book [of Euclid], and without producing the sides of the trapezium.

Let the [triangles] abc, ABC be equiangular, prove by the first Book [of Euclid], that axB = Axb; bxC = Bxc; and axC = Axc.[5]

I have completed my twelve geometrical Apostles,[6] and lest some of them might become traitors, I shall send you four others the moment our examination is over. If you have not received a letter of this date from your father, I am sure you will get one from him in a few days.

Excuse haste and let me say in the words of the inspired writer:

'I will not look on your like again'[7]

I am your ever faithful teacher | John Conwill

RI MS JT/1/11/3512
LT Transcript Only

---

1. *Ballinabrauna M.N.S.*: The National School at Ballinabrannagh, at which John Conwill taught, was attended by both boys and girls; thus 'M.' probably stands for 'Mixed'.
2. *Mr Butler's district*: probably Irish National School District 24, which covered the Baronies of Idrone and St Mullins Lower.
3. *the doctrine of Old Euclid*: Euclid (*c.* 325 BC–*c.* 265 BC), the Alexandrian mathematician whose book, generally called *The Elements* or *The Elements of Euclid*, passed through many modern editions and was the basic text for the study of geomety.
4. *in directum*: in a straight line.
5. *axB ... Axc*: a is the length of the side opposite ∠a; A is the length of the side opposite ∠A; etc.
6. *twelve geometrical Apostles*: the twelve mathematical problems that Conwill sent to Tyndall.
7. *I will not ... like again*: 'We shall never see your like again'; from Samuel Warren's novel *Ten Thousand-a-Year* serialized in *Blackwood's Edinburgh Magazine*, 48 (1840), p. 171.

From William Wright 1 November 1840 0018

Roscrea Nov. 1st 1840.

Dear John

I hope you will excuse me for not replying to yours of the –[1] (I am at a loss to know <u>when</u>) before this, but if you were acquainted with what interfered to prevent me these two last Sundays past any excuse would be superfluous. I was left alone with my own Sally,[2] her brothers being in Dublin and I thought it was my duty to make the best of my time at the expense of neglecting you, but I know John that you will as readily forgive me for it as I now ask you. I am happy to inform you that I am progressing with rapid strides in my lady's favour, But there is a most impassable barrier between us, that is the consent of her friends and under existing circumstances I cannot see the removal of it, and I am unable to gain the only alternative of running away, as these 'verilys'[3] have such an aversion to it. However, Jack, we must only hope for the best. I expect you are doing something for yourself and if you would allow a novice such as your humble friend to give you a little advice, you should endeavour to tread as lightly yourself as you can possibly do, although you should likewise make the fair lady feel as much as you could. For the state of uncertainty that I possess at present is dearly bought and bear me out in the simple advice I have offered to you, although it is unnecessary I know as you have too much good sense to implicate yourself in as hazardous a situation as I have. For if her brothers once knew of our intimacy I would on the instant loose my employment[4] and it is not for that I would give myself one moments thought, but I fear if I was separated, her friends would influence her to forget me and then would all be lost indeed to Billy – all that would be worth possessing. I hope you may never be placed in the like predicament. I expect you enjoy good health and spirits and the remnant of Old Leighlin who are with you in Youghal. By the bye I have something to say to you about Joe[5] – a few days preceding the receipt of your first letter from Youghal I had a line from Joe requesting a lone of £1. At the time it was not in my power to comply with his request in full. But I did send 10 shillings, all that I could then spare, and he never acknowledged the receipt of it. Now it is not for money that I care, as I never expected even at the time that it would be returned But that I thought common decency would entitle me to an acknowledgment of it – And another thing, a few weeks before that I received a rather impertinent epistle from him, requiring an explanation of my not acquainting him of my arrival in Carlow at the time that I had the pleasure of seeing you there. That letter I did not answer, as I did not feel called upon to give him an account of my conduct upon that or any other occasion. However I do not care one jack straw for the whole transaction, I would not like that you would let him know that you know any thing of the matter.

I will now take my leave of you and I will expect that you will not take pattern after me[6] in delaying to write. If your own statement be true, you have no excuse for a delay, for when a young woman is not in the case every other excuse is null and void.

Fare thee well Jack and may you always be as happy as Bill would wish you to be Yours truly | William Wright.

<*text missing*> a longer epistle <*several words missing*> for the present by my informing you that I am getting on pretty well with my business, but still there is the doubt continually present that we may be discovered and that is instant separation and goodness knows what may follow. About Joe I will have something more to say to you the next time which you may expect shortly

Adieu | I am yours truly | W. Wright[7]

RI MS JT/1/11/3885
LT Transcript Only

1. *yours of the* –: LT added a question mark here. Tyndall's letter to Wright is missing.
2. *my own Sally*: Sarah Maria Neale, whom Wright subsequently married.
3. *'verilys'*: good people.
4. *loose my employment*: see letter 0012, n. 3.
5. *Joe*: Joseph Payne.
6. *take pattern after me*: follow my example.
7. *a longer epistle … W. Wright*: It is not clear how this text, which was typed on a separate sheet of paper, relates to the preceding letter.

From John Tyndall, Snr 8 November 1840 0019

Mr John Tyndall | Ordnance Survey Office | Youghal
Leighlin Bridge | Nov. 8th, 1840

My dear John

I received your letter of the 4th instant[1] and am glad to hear from you. I wonder you did not let me know where Mr Reid lives as I want that information particularly. You were informed right with regard to the death of Mr Vigors,[2] his remains passed through this town on Friday last to be interred in the family vault in old Leighlin. Colonel Bruen was early in the field, having started before the body left England, but no candidate has yet appeared in the radical interest though they have applied to several to come forward. Some say it will be a walk over with the Colonel others that it will not. My opinion is that it will not. And I was speaking with Mrs Steuart on yesterday evening about you being employed as a writer during the contest,[3] she told me that all her interest should be exerted on your behalf. Mr William Duckett[4] is expected at Mr Steuarts' this evening, and I am to be with him tomorrow morning, when I hope to have everything settled to your satisfaction. I was also speaking to Matthew Griffith on the subject, he told me that he would speak to Captain Carey[5] tomorrow in Carlow, and at the same time for me to speak to Mr John Alexander of Milford whose interest is all powerful with Captain Carey – which I intend doing on Tuesday next. As soon as the second candidate starts and the writ of election issued I will send you word so as to afford you time to be at home. We are all well and send our love to you.

I remain Your affectionate Father | John Tyndall

Write immediately and dont forget to let me know where Mr Reid lives[6] or in what street he lodges.

RI MS JT/1/10/3193
LT Transcript Only

1. *your letter of the 4th instant*: letter missing.
2. *death of Mr Vigors*: Nicholas Aylward Vigors (b. 1785 at Old Leighlin), the MP for the County Carlow, died at his London residence on 26 October 1840. An ultra-Liberal, who often voted with the Radicals, Vigors also contributed to zoology, was a member of several scientific societies, including the Royal Society of London, and was one of the founders of the Zoological Society in 1826.
3. *employed as ... the contest*: poll-clerk, the official with clerical duties during the election (*OED*).
4. *William Duckett*: William Duckett (1796–1868) was a prominent Tory magistrate, agriculturalist and landowner of Russellstown Park, County Carlow, an estate of 1,232 acres.
5. *Captain Carey*: Captain Henry Cary was Henry Bruen's estate agent.
6. *where Mr Reid lives*: Christopher M. Read had resigned from the Survey on 20 October 1840 (NAI OS/1/17). Tyndall's father would have known him when he worked on the Survey at Leighlin Bridge.

From Maria Payne [8] November 1840 0020

Bagnalstown Sunday night | Nov. 10th 1840.[1]

My dear John

I was almost beginning to think you had quite forgotten your B'town[2] correspondent, however I am glad to find that you at last remember bygone hours. You desire a description of the Races.[3] I am a bad hand at descriptions. The weather was very cold and wet, still there was a very large meeting and great sport. In July there will be another,[4] which will be pleasant – perhaps you would grace it with your presence. I hope you enjoyed yourself hallow ee'n.[5] we commemorated that night twelvemonth very imperfectly indeed, as there was some highly prized friends absent.[6] however that was unavoidable, though still we could not help regreting them. You will come home at Christmas of course. I assure you we will be delighted to see you. As to any thing in the shape of news there is nothing just now, only the death of our County member Mr Vigors, which will cause an election. I believe he died in England but was buried in Old Leighlin with his ancestors.[7] George[8] has not suffered any change from the smallpox. I am sure you are a famous dancer ere this. We danced until morning hallow-ee'n, only think of that – are we not a good set, you know the next day was Sunday. Dear John, I am the worst hand in the world at writing long letters, you must consequently excuse the shortness of this and indeed all my epistles. Emma[9] was quite incensed that Mr Roberts should shorten your letter, she says if he has thought proper to refuse us even in place a his memory he should not teach you to forget your old acquaintances though he found the task an easy one.

Emma sends her love to you and Ann McGhee and my mother, also the Miss Cummins.

I am getting very sleepy and must conclude by subscribing myself

Your affectionate cousin | Maria Payne

P.S. They say a woman's postcript contains the most important part of her letter I believe that it is true Write soon remember I will be expecting to hear from you daily

Farewell | M.Payne

RI MS JT/1/11/3838
LT Transcript Only

---

1. *Nov. 10th 1840*: 10 November was a Tuesday; hence the letter may have been written on 8 November.
2. *B'town*: Bagenalstown
3. *the Races*: held at the Ballybar course on 27 and 29 October (*Freeman's Journal*, 29 October 1840, p. [2] and 30 October 1840, p. [2]); see letter 0015, n. 1.
4. *In July there will be another*: The race meeting was held on 2 August 1841 (*Bell's Life in London and Sporting Chronicle*, 8 August 1841, p. 2).
5. *hallow ee'n*: Halloween – the night of 31 October – has strong Celtic, especially Irish roots, marking 'Samhain' the end of summer.
6. *some highly prized friends absent*: including 'my uncle John and family and Tom Wall', mentioned in letter 0015.
7. *buried in ... his ancestors*: On 26 October 1840 Vigors died at his London residence, Chester Terrace, Regent's Park, and he was later buried on St Laserian's Cathedral, Old Leighlin; see letter 0019, n. 2.
8. *George*: probably George Payne; see letter 0015, n. 7.
9. *Emma*: probably Tyndall's sister, but possibly his cousin, Emma, daughter of Caleb Tyndall.

From William Wright 15 November 1840 0021

Roscrea Nov. 15th 1840

Dear John

I received yours of the 4th inst.,[1] in due time. I was not at all surprised to find that you were disappointed at my neglect, nor at your readiness to forgive me for it. Now I hope you will not be displeased at the brevity of this, it wants only a few minutes of post time and to tell you the truth I have been in the same delightful predicament[2] all day as that which deprived you of my last scrawl, were it not for that I would have written a longer epistle unfitted as I am for the task. You must be satisfied for the present by my informing you that I am getting on pretty well with my business, but still there is the doubt continually present that we may be discovered[3] and that is instant separation and goodness knows what may follow. About Joe I will have something more to say to you the next time which you may expect shortly.[4]

Adieu | I am yours truly | W. Wright

RI MS JT/1/11/3886
LT Transcript Only

1. *yours of the 4th inst.*: letter missing; 'inst.' is abbreviation for 'instante mense', Latin for this month.
2. *the same delightful predicament*: presumably, being in love.
3. *we may be discovered*: his relationship with (possibly engagement to) Sarah Maria Neale.
4. *About Joe ... expect shortly*: re Joseph Payne, see letter 0018.

To John Tyndall, Snr 10 November 1840 0022

Youghal, | Tuesday morning | Nov. 10th, 1840

My dear Father

I received yours of the 8th[1] about five minutes ago. I totally forgot the request contained on your last letter respecting Mr Read's residence.[2] I lose not a moment's time in now communicating that to you as he is on the point of leaving us, in fact he has left us tho' he is still in town but he expects every day to be leaving it – he lives at Mr McMahon's North Main Street,[3] where if you address a letter to him he will receive it. If you have any communication to make you had better lose no time in making it as he'll leave town shortly. I hope your efforts to get me employment[4] may be crowned with success. These few minutes are stolen I have time to say no more farewell

Your affectionate son | John Tyndall

RI MS JT/1/10/3194
LT Transcript Only

1. *yours of the 8th*: letter 0019.
2. *your last ... Mr Read's residence*: see letter 0019, n. 6, although the request appears to have been made in an earlier letter, now missing.
3. *Mr McMahon's North Main Street*: *Slater's* lists a John M'Mahon, grocer, 123 North Main Street, Youghal.
4. *get me employment*: see letter 0019, n. 3.

To John Tyndall, Snr 17 November 1840 0023

Youghal | Nov. 17th, 1840.

My dear Father

It is not because I have any particular news to communicate that I at present write to you, it is something concerning the election that I would wish to say to you. I have been told that Col. Bruen has been opposed by a Mr Ponsonby from the Co. Kildare.[1] I have also heard that there is every prospect of Col. Bruen's return and that the election wont last longer than two days at the furthest, if this be the case it would be mere loss of time on my part, even if employment was certain, to go to Carlow;[2] however this does not lesson my gratitude to my friends there one whit, to Mrs Steuart I feel extremely obliged, her kindness shall not be easily forgotten.

I hope you received the desired information respecting Mr Read's residence,[3] for fear the communication might have been mislaid I will here repeat it – he lives at Mr M'Mahon's North Main Street.

Nothing particular has happened since you last heard from me. I enjoy excellent health, three of the members of my family[4] have just departed to the field so that it is now reduced to its pristine number I'm not sorry for this movement as it will afford me an opportunity of studying during the long winter's nights which would be impossible if things had continued as they were.

I received a letter from Corporal Davey lately he is employed at the Ordnance Works of Woolwich[5] along with Newton the Sapper[6] who left us in Leighlin, from the tone of his letter I could gather that he would prefer being amongst us at present to his situation in England. Be so good as to let me have your opinion respecting the election if you think that it wont last more than two days I think it would not be judicious to continue your efforts to get me employment. When you next write send me word how my Mother is you seem to forget her totally in all your epistles and in so doing you forget that which would be most interesting to me, tell Emma she may expect a letter from me one or two days after you receive this.

I remain your affectionate son | John Tyndall

RI MS JT/1/10/3195
LT Transcript Only

---

1. *a Mr Ponsonby ... Co. Kildare*: The Hon. Frederick George Brabazon Ponsonby, whose family owned the Bishop's Court Estate, County Kildare.
2. *mere loss ... to Carlow*: Tyndall's father had suggested that he be 'employed as a writer during the contest' (letter 0019).
3. *information respecting Mr Read's residence*: see letters 0019 and 0022.
4. *three of the members of my family*: Tyndall shared accommodation with Evans, Ginty and Tidmarsh.
5. *the Ordnance Works of Woolwich*: The Royal Arsenal, Woolwich, had been established in 1671 and was the main site in London for the development, manufacture and testing of ordnance. The barracks of the Royal Sappers and Miners was located close to the Arsenal. Davy's letter is missing.
6. *Newton the Sapper*: James W. Newton, a Private in the 13th Company of Royal Sappers and Miners, joined the Survey in February 1834 and worked in the 5th Division, C District under Lieutenant Wynne. In August 1839 he was transferred to the Survey's head office at Mountjoy, Dublin (NAI OS/1/14–16).

To Emma Tyndall 21 November 1840 0024

Youghal | Saturday night | Nov. 21, 1840

My dear Emma

Weary and worn after my week's work I have just sat down snugly, contentedly, and comfortably by a roaring fire to pay the debt which has been so long owed to you I mean that of a letter. You no doubt expected a doleful catalogue of calamities when you read the commencement of this letter, however I have no such catalogue to relate. I am quite contented and comfortable, the whistling of the wind out-side makes me doubly so, you know it is very sweet to listen to the howling of the storm when you have nothing to dread from its effects; such is my present situation – the blazing light of a cheerful fire is a pleasant contrast with the howling dreariness of a November night. I got a letter from my father this morning,[1] it precludes all possibility of my seeing you during the ensuing election. I was surprised to find that Mrs Payne intended to leave Bagnalstown so soon, as she did not intimate to Joe[2] her intention of doing so. I received the letter just a little before breakfast time but judge of my surprise when I was just leaving the office to come to breakfast when I saw the self same Mrs Payne passing by on the Dungarven Car.[3] She looks remarkably well, she appears to have got even fatter than she was when I saw her last. She is now settled quite comfortably and I think if she manages properly she ought to be able to continue so. Joe's and Tom's pay together amounts to £1 a week[4] this I'm sure ought to support her well. She was telling me that Jim Walsh[5] and my father had a great fight lately tell Jim I'll trash him when I go home if he does not be quiet. Has he the medal yet? I hope he uses my old hat well. Mrs Payne tells me that he has preserved it like a sainted relic, she also told me that my mother was knitting some stockings for me, when I see her I'll let fall a shower of thanks and blessings on her, however my stock continues very good. Why did you not go to Bagnalstown on Halloweve[6] as I've been told you were invited there. I suppose you think that I have forgotten the subject which was the chief cause of our present correspondence – I mean your history. Though I have kept silent on that subject to the end of my letter still if the first part should be appropriated to the most important subject it should be placed there, however I think the end of a letter is as good a place as can be to impress a correspondent with the importance of any one thing. This is the reason I have reserved it for that place. Persevere! Persevere!! Persevere!!!

Your affectionate brother | John Tyndall.

RI MS JT/1/10/3196
LT Transcript Only

1. *a letter ... this morning*: letter missing.

2. *Joe*: Joseph Payne.
3. *Dungarven Car*: the coach which ran every day except Sunday to Dungarvan, which is on the coast about 16 miles north-east of Youghal.
4. *Joe's and Tom's ... £1 a week*: At that time Joseph Payne was earning 2*s*. per day and Thomas Payne 1*s*. Therefore together their weekly income was 18*s*.
5. *Jim Walsh*: Walsh was apprenticed as a shoemaker to Tyndall's father.
6. *Halloweve*: All Hallows' Eve (or Halloween) falls on 31 October, the evening preceding the Feast of All Saints, otherwise All Hallows' Day.

From William Wright 29 November 1840 0025

Roscrea Nov. 29th 1840.

My dear John

Again I have trespassed upon your kindness and once more have to sue for your pardon to forgive me for not writing to you before this, but it solely was owing to my being engaged with my fair lady[1] and John you know that no business no matter how pressing would justify me in deserting my post for one minute – and another reason, if I had not a friend like you so ready to forgive me for my negligence I perhaps would be more cautious of risking your friendship. I am still keeping up the steam as well as I can with the mistress and I believe not in vain – if we can believe what women say, and we are apt to believe what we hear in favour of ourselves. Jack, I hope you are not letting the time slip over but have your hook baited although you and I were not always the most successful fishermen. You will get a good take yet and I am sure no one would rejoice more than Bill. I will expect a line from you shortly. Hoping that the Youghal air will have the power of adding to your appetite which by this time, according to your own account, ought to do credit to Billy Henry.

I hope when you last heard from ever-to-be-remembered Lbridge,[2] that your Father and Mother were well. You I suppose are aware of the struggle which takes place in Carlow for the representation of the county on tomorrow. I sincerely hope it may end in the return of the Colonel, although I doubt it much.[3] To you or me it is very little matter who is in or who is out, but still I would like to see the gentry support their rights and protestantism and long has Carlow been <*words missing*> of in the hands of the demagogues.[4] Write to me soon for I long to hear from my old friend and shipmate of the Red Cot.[5]

Believe me ever yours | William Wright

RI MS JT/1/11/3887
LT Transcript Only

---

1. *my fair lady*: his fiancée Sarah Maria Neale.
2. *Lbridge*: Leighlin Bridge.
3. *the struggle ... doubt it much*: In the County Carlow election the Tory and Protestant

Colonel Henry Bruen triumphed over the Radical candidate the Hon. Frederick George Brabazon Ponsonby.

4. *the hands of the demagogues*: Carlow County had previously been held by the Liberal John Ashton Yates and the Liberal-Repealer Nicholas Aylward Vigors, whose recent death had caused the by-election. In describing Yates and Vigors as 'demagogues', William Wright portrayed them as unprincipled and self-serving popular agitators.
5. *shipmate of the Red Cot*: not identified.

From John Tyndall, Snr 8 December 1840 0026

Mr John Tyndall | Ordnance Survey Office | Youghal
Leighlin Bridge | Dec. 8th. 1840.

My dear John

I send you the glorious intelligence of Col. Bruen's triumphant victory over his Radical opponent,[1] after a tremendous conflict of five days. They commenced the battle on Monday the 30th of Nov. and ended on Saturday the 5th, of Dec. Ponsonby, the Priests' darling, sent in his resignation on Friday, bid goodbye to the mendicants of the Radical Club[2] and left the Colonel in possession of the field with a majority of one hundred and sixty seven amidst the roars and huzzas of the Conservatives. I was in Carlow three times during the contest. Monday, Thursday, and Saturday. I would not have went so often but Mr Steuart brought me with him in the Gig.[3] The Priests are really out of their reason and the Roman Catholics equally so. They threaten us with slaughter, but we laugh at their nonsense. John, you would be delighted to see the noble Band of Protestants that filled the Court House of Carlow on Saturday evening when the Colonel was declared.[4] Their shouts with 'hear him, hear him,' and the playing of a Kent Bugle[5] 'The Conquering Hero Comes,'[6] was truly gratifying. I wont bother you with more at present, as your mother, who, is in good health, expects you home at Christmas, but conclude

Your affectionate Father | John Tyndall

P.S. | Jim Walsh desires you to let Mrs Payne know that the person who promised to put up the partition[7] has not done so, he sent word to George[8] about it but received no answer. If she allows him in the rent he will have it done himself | J.T.

RI MS JT/1/10/3197
LT Transcript Only

---

1. *triumphant victory ... Radical opponent*: the voting to Thursday night was 586 for Colonel Bruen and 443 for the Honorable Frederick George Brabazon Ponsonby; as of Friday night the votes were 719 and 555 respectively. Contemporary accounts claim that intimidation was practiced by the supporters of both candidates.
2. *Radical Club*: The Radical Club was founded in London in 1833. Although Tyndall, Snr

claimed that Ponsonby was a Radical, contemporary newspapers generally described him as a Liberal. Daniel O'Connell considered that Ponsonby lost the election because he refused to align himself with the Repealers: 'It proves that the Whigs cannot prosper without the Repealers'. (O'Connell to P. V. FitzPatrick, 8 December 1840: M. R. O'Connell (ed.), *The Correspondence of Daniel O'Connell*, 8 vols (Tallaght, Dublin: Blackwater, 1972–80), vol. 6, p. 389. See also O'Connell letters 2762, 2765 and 3775.)

3. *Gig*: a light two-wheeled one-horse carriage (*OED*).
4. *the noble Band ... was declared*: The final stage of the election and Henry Bruen's triumphant speech after he had been declared the winner are reported in *CS*, 12 December 1840, pp. [1–3].
5. *a Kent Bugle*: a curved bugle, having six finger keys or stops, by means of which the performer can play upon every key in the musical scale. Also known as the keyed bugle, it was probably named after the Duke of Kent.
6. *The Conquering Hero Comes*: The chorus and march 'See the Conqu'ring Hero Comes!' is sung to mark the victory of the Jewish people in George Frederic Handel's oratorio *Judas Maccabaeus*.
7. *put up the partition*: presumably at Mrs Payne's house in Bagenalstown.
8. *George*: George Payne.

From John Tyndall, Snr 14 December 1840 0027

Mr John Tyndall | Ordnance Survey Office | Youghal
Leighlin Bridge | Dec. 14th, 1840.

My dear John

I received your letter of the 12th instant.[1] Its contents fills my heart with delight, I read it for Mr Stewart and Mr Beauchamp Newton, who expressed their utmost pleasure at your performance[2] and said I was happy to have a son able to do the like I intend showing it to the Dean[3] and Mrs Stewart the first opportunity. I was in bed when your mother brought it to me, I read it there and nothing I have read this good while has given me so much pleasure. And now, for your further information, I have to let you know that there was so much interest taken with regard to the Election of Carlow that in Liverpool six thousand good men and true assembled on the beach waiting for the Packet which was to bring the result. The moment they heard it announced, their shouts rent the air and one cheer more for the Carlow Boys[4] was given. Mr Mahon in this town is a complete Bedlamite,[5] the Sunday after the Election he told as much lies as would disgrace a cannibal, but it is by lies they live, move, and have their being.[6] Mr Conwill read a beautiful letter[7] from you last night, your mother was very well pleased, he said he would answer it this day, and as we expect you home at Christmas I will conclude

Your affectionate father | John Tyndall.

RI MS JT/1/10/3198
LT Transcript Only

1. *your letter of the 12th instant*: letter missing
2. *your performance*: probably a poem or vignette written by Tyndall and included in the missing letter.
3. *the Dean*: Dean Richard Barnard.
4. *the Carlow Boys*: the defenders of Protestantism in Carlow.
5. *a complete Bedlamite*: utterly mad; Bedlam was the popular name of the Bethlem Lunatic Asylum, Beckenham, London. Father John McMahon – here 'Mr Mahon' – was the Catholic curate of Leighlin.
6. *they live ... their being*: Cf. 'For in him we live, and move, and have our being' (Acts 17:28).
7. *a beautiful letter*: letter missing.

From John Conwill 21 December 1840 0028

My morning thought, my nightly dream,
The subject of each prayer;
Have I a wish, a hope, a scheme,
Which Tyndall does not share.

Monday morning[1]

'My dear Tyndall'

'My motto', to use your own words 'needs not be doubted' for I am convinced that it has been used by some love-lorn creature when he greeted the fair object of his affection. This circumstance does not render the 'motto' derogatory to the 'present occasion.' I doubt not but a corresponding feeling exists in your breast, and I expect that you know by this time that you have been premature in your animadversions on my unkindness.[2] You are very happy in your figures of speech. I defy all the powers of exaggerated hyperbole to transcend your figurative epithets with regard to the adamantine feelings that exist in my heart. However I must be a little egotistical – I fearlessly affirm that better feelings never strung the heart of an Irishman than the feelings are by which my heart is actuated. I do not suggest that you challenge my integrity, but I say that your comments as to my being apathetic are already disproved by the letter which contains an account of Professor Hennessy's expanded hints to me on the definition of Number.[3] Ah! My dear Tyndall, I am well aware that it is your overweening anxiety for my felicity, which urges you to judge rashly of my conduct towards you, and therefore I shall bear a thousand allegations, from my friend. Disembodied creations cannot write, consequently this letter shows I have not gone to that home whence no traveller returns.[4] The rich man's talent, in modern times, is far superior to the widow's mite.[5] It matters not what your abilities are if you possess an abundance of the ryno;[6] it will screen all your imperfections, how glaring soever they may be, and hence I care little about the political cant of Bruen's unshrinking Tory principles or scapegoat Ponsonby's Whiggish recreancy.[7] I deem the whole clan of Whigs and Torys a peculating[8] clique. I shall now conclude in the words of Goldsmith:–

> Wheree'er I go, whatever scenes to view.
> My heart untravell'd shall fondly turn to you.[9]

Your ever faithful teacher | J. Conwill

P.S. My Mother's respects to you.

'My valued Tyndal'

There is a peculiarity in the coincidence of our writing. I wrote this letter in the School; and when I came home, what had I for lunch but your sarcastic one on Hennessy's audacity. You may rest satisfied that I would not undervalue myself so much as to break a lance with such a miscreant, but indeed I made Ford give him a sound snubbing.

I will write in the course of a month, God willing.

J. Conwill.

RI MS JT/1/11/3513
LT Transcript Only

1. *Monday morning*: LT gives postmark 'Leighlin | Dec. 21. 1840'.
2. *'My motto' ... my unkindness*: Conwill was responding to and quoting from a letter of Tyndall's which is missing.
3. *the definition of* Number: An extensive literature exists on the definition of number. For example, Euclid defined number as 'a collection or assemblage of several units, or several things of the same kind; as 2, 3, 4, &c, exclusive of the number 1', whereas Stevin defined 'number as that by which the quantity of any thing is expressed: agreeably to which Newton conceives a number to consist, not in a multitude of units, as Euclid defines it, but in the abstract ratio of a quantity of any kind to another quantity of the same kind, which is accounted as unity' (C. Hutton, *A Philosophical and Mathematical Dictionary*, 2 vols (London: Hutton, 1815), vol. 2, p. 113). Issues relating to number occur in other letters, especially 0039 and 0060.
4. *that home ... traveller returns*: adaptation of 'The undiscovered country from whose bourn | No traveller returns' (W. Shakespeare, *Hamlet*, III.i.81–2).
5. *The rich man's ... widow's mite*: The ability of each to give according to their circumstances was a frequent theme of sermons. Cf. 'the rich men casting their gifts into the treasury ... [and] a certain poor widow casting in thither two mites' (Luke 21:1–2).
6. ryno: money (*OED*).
7. *recreancy*: treachery (*OED*).
8. *peculating*: embezzling (*OED*).
9. *Wheree'er I go ... to you*: 'Where'er I roam, whatever scenes to see | My heart, untravelled, fondly turns to thee' (O. Goldsmith (1730–74), 'The Traveller', lines 8–9).

From Maria Payne 20 December 1840 0029

Sunday night 7 o'clock | Dec. 20th 1840.

My dear John

I flatter myself that you will forgive me all my offences when I tell you it is out of a sick bed that I am addressing you labouring under the effects of a severe cold. Your good nature will I am sure excuse the numerous defects of this scrawl. The charge of suffering absence to obliterate your idea from my mind is one to which I plead not guilty – if you knew the disposition that I am of you would not accuse me of forgetting an old friend – the very name carries with it a recommendation and believe me your friendship, far from being unrequited, is returned with interest and one that I shall always be proud to acknowledge. You say that some gay scenes occupy my hours I will answer you in the words of an old ballad:–

> And tho' I mingle 'mid the throng
> Where pleasure twines her wreaths of flowers
> I cannot sing the same light song
> I used to sing in happier hours.[1]

You will perhaps think me a romantic fool – I own that I am partial to the spirit – I think you are romantic yourself – if that is the case, you will pardon me introducing poetry. As a correspondent I fear that I am not calculated to shine. Your goodness will overlook the many faults that are in my epistles. I have no news strange or uncommon to communicate The <*word missing*> at the Friary[2] are quite well and your favourite[3] as charming as ever – you knew who I know. I must now draw to a conclusion, however I must not omit giving you a long list of loves My mother, Ann McGhee, Emma, Miss Cummins and Bridget[4] desire their compliments

And now farewell

Your sincere friend and wellwisher | Maria Payne | Bagnalstown.

RI MS JT/1/11/3839
LT Transcript Only

---

1. *And tho' ... happier hours*: third stanza of 'Ballad' by B. H. G., *Dublin University Magazine*, 3 (1834), p. 123.
2. *the Friary*: The first monastery established in Ireland by the Carmelites was at Leighlin Bridge in 1271. In the seventeenth century the Carmelites were evicted by Sir Richard Bellingham, but in 1730 the Leighlin Friary was rebuilt on the outskirts of Leighlin Bridge on a site overlooking the River Barrow. By the time this letter was written the Friary was owned by James Thomas, whose family was wealthy and influential.
3. *your favourite*: possibly Ellen Wall or 'Jane' mentioned in letter 0033.
4. *Bridget*: Bridget Cummins.

From John Roberts, Thomas Charles Higginson, George Hamilton Wright and George Holmes 22 December 1840 0030

Castlemartyr,[1] Tuesday Night
Dec. 22nd. 1840

My dear Walter[2]

In considering on the subject of going to Cork on Friday, we have come to the conclusion that it would be a cold Christmas to us a lot of strangers in a strange place, – beside it would be attended with a good deal of heavy expense which in the end would produce no comfort – Another argument – We have a particular desire that you, Ginty, Evans, and Tidmarsh shall spend that day with me, Wright, Holmes and Higginson – here in Castlemartyr. We will give you a good round of beef – a good rice pudding – a couple of tumblers of punch plenty of tea and coffee and a comfortable turn in. We can then go up to Cork on Saturday morning and return when we Choose. Now Walter I know you are not altogether a man what will be obstinate like Ould Davy[3] – so agree to this our proposal and invitation. I know you and Ginty will come and also Tidmarsh. If Jimmy[4] shows any signs of disapprobation, castigate him à la Sam Weller.[5] I mean to say à la Higginson.

Don't forget to let us have a line by return

In Haste | Wat | farewell

till Thursday night when we will expect you

John Roberts | Thos. Higginson | G. H. Wright | & Geo. Holmes

RI MS JT/5/14/58
LT Transcript Only

---

1. *Castlemartyr*: about 10 miles west of Youghal and 20 miles east of Cork. A team that included Roberts, Higginson, Wright and Holmes was undertaking surveying work at Castlemartyr.
2. *Walter*: Walter Snooks, one of Tyndall's pseudonyms.
3. *Ould Davy*: James Davey.
4. *Jimmy*: John Tidmarsh.
5. *Sam Weller*: John Roberts' pseudonym (see letter 0109) and a character in Charles Dickens' *The Pickwick Papers* (1836–7).

1841

To John Tyndall, Snr 4 January 1841 0031

Youghal, Monday night

My dear Father

It's a long time since you and I had an epistolary converse, you no doubt attach the fault to me, but I think it rests with you. 'Two against one' says the proverb 'will kill a man'[1] but it's not the demands of two correspondents that I have to satisfy, but half a dozen. There's Emma, Mr Conwill, Maria Payne, Wm Wright and three or four more to each and every one of whom I have to offer the tribute of my pen, you appear to think that I have a right to send letter for letter with you I have stated the case as it stands and you cannot but see that you have a decided advantage over me, instead of half a dozen you have only my poor solitary self to correspond with, you also have more time than I have. I'm now confined from daylight to darknight. I dont intend to convey the idea that I'm discontented with my present situation, far from it; I doubt very much whether Queen Victoria[2] though a diadem[3] screens her brow is as happy as I am. I enjoy excellent health, have a good appetite, and a good pair of hands to satisfy its demands. So I see no reason at all why I should not be happy.

I'm glad to find that my last letter pleased you but had I the slightest idea that its contents would be made known to either Captain Steuart or Mr Newton[4] I would have produced one more worthy of their approbation I would have taxed my energies a little more and I doubt not that they would be equal to the task of producing a better letter than the one to which you allude, the reception it met with is however to me very flattering.

Great changes have taken place lately and we are on the eve of greater. Foy got married lately without leave for which he was broke from being Corporal, he has been reduce to-day and drafted to another division,[5] he was in charge of the field parties when this took place. Collins has replaced him in the field and I occupy Collins's post in the office. Our next destination is not Scotland[6] as we anticipated, it is already fixed upon that we're to go to England. I think we'll depart sometime about the beginning of summer. I'll see you before I go. I think the good men of the division will be more encouraged now than heretofore though I cant say much for them as yet having only got 2d a day of an increase within the last twelve months.[7] I had fifteen shillings leaving Leighlin, I have now sixteen. I'm sure of getting a rise next month as Corpl Mulligan the person now in charge told me that I would be specially recommended. How is my mother? I hope you spent a happy Christmas. Where do you think I spent mine? – In Cork. I was invited there by the father of a Cork lad[8] a fellow lodger of ours. Ginty came along with me. I spent Christmas day viewing the ruins of the far famed Blarnet Castle[9] rambled through the parade and dyke – these places are well known to my mother. I spent Christmas most pleasantly – they were very nice people that entertained us – We returned on the Monday following Christmas day and are now plodding away at the old work. I suppose you saw Phil Ryan. I searched the town before he went for a box of cephalic or Princess mixture[10] but could get neither to send you. I thought you would not relish the Cork snuff as it is of an inferior quality and so I declined sending it.

Send me word whether Emma continues to read or not. How is my uncle Caleb? Send me every bit of news you can collect.

Good bye | I remain your affectionate son | John Tyndall

RI MS JT/1/10/3199
LT Transcript Only

---

1. *'Two against one' ... 'will kill a man'*: This does not seem a very common proverb, although in David William Paynter's novel *The History and Adventures of Godfrey Ranger*, 3 vols (Manchester: R. & W. Dean, 1813), one of the characters remarks: 'Two against one is enough to kill a man, as the saying is' (vol. 2, p. 37).
2. *Queen Victoria*: Queen Victoria (1819–1901) had ascended the throne of Great Britain and Ireland in 1837.
3. *a diadem*: a crown; Victoria's coronation had been held at Westminster Abbey on 28 June 1838.
4. *my last letter ... Mr Newton*: letter missing but Captain William Steuart and Beauchamp Newton are referred to in letter 0027.
5. *Foy ... another division*: On 15 December 1840 Thomas M. Foy, who had been absent and married without permission, was demoted to the rank of Private, transferred to the 4th Division, under Lieutenant Fenwick, and his pay reduced from 1/6 to 6*d.* (NAI OS/1/16–18 and OS/2/18).
6. *Scotland*: The trigonometrical survey of Scotland had commenced in the 1810s, but had not been completed. The triangulation of the north-west of the country was recommenced in 1837, at the behest of Colonel Thomas Colby, who sent small parties of surveyors from Ireland to help with the work, which was completed in 1841.
7. *2d a day ... twelve months*: In December 1839, while still working at Leighlin Bridge, Tyndall's daily pay was increased to 2*s.* 6*d.* (15*s.* a week); in June or July 1840 his pay had risen by 2*d.* to 2*s.* 8*d.* (16*s.* a week) (NAI OS/1/16–17).

8. *a Cork lad*: John Tidmarsh.
9. *Blarnet Castle*: Blarney Castle, where the famous Blarney Stone is located, is 5 miles to the north-west of Cork. Although earlier fortifications were built on the site, the surviving stone dates from 1446 and the Georgian gothic house was constructed in the early eighteenth century.
10. *box of cephalic or Princess mixture*: varieties of snuff; cephalic (for ailments of the head). An advertisement for Collins' Cephalic Snuff Cordial recommended it for headaches, chronic deafness, weakness of the nerves of the head and dimness of sight (*The Traveller's Album and Hotel Guide* (London: Cassell, Petter and Galpin, 1842), p. 32).

To John Tyndall, Snr 9 January 1841 0032

Youghal | Jan. 9th, 1841.

My dear Father

I hasten to give you a receipt for your welcome present and to thank my mother for her's. I was just sitting down last night to write to you when I was hauled off to hear a controversial sermon preached by an English gentleman the subject was 'private judgement'[1] he was not what you might call a thunderer yet he treated the subject in a clever manner he went calmly and coolly to work and reasoned dispassionately on every subject he introduced he dragged the scarlet covering from the beast[2] exposed him to his hearers in his naked deformity but forbore further laceration, I suppose he was afraid of offending the Roman Catholic portion of his hearers.

The boots fit me capitally. I found it rather difficult at first to get on the right one but now it goes on quite easily.

Mr Ryan's pamphlet[3] is an interesting one, I suppose he has returned to America. I hope you did not neglect getting Robinson Tyndall's address from him it might prove useful at a future day.

We have just got an unexpected addition to our work which will detain us nearly a year longer in Ireland.

An order has just been sent out to the field directing a great number of surveyors to be discharged. Major Waters' letter was read for me this morning – the conclusion of it runs nearly as follows:–

'I expect that the remainder of the men in your division will exert themselves, as those who do so and those alone will have any chance of being selected for the Survey of Great Britain.'[4]

I was up in the room with Mulligan the other day on some business; he asked me would I go to England. I said I would not if I could get any better situation, he said that I would rise to good pay on the Survey, that my pay would be increased immediately on embarking for England and that I would not be there six months without receiving five shillings a day. This, however, is only conjecture.

I know I stand pretty high in Mulligan's estimation and I'm working hard to make myself worthy of his good opinion.

Has not your interest increased since the Colonel's[5] return? What do you think of an application for a situation in the Revenue?[6] I think it would be a surer speculation than the one in which I have embarked. The Dean's[7] and Captain Steuart's interest is I'm sure very powerful at present.

I met Jeffries the ganger[8] on the Parade in Cork I looked in his face of course but no sign of recognition passed between us

My situation in the office is at present very onerous, on me rests the whole responsibility of the most important part of it, they have already begun to call me Corporal Tyndall.

You may tell Mr Murray that I saw Wm. twice last week the first time I saw him I made him promise to write home, when I saw him last he told me he had done so, he is at present in the town of Middleton[9] just half way between this and Cork a letter addressed to J. Collins R.S. & M.[10] for Wm. Murray will be received by him. How is Mr McGee? Remember me to both him and Miss McGee.

Don't read what follows aloud – I wish when Emma is writing to me that you would look over her letter. I always thought that Emma was a better speller than you, but I now think you are able to beat her at it, I make this request because I know if you do so it will serve her greatly. In her last letter she spelt Phil Ryan's name Fill and in a former one she commenced Billy Henry's name with a small b thus – billy.[11] It would serve her much if you correct these mistakes – dont let her see this on any account. This is a long letter, I have sent you all the news I could collect. I leave it to yourself to judge whether its worthy of a reply or not. I think it is, it's a long time since you wrote to me, dont delay any longer. Remember me to Miss Clare.[12]

I remain your affectionate trooper | John Tyndall

Don't forget to send me word about R. Tyndall's address.

Note – Mulligan is the person in charge of the office – he occupies Davey's place.

RI MS JT/1/10/3200
LT Transcript Only

---

1. *controversial sermon ... 'private judgement'*: The issue of private judgement concerns the freedom of Christians to change their beliefs in the light of experience. This was a recurrent topic for sermons and lectures. For example, in April 1840 a Rev. Maguire delivered a lecture on 'Private Judgment' at the 'Adam and Eve'/New Church, Merchant Quays, Dublin (*Freeman's Journal*, 3 April 1840, p. [1]).
2. *he dragged ... from the beast*: 'I saw a woman sit upon a scarlet-coloured beast, full of names of blasphemy, having seven heads and ten horns' – Revelation 13:3. Opponents of Catholicism, like this lecturer, often treated the beast as a symbol of the Catholic Church.
3. *Mr Ryan's pamphlet*: not identified.

4. *I expect ... Great Britain*: original text not located in Ordnance Survey archive in NAI.
5. *the Colonel's*: Colonel Henry Bruen.
6. *the Revenue*: The Inland Revenue, which was responsible for collecting duty on taxed products such as malt.
7. *The Dean's*: Dean Richard Barnard.
8. *Jeffries the ganger*: not identified.
9. *Middleton*: Midleton, a town about 13 miles to the east of Cork.
10. *R.S. & M.*: Corps of Royal Sappers and Miners.
11. *In her last letter ... billy*: both letters missing.
12. *Miss Clare*: not identified.

From Maria Payne 10 January 1841 0033

Bagnalstown, January 10th 1841.

My dear John

You were not wrong in your opinion when you conceived that your expressive motto[1] would stir up my feelings I assure you that I gazed on it with unfeigned and heartfelt pleasure. Yes I do indeed love to think of bygone pleasure hours and the days of auld lang syne[2] will always contain a charm for me. Memory presents a faithful and sometimes a painful picture of the past and though its flowers are indeed fragrant we cannot always cull them without being pained with the thorns. I am glad to hear you say that you are romantic you will the more readily pardon the nonsense that frequently flows from my pen dear John how is it that you never speak of Jane[3] is it possible that you are not friends she never spoke of you but once I cannot think that you are in fault as I know that she is changable and those that are it is hard to account for them. Your advice with regard to the soft passion is indeed excellent and I must not forget that you have acknowledged that you are yourself a worshiper at Cupid's shrine. Come now be candid and in your next tell me who has succeeded in winning your valuable heart I am quite sure that she possessed something more than common to win such a heart as you are possessed of and I hope she will merit your warmest affection. I would have written to you sooner only in consequence of the new year I was very busy. Do you remember last year when you wrote the Reports for me?[4] I thought of you on this occasion how many things happen in one short twelve month Tell me how is it that we did not see you at Christmas I can assure you that we got disappointed You have asked me to inform you who I meant by your favourite in my last Well then in two words Ellen Wall[5] was she not a favourite? When do you purpose coming to the County Carlow for surely you will not quit the Emerald Isle without paying us a visit I need not tell you that you are welcome We hear that your party are going to England in Spring is that the case? I have got nothing in the shape of news to tell you only the death of Mrs Chapman[6] after an illness of eight days She is deservedly regreted by every one that knew her.

And now I must conclude as it grows late

Your sincere and affectionate cousin | Maria Payne

Write soon and answer all the questions that I have asked you concerning <*words missing*>

Yours &c &c

RI MS JT/1/11/3840
LT Transcript Only

---

1. *your expressive motto*: possibly the motto discussed in letter 0028.
2. *auld lang syne*: 'long, long ago'; the title of a poem by Robert Burns and set to a traditional melody.
3. *Jane*: not identified, presumably an earlier female admirer.
4. *you wrote the Reports for me*: 'Reports' not identified.
5. *Ellen Wall*: possibly Ellen Wall (b. 1813), sister of Tom Wall mentioned in letter 0015 (Irish Church Records).
6. *Mrs Chapman*: Marianne Chapman of Wycham aged 48 was buried on 12 January 1841 (Irish Church Records).

From William Wright 12 January 1841 0034

Roscrea, Jan. 12th 1841.

Dear John

I did not receive your kind letter of the 5th[1] until this evening being from home last week which may account for not answering yours sooner and strive to refute the charge of neglect which you have brought against me. Too long did I allow that fault to take hold of me and was and is determined that it never shall again If my memory does not fail me twice have I written to you[2] since I received yours of the I dont know when,[3] I expected that I had nothing to do but lay my hand upon it but when I went to search it had fled I know not whither. However Jack my old Trout,[4] I was just beginning to get a little uneasy at not hearing from you as I really did write twice since I received your last. But I attributed it to a change of quarters and perhaps was waiting to give Bill a description of them or perhaps – what was more likely to my imagination – that you had got into some love affair and I was expecting in your next epistle that I would read of the charms of your fair one and painted in those glowing colours which you are so well able to use without your ever going to Blarney.[5] Not that I would ever give you credit for the smallest particle of that accomplishment for I know my old friend's sincerity too well to think that time or place would make any change in him.

When reading of your visit to that far famed castle,[6] I wished almost to be along with you but the recollections of how I spent that day forbids me. I had the good fortune to be left in undisturbed possession of the being whom I love and although I can appreciate the delight and rapture I would have felt at viewing the ruins as well as the society of the companion of other days, which is only second to the society of my own one, I believe if both were within my choice I would select that which I possess. She[7] now joins me in congratulating you upon your sport and good appetite on your return and is obliged for the kind forbearance you have used towards me for her sake

I remain your old friend and now all but married man | William Wright

RI MS JT/1/11/3888
LT Transcript Only

---

1. *your kind letter of the 5th*: letter missing.
2. *twice have I written to you*: unless one or more recent letters from William Wright are missing, he is here referring to letters 0021 and 0025.
3. *yours of the I dont know when*: letter missing.
4. *my old Trout*: comrade.
5. *Blarney*: Kissing the Blarney Stone, located at Blarney Castle, County Cork, is said to endow a person with the power of eloquence, especially the ability to charm the fair sex.
6. *far famed castle*: Blarney Castle, which Tyndall visited on Christmas day; see letter 0031.
7. *She*: Sarah Maria Neale who married Wright on 7 February 1841.

From John Conwill 18 January 1841 0035

'In youth my scholar, and in age my friend,
Our love, dear Tyndall, but with life shall end;
Tho' sever'd now by intervening space,
Our souls and wishes join in sweet embrace.'

Ballinabrauna, M. N. S. | January 18th 1841.

'My dear Tyndall'

I possess an unbounded predelection for mottos, and therefore I have selected one, which in my humble judgment, conveys in the fullest terms my good wishes towards you; though some of the Munster sentence makers[1] might have the audacity to construe this motto into an effeminate harangue made by some unfortunate paramour to the fair portion of his fondest wishes, but I shall leave you to maintain the contrary. For the manly exposure of Hennessy's ingratitude and lying conduct,[2] I do sincerely thank you; It is what I anticipated from the pen of him whom I shall ever revere.

The other evening when Sol[3] had nearly completed his course o'er heaven's blue arch, I was crossing Frenchhorn-hill.[4] It was covered with a beautiful sheet of snow, which was rendered delicately white by the declining rays of the sun as he was sinking in the west: Here, said I in my soliloquy, Tyndall and I delineated some mathmetical diagrams when all nature wore its mantle of grey. Well, then I exclaimed in my enthusiastic reveries, I must commomorate the scenes of other days, so I commenced on my favourite spot to draw out geometrical schemes and eventually discovered the following beautiful theorems, which as an addenda to my apostles I send you;

In the trapezium ABCD, if $\angle DAB + \angle DCB = 2\angle rts$, to prove the $\angle ABC = \angle ADB$, within the limits of the 1st B of Euclid,[5] and without introducing the property of the circle.[6]

Having established the preceding theorem, you can by means of it and one of my geometrical apostles, demonstrate proposition B, Prop C, and Prop D of the VI B,[7] of Euclid within the limits of his 1st book.

I have nothing of a local character to inform you of save a melancholy accident that lately befel myself.

I was reposing the other night, whilst my thoughts were wandering about Youghal in pursuit of my Tyndall; Mary my wife, told me to begone; I was obliged to fly, though I am not an aliped.[8] My wonder at such conduct ceased not during the night; in the morning what news did I receive? I was told that Mary had given birth to a young daughter, she is named Catharine.[9]

I request that you may not raise your sarcastic pen to lash me for my inconsistency, if I have erred it is through love of you.

I remain | Your sincere teacher | John Conwill

RI MS JT/1/11/3514
LT Transcript Only

---

1. *the Munster sentence makers*: Conwill is presumably criticising the use of grammar by someone from County Cork, who Tyndall had mentioned in an earlier letter, now missing.
2. *Hennessy's ingratitude and lying conduct*: Michael Hennessy, the master of the National School at Graigue, whose views Tyndall disputed on both mathematics and grammar.
3. *Sol*: the sun.
4. *Frenchhorn-hill*: presumably one of the peaks to the west of Ballinabranah.
5. *1st B of Euclid*: First Book of Euclid; the first of the thirteen books of Euclid's *Elements*; see letter 0017, n. 3.
6. *In the trapezium ... of the circle*: probable reconstruction of LT's unclear transcription. $2\angle rts = 2$ right angles $= 180°$. The trapezium only enters into Book 1, Prop. XXXV.
7. *proposition B ... VI B*: These propositions pertaining to the Sixth Book of Euclid were added to later editions of Robert Simson's *The Elements of Euclid*.
8. *aliped*: wing-footed, like a bat (*OED*).
9. *Mary ... named Catharine*: John Conwill had married Mary Carty on 22 January 1840 and their first child, Catherine, was born on 27 December 1840.

To John Tyndall, Snr 23 January 1841 0036

Youghal, | Saturday night[1]

My dear Father

With feelings of the greatest pleasure I received yours of the 19th,[2] it was indeed a welcome epistle. I thought that you had either totally forgot me or that something unusual had happened to prevent you from writing.

Man it is said is born for disappointment,[3] it is a melancholy truth. You congratulate me on my being held in high estimation by the Corpl. in charge;[4] I believe I may safely say that there was not one in the division held in as high, but the pleasant anticipations which that gave rise to are now totally blotted. Mulligan is to be removed from us on Monday next. He is to take charge of the field party and Collins is to return to the office. Mulligan is replaced by a man named Colder from another division.[5] He may be a good fellow, but it is with great regret that we part with Mulligan.

We however did not let him go without making an effort to detain him. When I heard of the Lieut's[6] determination to remove him I immediately thought that a petition might avert the circumstances. I therefore sat down and wrote what I now send you,[7] quite hastily, got it signed by all the civilians in the office, and sent it down to the Officer's lodgings. In the evening we got his reply and indeed every man of us was stung by it – He expressed his surprise at the liberty we had taken and hoped that it would not be repeated. We could hardly suppress a groan when this was read for us. However we must hope for the best.

When I spoke in my letter last of looking about a situation I thought a member of parliament could obtain it, and I also considered that the Capt. could get the Colonel's interest[8] for you. However we're not confined to the revenue;[9] a situation in the bank of Ireland[10] would be equally as good; the Dean[11] and Capt. Steuart I think could get that for me. I wish you'd try what the Dean thinks of it, I flatter myself that he would be my friend – I shall expect to hear from you shortly – Good night | Your affectionate son | John Tyndall

[Enclosure]
Youghal | Thursday[12] morning
Sir[13]

With feelings of the deepest regret we have heard of your intention to remove Corporal Mulligan from the Superintendence of the office duties – this Sir, induces us the Civilians of your division to take a step which under other circumstances might seem unjustifiable.

Since the time of Corporal Mulligan's being placed in his present situation he has discharged the duties imposed on him with energy and well directed zeal – This remark taken absolutely would be no doubt superfluous, but when taken relatively – when taken in connection with us Sir, it assumes a very different form. He, by manly conduct, his strict regard for justice, and his acknowledged impartiality, has endeared himself to every man in your division; he has imparted energy to their efforts, they hoped under his superintendence to see the deserving man duly rewarded; and the knowledge that none but those who were deserving would be so rewarded induced every man to emulate his fellow in rising in his officer's estimation.

We therefore earnestly but respectfully beg of you Sir, to consider these circumstances. It is not a matter of surprise that we should deprecate the idea of a separation.

Far be it from us, Sir, to urge anything which in its effects might be detrimental to the business of the Survey (and here we would beg leave to make a suggestion). The field work as carried on at present under the superintendence of Private Collins has given universal satisfaction; inefficiency in that department cannot be the cause of Corpl Mulligan's being sent out – Every man finds it his individual interest to promote the credit of the division at large, it would then be quite preposterous to petition for anything that would be to its prejudice.

We must here plead guilty to the charge of selfishness. We have not consulted Corpl Mulligan's feelings in this instance, he is quite ignorant of our present-proceedings coming forward on this occasion.

In conclusion, Sir, we would beg of you not to judge harshly of our present proceeding, nor deem our petition impertinent. We feel assured Sir, from your acknowledged kindness of character that you will pardon its numerous defects.

Should the result of your reflections on this subject be favourable to our wishes it shall be repaid on our part by increased assidulity and redoubled exertion.

Signed &c.

RI MS JT/1/10/3202–3
LT Transcript Only

---

1. *Saturday night*: LT gives date 'Jan. 23rd 1841'.
2. *yours of the 19th*: letter missing.
3. *Man … for disappointment*: Tyndall is probably paraphrasing 'Mankind … alone is born crying, he lives complaining and dies disappointed' (W. Temple, 'The Gardens of Epicurus' (1685), in *The Works of Sir William Temple*, 2 vols (London: A. Churchill, 1720), vol. 1, pp. 170–91, on p. 170).
4. *the Corpl. in charge*: Corporal James Mulligan.
5. *Colder from another division*: Corporal Alexander Calder had been in charge of the 3rd Division, whose numbers had declined.
6. *the Lieut's*: Lieutenant Paul Whittingham's.
7. *what I now send you*: the enclosure that follows.
8. *the Capt. … Colonel's interest*: Captain William Steuart could influence Colonel Henry

Bruen, the recently-elected Conservative MP for County Carlow.

9. *the revenue*: the Inland Revenue; see letter 0032.
10. *the bank of Ireland*: The bank had been created by the Parliament of Ireland's Bank of Ireland Act (1781) and began trading in Dublin in 1784.
11. *the Dean*: Dean Richard Barnard.
12. *Thursday*: 21 January 1841.
13. *Sir*: This letter would have been addressed to Lieutenant Whittingham, who commanded the 5th Division.

To John Tyndall, Snr 24 January 1841 0037

Youghal, | Jan. 24th, 1841.

My dear Father

You no doubt will be surprised at this hasty repetition of writing to you. I spoke in my letter of last night[1] about exerting yourself respecting a situation for me in the bank of Ireland. I have considered that subject calmly since, and have come to the conclusion that it would be better to suspend all efforts of that nature for the present. Both the Capt. and Dean[2] would very naturally ask why I was so very anxious to leave my present employment. I'm afraid I could not give a sufficient reason for it. When I wrote to you last night my spirits were greatly depressed on account of Mulligan's leaving us, this shed a gloom over the situation and every thing connected with it. I had great hopes of being well advanced with Mulligan and I considered these as destroyed when I heard of him being taken away from us, however when I come to consider things calmly I have no reason to despond – I'm not afraid of being outstripped by any one in the office as a workman, and I stand on an equal footing with them at present as respects our new Superintendent.[3]

I hope you have not spoken to either the Captain or the Dean it would be unwise to obtrude on their kindness at present – Time may come when it will be more needed, and until then I think it best to let every thing remain as it is. I have no reason to be discontented with my present situation, and I hope it's better it will be getting. Don't blame me for inconsistency when you read this, when I came to think over things quietly I was vexed with myself for suffering my impatience to manifest itself so much. Should you be harsh on me in thus acting I have no one in Leighlin to plead my cause except my mother; to her I will appeal as counsel for my defence and I trust that her interference will soften circumstances

I remain your affectionate son | John Tyndall

Give my love to Emma

RI MS JT/1/10/3204

LT Transcript Only

1. *my letter of last night*: letter 0036.
2. *the Capt. and Dean*: Captain William Steuart and Dean Richard Barnard.
3. *our new Superintendent*: Corporal Alexander Calder, who supervised the office party in the 5th Division.

To John Tyndall, Snr 31 January 1841 0038

Youghal, | Jan. 31st, 1841.

My dear Father

I received your letter[1] about two minutes and a half ago, the news it brings is a cause of both pain and pleasure to me, pain because my hastiness was the means of making you and my mother fret, and pleasure to find that my last letter had effaced your sorrow. When I wrote my first I was unconscious of the pain that in a calmer moment I knew it would cause. I am only sorry to find that my last did not arrive in time to prevent you from shewing my first to Captain Steuart. I'm afraid he'll consider me troublesome, however it will be a while before I trouble either you or him on such a foolish subject again.

The sun of contentment again shines on me, it has dispelled the cloud of unhappiness which for a while hovered over my way. Mulligan left us last week to attend at Foy's Courtmarshal.[2] Foy while he had the payment of the party in the field run about £20 in debt. It was for the purpose of making stoppages in his pay legal that the Courtmarshal was held. After it was over Mulligan returned here for the purpose of recommending some men for a rise of pay, it being impossible for the new Corporal to estimate their abilities. I yesterday invited Mulligan to dine with us. He at first assented but afterwards he came down stairs and called me out and told me that as he had made up his mind to recommend me his dining with us might be looked on with suspicion by some individuals. I admitted the truth of what he said and at the same time asked him did he think that it was through an interested motive that I asked him to dine – 'no! God knows I don't' was his reply – So my Journal goes off tomorrow.[3] You judge aright respecting Roberts not paying me. When Roberts was in Leighlin I firmly believe he was an honest man, but now I regret to say he is devoid of every vestige either of principle or honour. He has now departed for Dublin[4] but before he went I did not forget to impeach his integrity to his teeth and expose his want of principle. His intentions may be honest, but he has involved himself in such a manner here as to render them very questionable.

I don't think it would be the least use for John Mooney to write to John Hennessey, we expect his dismissal every day,[5] I however will speak to him on the subject.

I wrote to Debby[6] last week I have received no answer from her as yet. I am extremely glad to find that Thomas[7] is doing well.

I commenced this letter just before dinner. I am just now after despatching an enormous meal of corned mutton and turnips, we are all here now secluded from the bustle of the town and a bystander I am sure would not hesitate to pronounce us a very happy little family.

As my dinner has rendered my mental faculties a little torpid I must conclude my letter with a desire to be remembered to my mother and Emma.

Your affectionate son | John Tyndall

Tell Emma I'll write to her shortly I intended to do so this evening but the same cause that prevents me from lengthening this letter entirely precludes the idea of writing to her – I mean my dinner.

RI MS JT/1/10/3205–6
LT Transcript Only

---

1. *your letter*: letter missing, but presumably a response to Tyndall's letters 0036 and 0037.
2. *Foy's Courtmarshal*: Thomas M. Foy was charged with having 'embezzled or misapplied' the sum of £19 11*s*. 3*d*. His court-martial on 27 January 1841 led to his detention in Cork Barracks for 14 days (NAI OS/2/18 and OS/1/18).
3. *my Journal goes off tomorrow*: Tyndall's 'Journal of Progress and Weekly Report', together with the journals from other members of the 5th Division, would be used to compile the 'Month of January 1841, Progress Report of Letter C District of the Ordnance Survey of Ireland, Commanded by Major M. A. Waters Royal Engineers' (NAI OS/1/18).
4. *He has now departed for Dublin*: John Roberts was discharged from the Survey on 9 January 1841, owing to 'idleness' (NAI OS/1/18).
5. *John Hennessey... every day*: John Hennessy was a civil assistant who worked for a few months in the 5th Division, C District.
6. *Debby*: Deborah McAssey.
7. *Thomas*: not identified.

From John Conwill 11 February 1841 0039

> While life will remain in this bosom of mine,
> Every thought of my soul shall to Tyndall incline.

Ballinabrauna | Thursday[1] Morning

'My dear Tyndall'

Three times hurrah for mottoizing!!! (There is a coining secundus[2] for you)

I received your epistle of last Sunday[3] on yesterday evening, and not withstanding your determination to keep up the fire of correspondence with unceasing ardour, yet I find that you must, at intervals, deviate from such a resolution. This circumstance is very dejecting to me, though I must confess that

I am frequently constrained to desist from responding to many of your letters through intervention of unforeseen events; however one of the poets says:

Act well your part, there all the honour lies.[4]

Your will is my deed at any time, and as to my giving nonentity an existence and abstracting the little gentleman from the vortex in which he seems to be enveloped, it was only a puerile effort on my part to presume such a thing; and though poets may hurl their innuendos at such seeming impossibilities, yet Emerson, Simpson,[5] and Newton, despite the phrenzied ravings and mad ambitions of some silly rhymers,[6] established mathematically, a positive value for zero, and even gave quantities below zero an existence from the investigation of transcendental curves;[7] but as I did

Cast of manure a wagon full around
To raise a single daisy from the ground
Create a whirlwind from the earth to draw
A goose's feather and exalt a straw.[8]

I flatter myself that I have satisfied you that nonentity is equivalent to any given quantity, in a geometrical point of view.[9]

Believe me dear Tyndall, that we have been favoured with severe weather in Carlow,[10] so inclement indeed, that Frenchhorn[11] was taken by the peak in such a way that a Siberian, inured as he is to the penetrating blasts of the Arctic Ocean, could not pretend to cross Old Bawnree.[12] I kept very close to the fire for four days but I have resumed my usual occupations, as winter's grey mantle has been very gently removed. I am sure the County Cork presented a terrific appearance, as the poet has been so confoundedly mistaken. He would not have been in error had he said (so far as I am concerned),

'Tis there the bonny laddy lives
The laddy I loved best.[13]

Fear not your geometrical apostles,[14] they are faithful servants; they will duly perform their mission: they will make the loquacious become dumb, and cause the dumb to speak. I wish you may have benefited from the effusions of their great prototypes Peter and Paul.[15]

Your ever faithful teacher | John Conwill

P.S. The reason I expatiated so much on your poetical phrase of four lines, is, Emerson, Simpson and Newton were attacked by some poetical dolts;[16] I thank you for the eulogium which you have conferred on me through the medium of that poetical quotation.

RI MS JT/1/11/3515
LT Transcript Only

1. *Thursday*: LT gives postmark as 11 February 1841.
2. secundus: second (Latin).
3. *your epistle of last Sunday*: missing letter, which would have contained one or more of Tyndall's mottos.
4. *Act well ... honour lies*: A. Pope, *Essay on Man*, IV.193.
5. *Emerson, Simpson*: The eminent mathematicians William Emerson (1701–82) and Robert Simson (1687–1768).
6. *phrenzied ravings ... silly rhymers*: A number of poems were written celebrating Isaac Newton and his scientific innovations, including Edmund Halley's poem that prefaced Newton's *Principia* (1687) and James Thomson's *A Poem Sacred to the Memory of Sir Isaac Newton*(1727). Romantics, however, often attacked Newton's view of the universe; for example, in several of his poems William Blake portrayed the Newtonian universe as sterile. John Keats, Charles Lamb and other poets attended a dinner in 1817 and drank to 'Newton's health, and confusion to mathematics'. An account of that incident was only published in 1853 in T. Taylor, *Life of Benjamin Robert Haydon, History Painter*, 3 vols (London: Longman, 1853), vol. 1, p. 385.
7. *a positive value ... transcendental curves*: Euclid, following Aristotle, made the distinction between number and magnitude; and in his *Elements* he expressed magnitudes as ratios, not numbers. By contrast, numbers had to be positive integers, and therefore numbers did not include zero and 'quantities below zero'. With the development of calculus and analytic geometry the separation between number and magnitude was undermined. Thus for Newton (and later writers such as Simson and Emerson) zero and negatives were accepted as numbers.
8. *Cast of manure ... exalt a straw*: from John Wolcot (Peter Pindar), 'Unpublished Line on Dr Johnson', *Mirror of Literature*, 14 (1829), p. 248.
9. *that nonentity ... point of view*: For Conwill, who was steeped in geometry, non-entity lacks any specific quantity or magnitude; therefore any quantity can be attributed to it.
10. *severe weather in Carlow*: *Freeman's Journal*, 9 February 1841, p. [3].
11. *Frenchhorn*: Frenchhorn Hill; presumably one of the peaks to the west of Ballinabranah.
12. *Old Bawnree*: Bawnree is within the parish of Old Leighlin.
13. *''Tis there ... loved best'*: Conwill composes his own lines in the style of the eighteenth-century Jacobite ballad 'Highland Lad', which laments the departure of a 'Bonnie laddie' to take part in a Jacobite rising (*Jacobite Meolodies* (Glasgow: Richard Griffin, 1825), pp. 194–6).
14. *Fear not your geometrical apostles*: Conwill had sent Tyndall twelve testing mathematical problems that he believed would distinguish between deep and shallow mathematicians.
15. *great prototypes Peter and Paul*: Peter was one of the 12 disciples (apostles) chosen by Jesus to promulgate his message. Although Paul (Saul of Tarsus, *c.* 5 – *c.* 67) was not among the original 12 apostles, he is often referred to as 'Paul the Apostle' on account of his teaching of Christ's gospel, recorded in 'The Acts of the Apostles' and 'The Epistle of Paul the Apostle to the Hebrews'.
16. *some poetical dolts*: see n. 6.

From William Murray 17 February 1841 0040

Carrigaline,[1] February 17th 1841.

Dear Milton.[2]

I assure you it is not in consequence of being in a leaden lethargy that I did not write sooner but to tell you the truth I had no paper until I went to Cork on Monday last. I would then have bought a cap for you but that I thought it useless in consequence of not having any way to send it. I was very sorry to leave Middleton[3] as I was very comfortable and now we are going to a horrid country where we cannot get any thing to eat. A drop of milk if you put your eye out you would not get. We left Middleton at seven in the morning which was beautiful until ten the snow came on then and continued so until we arrived at Camden Barracks[4] then to comfort us we could not get a bed at any price which was very pleasant after walking seventeen miles in the snow. George[5] and I got a bed by great good fortune with some of the spalpeens.[6] Did you hear from Sam[7] since he went I think he is a very foolish fellow to form such an attachment with that Whitegate girl that has neither family nor fortune but depending on Mrs Fitzgerald that is supporting her and especially against his father.[8] I heard that he is to be married in May and then go to Australia. Mrs F. is to send them out. You did not see the desk the Tyril[9] gave him, it is a beautiful one. I hope he will not be so ill-natured as not to write. Is not the office going to Cork the 1st of March or April? I would be delighted at that – this place is only seven miles from it I would see your purly[10] face again – you dirty dog what do I care about you – I suppose Collins had an immensity to say when he went in. there was no man like him, if he'd stopped out one day longer my journal would have been sent off, tell him I was asking for him write and tell me all the news soon. Remember me to Jim Tidmarsh and all the darlings

Your ever affectionate | Murray

I have a request to make which I know you wont refuse Will you be so kind, so condescending and I so unworthy as to be after writing the names which I am about to tell you Mina Higginson and Maria in copperplate,[11] and I will be exceedingly obliged to you. Please write them on the sketch map paper, and as you are so pressing I will trouble you to write the capital letters for me, as I am going to practice very much but those will do any time when you are not better employed.

RI MS JT/1/11/3833
LT Transcript Only

---

1. *Carrigaline*: a small town on the Owenabue estuary about 9 miles south of Cork.
2. *Milton*: one of Tyndall's pseudonyms.

3. *Middleton*: see letter 0032, n. 9.
4. *Camden Barracks*: Fort Camden overlooks Cork harbour.
5. *George*: possibly George Holmes.
6. *spalpeens*: common labourers (*OED*).
7. *Sam*: John Roberts (aka 'Sam Weller'), who discharged from the Survey on 9 January 1841.
8. *Whitegate girl ... his father*: not identified, but see letter 0054.
9. *Tyril*: not identified.
10. purly: possibly pearly, meaning pale.
11. *Mina Higginson and Maria in copperplate*: neither person identified, although Mina Higginson may have been a sister of Thomas Higginson. The significance of writing their names is unclear, although he might be trying to woo them.

From William Wright 18 February 1841 0041

Dublin, Feb. 18th 1841.

Dear John

Once more I am obliged to sue for forgiveness for my neglect. If it were not for the apology I have to offer I would hardly have trespassed on your notice. I got the words said on last Sunday week, [1] since then I have been engaged looking out for a proper place to commence business and amongst the rest I was at Ln Bridge[2] I saw your Father, Mother and sister they are all in the enjoyment of good health Your Mother when she saw me was a little affected, knowing and as she remarked if you were at home how happy we all would have been at the meeting. I have been unsuccessful in procuring a proper situation or rather the rents of houses are so very high and trade very dull and unsettled that I would not like to undertake anything of the kind. So that I have entirely given up settling in this country and intend to embark for New York from Liverpool on or about the first of next month. I will expect a line from you as soon as convenient, although I cannot hardly expect you would comply with my request on account of my own negligence. But whenever you get yourself in the same hobble[3] I will cheerfully give you two months to think of nothing save the Fair Lady.[4] My Sally[5] desires me to say that it is not her fault for my not writing to you before this but you may believe as much of that as you please. I am sorry that you will not have an opportunity of seeing her before we go out but I expect if that is now denied it is only making room perhaps for a future meeting in distant land, for if my memory is correct I have often heard you express a wish to go to America.[6]

Yours truly | William Wright

Direct your next letter (which you must send before this day week) to 13 Lower Camden Street, Dublin.[7]

RI MS JT/1/11/3889
LT Transcript Only

1. *I got ... Sunday week*: William Wright married Sarah Maria Neale on 7 February 1841 at St Nicholas Parish Within, Dublin (Irish Church Records).
2. *Ln Bridge*: Leighlin Bridge.
3. *hobble*: an awkward or perplexing situation from which extrication is difficult (*OED*).
4. *the* Fair Lady: Tyndall's (unnamed) lady friend, who he should consider marrying.
5. *My Sally*: William Wright's wife Sarah (née Neale).
6. *a wish to go to America*: Tyndall thus appears to have shared with many of his Irish contemporaries the wish to emigrate. In the early 1840s some 50–60,000 Irish emigrated each year, mainly to America and Canada. During the Irish Potato Famine of the late 1840s the annual rate nearly quadrupled.
7. *13 Lower Camden Street, Dublin*: *Slater's* lists the occupant as Matthew Bredon, private teacher.

From Maria Payne 19 February 1841 0042

My dear John

After a long and provoking silence I will own on my side I am again going to trespass on your attention I am sure you will credit me when I tell you that you are still dear to me and that nothing gives me more real pleasure than your sweetly romantic letters they breathe the very language that I love pray my dear John have you written many Valentines[1] this time I know your talents in that way are of the first order. I am quite sure you are not heartwhole[2] all this time in the South such a gay and gallant Lothario[3] was never intended to blush unseen like the desert rose.[4] Come now make me your confidant as I have no secrets of my own to keep I shall be the more capable of keeping anothers. There is nothing in the shape of news here everything going on as usual. When may we expect the pleasure of a visit from you in the County Carlow, it is a long delayed and I assure you anxiously looked for honour and you know how hope deferred maketh the heart sick.[5] Emma[6] requests me to present her love and compliments to you

Adieu my dear John | Your sincere friend | Maria Payne.

RI MS JT/1/11/3841
LT Transcript Only

1. *Valentines*: Although of pagan origin, St Valentine's Day, 14 February, was declared a Christian feast in ad 486. By the middle of the eighteenth century the day was marked by sending love notes or poems or, by the time this letter was written, printed notes decorated with lace could be purchased. With the introduction of the penny post in 1840 (and thus the reduction in postal charges) the sending of Valentines gained popularity.
2. *heartwhole*: not emotionally attached; not in love (*OED*).
3. *a gay and gallant Lothario*: Lothario was character in Cervantes' *Don Quixote* (1605/15) and later in Nicholas Rowe's play 'The Fair Penitent' (1703) who was a persistent and successful womanizer.
4. *blush unseen like the desert rose*: adaptation of the line 'Full many a flower is born to blush

unseen, | And waste its sweetness on the desert air', from T. Gray, 'Elegy Written in a Country Churchyard' (1751), 70.

5. *hope deferred maketh the heart sick*: Proverbs 13:12.
6. *Emma*: probably Tyndall's cousin Emma, daughter of Caleb Tyndall.

To Emma Tyndall 20 February 1841 0043

Youghal | Saturday[1] night

My dear Emma

Its almost an age since any thing in the shape of a letter has past between you and me, the blame of this does not rest with you – no – I acknowledge my transgression and like the prodigal son of old[2] I now return to a sense of my duty.

I however am not so much to blame as you may imagine. I would have written to you long before now had I not been prevented by circumstances, these circumstances the subsequent part of this letter will explain.[3]

On last Wednesday week I happened to go into the draftsmen's room where I met Corpl Colder our present Superintendent. He handed me a written order which specified that the Parish of Killeagh[4] was to be traced, examined on the ground, and drawn by me immediately.

This was quite unexpected by me. So I got the plan traced and on Thursday[5] morning at about 11 o'clock I found myself in readiness to depart for the field. I slung my sketching case across my back and with my bundle in one hand and an immense stick in the other I sallied forth for the village of Mount Uniacke.[6] I proceeded in this manner until I got outside Youghal, when I thought it more handy to carry my bundle in true traveller fashion. No sooner thought than done. Behold me now with my bundle on the end of my stick and my stick across my shoulder trudging along a miry road as happy as any little peddler[7] boy in the kingdom; ever and anon striking up a lilt of a song in order the more effectually to drive dull care away. After travelling about 7 miles, I came in view of my place of destination – a comfortable sight truly for a fellow who was obliged to make it his quarters for some time – a group of huts whose mud walls and smoky thatch were in perfect consonance with the dreariness of the weather.

'Shigintha Bierlagh'[8] was pronounced by me several times and as often the unwelcome negative 'ne Stiggum'[9] was the reply, at length I was fortunate enough to meet with one who could jabber a little English. From her I learned where the Surveyors stopped. I went to the door, knocked, and was let into a domicile the floor of which was most fancifully stuccoed[10] with countless footprints of pigs and poultry. Upon enquiry I found that the Surveyors had deserted the place about 6 weeks before. *<words missing>*[11] Having seen enough of the place to make me disgusted with it, I sallied forth once more in search of a better. At the opposite side of the road I saw an apology for a public house; into it I entered, called for a pint of beer, and while swallowing it spoke to the landlord about a bed. After sundry preliminaries which would do honour to a diplomatist when about to conclude a treaty, I agreed to become a lodger.

Here I must make a gap of 3 or 4 hours in my narrative at the end of which time behold me stuck at the right side of the hobstone[12] working with my fingers at a plate of parboiled[13] cabbage with might and main.[14] This was the gift of an old woman, and though I did not much relish it, yet I was afraid of offending the old lady if I refused. Shortly after I got an enormous dish of fat pork and some of the aforesaid greens the greater part of which was soon despatched by me, as at the time I think I would have graced the character of Esau and sold my birthright for a mess of pottage[15] – hunger is an arbitrary monster.

Shortly after this feed I retired to bed which was a very comfortable one. The sheets, which in whiteness would rival Mont Blanc's whitest coat,[16] were well aired before they were put on the bed. So that in this most essential point my most sanguine expectations were exceeded.

Friday being a fine day I started early to my work, tugged away as hard as I was able until dusk. The shades of evening found me in a remote corner of my work and as hungry as a hawk. I however bent my course homewards, begged for a cold potato at half a dozen houses,[17] but in vain, at length with a belly a perfect vacuum I met an old man who when he learned my condition requested me to go in and that he would get potatoes boiled for me. I declined his offer but succeeded in getting a drink of milk. I shall never say that the belly is a useless member after that evening, as the root sends the sap to the branches so did it when I had swallowed the milk convey such a vivifying influence to my legs that I bounded buck-like over hedge and ditch till I reached home.

Saturday being exceedingly wet, I did not stir out until about 2 o'clock, when I lapped[18] an old India rubber[19] which I had borrowed from Ginty about me and dashed off again to the ground. Having reached it, to my utter disappointment a tremendous shower commenced, and I had to seek the shelter of an old ditch where I remained as snug as a thrush lapped up in the cloak for a couple of hours. Here my mind took a wheel and I turned poet. Having rambled through the fields of imagination for a while, I began to think that if I did not substitute substance for shadow the night would likely overtake me in the situation that I have described, and so I commenced plodding the miry[20] fields of reality which I must say were strikingly contrasted with the flowery ways of fancy which a few moments before I had pictured to myself.

Nothing remarkable happened on Sunday excepting that we had <u>mutton</u> for dinner!!!!!

It would exhaust your patience to read all I could say respecting my recent pilgrimage in the field. Suffice it to add that I have returned to Youghal this day and am again comfortably reinstalled in my accustomed corner by the fire side.

I have however to sing the requiem of my old strong boots that Jem Walsh made me, my recent excursion proved a brain blow to the poor fellows.

This letter will I trust be some atonement for my former neglect it is nearly as long as half a dozen ordinary letter

Good night | Your affectionate brother | John Tyndall

What are you reading now?

I hear that we are going to remove to Cork on the 1st of April.[21]

Tell my father to write to me. He owes me a letter

Give my love to my mother I was a regular trooper on foot while I was out in the field.

RI MS JT/1/10/3207–9
LT Transcript Only

---

1. *Saturday*: LT gives postmark as '21 Feb. 1841'.
2. *the prodigal son of old*: The 'prodigal son' is based on the story in Luke 15:11–32 and refers to someonewho has lived a reckless or extravagant life away from home, but subsequently makes a repentant return (*OED*).
3. *these circumstances ... will explain*: This letter contains a slightly shortened version of Tyndall's 'Mount Uniacke Journal' covering the period 12–16 February 1841 (Journal, vol. 1, pp. i–ii). See also account (in another hand), JT/2/1, ff. 1a–4v.
4. *Parish of Killeagh*: The parish includes the village of Killeagh, 6 miles west of Youghal in the barony of Imokilly, County Cork. In 1841 its population was 2,815 (*The Parliamentary Gazetteer of Ireland*, 3 vols (London, Dublin & Edinburgh: Fullarton, 1846), vol. 2, pp. 464–5).
5. *Thursday*: 12 February.
6. *Mount Uniacke*: Three miles north of Killeagh, the seat of the Uniacke family during the eighteenth and nineteenth centuries.
7. *peddler*: an itinerant trader or dealer in small goods (*OED*).
8. *'Shigintha Bierlagh'*: 'dtuigeann tú Béarla' = 'you understand English?' (Gaelic)
9. *'ne Stiggum'*: 'Ní thuigim' = 'I don't understand [English]', or, possibly 'I don't understand your question' (Gaelic).
10. *stuccoed*: plastered, in the manner of fine Italian plaster-work (*OED*).
11. : 'Mount Uniacke Journal' reads: 'When I asked could I get a room, the reply was "och musha, in troth I dont know Sir, my mathur is at the berrin [burial ground?], but she'll soon be home"'.
12. *hobstone*: a level stone grate on which things may be set to warm (*OED*).
13. *parboiled*: partially cooked by boiling (*OED*).
14. *might and main*: utmost or greatest possible power or strength (*OED*).
15. *Esau ... mess of pottage*: Genesis 25:21–34 contains the account of the birth of Esau and Jacob, the twin sons born to Isaac and Rebekah. Esau, who is the first-born grows into a coarse labourer. One day, being very tired after working in the fields, Esau was ravenous and asked Jacob to give him a meal of 'red pottage'. Jacob struck the bargain with the exhausted Esau: Esau traded his birthright – the status in the family accorded to the first-born – for this mess of pottage.
16. *Mont Blanc's whitest coat*: Tyndall climbed Mont Blanc – the highest mountain in the

European Alps – on three occasions, the first in 1857.

17. *begged ... dozen houses*: By the 1840s potato was the staple crop in Ireland; the poorer the family the greater the dependence on potatoes. Some three years later the potato crop was blighted, leading to the Great Famine.
18. *lapped*: folded or wrapped garment, or anything supple (*OED*).
19. *an old India rubber*: a rubberized waterproof coat, manufactured using a process invented by the Scottish chemist Charles Macintosh.
20. *miry*: of the nature of mire or swampy ground (*OED*).
21. *we are going ... 1st of April*: instead they left Youghal for Kinsale on 30 June 1841.

From William Wright 25 February 1841 0044

Dublin, Feb. 25th 1841.

My old Friend

For the last time I address you from Irish ground I hoped that I would have a line from you before my departure for America but I begin to think that I did not give you my address in my last letter I will not leave Liverpool until Monday – if you direct a few lines to 15 Duke Street, Liverpool I will receive them. I leave this tomorrow for Liverpool and from thence on Monday for New York[1] as soon as I arrive I will write to you, till then farewell God be with you my old fellow

Your sincere friend | William Wright

The mistress sends her best respects to you remember me to the LnBridge[2] people particularly S Hennessy[3] and John, Joe, and Tom Payne.

RI MS JT/1/11/3890
LT Transcript Only

1. *on Monday for New York*: Wright sailed instead on Wednesday 3 March; see letter 0046.
2. *LnBridge*: Leighlin Bridge.
3. *S Hennessy*: not identified.

From John Tyndall, Snr 25 February 1841 0045

Mr. John Tyndall | Ordnance Survey Office | Youghal
Leighlin Bridge | Feb. 25th, 1841.

My dear John

I would have written to you long before this but hearing that you left Youghal for the field duty I thought it better to wait till your return lest the letter or letters sent by me might be mislaid and you never get them. I read a letter from you[1] that Emma got yesterday morning. Its contents pleased me very much as it gave an excellent account of your late pilgrimage through the country. I think an interpreter would be of service to you the next time you go to the fields as it is difficult to know how to act when the language is not well understood at both sides. However I hope your health has improved by the exercise, if it has you are very well paid for your labour. The last letter I received from you contained an expression that I dont approve of, that is that you would not trouble me nor Mr Steuart any more with a letter about such nonsense.[2] This is what I dont like, because it is to me you should write when any thing is amiss with you, for I am of course with the exception of your mother the most sincere friend you have on earth and expects at a future day to make interest enough to get a good situation for you. I am glad to hear that your Journal[3] has went off with a recommendation for a rise of pay – from everything that I can learn you ought to have got that before this, but better late than never.

I send you a pamphlet containing Doctor Cook's speech.[4] It is a great expose of that imposter O'Connell and will amply compensate you for the trouble of reading it. The postage for it is 2d which you wont neglect sending by return of post. A tremendous fire broke out at Magrath's of Nurney[5] on Saturday night the 6th inst. about 11 o'clock and consumed 13 fine Dutch cows in calf with a great quantity of thrashed oats and oat straw – they say to the amount of five hundred pounds worth. I believe it to be accidental and that is the general opinion, though they have sworn the contrary and has put in their claim against the assizes for the amount of the above sum. I sent you a pair of boots which I hope fits, the heels I know were too low but that is easy changed to high. Willy Tyndall wrote to you about the plan of a school house,[6] he is very anxious to have it done if you have any time at all to spare do it for him.

The Dean[7] was in the shop with me on Monday last enquiring how you were; when I told him I saw a letter from you to Willy and that I believed you were well he expressed great pleasure to hear so and that you had his best wishes. John McGee's and Cecilia's affairs are very bad, it is reported here this day that the latter has been served with three <*word(s) missing*> one from Mr Jos. Watson the other from a <*word missing*> man and the third I have not heard his name.[8] On the whole they are in a bad state. You see John what extravagance will bring people to.

I hope you are reconciled to your new Superintendent.[9] Always act the man and let not trifling things cast you down. The songs you sung leaving Youghal[10] and the manner you carried your bundle was truly laughable and shows a manly mind. I will now conclude with letting you know that we are all well and join in sending our love to you.

I remain your | affectionate father | John Tyndall

RI MS JT/1/10/3210
LT Transcript Only

---

1. *a letter from you*: letter 0043.
2. *The last letter ... such nonsense*: letter 0038 in which Tyndall stated: 'I am only sorry to find that my last did not arrive in time to prevent you from shewing my first to Captain Steuart. I'm afraid he'll consider me troublesome, however it will be a while before I

trouble either you or him on such a foolish subject again'. The foolish subject was the prospect of Tyndall getting a position at the Bank of Ireland.

3. *Journal*: Tyndall's 'Journal of Progress and Weekly Report'.
4. *a pamphlet ... Cook's speech*: An eminent Presbyterian evangelical, Henry Cooke (1788–1868), the minister at the May St Presbyterian Church, Belfast, had challenged Daniel O'Connell to debate the latter's call for the repeal of the union between Britain and Ireland. O'Connell, the foremost proponent of Catholic emancipation and known to his supporters as 'The Liberator', visited Belfast in January but refrained from debating with Cooke, whose attack on O'Connell was published as *The Authentic Report of the Rev. Dr. Cooke's Speech at the Great Conservative Meeting Held in the Circus, Wellington-Place, Belfast, on Thursday, January 21, 1841* (Belfast: W. M'Comb, 1841).
5. *A tremendous fire ... Magrath's of Nurney*: 'On Saturday night last, a dreadful fire broke out on the premises of Mr. William Magrath of Nurney ... farmer .... Thirteen cows, we understand, were burnt to death, and one hundred barrels of oats, which lay in the barn ready for market, were also destryed. All the out-houses were burnt to the ground, and a portion of the dwelling house.' (*Freeman's Journal*, 12 February 1841, p. [1]).
6. *the plan of a school house*: It is not known whether Tyndall produced these plans for his uncle William Tyndall, but he did design other buildings, including the porch at Steuart Lodge, the home of his patrons Captain and Mrs Steuart.
7. *The Dean*: Dean Richard Barnard.
8. *John McGee's ... his name*: report not traced.
9. *your new Superintendent*: Lieutenant Alexander Calder.
10. *The songs you sung leaving Youghal*: see letter 0043.

From William Wright 2 March 1841 0046

Liverpool, March 2nd. 1841.

Dear John

I received both your last letters.[1] I have just time to spare to inform you of it and of our intention to sail on tomorrow.[2] we did expect to set out on yesterday but the ship had not her stores on board. It is needless to tell you how much I prized your letters, arriving as they did on the eve of my departure from all I hold dear on earth except my Sally.[3] Were it not for her presence I must confess I would hardly leave old Ireland, associated as it has been with every thing that I love. You Jack know little of the pang it costs But with my Sally I will or would find a home and a happy one in the deserts of Arabia. She sends her compliments to you for the kind forbearance you have exhibited towards her in not blaming her for keeping me from writing to you. Farewell my ever loved and respected more than friend.

You may rely upon it if it pleases providence that we arrive safe I write to you by the first opportunity, hoping soon to have the pleasure of seeing you along with us

I remain | Yours truly | W. Wright.

RI MS JT/1/11/3890

LT Transcript Only

---

1. *both your last letters*: letters missing.
2. *to sail on tomorrow*: Wright probably sailed on the ship *Oswego* which left Liverpool for New York on 3 or 4 March 1841.
3. *Sally*: Sarah Wright (née Neale).

To John Tyndall, Snr 1 March 1841 0047

Youghal | Monday[1] night

My dear Father

At about 8 o'clock on Saturday night Ginty entered our little domicile. He immediately began to lug letters out of a capicious pocket which he has the happiness of possessing. Among the rest it gave me great pleasure to find one from you.[2] Your letters are so few and far between that I consider it a great treat to see your well known fist on the superscription.

I am glad to find that my account of my perigrinations among the savages[3] pleased you. I was thinking it would amuse you – but I did not tell you the half of it, one instance in particular that I omitted before I must here recount. You know that before Shrove Tuesday is always a great time of matchmaking.[4] It was the whole occupation of the country people during my sojourn amongst them; you'd imagine it was a species of traffic that was going on. When fortunes are in question, the parties differ sometimes about very trifling matters. A couple in the village where I was differed about a pot; a long contest ensued, which I saw was likely to terminate warmly. I thought it right to split the difference, they however could not be prevailed upon to break the pot.

I now proceed to justify the expression[5] that you allude to in your letter. In the first place you must admit that it was nonsense – the more ebullition of a disordered imagination – and besides I was so angry with myself for acting in such a hasty manner that I could scarcely conceive any words competent to atone for my error. be assured of it my dear Father there is no one on earth that I would apply to the thousandth part as soon as to yourself – in fact there is no one else I would apply to at all.

My rise of pay has come back it is a repetition of the first I got– namely 2d!!![6] This however does not dishearten me one bit. It is a consolation that I am at the head of the pay list at any rate.[7]

Words cannot convey my gratitude to the Dean[8] for his kindness in enquiring for me. It shall always be my highest ambition to preserve the good opinion of such a man.

I am quite reconciled to our new Superintendent[9] , though he is not so much to my taste as Mulligan. He is what a latinist would call obscurus.[10] It's very hard to pronounce an opinion on him. I believe however that he has as high an opinion of me as of any other in the office, I can defy him to think otherwise.

A thousand thanks for your present of Doctor Cooke's speech.[11] It is most excellent. I think Dan[12] never met a truer opponent his Kerry flesh[13] must partake of the nature of an elephants if he be impervious to the castigation which the Dr has given him. He certainly did strip the demon Agitation of this hypocritical garb. and exhibit him to a wondering public in all his native deformity. The Doctor's energy of expression surpasses <*word illeg*> Oh how soul thrilling it was to hear him call the martyred sons of the reformation from their blood-stained graves and exhibit them in all their awful panoply of reeking innocence as the effects of popish intolerance. Surely O'Connell must sicken at the sight, not in disgust at his Church's atrocity, but in very bitterness at seeing such indubitable testimony of his own secret falsehood. Well may it be said that O'Connell is the bane of Ireland, but Cooke will prove the antidote, the hardy Northerns[14] have reason to be proud of their hero.

I heard of the fire in Nurney[15] before I received your letter.

I'm extremely sorry to hear of the increasing embarrassment of John McGee's affairs. I wonder the poor man does not sink under such accumulated misfortunes.

Has any news been heard from John Treacy since he went to Australia?

Wm. Wright is on his way to America to night. He left Dublin on the 26th ult,[16] and he left Liverpool to day – poor fellow I was the last he wrote to, he and I carried on a constant correspondence ever since I left home. Did he tell you anything about his wife – are her friends reconciled to their marriage? I believe the Neills are very wealthy[17] and that Wm possessed but little of this world's treasures. I would not wonder if they disliked the union. How is my uncle Caleb and aunt Dolly?[18] I sent the plan to Bill, I'm afraid he did not like it. He seemed to be in such a hurry that I did not take time to finish it properly.

Give my love to my mother and Emma

RI MS JT/1/10/3211
LT Transcript Only

1. *Monday*: LT gives postmark as 'March 2nd. 1841'.
2. *one from you*: letter 0045.
3. *my account ... the savages*: Tyndall's survey of the Parish of Killeagh, related in letter 0044 to his sister Emma.
4. *Shrove Tuesday ...* matchmaking: Shrove Tuesday, which fell on 23 February in 1841, is the day preceding Ash Wednesday, the first day of Lent, the season of penitence. It is a feast-day, at which pancakes are consumed. It was a traditional day for Irish weddings because of the prohibition on celebrations during Lent.
5. *the expression*: 'an expression that I dont approve of'; see letter 0045.
6. *namely 2d*: Tyndall's pay rose from 2*s*. 8*d*. a day to 2*s*. 10*d*. a day.

7. *I am ... any rate*: Evans, another draughtsman, was then earning 2*s*. 9*d*. a day, whereas Ginty's pay was 2*s*. a day and Higginson, a field surveyor, earned 1*s*. 8*d*. a day.
8. *the Dean*: Dean Richard Barnard.
9. *our new Superintendent*: Lieutenant Alexander Calder.
10. obscurus: dull (Latin).
11. *Doctor Cooke's speech*: see letter 0045.
12. *Dan*: Daniel O'Connell.
13. *his Kerry flesh*: O'Connell was born in the village of Carhan in County Kerry.
14. *the hardy Northerns*: Henry Cooke was a Presbyterian minister in Belfast, County Antrim in the northern part of Ireland, and a stronghold of Presbyterianism and anti-Catholicism.
15. *the fire in Nurney*: see letter 0045, n. 5.
16. *ult*: abbreviation of ultimo (Latin), of last month.
17. *the Neills are very wealthy*: the family of Sarah Neale, whom William Wight had recently married.
18. *aunt Dolly*: Caleb Tyndall's wife was probably named Dorothea; a Dorothea Tyndall was buried on 15 May 1887, aged 86 (Irish Church Records).

From John Conwill 6 March 1841 0048

Great Tyndall still doth lead the van,
In every line he speaks the man.

Ballinabranna M.N.S.[1] | March 6th 1841.

'My dear Tyndall'

The sentiments which your letter of last Saturday[2] breathes, are those of a person deeply versed in the philosophy of nature. Yes, I hesitate not to declare in the face of the universe that Tyndall possesses a mind which for its vividness and effusions is not inferior to that of a Newton.[3]

Shakespear's[4] phrases are so cogent in themselves that any comment of mine would derogate from the merit of that great man, however I must say that the words which you have cited from his work, namely, "The world is a stage"[5] are words that I have introduced on various occasions to exhibit the frailty and inconstancy of this life; I have also used them to point out the various scenes through which man is ramified during his pilgrimage in this sublunary world.

It appears even in our own native isle of the ocean, there is another Juan Fernandez in which my beloved pupil has undergone the fatigues of an Alexander Selkirk.[6]

When I read your fanciful description of the brutes of whom you were Lord from the centre all round to the sea, I could not but question the boasted records of the literary acquirements of the peasants of Munster.[7] I could not but think that my pupil was as competent to recognize the talents of these people as any alien itinerant historian who has dared to proclaim that he recognized more classic lore among the inhabitants of the South than what could be displayed by the first literary characters in Cambridge or Oxford. Blast such bare-faced falsehoods – You have not suggested that the oases of literature exist in that place of well-contested fame.[8] No more of this until another time.

While I was reciting your letter in the presence of your mother, I frequently cast a glance at her to behold the big tear that, as it were, furrowed her cheek and, believe me, her tears caused no small emotion to pervade my bosom.

When she heard your exile read, she burst out into a most sorrowful strain, saying ah! John, my beloved John!! my darling Trooper!!! how truly may I say in the words of the same exile Alexander:–

Oh tell me I yet have a son
Tho' a son I am never to see.[9]

You want no stimulus to rouse you to send an epistle written in your wonted drollery to your mother, as she is very dejected in spirit because she had not the pleasure of your company last Christmas, a circumstance which she fondly anticipated, but alas! providence disappoints us when we least expect it.

Butler has been removed from Carlow district and he is succeeded by a man named Shean, who was in my school February 26th. I had to examine all the classes in his presence; during part of the examination he asked Ford[10] to parse 'Pompeii.'

The shroud of years thrown back thou dost revive.[11]

Ford said that revive was an active verb in this sentence. No English scholar can understand it. And Shean was never the active sort. Hold Sir, said I, not so fast. I am a mere English grammarian and I will account for the verb in all its bearings. In the sentence alluded to it is intransitive, being the same as if I said to you, Sir, Thy sickness being removed thou dost revive. The punctuation, Sir, led my boy astray, as there should be a comma at back. That it is an active verb parse this sentence.

'The doctor revived his spirits by his timely aid.'[12] Moreover, Sir, said I here is Walker[13] with common sense on my side. But again, Sir, give me leave to ask you is a knowledge of Latin, Greek &c, a guarantee for infallibility in the English language? Did you ever read old Billy Cobbett's grammar, which contains thousands of errors selected from Blair, Johnson and Watts,[14] any of these men knew more languages than all the Inspectors under the patronage of the National Board. I am convinced, Sir, said I, that you will not hold any antipathy against me for thus opposing you, because I deemed myself called upon in an especial manner to vindicate my own character. He said he was well pleased and gave me an excellent report.

Hennessy came to me lately and implored pardon. Well said I, 'To err is human, to forgive divine'.[15] I pardoned him. He showed me a construction given by one Kelly to the sentence 'Windmills are made use of to grind corn'[16] ; the construction runs thus 'Windmills are made to be of use to grind corn'[17] I told Hennessy to point out for me whether there was any difference in the following forms. 'John made use of meat.' 'John made meat be of use.' Another thing philologer Kelly[18] insinuated to him was, where was the authority for introducing 'used' for 'made use of'? I have no room here to recount the milling I gave him on the last remark, I shall do so in my next letter. I do wish that you should send me some of your observations on this subject.

I am your faithful teacher | J. Conwill

P.S. As to Evans farewell to <*word(s) missing*>

On my conscience, Tyndall, I thought that I gave a hop, step and a leap into Youghal the other night; that we conversed about Milton, Dryden, Homer[19] and other wits of antiquity; and that as we were conversing about the scenes of bygone days, I was attracted by the contortions of a person whom I imagined lay near me. I then interrogated you about what happened [to] him – you were about to answer me when he screamed aloud and I thought a cow burst forth from his bowels after which followed twelve black monsters that devoured the poor animal. Heavens! I cried, are we to have twelve years of famine. Just as I was pondering on Pharaoh's dream[20] I was started from my sleep by a shrill call from my mother. She said I was sleeping uneasily.

RI MS JT/1/11/3516
LT Transcript Only

1. *Ballinabranna M.N.S.*: see letter 0017, n. 1.
2. *your letter of last Saturday*: letter missing.
3. *Tyndall ... a Newton*: Conwill's uncanny appreciation of Tyndall's potential as a scentist. Isaac Newton (1642–1727) was widely hailed as the most outstanding natural philosopher (scientist) and mathematician.
4. *Shakespear's*: the eminent playwright and poet William Shakespeare (1564–1616).
5. *'The world is a stage'*: 'All the world's a stage, And all the men and women merely players.' (W. Shakespeare, *As You Like It*, II.vii.142–3).
6. *Juan Fernandez ... Alexander Selkirk*: Conwill compares Tyndall's lonely sojourn in rural Co. Cork to the experience of a Scottish buccaneer named Alexander Selkirk (1676–1721), who was marooned for four years on the Juan Fernandez Islands in the South Pacific, some 370 miles off the coast of Chile. Selkirk's adventures inspired a number of books including Daniel Defoe's *The Life and Adventures of Robinson Crusoe* (1719).
7. *the peasants of Munster*: In his letter 'of last Saturday' Tyndall had presumably sent Conwill a narrative of his visit to the Parish of Killeagh similar to that contained in letter 0043. Munster is an ancient province of Ireland comprising several counties in the south west of Ireland.
8. *place* of well-contested fame: may refer to Cork's turbulent history, including the Irish Rebellion of 1641 when it was a Protestant stronghold.
9. *Alexander ... never to see*: Conwill appears to be paraphrasing Nathaniel Lee's tragedy *The Rival Queens, or the Death of Alexander the Great* (1677), II.

10. *Ford*: one of Conwill's pupils.
11. *'The shroud ... dost revive'*: from the poem 'Pompeii', by J. Beldam, *Fourth Book of Lessons, for the Use of the Irish National Schools* (Dublin: printed by P. D. Hardy, 1835), p. 165. The punctuation centres on a sentence whose punctuation is deficient. Therefore, the pupil mistakenly parsed the verb 'revive' as a transitive verb (takes an object), rather than an intransitive verb (which does not take an object).
12. *'The doctor ... timely aid'*: The sentence is an example of the verb 'revive' being correctly used in an active (transitive) sense to govern the object word 'spirits'.
13. *Walker*: probably the lexicographer John Walker (1732–1807), whose works included *The Teacher's Assistant in English Composition* (1801) and *Outlines of English Grammar* (1805).
14. *old Billy Cobbett's ... and Watts*: The radical journalist and pamphleteer William Cobbett (1763–1835) published *A Grammar of the English Language* (New York: Clayton & Kingsland, 1818) in which he criticised the use of language by such eminent authorities as Hugh Blair (1718–1800), Professor of Rhetoric at the University of Edinburgh and author of *Lectures on Rhetoric and Belles Lettres* (1767), Samuel Johnson (1709–84), essayist and lexicographer, and Isaac Watts (1674–1748), hymnwriter and author of *Logic, or The Right Use of Reason in the Enquiry After Truth* (1724).
15. *'To err is human, to forgive divine'*: A. Pope, *An Essay on Criticism*, I.525.
16. *'Windmills are made use of to grind corn'*: This poorly-constructed sentence suggests that the purpose of the windmill is subsidiary to its existence.
17. *'Windmills are made to be of use to grind corn'*: This sentence by contrast links the construction of the windmill firmly to its function.
18. *philologer Kelly*: possibly C. Patrick Kelly, superintendent of the school at St Patrick's Monastery, Tullow (*Slater's*).
19. *Milton, Dryden, Homer*: John Milton (see letter 0009, n.9); the poet and playwright John Dryden (1631–1700); the ancient Greek poet Homer, author of the *Iliad* and *Odyssey*.
20. *Pharaoh's dream*: In Genesis 41 Pharaoh recounted his dream in which 'seven well favoured ... and fatfleshed' cows emerged from a river. Subsequently seven 'ill favoured and leanfleshed kine [cows] did eat up the seven well favoured and fat kine'. Joseph interpreted this as a prediction that seven years of plenty would be followed by seven years of famine.

From William Murray 8 March 1841 0049

Friday, March, 8th, 1841.

My dear Tyndall

When I received your long expected letter I had the paper before me going to pen another for I thought that you had not received the last. My dear Tyndall although I asked you to take the trouble of writing the names for me I did not mean that you should keep me in such anxious expectation of your letter. I wish I was near enough to receive some of your *<word illeg>* and not be in this horrible place. I suppose you heard of the wreck of the great Killarney Steamer,[1] I am lodging within 2 miles of the place.[2] I am sorry your opinion of Roberts is so deservedly bad. I assure you I never had a very good one of him from the time I had any intimate acquaintance with him, in Youghal, in Castlemartyr,[3] he squandered away his money playing cards and drinking and never had any for the mess, but I need not tell you of his goodness for you know it better than I do. I did not let him get into my debt much, he is in debt to Ferris[4] 9 shillings. I was sorry to hear poor Jim[5] was so near being turned out for his stiffness but he is well recompenced for it by having a Tyrett[6] to himself. Who do you suspect sent that new valentine, I envees you, why did not you tell them to send me one? But you did not answer my question about the office going to Cork. I wish it was, for I could see you often. I was asked from home would I like to be introduced to society in Cork. So that if you are there we may have some fun. I was in Cork several times since I came here. I would not have gone only that I had a little business, for it was no inducement for me. Why when you are such an hermit stuck up in that dark domicile. why dont you be something like a chieftain and go and pay the fair ones a visit -- such as Miss <*word illeg*>? If I was near you you would not be so. if you were as bad as I am separated from the only person that I cared anything about (G Holmes) – he was sent to Middleton[7] and I left Bird[8] alone. Remember me to all the Dashent[9] fellows

Yours ever | Murray

If you were here the other day when I reclaimed my character by mounting part of Camolin[10] barracks where none of the other fellows would follow.

RI MS JT/1/11/3834
LT Transcript Only

---

1. *the wreck ... Killarney Steamer*: The steamship *Killarney* was wrecked on 20 January 1838 at Barry's Head, 5 miles east of Kinsale, with the loss of fifteen lives.
2. *I am lodging ... the place*: possibly at Nohoval.
3. *Castlemartyr*: Wright, Roberts, Holmes and Higginson had earlier surveyed the Castlemartyr area; see letter 0030.
4. *Ferris*: Thomas Ferris was a civil assistant who undertook surveying work in the 5th Division, C District.
5. *poor Jim*: Phillip Evans.
6. *Tyrett*: not identified, possibly a transcription error by LT.
7. *Middleton*: see letter 0032, n. 9.
8. *Bird*: possibly Joshua Boyd, a civil assistant who worked in the Examinations Office at the C District headquarters in Cork.
9. *Dashent*: decent?
10. *Camolin*: in County Wexford, about 20 miles north of Wexford.

To Emma Tyndall 15 March 1841 0050

Youghal, Monday[1] evening

My dear Emma

I see yourself is the only person I can depend on for a return to my epistles I must confess that tho' you are not generous you are just in your epistolary affairs. You always send me one for one. I thought I could extract something more from you, but I find that impossible, so I must only make myself as good a fellow as you and write letter for letter with you. I dont forget the quizzing (as Frazer[2] used to say) you gave me in your last letter about the keenness of my appetite, oh my girl if you were out in Gurtnagopple[3] with me you would never have need of Camomile flowers.[4] I think if I met a child out there with a cold potato in his hand I would act my father's part and become a robber of the innocent. To tell the plain candid truth my appetite is something off the common. Evans and Tidmarsh swear that I must pay two thirds of the mess, as I am able to eat twice as much as either of them. However one comfort is that I am able to beat them both, so that if ever they become headstrong I'll thrash them into obedience.

I am just after exhausting my purse in purchasing a suit of clothes from head to foot. Tidmarsh counted 13 patches!!! on the trousers that I wore yesterday (Sunday!!) and yet strange to say notwithstanding all this there was something aristocratic in it, it was a perfect specimen of the relics of auld decency.

Tell me Emma is there anything amiss in Bagnalstown? I wrote two letters there and received no answer to neither. I believe Mrs Payne has not heard from them this good while. I am afraid something is the matter or Maria would have answered my letters. Tom Payne has an increase of 4d a day lately he has now 10 shillings a week. Mrs Payne is worth £70 a year at present, but I think if she had twice as much she would contrive to spend it without being comfortable.

Poor Wm Wright is on his way to America now, he wrote to me from Liverpool.[5] He appears to be greatly attached to his young wife. When I was in the country I lodged in the same house that Francis Heydon[6] lodged in two months before, they appeared to like Frank exceedingly as whenever his name was mentioned they were loud in their praise of him. Evans was out with me to day making my bargains.[7] I'm twice as good a hand as he is at cutting them down, he was surprised at the brass of face[8] that I exhibited. Among other articles I bought a couple pair of soles, it would not be easy to deceive me in them. John Murphy[9] when purchasing from myself was never half so great a shuffler[10] as I am. I think if I was dealing with my father he'd turn me out of the shop.

Give my love to my mother, ask her if she did pay my father the 2d for which he billed me, if not I'll take out a process and get Woodhouse[11] to serve it on her for my 11½d. I sincerely hope however that she'll see that it's her own interest to pay the required sum.

Do you read anything now Emma? I hope you have not paid it by, you will find it of infinite service to you. I have read a good deal myself since I came to Youghal.

Send me every little bit of news you can collect, how Mr and Mrs McGee are, how my uncle Caleb's family are. Send me word how the plan I sent Bill[12] answer. I saw a caution from him in the Carlow Sentinel[13] respecting some affairs between him and James Treacy. How is my uncle Ned[14]

My father owes me a letter, tell him to write to me

Good night | I remain Your affectionate brother | John Tyndall

RI MS JT/1/10/3213
LT Transcript Only

---

1. *Monday*: LT gives postmark as 'March, 16th, 1841'.
2. *Frazer*: not identified.
3. *Gurtnagopple*: 'field of horses' – probably reference to Tyndall's description in his previous letter to Emma – letter 0043 – of his visit to the Parish of Killeagh.
4. *Camomile flowers*: to treat indigestion.
5. *he wrote to me from Liverpool*: see letter 0046.
6. *Francis Heydon*: Francis Heydon was from Leighlin Bridge and one of Edward Heydon's sons. Unlike his brother William, he appears not to have worked for the Ordnance Survey of Ireland.
7. *my bargains*: purchase of the new suit.
8. *brass of face*: audacity.
9. *John Murphy*: not identified.
10. *shuffler*: *OED* defines shuffler as 'One who acts in a shifty or evasive manner; a slippery, shifty person'; but Tyndall probably meant someone who strikes a hard bargain.
11. *Woodhouse*: not identified.
12. *the plan I sent Bill*: see letter 0045.
13. *a caution ... Carlow Sentinel*: William Tyndall warned the public not to accept any bills from James Treacy, who was insolvent (*CS*, 6 March 1841, p. [1]).
14. *my uncle Ned*: Edward Styles.

From John Tyndall, Snr 17 March 1841 0051

Mr John Tyndall | Ordnance Survey Office | Youghal
Leighlin Bridge | March 17th, 1841.

My dear John

I was very much disappointed this morning when your mother brought your letter to Emma[1] to the bedside to me. Mr Winnet[2] who is now the letter carrier called to the house with it and said 'here is a letter for Mr Tyndall from his son.' Your mother ran up with it to me without looking at the superscription and said 'here is your Patrick's pot[3] for you.' I put on my spectacles and took a view of the cover. I immediately saw it was for Emma. The moment she heard it was for her she demanded it immediately. I gave it to her – she read it then handed it to me – I read it and was highly pleased with its contents, particularly that part of it where you say you can eat as much as your two comrades this was delightful intelligence to me and your mother, as I know you cant be such a cormorant for eating[4] without being in very good health, Your mother says she paid you the 11½d long ago when you were going to the fair of Orchard[5] from Nurney and also to Leighlin Bridge, and as I am willing to forgive you the debt I thought legally due, do you compromise the matter with your mother. I am just after seeing about five hundred teetotlers pass.[6] Hughes comes this moment from the upper window not thinking it worth while to go to the door, they assembled at the Priest's house,[7] marched with the band before them through the Royal Oak Bagnalstown, and round to Leighlin Bridge amidst the most tremendous rain – Priest McMahon Commander in Chief and Paddy Malloy[8] from Old Leighlin second in Command, for I saw him at the head of the division with a white rod in his hand a little behind McMahon who rode on horseback in front. If protestants acted so what an outrage on the majesty of the people would be the eternal cry of the repealers.[9] The plan you sent Willy[10] done very well but their funds is at such a low ebb at present they cannot go on with the Minister's house. John, if William Heydon is in Youghal I would be glad if you would speak to him and let him know from me that his father is not well off at present, being in no sort of business and that it is a very great shame for him to say he never sent one shilling to his father since he left Leighlin; now be the amount never so small you know he could from time to time save something that would be of service to him. I am glad to hear you have bought a suit of clothes; your shoes and boots I suppose are nearly worn out as you bought two pairs of soles the other day. If you had come home at Christmas I would have got made for you two or three pairs of shoes, but I think you dont intend coming any more. There was a great Christening at Caleb's on Sunday last,[11] the house though large I hear was filled with people from different parts of the country and some of them very decent. Amongst the number Mr Lee from Milford[12] and several others too numerous to mention. I had a letter from you to Mr Conwill until evening yesterday, waiting for a messenger, when I got one. I expect him here this night to read your letter and his answer but he has not come yet. I now conclude with sending our love to you and

remain your affectionate father | John Tyndall

I was not well writing this letter you'll therefore not mind any faults in it.

RI MS JT/1/10/3214

LT Transcript Only

1. *your letter to Emma*: letter 0050.
2. *Mr Winnet*: not identified other then being the 'letter carrier'.

3. *your Patrick's pot*: St Patrick, the patron Saint of Ireland, is said to have persuaded a miserly innkeeper to fill his customers' glasses to overflowing. The receipt of Tyndall's letter on St Patrick's Day – 17 March – was celebrated by his mother as a similar sign of gladness and generosity.
4. *a cormorant for eating*: As cormorants are fish-eating birds that dive on their prey, Tyndall is likewise portrayed as swooping on his food.
5. *the fair of Orchard*: an agricultural fair held in various Irish counties.
6. *five hundred teetotlers pass*: The Irish temperance movement, led by Father Theobald Mathew, was strongly supported by many Catholic priests. It held marches on St Patrick's Day 1841 in many cities and towns, including Tullow (*Freeman's Journal*, 24 March 1841, p. [3]).
7. *Priest's house*: probably the Priest's House in Dunleckny, close to Dunleckny Manor, about a mile north-east of Bagenalstown.
8. *Paddy Malloy*: not identified.
9. *the repealers*: the movement, spearheaded by Daniel O'Connell, that sought the repeal of the union between Britain and Ireland.
10. *The plan you sent Willy*: see letter 0045.
11. *a great ... Sunday last*: not indentified, but it would appear that neither Tyndall nor his father was invited.
12. *Mr Lee from Milford*: Peter Lee worked on John Alexander's Milford estate, where he was employed as a clerk or possibly steward.

From John Conwill 20 March 1841 0052

Give to Caesar the things that are Caesar's
And to God the things that are God's.[1]

Ballinabrauna M.N.S.[2] | March 20th 1841.

'My dear Tyndall'

I perceive by your last interesting epistle[3] that you possess fortitude enough to repel the bold attempts of the presuming critic on the insiduous intrusions of the atrocious parasite; but that you are completely unmaned by my extravagant eulogiums, because you are convinced, that I pay no respect to those persons who endeavour to raise the edifice of their future felicity on the destructive remains of another's good name; and that I inherit not the sly artifice of the sordid flatterer who basely exercises his base cunning to carry his objects into effect. Oh! you are perfectly aware the one non-descript of persons is as odious in my view as the other. And while I breathe vital air or float on the stream of existence, I shall use all my energies to subvert the sinister manoeuvres of the malignant critic and the crouching movements of the parasite.[4]

But again as to your being unmaned by my panegyrics; while Conwill can elevate his pen and bring it into collision with paper, the merits of my beloved pupil shall never sink into oblivion. The praise that is Tyndall's Tyndall must get as well as my limited capacity empowers me to portray it.

Never shall the expression of a refined Poet be verified in the talents of my far-famed Tyndall whilst the humble abilities of his teacher can proclaim these talents to the world, No! no!! no!!! never shall it be said of Tyndall's capabilities that they sink into the shade unnoticed. The words of Gray shall not be fulfilled in that respect:–

> Full many a gem of purest ray serene
> The dark unfathomed caves of ocean bear;
> Full many a flower is born to blush unseen,
> And waste its sweetness on the desert air.[5]

I had another interview with Mr Sheean, we discussed freely on the verb, revive.[6] I put my argument in this manner. A mere English scholar cannot understand the verb, should he construe it as an active verb; therefore Johnson Elphinson, Kenrich, Sheridan, Nares, Entick &c[7] are mere English scholars, because they have given it as their opinion that the verb in various instances is truly active. Now, Sir, said I, go forth and tell the literary world that Walker[8] Johnson &c had not a knowledge of Latin Greek &c and you will render yourself a butt for the finger of scorn to be pointed at. Believe me this argument had such an ascendancy that the doctor[9] became electrified. He received such a shock that his batteries[10] were hurled to the ground[11] and he departed evidently ashamed and confused.

I knew I was not astray when I formed my idea of the want of refined knowledge that existed in the south of Ireland. The mode by which the prodigies of Munster[12] obtained so much celebrity is this. They purchased the periodicals and other works published in England, from which works these would-be satellites in mathematics gleaned all that was concise and beautiful and gave that as the result of many a weary hour of their own study, to every satchelier[13] that through prejudice, or some other fantastical notions, moved up to that focus of plagiarism & imposition in which you have so long subsisted. Your observations on Professor Kelly's[14] modest paraphrasing of our excellent grammatical jaw breakers, proved to be an invulnerable stumbling block to that puerile creature's comments. I proved our construction of the expression thus, 'Windmills are made use of to grind corn': 'Windmills are used to grind corn': the expressions being equivalent, if we take away equals from equals, the remainders must be equivalent wherefore 'made use of' is equivalent to 'used'. I modified the sentence by hints of minor importance that are not worth noting in this place.

I read your last luminary[15] for your mother, whose bosom heaved like the agitated billows of a stormy sea, until I came to that part of the epistle that specified your determination to come home next summer and then her breast heaved no longer, it became like the calmest seas but to our surprise, if I may say so, she made the welkin[16] ring by hurrah! hurrah!! hurrah!!! My Trooper! I will see you once more. Yes I am consoled in the happiness of my contemplating

That I still have a son
And that son I will certainly see.

I am your ever faithful teacher | John Conwill.

Arrah! Jack – at a push, how bad you are for a possession of some of the old witch's incantations, to enable you to develope the mystery of my dream, on my soul you know the drift of that fancied vision with the very glance of your eyes. I had a comical scene the other night which you will hear from me another time. I have laboured this fortnight under a severe attack from cold.

You will no doubt be amazed at the delay of this letter but a terrific hurricane blew from the direction in which Youghal is situated and therefore it was utterly impossible for me to venture into Leighlin Bridge until Monday evening to post this letter to my revered friend and pupil.

RI MS JT/1/11/3517
LT Transcript Only

1. *'Give to ... are God's'*: 'Render therefore unto Caesar the things which are Caesar's; and unto God the things that are God's' – Matthew 22:21. In this and many other contexts this biblical verse is taken to emphasize the distinction between the religious and the mundane.
2. *Ballinabrauna M.N.S.*: see letter 0017, n. 1.
3. *your last interesting epistle*: letter missing.
4. *the sinister ... the parasite*: probably the unnamed controversialist attacked in letter 0060.
5. *'Full many ... desert air'*: T. Gray (1716–71), 'Elergy Written in a Country Churchyard' (1750), 53–6.
6. *the verb, revive*: see letter 0048, nn. 11–12.
7. *Johnson ... Entick &c*: Samuel Johnson (1709–84), the educationalist James Elphinston (1721–1809), the writer William Kenrick (1729/30–79), probably the Irish clergyman and schoolmaster Thomas Sheridan (1687–1738), the clergyman-philologist Robert Nares (1753–1829), and the schoolmaster, writer and grammarian John Entick (*c.* 1703–73).
8. *Walker*: the grammarian John Walker (1732–1807).
9. *the doctor*: Conwill uses title 'doctor' sardonically to empaphasise Mr Sheean's ignorance.
10. *electrified ... his batteries*: Conwill draws on the vocabulary of electricity, which he extends to the human subject to whom he attributes 'batteries' and the ability of being 'electrified'. Shocks could be produced by lightening and by powerful electrical machines, in which a glass plate or disc is rubbed while being rotated.
11. *hurled to the ground*: a bolt of lightning or a discharge from a powerful electrical machine would have that effect.
12. *Munster*: see letter 0005, n. 6.
13. *satchelier*: possibly the bearer of a satchel, as used by students.
14. *Professor Kelly's*: possibly C. Patrick Kelly, superintendent of the school at St Patrick's Monastery, Tullow (*Slater's*).
15. *your last luminary*: letter missing.
16. *welkin*: heavens (*OED*).

To John Tyndall, Snr 22 March 1841 0053

Youghal, | Monday[1] night

My dear Father

I was just sitting down this time last night in order to answer your letter, when I suddenly recollected that I had not fulfilled your request respecting Wm Heydon.[2] This caused me of course to lay pen and ink aside and defer writing till this evening. I brought your letter with me this morning to the office and after calling him out of his room read that part which related to himself. I said a good deal to him. I plainly told him that he possessed not one of the feelings of a child. He made not a word of answer and so I left him without requesting him to make any promise of future amendment. This I thought would have more effect at another time; accordingly when leaving the office this evening I called him aside and asked him what answer I would send to you respecting him. I asked him how much he was able to save in the week; he told me, and after some time he promised to be a good boy in the future and save all he could for his father. Being afraid that he might not be able to keep his resolution, especially when surrounded by a parcel of young lads like himself, I adopted the following plan – I'm to see him paid every Saturday night[3] and then on the spot he is to deposit three shillings each week in my hands. When it amounts to anything worth while I'll send it to his father. He cannot begin to save until next Saturday week, <*word(s) missing*> having a scale to pay for this week. So I have done all in my power with him.

I think you must have formed some very erroneous notion respecting my appetite – why I cant finish a fourpenny[4] all out at a time; surely I don't deserve to be called a Cormorant for that. However I'm half starved to night. Tidmarsh has turned messman,[5] vice Evans, he has most extravagant notions of economy and the moment he assumed the functions of his new office he drew up a declaration and made us sign it, this made us promise that we are to pay implicit obedience to him and never murmur against any measures adopted by him. The consequence of all this is that he has made out a bill of fare in which I must confess the luxuries of life are excluded, one article in it is in the shape of a postscript in which the use of bread is totally prohibited. But I intend to take a copy of this document and exhibit it to you in Leighlin when I go there. It's well for us that he will remain in office no longer than one week.

I'll match my uncle Caleb for not inviting me to the Christening.[6] I was usually a guest on such occasions. Surely the hop-step and leap that separates us could not prevent him from sending Emma or Billy over to Youghal to invite me. I'll give him a tongue thrashing when I go to Carlow.

Emma owes me a letter she need not think of getting over writing it, if I don't get what's justly due by fair means I must by foul I'll process her and bring her to the Mall house[7] if she does not pay me my just and lawful due – tell her to answer my questions relative to the Bagnalstown people.[8]

The only part of your last letter that I dislike was the postscript, this informed me that you were not well in writing it. Avoid getting cold. I hope however you will need no other doctor than my mother to cure you. I'm extremely glad to hear that she is still able to bounce up stairs.

I was out in the country the whole of Friday and Saturday. I was sent by Lieut Whittingham to seek information respecting the boundaries of gentlemen's demesnes.[9] I visited my old host[10] while I was out, he informed me that Wm Murray had slept a night there (i.e. in Mount Uniacke) since I left it.

I'm working away at the drawing in the office and altho' I was a year without any practice still I think that my plans at present are as highly esteemed as those of any other person in the office. Collins and I work in the one room I believe there never was a man on the Survey that he was so obliging to as he is to me we're as great as two brothers. An angry word never drops between us.

We had no display of 'de boys' here on Patricks day.[11] I suppose Jim Walsh formed one of the glorious band that got themselves ducked marching through Leighlin, if we could take the will for the deed I'd stake a day's pay that many among those very lads should plead guilty to the charge of intoxication.[12] I'll engage more than one of them wished to have a 'Johnny' on his tongue[13] when the gentle influence of a March shower had damped their overalls

Warmet[14] won't disappoint you in this

Good night | Your affectionate son | John Tyndall

RI MS JT/1/10/3215–6
LT Transcript Only

---

1. *Monday*: LT gives postmark as 'March 23rd. 1841'.
2. *your letter ... Wm Heydon*: see letter 0051 in which Tyndall, Snr, had asked Tyndall to intervene with William Haydon – a civilian assistant in the Youghal office – to send money to his father who was in financial difficulties.
3. *I'm to ... Saturday night*: William Heydon's weekly pay was 9*s*.
4. *a fourpenny*: a very substantial loaf of bread costing four pence.
5. *messman*: cook (*OED*).
6. *the Christening*: see letter 0052.
7. *the Mall house*: where the Petty Sessions were held in Youghal.
8. *the Bagnalstown people*: probably the Payne family, who lived in Bagenalstown.
9. *demesnes*: private estates (*OED*).
10. *my old host*: see letter 0043.
11. *'de boys' here on Patricks day*: St Patrick's Day falls on 17 March; St Patrick being the most widely acknowledged patron saint of Ireland. According to Tyndall, the Nationalists in Youghal did not celebrate it publicly in 1841.
12. *guilty to the charge of intoxication*: the march was in support of the teetotal movement.
13. *'Johnny' on his tongue*: a taste of whisky.

14. *Warmet*: not identified.

To John Tyndall, Snr 10 April 1841 0054

Youghal | Saturday[1] night

My dear Father

About half an hour ago a letter from you was handed to me by John Hennessy.[2] So you see that I have lost no time in answering it. You complain of my expecting letter for letter from you, but I assure you this was not the cause of my delaying to write to you. It was something else which I won't tell you at present.[3] I'm sure you'll laugh when you hear it from my own lips when I go home. I was just on the point of sending to you for Poole's 'Nullity' or his 'Dialogues'.[4] This is just a hint. You draw in your letter a comparison between our ages – you complain of a palsied hand. Now with great deference to you, father, your writing contradicts your words. I positively think I never saw better writing after you than your last letter contained. This does not argue much in favour of a trembling hand. The steel pen[5] must be valuable, but I'm afraid if the hand that guided its movements was not steady its own efforts must fail in producing good writing.

Never dread my health, if you could but see the mess of herrings and butter that I despatched this day you would not be over anxious concerning that matter. I however will write to you weekly from this out.

I'm extremely sorry to hear of Mr Conwill's health being delicate, it gives me great pleasure however to find that he is recuperating, tell him to give up Diophantus and Descartes[6] for a while, they are bad medicine for a delicate frame.

There is one portion of your letter in particular which gave me great concern and that is where you tell me that the times are very bad. Are they very bad with you? If I thought so I'd starve myself to keep you independent. Have you a good stock at present? Do you keep Jem[7] constantly employed? Egad[8] this is the worst news I heard since I came to Youghal.

But after this unpleasant news you have penned a little palliative, it gives me great pleasure to hear that you were able to afford the girls some fun there is an amazing spirit contained in two or three *[*tumblers*]*.

Did you hear how was Sarjent McKim[9] from Jacob Little?

I suppose I need not tell you that my feeling with respect to Mrs Steuart is as strong as your own. I believe in her loss I would have to deplore that of one of my best friends.

I'm just after extracting the first halfcrown[10] from Wm. Heydon.[11] I would have got more from him but being prevented from working on Good Friday and not being paid for it diminished his week's pay something, he promises well, I'll make him keep up to his trumps. I have John Murray also under contribution[12] I was afraid that he was spending the overplus of his pay foolishly, so I make him give it up to me now on every Saturday night I have six and six[13] of his at present when it amounts a little higher I'll buy some clothes for him.

I saw Wm Murray a couple of times lately he's as strong as an ox and comfortably dressed. I wanted him to come and eat his dinner with me the other day and after much pressing he promised to follow me up but he did not come.

I like our present Corporal[14] better and better every day. He's a very fair man I think.

Things are going on in their old way here. Roberts is to be married on the 1st of May to a girl for whom he farmed an attachment while employed in the field.[15] She's a young woman that a Mrs Fitzgerald a lady here has taken under her protection. Mrs F. is to give him a hundred pounds, and to send both of them out to Australia. Roberts's father is very averse to the match I believe however the son is determined to persist in his choice. I owe I suppose half a dozen letters, the same cause that prevented me from writing to you operated in like manner with respect to the rest of my correspondents. Its nearly 12 o'clock so I'll have my Easter's egg[16] on you and bid you Good night

Your affectionate son | John Tyndall

RI MS JT/1/10/3217
LT Transcript Only

---

1. *Saturday*: LT gives postmark as 'April 12th, 1841', a Monday.
2. *letter from ... John Hennessy*: letter missing. Hennessy was a civil assistant who had worked for a few months in the 5th Division, C District, but had now left the Survey.
3. *something else ... at present*: possibly the invitation to debate 'Popery'; see letter 0057.
4. *Poole's 'Nullity' or his 'Dialogues'*: The Nonconformist minster Matthew Poole (1624–79) was the author of the two anti-Catholic works, *The Nullity of the Romish Faith* (1666) and *A Dialogue between a Popish Priest and an English Protestant* (1667).
5. *steel pen*: a pen with a steel nib, split at the tip like a quill (*OED*).
6. *Diophantus and Descartes*: the demanding writings of the Greek mathematician Diophantus (third century AD) and the French philosopher René Descartes (1596–1650).
7. *Jem*: James Walsh, who was Tyndall's father's apprentice boot and shoemaker.
8. *Egad*: a softened oath originally derived from A God! (*OED*).
9. *Sarjent McKim*: not identified.
10. *halfcrown*: 2*s*. 6*d*.
11. *Wm. Heydon*: William Heydon; see letter 0051.
12. *John Murray also under contribution*: Murray was then earning 1*s*. 4*d*. per day (NAI OS/1/18).
13. *six and six*: 6*s*. 6*d*.
14. *our present Corporal*: Corporal Alexander Calder.
15. *Roberts ... in the field*: see letter 0040.
16. *Easter's egg*: the celebration of Easter by eating eggs; the egg symbolises Jesus's empty tomb, and thus his resurrection.

To John Tyndall, Snr 25 April 1841 0055

Youghal, April 25th, 1841.

My dear Father

On Friday evening I received your letter.[1] By it I perceive you are an expert man of business. Your despatch in going to Dublin and coming back brings to my mind a picture I saw somewhere the title of it was 'Travelling in the year 1900.'[2] It represents a fellow with a steam kettle under his arm taking a race from one end of Europe to the other every step or stride the fellow made would lead a person to imagine that he was possessed of 7 league boots[3] or some other mode of conveyance like it, the truth was he went by steam. I see you have improved the hint I gave you relative to what prevented me from writing to you.[4] I'll not tell you the circumstances at present neither will I undeceive you, you may indulge in your speculations for a couple of months and then you shall hear the whole story from my own lips. Your advice respecting my acting on the defensive shall be strictly attended to. My motto is

'I neither seek the combat nor shun it when it comes.' I agree with you perfectly that a little sophistry is quite needful in order to meet, with success, a subtle opponent; the less of it that is used however the better. It should be the business of a man with truth on his side to expose sophistry but not to use it. In order to expose successfully I admit that it should be well understood by even the champion of truth. While you were in Dublin it appears that you did not forget me. I received the pamphlet you purchased last night. I believe old 'Nullity' is a very scarce work. I can however get a loan of Poole's Dialogues[5] here so that renders it unnecessary for you to send me yours.

Your last letter conveyed very cheering news to me. I don't deserve one bit of praise for the sentiments which my last letter[6] breathed. I would be a monster if I acted otherwise than I there professed.

I made Wm Heydon send his father five and sixpence[7] last night. When I make use of the word made I don't mean to intimate that he did it with reluctance. He would have sent a little more had he not been sick for two days last week, which prevented him from laying his usual instalment in my hands. It was he sugested the propriety of sending even that small sum, for fear his father might want it. I have by my side at present Challoner's 'Catholic Christian'[8] I am after reading, marking and inwardly digesting 30 pages of the strangest arguments I ever saw brought forward in support of Transubstantiation.[9] I have taken these arguments one by one and I humbly conceive that I am fully able to refute each and every one of them. I find logic of infinite use to me where he draws distinction between substance and accident.[10] He's the most knotty fellow I have met as yet on the subject.

I am glad to find that Mr Griffith is well, you have not told me this but I judge it from your taking the two tumblers together

Good bye | Your affectionate son | John Tyndall.

RI MS JT/1/10/3218

LT Transcript Only

---

1. *your letter*: letter missing.
2. *Travelling in the year 1900*: not identified.
3. *7 league boots*: boots, which appear in a number of folktales and novels, that enable the wearer to make great strides at each step; a league is three miles.
4. *improved the ... writing to you*: see letter 0054.
5. *old 'Nullity' ... Poole's Dialogues*: see letter 0054, n. 2.
6. *my last letter*: letter 0054.
7. *Wm Heydon ... sixpence*: 5*s*. 6*d*. from his weekly pay of 9*s*.
8. *Challoner's 'Catholic Christian'*: *The Catholic Christian Instructed in the Sacraments, Sacrifice and Ceremonies of the Church*, first published in 1737 and frequently reprinted, by Richard Challoner (1691–1781), the Vicar Apostolic of the London district. This work consists of a series of questions and answers, and includes a section entitled 'Transubstantiation proved. Objections answered'. Tyndall had earlier attended a lecture by visiting preacher, James Godkin, who rejected the doctrine of transubstantiation 'most irreverently' (RI MS JT/8/1/2, p. 6).
9. *Transubstantiation*: the conversion in the Eucharist of the whole substance of the bread into the body and of the wine into the blood of Christ, only the appearances (and other 'accidents') of bread and wine remaining: according to the doctrine of the Roman Catholic Church (*OED*).
10. *distinction between substance and accident*: '... the catholic church, believes that the bread and wine are converted into the body and blood of Christ, so that there remains nothing of the inward substance of the bread and wine after consecration, but only the outward appearances or accidents' (R. Challoner, *Catholic Christian Instructed* (Derby: T. Richardson, 1843), p. 74).

From John Tyndall, Snr 30 April 1841 0056

Mr John Tyndall | Ordnance Survey Office | Youghal
Leighlin Bridge | April 30th, 1841.

My dear John

I received your letter[1] and am happy to know you are in good health. I perceive from the letter you sent Mr Conwill,[2] also from the one I received myself, that you have entered the knotty field of controversy with some bigot in Youghal[3] who when he was not able for you at grammer should attack your religions creed. His first onset was an active one by placing in your hands Challoner's Catholic Christian Instructed,[4] one of the ablest and most sophistical productions on the popish side of the question. I read it myself twenty seven years ago. It was given to me with the fullest confidence of my becoming a convert to his reasoning, but so far from it having the desired effect that it confirmed me more strongly in my former belief than ever; for what was this popish champion striving to do – to make it appear that two and two made six, or – what is equivalent to it – that a piece of bread is the body of a man. He might as well, as the great Tillotson says, go about to prove that an egg is an elephant or that a musket bullet is a pike.[5] I would advise you to get Doctor Jeremy Taylor's Confutation of Transubstantiation[6] or Carson on the same subject.[7] You understand Euclid[8] yourself, bring it to bear against it as Daniel MacFrony has done before you,[9] for it is an outrage on common sense Mr Conwill says he will send you some jawbreakers for the Youghal bigot[10] in his next letter to you on the whole I would much rather have your mind employed on something more useful to you, as it often creates very unpleasant feelings between man and man, and ends in a bottle of smoke.[11] I have now a very unpleasant task to perform that of letting you know Edward Styles having lost the situation he held in the loan fund.[12] When the books were examined, he was found a defaulter in the enormous sum of one hundred and seven pounds. At the same time there is but little blame attached to Ned, but all the blame laid at the children's door. Now I think he should have minded his children's conduct better than to allow them the handling of the people's money and making a bad use of it. His son John[13] – you know the sort he is – I met with him at a late hour in Nurney[14] about 8 months ago with two blackguards along with him and three tame cocks in their hands. I checked John for his acts, the consequence was the whole pack was in fully cry at me nearly ever since. Ned and I have not spoken to each other since that period. Jas.[15] Treacy has the situation, and indeed no man could get it that wanted it worse though I wont at present describe his situation. Still I am sorry for Styles though he does not deserve it from me.

John Kehoe[16] that lived at the Bridge is dead and buried. We are all well at present and join in sending our love to you

I am your affectionate father | John Tyndall

Get Tillotson's Discourse against Transubstantiation with Pool's Dialogues.[17] I am tired and the steel pen[18] no good.

Jeremy Taylor's Dissuasive from Popery,[19] a first rate work.

RI MS JT/1/10/3219
LT Transcript Only

---

1. *your letter*: letter 0055.
2. *the letter you sent Mr Conwill*: letter missing.
3. *controversy with some bigot in Youghal*: This controversy was presumably discussed in the letter to Conwill. The 'bigot' may be Anthony Lebert mentioned in letter 0058.
4. *Challoner's Catholic Christian Instructed*: see letter 0055, n. 8.
5. *the great Tillotson ... a pike*: In his *A Discourse against Transubstantiation* (1684), John Tillotson (1630–94), the Dean and later the Archbishop of Canterbury, wrote 'It might well seem strange if any man should write a book to prove an egg is not an elephant, and that a musket-bullet is not a pike' (p. 2).
6. *Doctor Jeremy ... Transubstantiation*: *The Real Presence and Spiritual of Christ in the Blessed Sacrament Proved Against the Doctrine of Transubstantiation* (1653) by the Anglican clergyman Jeremy Taylor (1613–67), who served as Bishop of Down and Connor.
7. *Carson on the same subject*: *The Doctrine of Transubstantiation Subversive of the Foundations of Human Belief* (Dublin: Tims, 1825) by Alexander Carson (1776–1844), who was ordained into the Presbyterian Church of Ireland but later became an independent and then a Baptist. Carson rejected the doctrine as incapable of proof and contrary to Common Sense.
8. *Euclid*: see letter 0017, n. 3.
9. *Daniel MacFrony has done before you*: possibly Daniel Featley's *Transubstantiation Exploded* (1638).
10. *the Youghal bigot*: see n. 3.
11. *a bottle of smoke*: a deceit or falsehood.
12. *Edward Styles ... loan fund*: The fourth annual report of the Leighlin-Bridge Charitable Loan Fund, for the year ending 31 December 1840, names Edward Styles as clerk and John Styles as messenger, for which they received £25 and £5 respectively. No financial irregularities were reported. (Catalogue – Small Private Collections (1), Carlow County Archives Service, Document P2/0006.)
13. *His son John*: John Styles (b. 1830) was one of Tyndall's cousins and the son of Edward and Mary (née Tyndall).
14. *Nurney:* a village about 3 miles east of Leighlin Bridge on the Tullow Road.
15. *Jas.*: James.
16. *John Kehoe*: not identified.
17. *Tillotson's Discourse ... Pool's Dialogues*: *A Discourse against Transubstantiation* (1684) by John Tillotson (1630–94), who served as Archbishop of Canterbury (1691–4); *A Dialogue between a Popish Priest and an English Protestant* (1667) by the Nonconformist minster Matthew Poole (1624–79).
18. *steel pen*: a pen with a steel nib, split at the tip like a quill (*OED*); see letter 0054.
19. *Dissuasive from Popery*: Taylor's *A Dissuasive from Popery* was first published in 1664.

From J. Holmes [April] 1841 0057

My dear Sir[1]

Permit me, at the earnest solicitation of some particular friends,[2] who with myself have entertained a very high opinion of your ability, to request you as a great favour to lend us your kind assistance by taking part in our humble endeavours against that Curse of our Country – Popery: provided it does not interfere with your situation, as in that case we should be very sorry that any unpleasant circumstances should ensue or that you should at-all suffer any inconvenience by your compliance with the request of

Yours sincerely | J. Holmes.

RI MS JT/5/16b/25
LT Transcript Only

1. *My dear Sir*: LT dates this letter 'Youghal. April and May 1841' and entitles it 'Invitation to speak at public debating society'. As letters 0055 and 0056 contain possible references to this letter, it may predate both of these letters. However, as the sequence of letters is uncertain, this letter and the two that follow (which also concern the public debate) have been kept together.
2. *some particular friends*: unspecified but probably including William Ginty, John Tidmarsh, and Phillip Evans with whom Tyndall shared lodgings. According to Tidmarsh, Tyndall and the other junior employees of the Ordnance Survey in Youghal often engaged in vociferous debates at their lodging house. Hearing of these debates, the local people 'requested [us] to give one of our "entertaining" discussions in the large room in the town'. (J. Tidmarsh, 'Reminiscences of John Tyndall', *Southern Australian Register*, 24 January 1894, p. 6).

From J. Holmes [April] 1841 0058

My very dear Sir[1]

Shall I congratulate you on your triumphant display of powerful talent and unrivaled ability in defence of Error?[2] Or shall I not rather regret that you did not take up the opposite side. But Sir such has been the general interest created by your unexampled and learned disquisition that I am induced once more to intrude myself upon your kind attention by most earnestly requesting that you will take it the next night upon the INVOCATION OF SAINTS AND ANGELS,[3] when you will be met by Mr LeBert,[4] who has proposed to meet you on either side; thus leaving you to your choice.

Now I hope you will pardon me for being so troublesome.

Your sincerely | J. Holmes | Thursday morning.

RI MS JT/5/16b/25
LT Transcript Only

1. *My very dear Sir*: LT dates this letter 'Youghal. April and May 1841' and entitles it 'Invitation to speak at public debating society'; see letter 0057.
2. *defence of Error*: In the public debate between Protestantism and Catholicism, Tyndall took the Catholic side. According to Tidmarsh, 'Tyndall led off splendidly ... and closed his twenty minutes in a rare flow of most telling language, and the room rung with applause. Ginty was on the other side. Poor Ginty! I shall never forget that scene; though a clever man and possessed of the finest abilitie she had not the nerve of Tyndall. He had everything well prepared, studied, and noted.I think if he had led off his chances would have been much better for going through it, but Tyndall's masterpiece was too much for him. He spoke for about five minutes, then suddenly sat down. Thus ended our public discussion.' (J. Tidmarsh, 'Reminiscences of John Tyndall', *Southern Australian Register*, 24 January 1894, p. 6).
3. *INVOCATION OF SAINTS AND ANGELS*: This important Catholic doctrine that attributes divinity to the saints and angels had been defended in 1838 by Father Thomas Maguire in a public debate in Dublin with the Protestant clergyman Tresham Gregg (*Authenticated Report of the Discussion which took place between the Rev. T. D. Gregg, and the Rev. Father Thomas Maguire* (Dublin: R. Cloyne, 1839)).
4. *Mr LeBert*: Anthony Lebert, schoolmaster, Friar Street, Youghal (*Slater's*).

From J. Holmes [April] 1841 0059

Dear Sir[1]

Now that we are about to discontinue our Wednesday evening meetings after the next; and that we are desirous to conclude with something that will sum up the whole, I hope you will not refuse this my last request, for which I shall personally feel exceedingly obliged and I am convinced with which the public will be highly pleased. If you come forward on either side of the question proposed. We are endeavouring to have four speakers and you know how much gratified all will be if you come. Any book that you want that I have you shall get with pleasure.

Say: 'Yes' or 'NO', but don't say 'No'.

Yours sincerely | J. Holmes.

RI MS JT/5/16b/25
LT Transcript Only

1. *Dear Sir*: LT dates this letter 'Youghal. April and May 1841' and entitles it 'Invitation to speak at public debating society'; see letters 0057 and 0058.

From John Conwill 1 May 1841 0060

Alas! for the days that are gone.
When I was in danger brave Tyndall stood nigh;
In him I always had one
Who would Butler and Sheehan[1] defy.

May 1st 1841.

'My dear Tyndall'

Your truly pathetic motto[2] would launch me into the debilitated state from which I had so lately recovered, were it not that I cited to my aid some portion of fortitude, at the moment that retrospection caused the scenes of by-gone days to flit, in glowing colours, across my mind.

So it appears you have emerged from your caterpiller state and resumed the proud position of a butterfly in your airy rounds of epistolary correspondence: Oh! happy abstention from your embryo state, for by such a transition, you have been protected from the ill-fated sting of the Munster ichneumon controvertist,[3] who once proudly sparkled as a luminary of the first magnitude, in his grammatical orbit, but being constrained to hide his diminished rays in this trajectory, he assumed the projectile force of a polemical hero, hoping thereby to retrieve his lost name and to trace out a course never before trodden in the religious world.

What does this Munster pratler mean. Does he intend with his controversial club[4] to level low whole phalanxes of his fellow men, for adoring our eternal Father conformably to the dictating of their consciences? Will he have the audacity to pronounce the sentence of eternal reprobation on his fellow man for not kneeling before the same altar with him.

'Curse on the heart,' says Moore, 'that would try truth by such a standard as this.'[5]

I despise such litigations altogether. Shame on the dastard that has promoted them. Ask the satanic ichneumon what was Tom MGuire's opinion at Glasgow about those persons who differed from him on points of belief.[6] Leave those persons who hold the green acres and those who have their hands deep in the pockets of the peasantry, to dispute about articles of faith, and live in peace and amity with your neighbour. This suggestion is not directed at you, but towards, if I anticipate truly, that dolt[7] who is already writhing beneath the poignancy of the irrefutable arguments that you have adduced in favour of those tenets which you profess.

I can safely throw the detention of my last letter on the back of fate, and if I guess right your controversy with the sentence-making doctor[8] protracted your answer until the 18th of April. I delayed writing this till I could assure you that I am completely retrieved and that I have concurred with your request[9] in setting aside my studies of the musty pages of Diaphantus[10] and Newton together with the elaborate pursuits of the Differential and integral calculus, as given by Lacroix.[11] This dereliction of study tended more to ameliorate my constitution than all the other medical operations I underwent; so that it is obvious to me that too close an adherence to study preys heavily on the human frame.

As to the finding of the logarithm of a decimal, I believe that you are pretty well aware of the method of determining that of a whole number, therefore find the log of the decimal just as if it were a whole number, and then prefix a negative sign and you will have the logarithm of the decimal sought. Hurrah for the great utility of nonentity[12] and the dragging of it into existence. Example:– required the log of 1000 or 001. You well know that the logarithm of 1000 is 3; and that division is performed by subtraction of logarithm hence 3–1 = 2 = log of (1000 ÷ 10) = 100. 2–1 = 1 = log of (100 ÷ 10) = 10. 1–1 = 0 = log of 10 ÷ 10 = 1. 0–1 = –1 = log of (1 ÷ 10) = 0.1.–1–1 = –2 = log of (0.1 ÷ 10) = 0.01. –2–1 = –3 = log of (0.01 ÷ 10) = 0.001;[13] which shows that if we prefix the negative sign to the logarithm of any whole number, we will have the logarithm of the decimal corresponding to the same number. If you desire it I shall, in my next letter, send the calculation of a complicated decimal, but I flatter myself the preceding illustration sufficiently simplifies the subject for you.

I had nearly forgotten to inform you that the Reverend Thomas C. Dixon[14] paid me a visit the other day. He listened on while I examined all the classes, after which he wrote a most excellent comment on my capabilities in my Report Book. Furthermore he has corresponded with the Board of Education and stated to them in the most flattering and eulogistic terms the efficiency and literary acquirements of your old teacher. I have no more at present to communicate to you save my mother's best respects, who earnestly solicits your presence among us this summer.

I remain your ever | faithful teacher | John Conwill.

RI MS JT/1/11/3518
LT Transcript Only

---

1. *Butler and Sheehan*: inspectors of schools.
2. *Your truly pathetic* motto: not identified, but possibly the motto in letter 0055, if Tyndall's father had shown that letter to Conwill.
3. *Munster ichneumon controvertist*: see letter 0005, n. 6 for Munster, although Conwill is here referring to County Cork. An ichneumon is a small parasitic hymenopterous insect (family Ichneumonidæ), which deposits its eggs in or on the larva of another insect (*OED*). Conwill is dismissing his unnamed opponent as a parasite, probably the 'malignant critic' mentioned in letter 0052. It might also refer to Tyndall's opponent in the public debate, see letters 0057–9.
4. *his controversial club*: possibly relates to the debate mentioned in the three previous letters.
5. *'Curse ... a standard as this'*: adaptation of '... perish the hearts, and the laws that try Truth, valour, or love, by a standard like this!' from 'Come, Send Round the Wine' (1820) by

the Irish poet Thomas Moore (1779–1852).

6. *Tom MGuire's ... points of belief*: At a public dinner in Glasgow in 1837 Father Thomas Maguire 'delivered a powerful speech against religious anti-Catholic prejudices' (*Freeman's Journal*, 6 December 1847, p. [2]).
7. *that dolt*: possibly either Anthony LeBert (letter 0058) or the 'Munster ichneumon controvertist' (n. 3).
8. *the sentence-making doctor*: see letter 0052, n. 9.
9. *your request*: see letter 0054.
10. *Diaphantus*: Diophantus (see letter 0054, n. 6) made important contributions to algebra and the theory of numbers.
11. *Lacroix*: The French mathematician Sylvestre François Lacroix (1765–1843) produced several influential textbooks including his three-volume *Traité du Calcul Differential et du Calcul Intégral* (1797–1800) and his *Traité Élémentaire du Calcul Differential et du Calcul Intégral* (1802).
12. *the great utility of nonentity*: For Conwill's views on non-entity, quantity and the theory of numbers, see letters 0028 and 0039, including notes.
13. *hence ... 0.001*: LT's typescript contains incorrect numbers and mathematical signs. The text has therefore been reconstructed by the editors.
14. *Reverend Thomas C. Dixon*: not identified.

To John Tyndall, Snr 2 May 1841 0061

Youghal, Sunday[1] evening

My dear Father

I received yours of the 30th[2] on last night. I was just after returning after taking a walk with Cuddy up the Blackwater.[3] As I entered the room I was agreeably surprised to find a letter from Mr Conwill[4] for me on the table, this brings me in debt to both you and him, and as I think one creditor in the line of correspondence quite enough, I hasten to discharge your bill. You are astray in your conjectures respecting my being engaged in controversy. I admit that I am a little immersed in it,[5] but not in the manner which you imagine. As I stated in my last letter[6] you shall hear the whole account from myself shortly. Why the names of the authors which you advise me to procure would in themselves act either as a spell in paralyzing the efforts of an opponent, or perhaps on the contrary would act the part of an African tornado in hurling the puny batteries of the man of sin and in ruin to the ground. I have considered Challoner's proofs[7] very calmly and I think in an unbiased manner. They only prove that a man may be a Newton in either the political or mathematical world and still be a child in the ways of religion. Surely if ever the apostle Paul's prediction[8] was fulfilled it is in such men as Challoner. His talent must be admitted but 'God has sent him strong delusion that he should believe a lie.'[9] I'm sure if it was possible for his shade to look on me at present it would frown indignation at my presumption. To see the opinions of his mitred majesty impugned by a stripling is strange and passing strange.

Mr Conwill's letter contains no small share of invective against my unfortunate grammatical opponent. He is not however deserving of it all. He tells me he is quite recovered, this I am extremely glad to hear.

I regret much to find that Ned Styles has lost the loan fund;[10] poor man, he must have an aching heart at finding himself the victim of his children's misconduct. My opinion of John[11] was never very high but indeed from what I have heard from you he is even worse than I conceived him to be his father made a spoiled child of him – was it because you checked John that he and you fell out?

I fully admit the wisdom of your remark respecting my time being employed in something more useful than controversy – though it must be admitted that every man ought to be able to produce a reason for his belief. You may depend on it that it is with reluctance I enter the field I act in full conformity with your advice that is I act on the defensive.

I am extremely glad to find that you are all well give my love to my Mother and Emma

Good evening | I remain Your affectionate son | John Tyndall

RI MS JT/1/10/3220
LT Transcript Only

1. *Sunday*: LT gives date as 'May 2nd, 1841', which was a Sunday.
2. *yours of the 30th*: letter 0056.
3. *the Blackwater*: Youghal is located on the estuary to the River Blackwater.
4. *a letter from Mr Conwill*: letter 0060.
5. *I am a little immersed in it*: Tyndall is probably referring to the public debate over Catholicism; see letter 0058.
6. *my last letter*: letter 0055.
7. *Challoner's proofs*: see letter 0055, n. 8.
8. *the apostle Paul's prediction*: presumably 2 Thessalonians 2:1–3 which indicates that true believers (like Richard Challoner) will be resurrected at Christ's second coming.
9. *'God has sent ... a lie'*: 'And for this cause [encompassing unrighteousness and denying truth] God shall send them strong delusion, that they should believe a lie.' (2 Thessalonians 2:11).
10. *Ned Styles has lost the loan fund*: see letter 0056.
11. *John*: John Styles: see letter 0056, n. 13.

To John Tyndall, Snr 22 May 1841 0062

Youghal, | Saturday[1] night

My dear Father

I believe a week has elapsed since you last heard from me[2] and as I promised that you should hear from me weekly I now raise the pen in order to fulfil that promise.

Silence I think has spread her wings over the County Carlow people. I wrote to Mr Conwill a good while ago and have received nothing as yet in return. You must recollect that you yourself are already indebted to me a letter so that on the postage of this I may set you down on the debt side of my account for two.

Things are going on in their old routine here I'm as strong and have as good an appetite as ever. I found a twopenny[3] quite insufficient for breakfast. The lads complain dreadfully of my undue share in the consumption of the mess.[4] One thing in particular they protest against loudly and that is my inroads on the sugar bowl. I am sure my mother must feel the benefit of my absence with respect to this portion of the furniture of the tea table.

You may recollect that we calculated on leaving this country about the end of September. We counted on having all the work with which we were furnished totally cleared up by that time. We had no idea whatever of getting any more in Ireland. Our views on this subject proved fallacious and yesterday we received an additional supply of 29000 acres,[5] this will detain us about 5 months longer in Ireland. I am not sorry for this though I believe our pay in England will be superior to what it is here.

Our present division is but the skeleton of Mr Wynne's. It has undergone a complete purgation. Roberts, Mackay, Packer[6] and several others have left us since we came here. Linedale took to his heels on Sunday morning last[7] – scarcely any one know anything of his going away, he appeared studious to conceal his exit as he has left the town greatly in debt. He was sick for a good while before he left this, and during his sickness he was well attended by the person with whom he lodged. He ran to the amount of nearly £2 and when he recovered he took no measures whatever towards the liquidation of the debt. He went in debt with other persons also, he took some drawing instruments from a person who lodged in the same house with him. You are aware that we keep a Journal in which the daily work of each of us is entered,[8] before Linedale went away he crammed this Journal with every sort of filth and vagabondism. In fact the production would be a disgrace to Paddy Molloy.[9] During the time we have been in Youghal he has conducted himself very badly; he always kept low company, had several fights with the townspeople, had to appear at the Sheriff's Court in the Mall house – In fact he has left behind him the character of a depraved and immoral lad. He boasted I believe before he went that he was going to Bagnalstown. Mrs Payne and I were speaking together the other day about him. I was astonished to hear from her that Miss McGee was favourably inclined towards Linedale. I laughed at the idea. However, at the time, but she appears to be so serious in her declaration that I know they must have some foundation. I know that reports are exaggerated but for Miss Ann McGee's sake I could wish them to be totally groundless. You will much oblige me if you let me know whether Linedale has gone to Bagnalstown. The idea of the thing however is so absurd that I cannot bring myself to entertain it for an instant certainly nothing could astonish me more than to hear of Miss Ann McGee fixing her thoughts on such a scapegoat as Linedale.[10] Evans is snoring like a pig, the harmonious notes of his nose appear to me as a warning voice which tells that bed time has arrived, write to me as soon as possible.

Good night | Your affectionate son | Jack Tyndall
How are my Mother and Emma?

RI MS JT/1/10/3221–2
LT Transcript Only

---

1. *Saturday*: LT gives postmark as 'May 23rd, 1841', a Sunday.
2. *since you last heard from me*: letter missing.
3. *a twopenny*: a fairly substantial loaf of bread.
4. *mess*: a serving of food (*OED*).
5. *an additional supply of 29000 acres*: By this time C District had completed surveying approximately 489,890 acres in County Cork. 122,610 acres remained, of which 19,670 acres were allocated to 5th Division – 'General Statement, Showing What Part of the Country Is Ready for the Officers to Take Up when the Work They Have in hands Is Completed', May 1841 (NAI OS/1/18). The figure of 29,000 acres in this letter may be a mistranscription of this remaining acreage for the 5th Division.
6. *Roberts, Mackay, Packer*: John Roberts had been dismissed for 'idleness' on 9 January 1841 (NAI OS/1/18); Thomas C. MacKay, another civil assistant in the 5th Division, C District, had been discharged on 24 July 1840, his work having been deemed 'inadequate' by Major Waters (NAI OS/2/16 and OS/1/17); another civil assistant, Walker G. Packer, appears to have gone on leave in December 1840 but did not return to the Survey (NAI OS/1/18).
7. *Linedale ... morning last*: Following a period of absence from duties owing to illness, Anthony Linedale absconded on 17 May 1841 (the day following Sunday, 16 May) – Monthly Report, May 1841.
8. *a Journal ... is entered*: One of Tyndall's Survey Journals, covering the period May 1839 to March 1840, is deposited in RI MS JT/3/36.
9. *Paddy Molloy*: presumably a disreputable character; possibly the Paddy Malloy mentioned in letter 0051.
10. *Miss Ann McGee ... Linedale*: Six days later, on 29 May, Anthony Linedale (St Peters, City of Dublin) married Anne MacGhee (Bagenalstown) at Dunleckney (Irish Church Records).

From John Tyndall, Snr 6 June 1841 0063

Mr John Tyndall | Ordnance Survey Office | Youghal
Leighlin Bridge | June 6th, 1841.

My dear John

I believe you think that I have totally forgotten you I not having written to you for so long a time. I have two reasons for the delay, the first is I was very busy for several days farming and when I came home at night I was so fatigued that I could not bring myself in humour to write a line; my second is I heard William McAssey of Nurney say that he saw a letter from you to Debby[1] and that in it you desired her not to answer it as you intended to be home in a few days I then thought it foolish to write and I had a watch out ever since for every coach that passed through from Kilkenney[2] expecting that you would be with it. I now have given up all hopes of that nature and think it high time to discharge a portion of my debts. I received your letter[3] in due time and am happy to hear that you have so good an appetite, I am afraid you overrate it, thinking to please me and your mother, If not, so much the better. You mentioned in your last that the Division is but the skeleton of Mr Wynne's and mentioned several that had left the office I was partly in receipt of that news myself as Fahy[4] passed through this town with his wife not long ago and left an excellent character[5] of you, both for abilities and good conduct after him. He did not tell me that he had left the office, he gave me to understand he was going home to see his friends before his departure for England. I hear himself and his wife walked from Youghal and Mr Conwill met them both the next day at Rathamore on their way towards Carlow walking.[6]

You mentioned that Linedale took to his heels, also that I knew full well, for he came to Bagnalstown and got married to Anne McGhee;[7] after their marriage they got off to Dublin, but it is reported he will return to Youghal and pay all his debts. I hear he gave a very impartial account of himself when he came to Bagnalstown – told how he was in debt but that he would discharge all in a short time. I know that you have heard that we are on the eve of a general Election. There are three radical candidates in field for this never to be forgiven County[8]– Dan O'Connell Junr and Ashton Yates for the County and Gisbourne for the Borough.[9] It is reported that old Daniel himself will agitate our beautiful County in person,[10] at the head of forty thousand men. This is a powerful body; but if the big Beggarman[11] brought double that number at his tail he will be met with a bold front by the hardy veterans[12] of the County Carlow, and his nominees driven with disgrace from the hustings as old Ponsonby[13] was before them. The Ministry are on their last legs,[14] and before tomorrow night may not have one to stand upon

I will now conclude with letting you know that we are all well and join in sending our love to you

I remain your affectionate father | John Tyndall

Dean Bernard is up in the County Cork at present.

RI MS JT/1/10/3223
LT Transcript Only

---

1. *a letter from you to Debby*: Deborah McAssey; letter missing.
2. *coach that passed through from Kilkenney*: the daily coach from Cork passed through Kilkenny on its way to Leighlin Bridge.
3. *your letter*: letter 0062.
4. *Fahy*: Michael Fahy had worked as a civil assistant in the 5th Division, C District, of the Survey since July 1836, but had been discharged for 'idleness' on 29 March 1841 (NAI OS/1/17–18).
5. *character*: character reference.
6. *from Youghal ... Carlow walking*: a distance of about 90 miles. Possibly Rathmore, County Carlow, which is about 8 miles west of Carlow, but is not on a direct route from Youghal to Carlow.
7. *married to Anne McGhee*: see letter 0062, n. 7.
8. *this never to be forgiven County*: probably because Henry Bruen gained the seat for the Conservatives in the December 1840 by-election.
9. *Gisbourne for the Borough*: At a by-election in February 1839 the Conservative Francis Bruen won by 2 votes but he was then unseated on appeal and replaced by the Liberal Thomas Gisborne (*c.* 1790–1852). Although Gisborne, who earlier represented Stafford and North Derbyshire, was mentioned at the election hustings in 1841, he did not contest the Carlow Borough seat, which was won by the Liberal Captain Brownlow Layard (*CS*, 22 May, p. [2] and 12 June 1841, p. [2]).
10. *old Daniel ... in person*: 'I will of course go down to Carlow at once when wanted and go from parish to parish *agitating*'. (Daniel O'Connell to John O'Connell, 26/28 May 1841: D. O'Connell (ed.), *Correspondence of Daniel O'Connell*, vol. 7, pp. 74–6). Owing to the importance of winning County Carlow from the Conservatives (and also maintaining his investment in his son Daniel's campaign), O'Connell spent a month agitating in the county. See O. MacDonagh, *The Emancipist: Daniel O'Connell, 1830–47* (London: Weidenfeld & Nicholson, 1989), pp. 195–8.
11. *the big Beggarman*: a Protestant nickname for Daniel O'Connell.
12. *hardy veterans*: Protestant defenders.
13. *disgrace from ... old Ponsonby*: The Liberal candidate, Frederick Ponsonby, had been defeated at the December 1840 by-election.
14. *The Ministry are their last legs*: Lord John Russell's Liberal administration suffered defeat on many issues.

To John Tyndall, Snr 9 June 1841 0064

Youghal | June 9th, 1841.

My dear Father

I received yours of the 6th[1] on last night. I knew full well that you had cogent reasons for not writing to me so that there was not the least fear of it being attributed to forgetfulness by me. The information you received from Wm McAssey was not correct, he must have mistaken Debby's's[2] words. When I have settled about going home you are the first that shall hear of it. A circumstance however happened to day which was very near sending me home sooner than I expected. This morning after breakfast Corpl Colder came into our room and judge of our surprise when he told us that the 5th division (ours) was to proceed to Kinsale on Monday fortnight next. This intelligence was quite unexpected we had heard nothing about it until it was divulged this morning. A short time afterwards I was thinking that as I was about to move it would be best to get a month's leave and set out to Leighlin – at the same time I went up to Colder to consult with him on the subject (he and I are great friends). He however advised me to defer it until after I have been a while in Kinsale. I asked him was he sure that I would get leave at that time, he said there was not the least doubt but I would. I therefore concluded that it was best to take his advice and so have deferred my going home for some time longer. Kinsale is about 50 miles from this, but I believe its distance from Leighlin is much the same as that of Youghal,[3] they not being in a direct line. The moment I muster up a little cash after I arrive there I will start for Leighlin. I won't be long doing that, I promise you. You mentioned Fahy in your letter. He is discharged.[4] He, his wife, and a fellow named Pilkington[5] who is at present with us had a row in the street one day. Mr Whittingham rode up and was an eye witness of the scene. The consequence was that Fahy was dismissed on the spot and summoned by Pilkington. They appeared before the Magistrates and Fahy got a fortnight's confinement in Cork Jail, and was on his way homeward when you saw him. You speak also of Linedale's marriage: could I concentrate all the energy of language into one word and speak with respect to the part that Miss McGee acted in that farce, I would say that she was a <u>fool</u> – <u>a most egregious fool</u>. I have a shilling bet already on the result of the expected election[6] with Phil Ryan. He says that Bruen will be thrown out and he founds his opinion on the influence which he knows the priests possess over the people.[7] I think however they have been too often gulled by those very same priests to be again imposed on by them

I forward another letter to Mr Conwill along with this

Good bye | Your affectionate son | John Tyndall | Give my love to my Mother and Emma.

RI MS JT/1/10/3224
LT Transcript Only

1. *yours of the 6th*: letter 0063.
2. *Debby's*: Deborah McAssey.
3. *distance from Leighlin ... of Youghal*: Rather surprisingly, Tyndall underestimated the difference in distance; by road the distance from Youghal to Leighlin Bridge is approximately 85 miles, whereas the distance from Kinsale is about 125 miles.
4. *Fahy ... is discharged*: see letter 0063, n. 4.
5. *a fellow named Pilkington*: Macdonald Pilkington had early been employed by the Survey but had been removed. He had then been rehired and was a member of Mulligan's surveying party in the 5th Division, C District.
6. *the expected election*: see letter 0063.
7. *the influence ... the people*: 'The people are rousing, and the Catholic clergy are, for the first time for years, taking their station'. (Daniel O'Connell to ?, 18 June 1841: D. O'Connell (ed.), *Correspondence of Daniel O'Connell*, vol. 7, pp. 94–5.).

From Thomas Charles Higginson11 June 1841 0065

Millstreet,[1] Thursday | June 11th 1841.

My dear Tyndall

On Friday I set out from Nohoval[2] to this when arrived in Cork the first to be done was to look for Miss Duble but that was easier said than done when after walking my feet off looking for Orrory Hill[3] I was returning from the search when I met a peeler[4] asked if he knew and where she lived he asked me was it Mr Duble the butcher.[5] He said if I returned with him he would show me where Orrory Hill was and where he lived, but being rather tired I declined. So ended the search On Saturday I set out for Macroom 24 miles along the river Lee, the scenery of which is splendid. in Macroom there stands a fine old abbey.[6] I was so tired when I arrived that I could not go to see it. On Sunday I marched to Millstreet what a different scenery to the last days march. Instead of the Lee winding its way through glens wooded to the water's edge, I now came in view of a country covered with mountains like the waves of the troubled ocean one rising higher than another but our work is very level. I suppose you're delighted at going to Kinsale to be under Lieut. Fenwick as Whittingham is going away.[7] I have no more to say at present give my love to Jim, Tid, Cuddy and the blackguards

Yours ever | T.C.H.[8]

RI MS JT/1/11/3728
LT Transcript Only

1. *Millstreet*: a market town 25 miles north-west of Cork.
2. *Nohoval*: 5 miles east of Kinsale, close to the spectacular Nohoval Cove.

3. *Orrory Hill*: Orrery Hill is to the north of Cork.
4. *a peeler*: a member of the Irish constabulary (*OED*).
5. *Mr Duble the butcher*: possibly William W. Deeble, butcher, in the Corporation Market, Cork.
6. *in Macroom ... old abbey*: Macroom is a market town in County Cork and is situated on a tributary of the River Lee. As there is no abbey in Macroom, Higginson may have been referring to St Colman's church.
7. *Whittingham is going away*: Whittingham was briefly stationed in Kinsale before leaving the Survey on 11 July 1841 after 2 years and 168 days' service.
8. *T.C.H.*: Thomas Charles Higginson.

From John Tyndall, Snr 23 June 1841 0066

Mr John Tyndall | Ordnance Survey Office | Youghal
Leighlin Bridge | June 23rd, 1841.

My dear John

I received your last letter[1] in due time and am happy to hear from you, and that you are well. I have now to let you know the present state of the County Carlow[2] as well as I am able to perform such a task. We have been cursed for the last nine or ten days with all the leading agitators from the Corn Exchange Dublin[3] such as Danl O'Connell, John O'Connell, Tom Steel, Tom Reynolds,[4] with several other mountebanks like them. They marched into this town on this day week and the Chapel Band playing before them, with a shirtless rabble surrounding the vehicle that carried them. They set up a platform in the inn yard and decorated the Butter Crane House with green bows.[5] As soon as your uncle Caleb heard of the latter transaction, he darts down amidst the whole band of jugglers, mounts the house and throws every green bow to the ground, when the whole band of thimblemen[6] groaned and hissed him till they were tired. The house belonged to Caleb himself, therefore he had a right, if he wished, to take them down. Don't you wonder they did not dash his brains out. If they gave one blow its my opinion lives would have been lost. They then commenced spouting out the most filthy abuse against the Tories and Orange-men[7] amidst the partial cheers of the shoeless and shirtless crew that surrounded the platform. There was not a man of the least respectability amongst them unless Terence O'Neill, Monk Byrne, and Woodcock.[8] Nothing can surpass the excitement here and through the whole County from this rebel gang. Old Danl[9] with the rest of the minor fry marched into Borris[10] on Monday last, at the head of (as report speaks) forty thousand men. Mr Doyne, the worthy agent of the Kavanagh Estate, took time by the forelock and had the chief part of the freeholders at Borris House with the outward gates locked. They did not attempt an entrance, but their abuse of Mr Doyne and the Kavanagh family was abominable.[11] There is many bets here with regard to the result of the coming Election. Some of the protestants are in high spirits while some, knowing the treachery of popery, are not so sanguine. My opinion is that Bruen will be returned along with his colleague Mr Bumbury. Nothing can surpass the rage of the surpliced ruffians[12] and their flocks equally so. I still believe, notwithstanding their threats of extermination, that Bruen will gain the victory. I am very glad that you will not be at home during the contest because there will be mobs from the Countys Kilkenny, Wexford, Kildare, and Queens, to gain their point if possible by brute force, because they know full well they have no chance any other way. If the Barony boys[13] prove true on this occasion they will gain for themselves immortal fame after all the threats and curses that has been heaped on them by all the priests in the County. I am barely able to write this letter my mind is so uneasy. Still I will say Bruen will be the victor. God grant he may! is the sincere prayer of your affectionate father. John Tyndall

N.B. | We are all well, write before you go to Kinsale I saw a letter from Francis Heydon. He is discharged[14]

RI MS JT/1/10/3225
LT Transcript Only

1. *your last letter*: probably letter 0064.
2. *present state of the County Carlow*: see *CS*, 3 July 1841, p. [2] for events at Leighlin Bridge on 27 June; letters 0073 and 0074.
3. *Corn Exchange Dublin*: A number of prominent meetings of the Loyal National Repeal Association of Ireland had been held at the Corn Exchange, Dublin.
4. *Tom Steel, Tom Reynolds*: Despite being a Protestant and a landowner, 'Honest' Thomas Steele (1788–1848) had been one of the most prominent advocates of Catholic emancipation and continued as a staunch supporter of Daniel O'Connell. A merchant and Vice President of the National Trades Political Union, Thomas Reynolds (b. 1793) was active in O'Connell's Repeal movement.
5. *Butter Crane House with green bows*: Caleb Tyndall was the Weighmaster of the Butter Crane in Leighlin Bridge and since 1819 had been employed in that capacity by Captain William Stewart. Green bows were a symbol of the Irish nationalists.
6. *thimblemen*: tricksters – derived from the shell game (thimblerig) often used by swindlers to defraud players.
7. *Orange-men*: members of the Protestant Orange Order founded in 1796 in County Armagh.
8. *Monk Byrne, and Woodcock*: possibly Denis Byrne, who had been trained for the priesthood at St Patrick's College, Carlow, 1824–30; in 1840 Captain Thomas Woodcock (probably of Coolnakisha) faced criminal charges for causing an affray (*Freeman's Journal*, 24 April, p. [3] and 2 June 1840, p. [3]).
9. *Old Danl*: Daniel O'Connell.
10. *Borris*: a small town on the River Barrow about 8 miles south of Leighlin Bridge, where the Kavanagh family estate was located.
11. *Mr Doyne ... was abominable*: Charles Henry Doyne (1809–67), whose family possessed a large estate in Tullow, was the land agent for the extensive and wealthy Kavanagh Estate at Borris; 'with such a large block of tenant farmer votes under his control, he wielded

enormous power' at times of political turmoil – J. O'Toole, *The Carlow Gentry* (Carlow: n.p., 1993), p. 24. Doyne was deemed a hero at the December 1840 election for saving the Kavanagh Estate and again on 14 June 1841 for confronting Daniel O'Connell when he tried to initiate agitation in St Mullins (*CS*, 19 June 1841, p. [3]).
12. *surpliced ruffians*: Catholic priests.
13. *Barony boys*: a local gang or society of Protestants who opposed Daniel O'Connell and the Catholic Repealers.
14. *Francis Heydon ... discharged*: Although the name Francis Heydon is given here, it was his brother William who was discharged from the Survey on 12 June 1841 (NAI OS/1/18).

From John Tyndall, Snr 25 June 1841 0067

Mr John Tyndall | Ordnance Survey Office | Youghal
Leighlin Bridge

My dear John,

I think you received a letter from me[1] about the time I received one from you and that was this morning. I am glad to perceive that you are not sorry for the removal to Kinsale under an Officer[2] that will reward the man of merit. You mention in your letter that the Heydons are discharged and that William has enlisted[3] it gives me little concern what becomes of them as we have too many of the breed here.[4] Their father is out every night with mobs headed by Thimblemen priests[5] forcing Colonel Bruen's freeholders most of them against their wills to quit the country.[6] Nothing can exceed the alarm in this County. Several of the electors have been dreadfully beaten by these nocturnal mountebanks. I would not wish – as poor as I am – for a hundred pounds you were here. Men that was on friendly terms with each other now pass by without speaking. Still I say Bruen and his Colleague[7] will be the victors I will stop up on Saturday night till the coach comes in. I am not able to write more as the office will be closed

I am your affectionate father | John Tyndall.

RI MS JT/1/10/3226
LT Transcript Only

1. *a letter from me*: letter 0066.
2. *an Officer*: Lieutenant Fenwick, although Lieutenant Whittingham transferred to Kinsale briefly, but left the Ordnance Survey on 11 July 1841.
3. *William has enlisted*: William Heydon, who had recently left the Survey (see letter 0066, n. 14), had joined the British army.
4. *too many of the breed here*: the Heydons were Roman Catholics.
5. Thimblemen priests: scheming, dishonest priests; see letter 0066, n. 6.
6. *Colonel Bruen's freeholders ... quit the country*: A freeholder is one who owns a property which if worth £10 or more permits that freeholder to vote in an election. As Bruen expected his freeholders to vote for him (and against the Repeal candidates) they were allegedly being forced to absent themselves from County Carlow during the election.

7. *Bruen and his Colleague*: Henry Bruen and Thomas Bunbury, the Conservative candidates at Carlow.

From John Tidmarsh 27 June 1841 0068

Mr McGrath, Cork Lane, Youghal
Cork, June 27th 1841.

Dear Spigadurry[1]

I just arrived this minute and am after laying in a queer lot of tea. I got a queer squeezing on the coach between two angels. I went to hunt for carriage at Middleton[2] and did not think I would be in time for the coach so I gave the note to one of the little girls who were at the head of the street and she took it to the house. I am requested to make you come here on Monday at four o'clock and you will come down to Cove[3] on Tuesday with my sisters[4] and some more of em – now dont neglect coming, you will be accommodated here and I will give you a side of my bed. You varmint![5] pon my rankins[6] now if you dont come I will choak you when I catch a hold of you. But putting all jokes aside | COME.

JACK.

RI MS JT/1/11/3869
LT Transcript Only

1. *Spigadurry*: presumably a term of comradeship.
2. *Middleton*: see letter 0032, n. 9.
3. *Cove*: now Cobh, a seaport town on the south side of Cork Harbour, some 9 miles from Cork.
4. *my sisters*: Christina and May Tidmarsh.
5. *You varmint!*: derived from vermin – an obnoxious creature – the term is used here to make a friendly threat.
6. *pon my rankins*: presumably an oath.

From John Conwill 27 June 1841 0069

'The Villager'
Happy the man who far from strife
In shades paternal leads his life
Stranger to ills that fortune brings,
From bed at early twilight springs
Before the cock salutes the morn
Or hunters sound the echoing horn
Ere thro' the canvas shuttles play[1]
Or lark salutes the rising day
By health and vigour taught to toil
To drive the team or turn the soil
Or by his task yon towering oak

Bows its tall head beneath his stroke
And thus employ'd with double force
Time urges on his rapid course
Till sol[2] has lost his mighty power
Of gilding o'er the cloud-top'd tower
Then without guide he measures o'er
The path so often trod before
By fire or candle light made gay
He sits and chats the night away
Like Joseph[3] faithful to his trust
Honest himself thinks all men just
From bribes and fell corruption free
No gold charms him to infamy
Tho' boys well versed in letter'd lore
Say blessings may be found in schools
And weak as water pen down rules
He wisely has the truest wealth
Possessing competence and health.

June – 27th. 1841.

'My dear Tyndall'

When I read your poetical effusion[4] I felt at a loss for a topic on which I might compose some lines in verse, till the present state of our ever-to-be-distracted and ill-fated country[5] presented itself to my views. Alas! I said, are the natives of old Erin[6] to be eternally pitted against each other by the crafty wiles and cunning tricks of those political knaves, who are enjoying the loaves and fishes,[7] whilst the peasantry are labouring under the most heart-felt privations in many parts of the Island. I shall never broach this subject again, as I hold in abhorence all notions of politics. The reason that compelled me to mention the subject in this letter is, that you may understand the motives which induced me to compose a few paltry lines on the contented villager.

I am a very bad poet and were it not that there was something of magic in your splendid lines, I would not presume to write one word in meter; I deem myself no adept in prose, much less in rhyme. Pardon any thing incongruous in the poetical prelude to my letter.

I informed you in my last letter that Ford and Whealan[8] were determined to be present in Professor Murphy's school[9] whenever the public examination should take place. The examination has passed over and my boys were spectators. But observe, Murphy thought to keep Ford out. The Revd Mr Murphy, who is my friend, linked Ford to the Professor's[10] desk and suggested to the doctor not to presume to preclude any person from appearing in a school instituted by the public and the government.

Sheehan went through his usual routine of business; so much so that he pronounced the whole clique a set of dan-dra-heads.[11] And why not? read the following –

Sheehan asked the grammar class was the following sentence correct English: 'I intended to have come last week.' The lads proved to be very idiots on the occasion and surely so they ought, for their learned and plumbean professor[12] insinuated to them when they failed that conjunctions connect the same moods and tenses of verbs. Now by what kind of insertion Murphy could introduce a conjunction into the sentence I leave you to judge. I must now wish you farewell for some time

Your ever faithful teacher | John Conwill

RI MS JT/1/11/3519
LT Transcript Only

---

1. *thro' the canvas shuttles play*: the villager is engaged in weaving canvas – a strong or coarse unbleached cloth made of hemp or flax (*OED*) – on a handloom in his cottage.
2. *sol*: the sun.
3. *Like Joseph*: Despite being sold into slavery by his brothers, Joseph subsequently remained true to his family (Genesis 37–50).
4. *your poetical effusion*: letter missing.
5. *the present ... ill-fated country*: This letter was written during the run-up to the bitterly fought election in Carlow at a time of social unrest when Tyndall's uncle Caleb fired at the radicals agitating in support of Daniel O'Connell, Jnr.
6. *old Erin*: romantic name for Ireland.
7. *the loaves and fishes*: Although this phrase, which is based on several passages in the gospels, usually refers to feeding the multitude, Conwill here uses it to refer to the luxury of the political elite contrasted with the povery of the multitude.
8. *Whealan*: a pupil at Ballinabranna School.
9. *Professor Murphy's school*: The teacher at the National School for Boys at Leighlin Bridge was James Murphy (d. 1845), the Catholic curate of Leighlin parish, 1839–42 (M. Nevin, *A Watershed in the Ecclesiastical History of Leighlin Parish* (Carlow: M. Nevin, 2009), p. 41).
10. *the Professor's*: presumably Professor Murphy.
11. *dan-dra-heads*: presumably dunderheads, thus stupid people.
12. *plumbean professor*: plumbean means leaden; thus Conwill dismisses the professor as possessing a dull intellect. There is also a word-play on 'Plumean Professor', being the title of the Professor of Astronomy and Experimental Philosophy at the University of Cambridge.

From John Tidmarsh 28 June 1841 0070

Cork | June 28/41[1]

You blather[2] of a fellow! I received your note[3] this minute and I read it for them all and they were all greatly disappointed, however be here in the morning on Tuesday as they had a few friends invited and dont disappoint again. I am now going to finish the little plan The methodist house here[4] was crowded to excess yesterday I suppose there was over 3 or 4 thousand persons there. There are a great deal of improvements here since you were here last. There is an ordination of methodist preachers to take place here. So be in time for it and several other little things too numerous to mention. I did not see any of the office beggars walking about here yet. I suppose you did not hear from the little beggar[5] yet You mentioned that it was with a bit of stick you wrote your letter, why it is with a town pump[6] I am writing this | Your varmint[7]

RI MS JT/1/11/278
LT Transcript Only

---

1. *Cork | June 28/41*: postmark given by LT.
2. *blather*: voluble talk void of sense (*OED*).
3. *your note*: letter missing.
4. *The methodist house here*: Two Wesleyan chapels (St Patrick Street and Henry Street) and one Primitive Methodist chapel (French-church Street) are listed in *Slater's*. It would appear that Tidmarsh, who came from Cork, was a Methodist.
5. *the little beggar*: possibly Phillip Evans.
6. *with a town pump*: presumably referring to the breadth of Tidmarsh's handwriting; the pipe used on a town pump being perhaps 3–6 inches in diameter.
7. *Your varmint*: see letter 0068, n. 5.

To John Tyndall, Snr 30 June 1841 0071

Youghal | Tuesday 12 o'clock | June 30th, 1841.

My dear Father

I received yours of Friday night[1] on the following evening. According to your account the County Carlow is in a blaze. This is strikingly contrasted with the state of party feeling here. Mr Cavendish[2] the liberal candidate canvassed the electors here on yesterday accompanied by about 60 ragamuffins I could not mark 2 persons of any respectability among his train. No person has appeared as yet on the conservative side, it is generally thought that there will be no opposition, the gates of the town however are covered with placards stating that a conservative candidate will most assuredly come forward and at the same time telling the electors of Youghal to keep themselves unpledged. Mr Cavendish is a nephew to the Duke of Devonshire and the Duke possesses great influence here.[3] All however is quiet, not the least excitement. I was speaking to a conservative elector last night and one who plainly told Cavendish in the former part of the day that he would vote in his teeth[4] in the event of an election. He told him that a brother to the marquis of Waterford[5] was expected on the conservative side.

It appears from your letter that my presence in Leighlin would be far from being pleasant to you. Indeed I think if I was there that no angry feeling would exist between me and any one whom I ever considered in the light of a friend. I would no doubt uphold and extend my principles by every means in my power, but I think I would do it in such a manner as not to offend any man therefore I think that my presence in Leighlin would not be so dangerous as you seem to think. I must however candidly confess that if there was any fear of my uncle Caleb being injured for acting in the way he did[6] I would be the first person to stand at his side – aye and die in his defence. He acted I think as a firm consistent man should act. If the house were mine I would just do the same. This statement may seem to clash with another remark in this letter, that is where I say 'that I would uphold my principles without offending any man;' but I spoke there with reference to the means which I possess of upholding them, and that is with my tongue alone, my property in Leighlin being somewhat below the value of 0.[7]

You speak in no very commendatory terms of Ned Heydon[8] I would not however let the sin of the father operate in my breast against the child. Francis[9] I think is a very good little fellow, and one, I think, that were he capable of judging how things stand would condemn his father's present proceedings. Ned's present mode of acting I think may be accounted for if we take his circumstances into account. You know he has nothing to lose and in the destruction of property, the chaos of society, he may pick up something. This I think influences the man's actions.

Were I to take the former part of your two last letters as a premise, and from them draw a conclusion, that conclusion certainly would not be favourable to Col. Bruen. But your last resounding sentence that Bruen would be the victor destroys the premises and therefore the conclusion falls along with it to the ground. I could not repress a hearty 'amen' on reading your ejaculation in favour of Bruen's return. My heartfelt wish is that your predictions may be verified. It is most sickening to see a popish priest[10] rampant in his power and riding roughshod over the sycophant slaves of Rome. And this would most assuredly be the case could they but wield their magic wands to the extent that their desires would prompt. The zephyr's breath[11] would again flaunt the purple of Babylon.[12] The 7 heads and 10 horns[13] would again make their appearance in all their former hideousness the heretical offspring of the reformation would again writhe upon the rack or become fuel at the stake, and not content with thus treating the living, the mansions of the dead[14] would be profaned and the tragic farce of Bucer and Phagia[15] would be again reacted. Oh! the tender mercies of our holy mother!!!

Ginty and I are the only two persons connected with the Survey at present in Youghal. We start for Kinsale at 7 o'clock tomorrow morning. The luggage started yesterday morning. It went as far as Middleton yesterday, it will be in Cork to day, and in Kinsale tomorrow evening. I will get into Kinsale in one day therefore will arrive there by starting tomorrow morning as soon as they will. I got a very kind invitation to make Cork my head quarters for a day and yesterday I was promised a delightful sail to Cove[16] on to day. I declined the invitation, having made up my mind not to start till Wednesday morning.

This is the last letter you will receive from Youghal. Your future letters must also be directed to Kinsale. As soon as I arrive there I will write to you.

I remain your affectionate son | John Tyndall

RI MS JT/1/10/3227–8
LT Transcript Only

---

1. *yours of Friday night*: letter 0067.
2. *Mr Cavendish*: The Liberal candidate Charles Compton Cavendish (1793–1863), who had previously represented four English consituencies, was elected.
3. *the Duke possesses great influence here*: William George Spencer Cavendish, 6th Duke of Devonshire (1790–1858); Lismore Castle, County Waterford, was one of several estates owned by the dukes of Devonshire.
4. *vote in his teeth*: vote against him (Cavendish).
5. *brother to the marquis of Waterford*: John de la Poer Beresford (1814–66), brother of Henry de la Poer Beresford, 3rd Marquess of Waterford (1811–59). John Keily, Jnr, of Strancally Castle, Youghal, was nominated as the Conservative candidate, but it is not clear whether he resigned before the poll (*London Standard*, 9 June 1841, p. 2).
6. *acting in the way he did*: Threatened and intimidated by a mob on 27 June 1841, Caleb Tyndall, a fervent Orange-man, fired a shot that wounded a Mary McAssey. He was subsequent brought before the court, but was not found guilty having pleaded his innocence and that he acted in self-defence (*Freeman's Journal*, 24 July 1841, p. [4]).
7. *below the value of 0*: less than zero.
8. *You speak ... of Ned Heydon*: for his father's antagonistic comment on the Catholic Heydon family, see letter 0067. Ned (Edward) Heydon was the father of Francis and William.
9. *Francis*: Francis Heydon; see letter 0050, n. 6.
10. *a popish priest*: Father James Maher who aided the agitation in favour of O'Connell.
11. *The zephyr's breath*: a light breeze.
12. *the purple of Babylon*: The evil whore of Babylon in the Book of Revelation (ch. 17) was often identified with the Roman Catholic Church. 'And the woman was arrayed in purple and scarlet colour' (Revelation 17:4).
13. *The 7 heads and 10 horns*: I 'saw a beast rise up out of the sea, having seven heads and ten horns, and upon his horns ten crowns, and upon his heads the name of blasphemy' (Revelation 13:1). As in this letter, the beast has often been identified as the Roman Catholic church.
14. *the mansions of the dead*: catacombs.
15. *the tragic farce of Bucer and Phagia*: The Reformation theologians Martin Bucer (1491–

1551) and Paul Fagius (1504–49) were both condemned for heresy during the Counter-Reformation and were dismissed from their teaching posts at Strasbourg. Both fled to England, where they were well-received by Thomas Cranmer, and gained teaching positions at Cambridge, where they died. During Queen Mary's Restoration of Roman Catholicism their remains were exhumed and burned.

16. *Cove*: see letter 0068, n. 3.

To John Tyndall, Snr 1 July 1841 0072

Kinsale!!! July 1st. 1841.

My dear Father

When you receive this letter you may truly say that a voice from '<u>the world's end</u>'[1] salutes your ear, as I am at present siting about 8 or 10 perches[2] from a place which goes by the name. I started from Youghal on Wednesday morning at 7 o'clock, and arrived in Cork at 11. Here, after seeing our luggage removed or rather after removing it ourselves to the Kinsale Coach office, and paying our fare there, we (namely Ginty and myself) issued forth thro' the city of bragge and blarney.[3] Having been met by Tidmarsh at the coach office we were straightway conducted to his father's house, where the moment we got in we were set down to bread butter cheese and ale. After despatching a moderate quantity of that, and having some business to transact in town, we again sallied forth, and did not return until half past 2. On our entrance our noses were most agreeably saluted by the effluvia of a smoking roast leg of mutton. On this we dined most heartily and I took a tumbler of punch made out of the best Cork whiskey. After dinner, being fortified by this for the road, we again to coach and left Cork at a ¼ before 4. I was sitting beside two gentlemen whose whole discourse was about the coming election. I think they were radicals, for they had the Southern reporter[4] along with them, and tho' they did not actually express their opinions, still I could gather from their discourse that they were what I have stated them to be. Among other subjects the present state of Carlow was discussed. They seemed to take a great interest in it. They were fully acquainted with the disturbances that have taken place there. I think the eyes of Ireland are at present turned on Carlow. We arrived in Kinsale at about 7 o'clock and were met by Foy[5] at the coach office. The next job was to seek for lodgings. After rambling for a while thro' streets something less than 50 miles in breadth, we at length hopped on a place where we thought it likely that we might be accommodated. We were shewed up stairs by the landlord and after a few preliminaries we closed our bargain. Tidmarsh and I took two very good rooms at the rate of 5 shillings a week.[6] Here I intend to remain until Evans comes back, if he does not like our present location we have plenty of time to look out for another. I have not yet rambled through the town, so on that account I am not able to give you a full description of it in this letter. One great feature that I could not help remarking was the narrowness of the streets. A reasonable standing leap would carry me from the door of my lodgings into the shop at the opposite side. Our host is a Wesleyan[7] and an active conservative. He is up nearly every night securing votes for Col Thomas.[8] Our officer[9] is in the barracks and the 1st thing we heard when we went into it was an order read which prohibited any of our lads from meddling in the anticipated political struggle.

This is a great town for fish. The houses of the World's End are actually thatched with it, we had I believe about 3 dozen fresh herrings for dinner yesterday, they were however not much larger than pinkeens.[10] I received a long letter from Evans.[11] He tells me the conservatives of Carlow are determined to either stand or die. It is very near post time and for fear of being late I must conclude. Give my love to my mother and Emma

Your affectionate son | Jack.

RI MS JT/1/10/3229
LT Transcript Only

---

1. the world's end: World's End is on the coast road about half a mile south of the centre of Kinsale.
2. *perches*: an Irish perch measures 21ft.
3. *the city of bragge and blarney*: the City of Cork.
4. *Southern reporter*: The thrice weekly *Southern Reporter* (1807–22, renamed as the *Southern Reporter and Cork Commercial Courier*, 1823–71) was described as 'an out-and-out Radical journal' by the Tory *Morning Post* (29 April 1840, p. 3).
5. *Foy*: probably Private Thomas M. Foy of the Corps of Royal Sappers and Miners.
6. *rate of 5 shillings a week*: Tidmarsh was earning 11*s*. and Tyndall 16*s*. 6*d*. per week.
7. *Our host is a Wesleyan*: The principal Wesleyan Methodist meeting house in Kinsale was in St Patrick's Street.
8. *Col Thomas*: The sitting MP for Kinsale, the Conservative Colonel Henry Thomas (*c*. 1780–1858), who had a distinguished military career and served in several campaigns, initially stood as a candidate but withdrew before the election.
9. *Our officer*: Lieutenant Paul Whittingham.
10. *pinkeens*: minnows (*OED*).
11. *a long letter from Evans*: letter missing.

From John Tyndall, Snr 7 July 1841 0073

Mr John Tyndall | Ordnance Survey Office | Kinsale
Leighlin Bridge | July, 7th, 1841.

My dear John

I received a letter[1] from you on Monday morning last and a note this morning. I did not intend letting you know any thing about our present situation until after the Election would be over because I did not wish to have you fretting – not that I consider that you would have cause to fret when I tell you the truth of the whole matter.[2] I told you before that mobs from the Co[3] Kilkenny headed by the Leighlin mob went through the country by night taking the Colonel Bruen's freeeholders[4] most of them by force. On Sunday night week,[5] a party of the above villians went to Kildrenagh a short distance above Dunleckney, took them from their homes by force, and conveyed them into Leighlin Bridge. As they passed your uncle Caleb's house they gave three cheers for O'Connell. He being up at the time gave three cheers for Bruen. The rebels attacked his house, broke several panes of glass, and struck himself on the breast with a stone. He fired out, and the misfortune was that Patt McAssey's wife[6] received a ball or slug in the thigh. William Hackett from the Royal Oak,[7] one of the leaders of the gang got a black eye on the occasion. He says it was from a second shot fired by Caleb, but that story is a falsehood, as he had but one charge in his possession at the time. After the shot was fired the mob ran over the bridge and, as the story goes, consulted Priest McMahon, who advised them to secure him if they could. My opinion is he advised them to murder him, as they came back in a rage, particularly when they heard that he was gone. This occurred about half past 12 o'clock at night. But to give you a description of the people of Leighlin on this occasion is beyond my power, for they were all willing to steep their hands in his blood. Amongst the leaders in this tragedy was Jim Walsh, Tom Williams, and Ned Connell.[8] As for the former villain I have banished him,[9] he shall never enter my door. Nothing can exceed the state of feeling in this County at present. In Leighlin neither Roman nor Protestant speaks to each other and a system of exclusive dealing[10] is now in full vogue. No Roman Catholic dare put his nose inside my door. The fact of it is, neither life nor property is secure in this County at the present moment. Your uncle Caleb I believe is in jail at present, and that more for security than any thing else for Mary McAssey is not in the least danger, the wound being but a flesh one, and even if she died it would not effect Caleb, though the villians say they done nothing to cause him to fire; but the oaths of such rascals no one could believe. Caleb is I think bailed at this moment and will vote for the Colonel[11] next week. Mobs from the Countys Kilkenney, Waterford, Wexford, Kildare and Queens are to be in Carlow on Monday morning next – the day of nomination. O'Connell says he will have Caleb hanged from the rapper of his hall door, others say that he will be transported. The assizes commences on the 20th[12] therefore he must have his witnesses summoned this week, as the Election will keep every one busy the next. We expect an army here on Saturday if they do not come I will go to Carlow Sunday morning and remain there for the week, agitation and outrage has done its utmost in this County by placing man against man and woman against woman, Still I say that Bruen will be the victor.

I am your affectionate father | John Tyndall

RI MS JT/1/10/3230
LT Transcript Only

---

1. *a letter*: probably letter 0072.
2. *the truth of the whole matter*: A very different account appeared in several newspapers that described the incident as 'one of the most wonton outrages that every was perpetrated, even by Orangemen' – *The Examiner*, 3 July 1841, p. 7; *Bristol Mercury*, 3 July 1841, p. 4. According to the pro-Repeal *Leinster Reformer* (3 July 1841), Caleb Tyndall 'fired at some liberal voters that were driving by', while the Dublin-based *Morning Register* (3 July 1841, pp. 2–3) claimed that Mary McAssey 'who was wounded on Sunday night, is past all hope of recovery; and I have also been informed, upon unquestionable authority, that after the outrage was committed by Tyndall and his associates, some of his family were seen breaking his windows (they being inside at the time), so as to make it appear that the house was attacked by the people [outside], and to form a plea of justification for the cold-blooded attempt at murder that had been made'.
3. *Co*: County.
4. *Colonel Bruen's freeeholders*: see letter 0067, n. 6.
5. *Sunday night week:* 27 June.
6. *Patt McAssey's wife*: Mary McAssey.
7. *William Hackett from the Royal Oak*: Being built at cross-roads, the Royal Oak Inn near Bagenalstown became important as a coaching inn where horses were changed. Following the death of his more famous father, Cornelius, in 1821, William Hackett (d. 1841) became proprietor of the inn and he was also an avid follower of O'Connell.
8. *Tom Williams, and Ned Connell*: not identified.
9. *I have banished him*: Jim Walsh was Tyndall's father's apprentice.
10. *a system of exclusive dealing*: With the increasing separation of the Catholic and Protestant communities and distrust of each other, people only engaged with members of their own community.
11. *the Colonel*: Colonel Henry Bruen.
12. *The assizes commences on the 20th*: The trial took place on Wednesday 21 July before Chief Justice Bushe and was reported in *CS*, 24 July 1841, p. [2] and reprinted in part in *London Standard*, 28 July 1841, p. 2. A shorter report appeared in *Freeman's Journal*, 24 July 1841, p. [4].

From Phillip Evans — 8 July 1841 — 0074

Carlow, Thursday[1] morning

Turn over & read from A to B first

My dear John

I write to acknowledge the receipt of your letter of Sunday last[2] which I received in due time, it happened that it arrived in the morning and your humble servant did not get up till 10 o'clock. I had a most craving appetite and believe me I devoured the contents of your letter in the first fashion Oh, what an awful account have I to render at the last day when summoned before the tribunal of Heaven When you my learned and high minded friend or I would rather say Erin's[3] only child shivers at penning a few lines to an ould friend on Sunday Heaven I would say 'L--d incline our hearts to keep this day holy'[4] AMEN. You say that I have let the cat out of the Bag by my soul, I have not let the cat out of the bag yet, you thief let me know did you see any of the girls after I left town Jack how did you part with the Miss Whites.[5] I think I could answer this question myself for the next part of your letter supplies me with a sufficient answer. You parted well enough Jack future anticipations of meeting with those 'whose bright blue eyes and flowing locks would shake the saintship of an anchorite'[6] much less the heart of John Tyndall. This kept you from even casting a thought upon those from whom you were about to part, I am sure that instead of Ginty having one corner on his second visit to Cork I think he took the round of the table and done justice to the four, I have a pretty good idea of your abilities in tugging into a leg of mutton, thank G-d[7] you could not say on the occasion that I was mess-man for you never to befull savages. I suppose instead of saying I gave you tripes, it would now be: <*words missing*> Oh, ever dear and nonforgotten shades of Evans, roasted beef and York-pudding, spring from the bowels of the earth to 99 George's Street, Cork[8] and there bear testimony against your gormandizing and ill-natured mess-mates. You may well say eaten bread is soon forgotten. Jack I am not quite done with Cork yet. I doubt not the attraction of the joint of mutton, but come Jack for once be candid as I was, and let the cat out of the bag and say rather the attraction of the fair ones, from the description you hitherto gave me of your proceeding there, upon my s--l[9] I envy you very much. I think if the secrets of your heart when parting were known they were much of this description:–

> Now the time has come for parting
> Now my heart begins to swell
> Now the silent tears are falling
> Scarce can I say – farewell[10]

Jack I did not think you were such a blagyard.[11] you say the cohesion had great influence on you. I know very little about it, but I think the laws of cohesion say nothing about human bodies sticking to a chair, a table, &c &c, and now Jack what were you going to stick to? I'll leave you to guess at my ideas about the attraction which overpowered you. I think I am done with Cork. Your description of the town of Kinsale was very short and sweet.

A. Oh Jack! this moment my mother has brought me upstairs your unexpected letter of Tuesday evening[12] and now that I have finished reading it, I would tell you that something of the occurrence which you wrote about did occur. I am happy to be able to make your mind easy on that matter and that the paper from which it sprang set forth 7/8ths of that occurrence in the blackest lies that man could invent.[13]

John I have been at the scene of action during my visit at Leighlin. I will now commence & give it as far as it lies in my power without favour or affection. The transaction took place on Friday night last was a week[14] on that day the agitators were in Leighlin stirring up the minds of the people in the most horrid manner. All this passed pretty well, until about 11 o'clock that night, when a mob (headed by Hacket from the R. Oak)[15] from the Co. Kilkenny (having a few from Lr.[16] among them) was coming from Bagnalstown, they commenced shouting for O'C,[17] and groaning for Bruen. Well Sir this done very well until they arrived opposite your uncle's[18] door, some in the crowd said Tyndall is not in bed let us give him a groan. Your uncle was not in bed and came to the door immediately. On his appearing there was a volley of stones pelted at him which did not hurt him but had the desired effect on the windows, I think. It was now high time for him to defend himself so he let slip, hurted Hacket's nose and two small slugs entered a woman's thigh (the D---ls[19] care to her what brought her) her name is Macassey lives a little above your uncle's and would not say a word about it only for that ill looking savage Maher.[20] she has sworn information but much about her or the whole county and Dan[21] at the head, they cant do any thing. He will come off with flying colours, John you mention about bail being refused, I can tell you with confidence that he could be bailed if he wished what would prevent him? Doctor Roache[22] states that it was not more than the scrape of a pin. He is quite comfortable in the Turnkey's appartments[23] at the gate. He could go home if he wished but he prefers stopping until after the Election. All his family are at home I believe, minding their business. John I think this is as near the truth as possible, if you think I have neglected any thing send me word in your next; and now before I end this part of my letter I must tell you that it was partly by my advice that your father kept back from writing to you about it. I thought no matter how it happened that you would feel pain, and that it would be full time for you to hear it on my return, but as you have asked me about it I send you all I know. I would advise you not to mention it to your father but you now know how to act better than I can advise.

B.I will now return to Kinsale. I am happy that you hit upon so good a place on your first entering the town. The last part of your letter in some little degree surprised me – that portion about Mr Fenwick's plans.[24] I am proud in being able to join with you in saying 'we will in short acknowledge none as our superior'.[25] John I hope we will be trifled no longer – is Barney[26] gone Is there any talk of recommending John.[27] I must now hurry on to say something about the times. you I suppose have got my letter[28] in which I stated about the Bor.,[29] being lost, lost John by bad Protestants who wanted too much for their votes. no matter, England has done her duty, let Ireland go as it may, we will have a majority in the House. John it is the current report here that on Monday and the following days that we will have hot work, nothing expected in the town but open assassination.

the police are here from Meath[30] the military still continue coming in which will have Col. Jackson as commander he is a true blue.[31] Four pieces of cannon are expected in every moment eh boy does not this look like hot work. Some of Col. Bruen's freeholders were coming in to day. on arriving at the potato market the cars[32] of potatoes were drawn across to prevent them from proceeding (I was an eye witness). Well Sir in front of the brake you might behold 2 middle-aged determined men (at this moment you appeared to me telling me about those men who form the American Congress)[33] here Sir you might in miniature behold two of that description – true, bold and determined men they Sir without delay displayed 2 fine blunderbusses their appearance calmed the boys, the police arrived and after a little the car and men arrived safe at the Club House[34] the next moment you might behold 10 cars laden with the Newtown Barony Boys[35] escorted in front and rear by the 12 Royal Lancers[36] John it is expected that the beggar man[37] will be kicked out of Dublin and will then start for this County. We still hope to welt Dan himself. John rest assured that you will have the poll of every evening. Write to me soon any time I have any news I will not delay in writing to you. Give my love to Tid,[38] Ginty, Joe Payne, Corp. Colder, Mr Collins and poor Tom Foy,

John now that I have finished this part of my duty I go first with this to the post and thence in search of the newspaper Sentinel[39]

Yours dear John | P.D. Evans

Should you say anything about your uncle to your father let me know. Excuse this as I am in a hurry until I get it off.

RI MS JT/1/11/3563–4
LT Transcript Only

1. *Thursday*: LT gives postmark as 'July 8th 1841'.
2. *your letter of Sunday last*: letter missing.
3. *Erin's*: Romantic name for Ireland.
4. *incline our ... day holy*: 'Lord, have mercy upon us, and incline our hearts to keep this law' occurs several times in the *Book of Common Prayer*. One such occurrence refers to the fourth Commandment concerning the holiness of the Sabbath (Exodus 20:8–11).
5. *Miss Whites*: not identified.
6. *whose bright blue ... an anchorite*: Byron's 'Childe Harold's Pilgrimage', xi.3–4.
7. *G-d*: God.
8. *99 George's Street, Cork*: *Slater's* lists John Ford, cork cutter, at that address.
9. *s--l*: soul.
10. *'Now the time ... farewell'*: no known published source.
11. *blagyard*: variant spelling of blackguard, a person who behaves in a dishonourable or contemptible way (*OED*).
12. *your unexpected letter of Tuesday evening*: letter missing.
13. *the paper ... man could invent*: see letter 0073, n. 2.
14. *on Friday ... a week*: The attack involving Caleb Tyndall occurred during the night of

27–8 June; i.e. Sunday night–Monday morning.
15. *Hacket from the R. Oak*: see letter 0073, n. 7.
16. *Lr.*: possibly contraction of Leighlin Bridge.
17. *O'C*: Daniel O'Connell.
18. *your uncle's*: Caleb Tyndall.
19. *D---ls*: Devil's.
20. *that ill looking savage* Maher: Father James Maher who aided the agitation in favour of O'Connell.
21. *Dan*: Daniel O'Connell.
22. *Doctor Roache*: Dr Benjamin Roche of Bagenalstown, listed among physicians and surgeon in *Slater's*.
23. *the Turnkey's appartments*: the jailer's living accommodation.
24. *Mr Fenwick's plans*: In July 1841 Lieutenant Fenwick took command of the 5th Division, C District previously commanded by Lieutenant Whittingham.
25. *'we will ... our superior'*: probably a quotation from an earlier letter – that has not been located – from Tyndall to Evans.
26. *Barney*: presumably Lieutenant Paul Whittingham, whose middle name was Bernard and who left the Survey on 11 July 1841, after briefly commanding the 4th Division in Kinsale.
27. *John*: possibly the (unrealisable) suggestion that Tyndall himself should replace Lieutenant Paul Whittingham.
28. *my letter*: letter missing.
29. *Bor.*: Borough.
30. *police are here from Meath*: police reinforcements from County Meath, which is to the north of County Carlow, beyond County Kildare.
31. *a true blue*: probably a staunch Tory, although possibly a staunch Presbyterian (*OED*).
32. *cars*: wagons.
33. *at this moment ... American Congress*: relevance unclear.
34. *the Club House*: one of Carlow's principal inns.
35. *Newtown Barony Boys*: a gang from the townland of Newtown – about 3½ miles east of Leighlin Bridge – which, from the context, probably comprised Catholic Repealers opposed to Colonel Bruen.
36. *the 12 Royal Lancers*: 12th Regiment of (Light) Dragoons (Lancers), a cavalry regiment of the British Army then stationed in Dublin.
37. *the beggar man*: Among his opponents Daniel O'Connell was known as the 'Big Beggarman'.
38. *Tid*: John Tidmarsh.
39. *the newspaper Sentinel*: the *Carlow Sentinel*.

To John Tyndall, Snr 11 July 1841 0075

Kinsale, | Saturday evening | July, 11th, 1841.

My dear Father

With gratitude I have to acknowledge the receipt of yours of the 7th.[1] It relieved my mind of a great load of anxiety. Things are not half so bad as they were represented to be. I heard that there was a woman shot in the thigh and an eye shot out of a man's head and the whiskers shot off another. I heard also that the immense bail of £2000 was refused for him. This led me to conclude that his offence was a most aggravated one. I am extremely glad to find that my conclusion was a false one.[2] I commend your determination of going to Carlow highly. You would be a fit object for the mob to wreak its fury on (oh I wish I was at home!) they will scarcely think of injuring my mother or Emma.

We had an election here this week, there was a mob collected of course. The rival candidates were Mr Atwood conservative,[3] and Mr Watson radical[4] – both Englishmen. Mr Atwood had been beaten by Lord John Russell in London.[5] He was not here himself, his agent polled for him. The Radical was returned by a majority of 18. Had Atwood been here in time himself he would have been returned. The people here, even the radical voters, are not at all pleased with the manner in which Watson left town.[6] They say he departed very shabbily.

He will be petitioned against without doubt will be thrown out, it is thought that it will not be defended. One day during the election I was attacked by a party of fishermen with green boughs[7] in their hands a dozen of them gripped me, I thought to laugh them out of their holds, but to no purpose. I knew that it would be madness to strike one of them – that if I did so my clothes would be all torn – so I had to give them half a pint[8] to appease them. The town at present is quite peaceable all excitement has subsided. It is just post time so I must bid you good bye.

Your affectionate son | John Tyndall

RI MS JT/1/10/3231
LT Transcript Only

---

1. *yours of the 7th*: letter 0073.
2. *I heard ... a false one*: false rumours that circulated concerning the shooting of Mary McAssey and William Hackett by Caleb Tyndall. Some of the published accounts were equally overblown; for example, the Dublin *Morning Register* (3 July 1841, pp. 2-3) claimed that 'the poor woman Mackessy ... is past all hope of recovery', while *The Pilot* (30 June 1841, p.3) insisted that the 'poor man [Hackett] is dreadfully injured and is likely to lose the sight of his eye'.
3. *Mr Atwood conservative*: The previous MP, the Conservative Colonel Henry Thomas (see letter 0072, n. 8), had withdrawn shortly before the election and been replaced by Matthias Wolverley Atwood (1808–65), a banker from London who had previously represented Greenwich.
4. *Mr Watson radical*: The lawyer William Henry Watson (1796–1860) stood as the Reform candidate and defeated Atwood by 18 votes.
5. *Lord John Russell in London*: Four MPs had recently been elected to represent the City of London. Lord John Russell (1792–1878), the leading Liberal politician and future Prime Minister, had come fourth in the poll beating Atwood by just 9 votes (*Morning Chronicle*, 1 July 1841, p. 4).
6. *the manner ... left town*: After the result was declared there were complaints that a num-

ber of bad votes had been cast and that Watson was not qualified to sit in Parliament. Watson did however represent Kinsale for the next six years.

7. *green boughs*: Green bows were a symbol of Irish nationalism.
8. *half a pint*: probably either beer or whisky.

From Phillip Evans 13 July 1841 0076

Carlow, Tuesday[1] evening

Forget me not.

My dear John

Your motto I am sure from experience does speak the sentiments of your mind, and now Jack before I go any further allow me to tell you that I have just eaten my dinner after which I took a drop of the Barley[2] and upon my s--l[3] I am now 4s 11d <*word(s) illeg*> So Jack you must take pot luck in this epistle.

I received your wished for letter[4] this morning it afforded me much comfort in knowing that you were settled in mind, respecting your uncle Caleb Believe me when you feel pain I take a share in it, and when you rejoice I rejoice. I am glad that you did not let your landlady[5] run the old soldier on you, that's right boy nothing like 'shaving close' The election commenced yesterday the vagabonds commenced coming in very early but that brave and gallant officer (Col. Jackson) knew their movements. He is the boy that soon settled them no less than a band at the head of every County. The Queen's County boys were coming in they arrived at Graigue Bridge[6] but to their great surprise and mortification there they met a fine piece of cannon staring them in the face. Blast the one of them Jackson allowed to cross the Br.[7] So they had to return just like so many – what – I cant compare them to anything that will answer so well as the Big beggar man himself[8] on his arrival in Carlow after loosing the City[9] John he was in an awful way.

Well Sir, | The election was to commence at 10 o'clock and from the regulations made by the military and police all Dublin St and the vicinity of Court House[10] was as free from mob as if there was no such thing going on in the town everything quite peaceful tickets were given at both sides to the amount of 200 for admission into the Court House the gates were now opened and those and only those who had a ticket were admitted (fine regulation Jack) on this eve (Monday) the radicals sent a proposition requesting no speeches on either side, this was agree'd to, so the Members were proposed and seconded and the proceedings closed until this day. I was in court this day and saw all the Barony Boys 160 in number (all RC.[11] to a man, polling for Bruen and Bunbury. John I saw one glorious sight a voter coming up on O'Connell's tally and voting for Bruen and Bunbury, all from the looks and power of his landlord who was standing at my side. immediately after voting his landlord was crossing the table to fold him in his arms as the prodigal son)[12] when he was stopped by one of the opposite party – now Sir the fight commenced now you might behold John Alexander, Sir T. Butler, Newton,[13] to fighting like men for the cause, nothing serious occurred but the conservatives came off victorious. Hurrah for Carlow! In another Barony three voters came up on O'Connell's list and voted in their teeth,[14] every thing went on yesterday very quietly, there was a great number of people in town yesterday. Jack it is very near post time, so you must be content with a short letter in lieu of a long one I send 30 as compensation. I have seen Barlow[15] one Friday and delivered your message he said he would write to you. He has surveyed a foundry in Graigue. He and I go to examine it tomorrow. I believe I will give the *[word illeg]* a bat for him John the Sentinel[16] you wrote to me for I could not get for love or money in Carlow. they were all picked up as soon as they came out of the press. My uncle sent his to Scotland and I got him to write for it he got an answer that it was sent to America, but I now send you (per this post) the Sentinel of last Saturday.[17] John write soon and give my love to Tid,[18] Ginty and anyone else you like. It is post time.

Yours P.D.E.

RI MS JT/1/11/3565

LT Transcript Only

1. *Tuesday*: LT notes 'probably 13 July 1841'.
2. *a drop of the Barley*: whisky.
3. *s--l*: soul.
4. *your wished for letter*: letter missing.
5. *your landlady*: not identified.
6. *Graigue Bridge*: the bridge on the River Barrow linking Graigue (on the west) with the rest of the town of Carlow.
7. *Br*: Bridge.
8. *the Big beggar man himself*: Daniel O'Connell.
9. *loosing the City*: Daniel O'Connell had been defeated in the City of Dublin election a few days earlier.
10. *Court House*: Carlow's impressive Courthouse, built to a classical design, was the venue for counting the votes. As the ballot was public, not secret, the audience could see how each person voted and the opposing parties usually responded with cheers or jeers, and sometimes with violence.
11. *RC.*: Roman Catholic. Evans reported that contrary to expectation Catholics voted for Bruen and Bunbury, the Tory candidates. The 'Barony Boys' in this context appears to refer to the band of local Catholics.
12. *the prodigal son*: see letter 0043, n. 2.
13. *Sir T. Butler, Newton*: Sir Thomas Butler (1783–1861), 8th Baronet of Ballintemple, Tullow, Co. Carlow, was a Captain in the 6th Dragoon Guards, Colonel of the Carlow Militia and had served as High Sheriff of Co. Carlow. It is not clear which member of the Newton family was present; probably either Beauchamp Bartholomew Newton (1798–1850) of Rathwade, Bagnelstown or his younger brother Henry Newton (*c.* 1800–63) of Mount Leinster, Borris.

14. *voted in their teeth*: voted in public against the Repeal candidates.
15. *Barlow*: possibly Robert Barlow (1813–83), a private contractor who undertook surveying work for the Ordnance Survey before emigrating to Canada in 1855.
16. *the Sentinel*: the *Carlow Sentinel.*
17. *last Saturday*: 10 July 1841.
18. *Tid*: John Tidmarsh.

From Thomas Higginson 17 July 1841 0077

Friday, | Mill Street,[1] July 17th 1841

My dear Tyndall

When I was coming in from work I met the messenger and to comply with your request sat down to write to you not indeed that I had any thing particular to communicate only that if I did not write to you you would not write to me, There is a strong report here that there are five men to be discharged[2] at the end of this month and that the remainder are to go into the office when this work is done I suppose if I am not discharged that we will be soon together again but I think <u>that</u> bloody Collins will do all in his power to get me off, [3] but no matter. I hear that the conservative member is put out for Carlow they did not do like poor Dublin that gave <u>Dan</u>[4] the decent turn out. I wonder you did not go home with the others

Dear Tyndall, I have no more to say to you that have all the news. Write to me soon. Remember me to Foy and Collins bad luck to him

I remain yours ever attached | T.C. Higginson

RI MS JT/1/11/3729
LT Transcript Only

---

1. *Mill Street*: a market town 25 miles north-west of Cork.
2. *five men to be discharged*: Robert Martin and Archibald McLachlan may have been among the men considered for discharge as they were transferred to the English Survey in August 1841.
3. *to get me off*: Higginson remained on the Irish Survey until September 1841, when he joined the English Survey in Yorkshire.
4. <u>Dan</u>: Daniel O'Connell, who had lost the election contest for Dublin City, but with a substantial number of votes (2069).

From John Tyndall, Snr 19 July 1841 0078

Mr John Tyndall | Ordnance Survey Office | Kinsale |
Leghlin Bridge | July 19th, 1841.

My dear John

A second time I send you the glorious intelligence of Colonel Bruen's return for Carlow,[1] also that of Mr Bumbury's after one of the most tremendous struggles that ever took place in any County. The mobs that passed through this town on Monday last[2] on their way to Carlow, headed by the most hellborn agitators with long poles in their hands and concealed pikes in their bosoms, was enough to strike terror into the most chicken hearted. They had along with them three green flags[3] with mottoes on them and one of them was a representation of the devil with Bruen on his back. Below was printed letters:– 'The Devil running away with Bruen'. The reception they met with in Carlow was anything but pleasing to them. As soon as they arrived at Bunn Bridge[4] there was a very snug cannon there with a blazing match ready to receive them. Priest McMahon marched up to Colonel Jackson who commanded the military and told him that he would go bail for the peace of the town if the people, as he called them, were permitted in. Colonel Jackson asked him who he was and told him to be off about his business and that he, not the Priest, would secure the safety of Carlow at this time. Nothing can describe their chagrin; after all their swaggering and display they were obliged to return home in the midst of the most dreadful rain while starving all, without being able to rob a single baker of one loaf of bread. It was their intention to ransack the town and carry away their illgotten booty. This is no idle story but a positive fact as they brought cars and carts with them for that purpose. Nothing can describe the state of society in the Co Carlow at this moment and I believe the Romans[5] and Protestants will never be on good terms with each other. They threaten us with a rebellion but we laugh at them with scorn. Your uncle Caleb was bailed on Friday last. Mr Steuart and Mr Henry Newton were his sureties. The trial I think wont go on this assize though Priest McMahon and Mary McAssey with about forty witnesses has left this town this morning for Carlow to hang him, but they will see to their cost that it's all a mountain in labour.[6] I was in Carlow from Tuesday till Saturday, and in the Court House when the Colonel and Mr Bumbury were declered duly Elected by the High Sherif amidst the most tremendious cheering and kentish fire[7] ever beheld. The majority was small – only 9[8] – but if you knew our situation here you'd think we ought to be in a large minority. I told you that Bruen would be the victor amidst the most discouraging circumstances, and you cant call me a false prophet.

I am your affectionate father | John Tyndall

Caleb voted when bailed, I was in the booth.

RI MS JT/1/10/3232
LT Transcript Only

---

1. *Colonel Bruen's return for Carlow*: A late edition of *CS* was printed on 17 July in order to print the result of the election.
2. *The mobs ... Monday last*: On 12 July 1841 the mob in Carlow, which according to one observer consisted of 'more than 100,000 people', was kept in check by Colonel Jackson and his troops. 'Not a drop of blood has been shed, and that is invariably the result of magisterial firmness and timely preparation' (*CS*, 17 July 1841, p. [3]). This invasion of Carlow was widely reported in the national press (e.g. *London Standard*, 22 July1841, p. 2); see also letter 0076.
3. *green flags*: green is a symbol of Irish nationalism.
4. *Bunn Bridge*: presumably the bridge over the River Barrow linking Carlow and the suburb of Graigue (now Graiguecullen Bridge) which was built in 1569 and widened in 1815.
5. *Romans*: Roman Catholics.
6. *a mountain in labour*: In a story from the ancient Greek collection *Aesop's Fables*, a mountain groaning loudly in labour gives birth only to a mouse. The fable is usually interpreted as relating to grandiose claims or threats which prove to be empty.
7. *kentish fire*: protracted clapping in unison in order to drown out an opponent; a tactic thought to have originated in 1828–9 when the people of Kent opposed speakers who favoured the Catholic Relief Bill.
8. *The majority was small – only 9*: Bruen polled 705 votes, Bunbury 704, John O'Connell, Jnr, 696, Ashton Yates, 697. In asserting that there was only a majority of 9 votes, Tyndall's father appears to have been considering only the votes for Bruen and O'Connell and ignoring the votes for the other two candidates. 8 votes separated both Bruen and Yates and Bunbury and O'Connell.

From Phillip Evans 21 July 1841 0079

Carlow, Wednesday[1] evening 7 o'clock

My very dear John

I believe I stated in my last letter[2] that I would not write to you until I should first get one from Kinsale, but important business, business, which if possible is as dear to me as to you now calls upon me, and it is this, that your uncle Caleb has been duly and lawfully <u>acquitted</u>.[3]

His trial commenced at 10 o'clock this morning the rabble produced 9 witnesses, which I believe would swear that the houses in Carlow were not made of stones or morter. John they were drilled well, but truth is powerful and must prevail. Your uncle had your uncle Wm.,[4] his apprentice, and two policemen.[5] He had two of the best counsellors on the Circuit, McDonough and Walker,[6] these are the boys that laid forth the truth, his jury comprised of 10 good and stanch protestants and two Roman catholics. At 6½[7] the trial terminated, the jury retired and in ¼ hour, to the surprise of the vagabonds in court, they again made their appearance. I knew, I knew he was victorious, the jurymen's names were now called over – now boy, O, my ear was cocked, the next minute the Prisoner was NOT Guilty. Hurrah Hurrah for Tyndall I say. I saw your father in town to day. He told me he expected a letter from you

Fair and flowing be thy way
And bright the skies above thee
Nor grief nor gloomy care impart
Their cruel pangs to wound it[8]

John it is quite useless for me to say that the above few lines expressed the sentiments of my heart and that it is the fervent wish (I could say prayer for you know I would be telling a great lie for I never say one) of your humble servant that grief nor gloomy care may never make their appearance near you. John if the Sentinel contains any part of the trial[9] I will send it to you. Could I do anything for you before I leave this country?

'Sleep on now and take your rest'[10] | Give my love to all the young fellows

I remain your affectionate companion | P. D. Evans

RI MS JT/1/11/3566
LT Transcript Only

1. *Wednesday*: LT gives postmark 'July 21. 41'.
2. *my last letter*: letter 0076.
3. *uncle Caleb ... lawfully* acquitted: the trial at Carlow Assizes of Caleb Tyndall for wounding Mary McAssey; see letters 0071, 0073 and 0075.
4. *uncle Wm.*: William Tyndall.
5. *his apprentice, and two policemen*: William Cooley (apprentice) and constables John Macken and William Watters.
6. *McDonough and Walker*: probably Francis Macdonough, QC, Lower Dominick Steet, Dublin and Richard Cotton Walker (1782–1850), barrister, of Upper Gloucester Street, Dublin.
7. *6½*: half past 6.
8. *'Fair and flowing ...* wound it': based on lines 1–2 and 11–12 of J. H. Bedford, 'To Amelia', in *Wanderings of Childe Harolde: A Romance in Real Life* (1825).
9. *the Sentinel ... the trial*: report in *CS*, 24 July 1841, p. [2].
10. *'Sleep on ... your rest'*: Matthew 26:45.

To John Tyndall, Snr 22 July 1841 0080

Kinsale, Thursday[1] night

My dear Father

I now sit down to congratulate you on your recent victory[2] – well and bravely has Carlow done her duty – The news of this triumph sped like a meteor until it reached this. I had it on Sunday last. You may rest assured that I was no unmoved listener to the intelligence. An echo from the World's End[3] answered the loud hurrah! contained in Phil's letter[4] – the rocks of Scilly rung and the banks of the Bandon[5] responded to the Carlow cheer – hurrah!

Words could not convey an idea of the transport with which I received it. Three leaps off the ground was the immediate consequence – 'twas great – 'twas glorious. Dan kicked out from Dublin[6] and his namesake Whelp[7] rejected from Carlow.

Mail! ye venerable cloisters of Derrynane[8] again he seeks your protection, let nature sympathise with him and upland herds shew him a dumb compassion – mourn with him ye streams and Oh! ye turrets of his fathers' domicile reecho[9] his piteous wail.

What has that demon traitor Maher[10] to say now! the strongest conjuration of my fancy I fear would fail to depict him in his true colours. Me thinks I see him – his ghastly countenance displaying the fiercest passions that ever agitated the bosom of an arch devil – With what contortions he writhes! – he foams – he swells – he groans – curses froth upon his lips – the fellest anathema of his mother dwells upon his tongue – oh! for the stake of former times oh! for the power of a Bonner[11] – a Bonner's spirit is here but fancy has attained her limit and Maher recedes before my mind's eye a distracted phantom.

Sound the loud timbrel o'er Barrow's blue wave[12]
The Hydra is crushed that would Carlow enslave.[13]

Language could not do justice to the boys of St. Mullins;[14] they have erected for themselves a monument in the hearts of all honest men that shall long outlive the ephemeral popularity of those fiendish agitators that would coerce them to their ruin. The noblest privilege of man is liberty to assert his right of thought; the Barony boys[15] have not resigned it.

But what shall we say of Col Bruen – of him who has borne the brunt of the battle in so many hard-fought days – like the oak of the forest he has braved the fury of agitation's tempest – Victory's brightest banner now waves above him – may propitious breezes long continue to fan his ample volume

Let fame her golden trumpet sound
Let Erin[16] join the theme
And Barrow's beauteous banks resound
In praise of Bruen's name.
You deem me an enthusiast, but
'Breathes there a man with a soul so dead
That never to himself hath said
This is my own my native land.'[17]

Carlow is mine – the wildest furzebush that blossoms on her braes is dear to me – blame me not then for being warm in her cause. May heaven strew its choicest blessings on her.

Good bye | Your affectionate son | John Tyndall

RI MS JT/1/10/3233
LT Transcript Only

1. *Thursday*: LT gives postmark 'July 22nd, 1841'.
2. *your recent victory*: the success of the Conservative candidates, Henry Bruen and Thomas Bunbury, in the Carlow County election.
3. *the World's End*: see letter 0072, n. 1.
4. *Phil's letter*: letter 0079, from Phillip Evans.
5. *Bandon*: The River Bandon in west Cork flows through Dunmanway and Bandon before discharging into the sea at Kinsale.
6. *Dan kicked out from Dublin*: Daniel O'Connell lost to the Conservative John Beattie West, 2069 votes to 2128.
7. *his namesake Whelp*: Daniel O'Connor, Jnr, who failed to be elected by the voters in the Carlow County constituency.
8. *cloisters of Derrynane*: the ruined Derrynane Abbey in County Kerry.
9. *reecho*: resound with (*OED*).
10. *that demon traitor Maher*: Father James Maher who aided the pro-O'Connell agitation.
11. *Bonner*: Edward Bonner (d. 1659), bishop of London, who gained notoriety for hounding and persecuting heretics.
12. *Barrow's blue wave*: The River Barrow flows through Carlow and Leighlin Bridge.
13. *Sound ... enslave*: based on one of Thomas Moore's *Sacred Songs* (1816) entitled 'Sound the Loud Timbrel'.
14. *the boys of St. Mullins*: O'Connell complained that only one barony – Lower St Mullins, which includes the Kavanagh estate – was 'remarkable for the indisposition of the clergy to agitation' (Daniel O'Connell to ?, 18 June 1841: O'Connell (ed.) *Correspondence of Daniel O'Connell*, vol. 7, pp. 94–5).
15. *the Barony boys*: in this context the Protestants from the barony of Lower St Mullins who defended the Protestant cause.
16. *Erin*: romantic name for Ireland.
17. *Breathes there ... native land*: W. Scott, *The Last Lay of the Minstrel* (1805), VI.i.1–3.

From John Conwill 24 July 1841 0081

Ambition
What is that passion men call
Which holds dominion o'er great and small
What's that disturber of the human breast
That robs its vot'ries of nocturnal rest
What is that sin by which the angels fell
From heav'n's high portals to the gulphs of hell
And soil'd the happiness which promis'd fair
In Eden's bow'rs to crown the unborn pair

'Tis hard to say though almost ev'ry mind
In ancient or in modern times we find
Is subjugated by its conq'ring pow'rs
And held in bondage all life's fleeting hours
The man whose pride and littleness of soul
Attempts at governance without control
Whose mountain arrogance with pain I see
My fellow-creature, perhaps my enemy
Is prompted by ambition's luring call
To make himself despicable to all
Tho' low his station and tho's mean his birth
He thinks himself of consequence on earth
And though unskill'd in Nature's various laws
Affects to state and trace the wond'rous cause
He's quite astonish'd that all human kind
Don't always rev'rence his magnific mind
Don't crouch like slaves before an Eastern king
And from Golconda[1] richest presents bring
From Peru Chili and the Brazil shore
Don't send him mines of gold and silver ore
And yet amid these notions of his brain
Say is he happy does his breast contain
One heart-felt pleasure which he dares reveal
One beam of comfort does he ever feel
Ah no his aching heart and throbbing breast
Continued fears and anxious cares molest
These are the evils which ambition brings
On all its vot'ries mendicants or kings
And those who would avoid its fatal snare
Should of its soft but syren arts beware.

July 24th 1841.

'My dear Tyndall'

I perceive, though your brain had been rendered a complete system of jarring elements in consequence of your removal from Youghal, and the agitated state of Kinsale when you arrived in that place, that you have not forgotten your old friend, and that your cranium has become consolidated by the fortuitous concussion of its elementary particles, hurrah for Epicurus,[2] he was a queer sort of fellow, but away with him for the present.

You talked of silence's having spread her mantle over me. I discard the allegation; because there never existed an individual, with few exceptions, whom I admire more than I admire you.

My lines on ambition were crossed by the Electioneering manoeuvers carried on in the County Carlow. I will not give you a portrait of the conduct of either the radical or the conservative party lest I might be considered biased to either side. I am sure your father has satisfied you on this subject, if Mr Evans should have neglected to do so. Mr Evan's gratitude to me deserves the highest eulogium[3] as he never so much as enquired after my welfare. Oh shame, never mention the fellow to me again.

Your uncle Caleb is acquitted just as I anticipated, with regard to his conduct I have nothing to say, because I am totally unaware of his transaction altogether; As there are so many stories afloat about him,[4] I am sure his opponents exaggerated the circumstance of his firing very much.

The Sunday before the polling commenced in Carlow, I was going to see your father and mother. I met Emma and Miss Newsome at Mr Faulkner's cottage,[5] Emma told me your father was gone to Nurney, I then formed a resolution to go to your mother, and ere I had arrived at the market-cross I was assailed by a crowd of jack-asses roaring on every side. I wheeled round, perfectly disgusted, and never appeared in Leighlin since, except on two evenings for the purpose of getting my surveying chain[6] fixed.

Your story of the fishermen of Kinsale[7] flits across my mind at the moment – really I pitied you in my heart when I contemplated you in the midst of a rabble bellowing for whisky and who were ready to deprive you of existence had you not conceded to their wishes. Really my dear Tyndall, my ideas are so confused that I can scarcely write a correct sentence on any subject.

Your ever faithful teacher | John Conwill

Your poem on Friendship[8] deserves the most unbounded applause. In my next letter I hope to write some lines on the same subject, though I must confess I am a very bad poet.

RI MS JT/1/11/3520
LT Transcript Only

---

1. *Golconda*: the city in southern India at the centre of the diamond trade and associated with such magnificent and expensive stones as the Koh-i-noor diamond.
2. *Epicurus*: The Greek philosopher Epicurus (341 BC–270 BC) posited that all occurrences could be explained by the concourse of small particles – atoms.
3. *eulogium*: a laudatory discourse; a formal expression of praise (*OED*).
4. *so many stories afloat about him*: see letter 0073, n. 2 for some of the widely differing reports of the incident.
5. *Miss Newsome at Mr Faulkner's cottage*: The magistrate Hugh Faulkner lived in Rathvinden Cottage, Leighlin Bridge. Miss Newsome has not been identified.
6. *my surveying chain*: a chain of 66 feet in length used to make linear measurements of distance.
7. *the fishermen of Kinsale*: see letter 0075 for the fishermen with green bows who attacked Tyndall.
8. *Your poem on Friendship*: poem missing.

From William Ginty 24 July 1841 0082

Ball.[1] | July 24th 1841.

Dear Snooks[2]

As I promised to write to you a letter when I arrived here, it is now time for me to scribble these lines. I am sorry indeed for not doing so long before this I hope my dear fello you will excuse me this I know you will for you and I are good friends. I now commence with regards with the little Beggarman[3] and myself, we have bidden our last farewell to the Youghal lasses, particularly that lovely creature Miss Stoul[4] who peeped at me out of her door when the car passed it and I put my hand to my mouth and saluted her which she returned in the most ladylike manner, and of course by nature the tears rolled down my cheeks for a good while – this the Beggarman was laughing at. Well when we passed the long wooden bridge and arrived at that nice little bridge called – <*word(s) missing*> met another lady Miss I do not know, the governess,[5] my little man's sweetheart when the car stopped we jumped off the seat and shooked hands with her oh what a soft and delicate hand she has – she asked us in but we did not nor had we time the car just starting we bade her good bye and told her we hoped we soon see her again. I hope so said she, the poor thing lifted up her beautiful hands to wipe away the tide which was falling down her pale cheeks. We arrived in Carlow safe but I did not reach my quarters till 4 next morning. Carlow election is over, Bruen and Bunbury has gained the victory by a small majority of 8[6] – the work was tight. Hurrah – Hurrah and again Hurrah, my old boy we are all right; there were great numbers of mounted and foot soldiers and police I never saw such a sight of them in my life. There were also four pieces of cannon at this election which frightened the people very much, they did not know what they were at first and as they found them out they ran away like mad devils, roaring and howling and crying murther, murther we never can face the canons for they will blow our brains out. Col. Jackson was the commander of the military forces a fine and warlike fellow with a monstrous pair of mustachioes, he did his duty like a gentleman and soldier he kept the town in good order and as clear as possible, Major Galway[7] commander of the police at this election. There is no use to tell you any more as it will take two or three sheets more of paper and besides that I will soon be with you, I was glad to hear that your uncle[8] is acquitted. The little Beggerman will start before me, I am sorry I will not be with him. I was very very busy here ever since, surveying and spending days with the jolly boys in different parts of the country all those prevented me from writing to you sooner. Write soon and I will write another and let [me] know how you are getting on and let me know how does the examination of the two points that I plotted look and believe me

Dear Snook | Your affectionate friend | G[9] | alias | Whiskerando[10]

RI MS JT/1/11/3577
LT Transcript Only

1. *Ball.*: LT speculates that this is Ballickmoyler, in Queen's County (now County Laois), a village approximately 3 miles north-west of Carlow. However, as there are a number of other places that begin with 'Ball', this identification is difficult to confirm.
2. *Snooks*: Walter Snooks, a favourite pseudonym for Tyndall.
3. *the little Beggarman*: probably Phillip Evans.
4. *Miss Stoul*: not identified.
5. *Miss ... the governess*: not identified but possibly the author of letters 0091 and 0099.
6. *gained ... majority of 8*: see letter 0078, n. 8.
7. *Major Galway*: Major R. A. Gallwey/Galway (d. 1845) appears to have previously held the post of Provincial Inspector of Constabulary for Munster (*Champion*, 5 January 1840, p. 7), but it is not clear what position he held in July 1841.
8. *your uncle*: LT's transcription reads 'your father', but the trial reffered to here was of Tyndall's uncle Caleb Tyndall.
9. *G*: William Ginty.
10. *Whiskerando*: probably an allusion to Don Ferolo Whiskerandos, a heavily-bearded character in Richard Brinsley Sheridan's play *The Critic* (1779).

To John Tyndall, Snr [3] August 1841 0083

Kinsale, Thursday evening, Aug. 3rd, 1841.[1]

My dear Father

I believe I made a promise some time ago to write to you every week.[2] Well to prevent a breach of promise, and not to pay a debt, I now write. Evans arrived here last night. I met him outside the town. He came through Youghal and stopped there in our old quarters all day on Sunday. He appears to have been tired of Carlow before he left it; the calm succeeded the turbulence of the election was so strikingly contrasted with the previous excitement that the poor fellow was near sinking into lowness of spirits it will take at least a week's <u>thumping</u> to bring him to his natural feelings. He brought me Swift's 'Tale of a Tub'[3] he tells me I must bring it back to you again – well this I will most assuredly do. I was just glancing over it yesterday evening and I find that it has passed through your hands, for all the choice portions are marked. The moment Evans entered Kinsale he sent up a long petition to all the Saints in the calender[4] to send him a safe deliverance out of it. The rattling of cars over the rough pavements, the cry of the fishwoman, the bawling of children, and the chattering of the World's End Irish, all rung very discordant on an ear that was accustomed to the sounds of <u>human</u> voices for upwards of a month. Poor fellow, when he's as long on his own hands as I have been he'll learn to deem such things mere trifles. An acquaintance with the world is a fine thing. He has a great many very interesting anecdotes respecting the election; he found it impossible to transmit them to me by letter. He tells me that he saw you sharing a chicken with my uncle Caleb. This gave me great pleasure as I concluded from it that old sores are healed and the breach made up.[5] I was greatly surprised to find that you were about removing to my uncle Bill's house.[6] You never told me a word of this. This was rather uncourteous as in the event of a missing you from your old domicile when I had arrived in Leighlin I might likely go astray in the town and so lose both myself and you. – But, jokes apart, I think the movement will be for the better. It will be in a far better thoroughfare, but I suppose this wont be of much advantage to you in these perilous times; exclusive dealing[7] I hear it is carried on to a great extent there, and if this be the case surely the name of Tyndall will trebly strengthen the determination of the propagaters of the system.[8] Let me know how you are situated at present. Who have you got in Walsh's place?[9] I'm promising myself a great job at your accounts when I go home. I'll be looking out for a nice account book, if I can possibly get one I'll buy it. Have you e'er a heavy hide[10] for me to cut up? I'm the boy that could dissect a 50 pounder.[11] You have not told me every thing about Mr McGee[12] this long time how is he? how did my uncle Ned get over his arrears in the loan Bank?[13] Was the payment enforced? I wish when you go to Carlow that you would call to Jos Wright and enquire did he get any word from Wm[14] since he went to America and if he did to send me his address. Wm promised to write to me the moment he'd land, but I would not wonder if his letter was missent on account of my removing from Youghal.

I have no more to say | Your affectionate son | John Tyndall

RI MS JT/1/10/3234
LT Transcript Only

1. *Thursday evening, Aug. 3rd, 1841*: 3 August 1841 was a Tuesday, so either the day or the date is incorrect.
2. *I believe ... every week*: letter 0062.
3. *Swift's 'Tale of a Tub'*: Jonathan Swift's satirical novel *A Tale of a Tub*, first published in 1704.
4. *long petition ... the calendar*: Although Roman Catholics and Anglicans subscribe to somewhat different lists of saints' days, the term is used here to mean that Phillip Evans expressed his profound wish to escape from Kinsale.
5. *old sores ... made up*: The source of the rift between Tyndall's father and his uncle Caleb is not clear, but neither Tyndall nor his father was invited to the christening mentioned in letters 0051 and 0053.
6. *my uncle Bill's house*: In *Griffith's Valuation* (County Carlow, p. 179) 'John Tyndall' is listed as the occupier of 22 Bridge Street in the town of Leighlin Bridge. It is therefore likely that this was the property to which his father moved in 1841, his uncle William Tyndall (who had moved to Carlow) having previously lived there.
7. *exclusive dealing*: see letter 0073, n. 10.
8. *propagaters of the system*: it is not clear whether Tyndall is referring to exclusive dealing (see letter 0073) or the previous norm in which Protestants and Roman Catholics dealt

with each other.

9. *Walsh's place*: The apprentice James Walsh had been sacked by Tyndall's father for participating in the Repeal agitation during the recent election.
10. *a heavy hide*: leather (which his father used in shoe-making) that Tyndall hoped to use in binding the account book.
11. *a 50 pounder*: presumably a side of a cow weighing 50 pounds.
12. *every thing about Mr McGee*: on John M'Gee's bankruptcy, see letter 0004, n. 11. McGee was due to appear before the Court of Bankruptcy, Dublin, on 24 August 1841 in order to discharge a debt of £461 0*s*. 9*d*. (*Perry's Bankrupt Gazette*, 7 August 1841, p. 8).
13. *uncle Ned ... loan Bank*: for Edward Styles' demeanour, see letter 0056.
14. *Jos Wright ... word from Wm*: Joseph Wright was probably the father, or possibly the brother, of William Wright who had departed for America in early March 1841.

To John Tyndall, Snr 10 August 1841 0084

Kinsale, | Tuesday[1] Evening

My dear Father

This morning I received your letter[2] it was indeed long looked for. My correspondence is much more limited than formerly. Some of my friends have flagged and others have departed to another hemisphere.

I never was more surprised in my life than to find that John Mooney was taking an active part in preventing Roman Catholics from dealing with you,[3] there were few men in Leighlin that I liked better than John and none on whom I would sooner place dependence. How mistaken I was. It's hard to form a correct estimate of human character. Were this intelligence from any other source I would deem it incredible. I wish I was speaking to John I think I would bring him to a sense of his want of consistency. These certainly are perilous times, when a man's own familiar friend is the first to raise his heel against him.

I'm sure this system of exclusive dealing must have injured you. You may be in want of a little money perhaps. I have two or three pounds by me which I intended for my expenses going home and returning. If you are agreeable I'll send you this I willingly give up the pleasure I would experience on seeing you and my mother and Emma could the money that I have at present be useful to you. <u>If you want it</u> it's foolish to spend it on coach fare. Speak to my mother and whatever determination she and you come to that I will abide by. Send me word immediately what you would wish me to do

Phil[4] desires to be remembered to you | Your affectionate son | John

I'll be expecting an answer every day, write one – be it ever so short.

RI MS JT/1/10/3235
LT Transcript Only

1. *Tuesday*: LT gives postmark 'Aug. 10th, 1841'.

2. *your letter*: letter missing.
3. *preventing Roman ... with you*: see letters 0073 and 0083.
4. *Phil*: Phillip Evans.

To John Tyndall, Snr 15 August 1841 0085

Kinsale | Sunday[1] morning

My dear Father

I received your letter[2] on yesterday morning, I was very glad to find that you did not delay in sending an answer to my last question[3] the motives however which influenced me in writing my last letter are very very much mistaken. You appear to think that I am quite indifferent about going home. To this charge I plead not guilty. I declare most solemnly nothing would give me greater pleasure than seeing you all. This pleasure I was willing to sacrifice, and it gives me no small concern to find it attributed to a comparatively base motive to that which really influenced me. 'I promised to go home often' – true, but you know I have not the command of my destiny; that promise was to be qualified by circumstances. Last Christmas it was utterly impossible for me to go.[4] I told you before how things stood and I'm sure Phil Evans verified what I said, if further proof was necessary, Well the hurry of the work continued thro' the spring and summer – in proof of this I will just say that three or four of the best draughtsmen had to work from 5 o'clock until 8 in the evening[5] for the purpose of getting the work forward – this was caused by Mr Wynne leaving it in a very backward state. Well then when we were about to leave Youghal I spoke to Colder about getting me leave and he strongly advised me not to apply until that quarter's work was finished, this I made you acquainted with and you will remember that you told me in answer that you would not wish for a hundred pounds that I was at home.[6] This you must confess was sufficient to make me defer my purpose. I think I have satisfactorily proved that it was not neglect or indifference which prevented me from appearing ere this in Leighlin Bridge. And as to my last proposal I have only to say that you wrong me.

But to the point, the moment I read your letter I went to Colder and asked him how he would advise me to act, he told me to go myself to Lt. Fenwick. I did so. I found him in his own room in the office. I made known my business and urged my request in the strongest terms I dare. He told me that he was expected to finish Lt. Whittingham's work[7] without delay and if the draughtsmen left him this would be out of the question. I told him that the work was in a very forward state and that there were hands enough to take up all the plans, he told me he was not aware of that and that he would not give me a decisive answer either affirmative or negative until he had found how the work stood. Well, I left the room disappointed, of course. During the day he came into the office and after looking over some of the work he came over to me and uncovered my plan. It is a very large one and I was just after commencing it. ‘There’ says he ‘how can you expect leave while you have this to draw.’ I told him that Evans could take up my plan. ‘No’ was the reply, ‘he has one of his own to attend to.’ I’ll work hard as I am able at the plan till I have it finished, and then I’ll apply again. The number of our draughtsmen has been reduced lately by a circumstance which you would not expect, and that is the dismissal of Mr Carroll.[8] He was drinking one evening last week at a village about half a mile from this; whiskey it appears makes him very quarrelsome; an extra quantity makes him a perfect madman. Well, a bickering commenced between him and one of the townsmen, it rose at length to such a pitch that they came to blows, a man rushed between them and as townsman had the advantage with respect to strength, he was secured, Carroll being thus set at liberty drew his penknife and stabbed his opponent in the arm. He escaped from the vicinity by a miracle, as he was waylaid on all sides by the sailors and had he been caught they would undoubtedly have killed him. This came to Mr Fenwick’s ears and he discharged Carroll on the spot, the fellow he stabbed I believe is as sorry as any of us for his dismissal, he freely forgave him tho’ the wound prevents him from working and will do so for some time. I have now stated everything. You see how things stand, and I’m sure the conclusion you must draw from these premises is that I am not guilty of the faithlessness attributed to me by my mother. I have relieved my mind of no small weight of anxiety

Good bye | Your affectionate son | John Tyndall

RI MS JT/1/10/3236–7
LT Transcript Only

1. *Sunday*: LT gives date ‘Aug. 15th, 1841’.
2. *your letter*: letter missing.
3. *my last* question: whether his father ‘may be in want of a little money’, which Tyndall could supply by sending savings set aside for his coach fare to Leighlin Bridge; see letter 0084.
4. *Last Christmas ... to go*: In a missing letter Tyndall would have explained to his father the reason why he was unable to travel to Leighlin Bridge the previous Christmas. Presumably Lieutenant Paul Whittingham had refused leave owing to pressure of work on the Survey. Tyndall instead took a short Christmas break; see letter 0031.
5. *5 o’clock ... the evening*: 5 o’clock in the morning until 8 o’clock in the evening.
6. *I spoke to Colder ... at home*: see letter 0064.
7. *finish Lt. Whittingham’s work*: Lieutenant Paul Whittingham had left the Survey on 11 July and been replaced by Lieutenant Robert Fenwick (NAI OS/1/18).
8. *the dismissal of Mr Carroll*: A civil assistant attached to the 5th Division, C District, James Carroll had been employed by the Survey since May 1838, principally as a writer. His dismissal is not recorded as the records for August 1841 are missing.

To John Tyndall, Snr 24 August 1841 0086

Kinsale, Aug. 24th, 1841.

My dear Father

This moment I'm after reading your letter.[1] I lose no time in sending off the second moiety.[2] I'm glad to hear that Mr Conwill intends writing to us as my correspondence is rather limited lately. The news you sent me respecting my mother's illness is a cause of sorrow and pleasure to me – sorrow to find that the old complaint has not yet forsaken her and joy to find that she has once more overcome it. May she always thus triumph over her enemies, tho' I believe the number to be very few. Father Matthew[3] was indeed a poor resort in such an emergency I'm afraid that even the magic wand or demoniac incantation of a Hohenlohe[4] would fail to affect the stubborn complaint.

I took a lovely row out to the mouth of the harbour yesterday evening. I was invited to the trip. You see I have made acquaintances in Kinsale tho' short I'm in it. There were six or seven of us together. We had a beautiful gig and four oars, and with these we made her skim the water in good style. We had a fine view of Charles Fort.[5] We ceased for a moment while it frowned on our track, to view the thundering tubes[6] that peer thro' its gaping portholes. We tugged on a little farther, the liquid plain was severed with the quickness of a sea bird, but soon the wound was healed and nothing marked our track save a line of foam that ever announces in calm water the furrow of a keel. Suddenly all eyes were directed towards the shore; between us and it an immense porpoise was rolling about. We pulled towards him and came quite close. Here we had an opportunity of viewing his antic[7] gambols. Sometimes he would pitch his huge body completely over water, and at others he'd rear himself and seem if I may compare great things with small to awkwardly imitate the gambols of a kitten. I have seen hundreds of them in Youghal harbour but I was never so near one of them before. It was just merely to fill the page that I have described this. A little narrative must supply the place of more important matter. I could however fill half a dozen sheets if I were to recount to you the many little incidents that have occurred even since I came here. I intended to go to see Bandon[8] one of these days. I was told the other day that Dean Bernard preached in a church about 3 miles from this a few Sundays ago. It's within a few minutes of 9 o'clock this is the hour the post office closes so I must conclude

Your affectionate son | John Tyndall

RI MS JT/1/10/3238
LT Transcript Only

---

1. *your letter*: letter missing.
2. *the second moiety*: 'the second halves of two thirty shillings notes'; see letter 0089.

3. *Father Matthew*: Theobald Mathew (1790–1856), a Capuchin priest known as the 'Apostle of Temperance', who delivered lectures throughout Ireland urging his audience to take the pledge of abstinence.
4. *Hohenlohe*: Prince Alexander Leopold Franz Emmerich of Hohenlohe-Waldenburg-Schillingsfürst (1794–1849), a German priest who reputedly effected a series of miracle cures in Ireland in 1823–4. See C. Connolly, 'Prince Hohenlohe's Miracles: Supernaturalism and the Irish Public Sphere', in D. Duff and C. Jones (eds), *Ireland, Scotland and the Romantic Aesthetic* (Lewisburg, PA: Bucknell University Press, 2007), pp. 236–57.
5. *Charles Fort*: the star-shaped fort that overlooks the mouth of the River Bandon and the entrance to Kinsale.
6. *thundering tubes*: cannon.
7. *antic*: bizarre (*OED*).
8. *Bandon*: a town on the River Bandon, about 7 miles north-west of Kinsale.

From John Conwill 4 September 1841 0087

Since independence is the thing
Which guides friend Tyndall's mind
He will not crouch to slave or king
Tho' to his desk he is confined
May independence urge the soul
Their proud Fenwick ne'er can bend
Fam'd Tyndall – tho' by fierce control
To his desk he is confined.
But Tyndall said the other day
That independence he resigned –
That – phantom like – it flit away –
And left him to his desk confin'd –
In logic too he shows his skill
Proves that matter ne'er can bind
The power of man's creative will –
Yet Tyndall's to his desk confined.

September 4th 1841.

'My dear Tyndall'

I was almost consumed in the flame of your asserted independence. I was almost annihilated by the electric fluid[1] of your logical deductions, excellent premises and inevitable conclusions, which you have so well produced in order to maintain that though 'the right of thought cannot be denied, the right of action may'. Did I say in my last letter,[2] my beloved disciple, that matter and mind can be acted on alike? If I did, I am a very idiot. But as well as I can remember I used two or three interrogatives about independence that I might bring the engines of your mighty mind to act upon that fourteen stone of unmanageable substance which has so decided to control your bodily movements. I feel great pleasure at your explanation of Mr Fenwick's hospitality to those placed under his superintendence.[3] So you tying Mr Fenwick by the legs on a pole is as feasible as for a mouse to tie a bell about a cat's neck – well to be sure, the poor mouse dare not come within the precincts of Mr puss's hemisphere, nor can poor Tyndall look upon rattle-snake Fenwick, without being rivetted to his desk. For heaven's sake spare your poor old teacher – do not bring the artillery of your Ethics to bear on him, pardon whatever inconsistencies may breathe through his epistles to you – because such may happen from his anxiety to behold you face to face.

If action and reaction continue the same,[4] you are like to remain stationary. I flatter myself that you will add such an impetus to Mr Fenwick's mind as will compel him to concede to you the privilege of coming to see a beloved mother whose only hope is centered in you. Nero has encountered Paul[5] but Paul had a firm supporter[6] who, by his ingenuity, rescued him from the fell fangs of his vile persecutor, and as his generous deliverer has, by a particular investigation, given him a longer day to live, I shall shed further luster round his path by a general illustration.

Assume [triangle] abc – erect gh –gd & *<word(s) excised>* to df join dh. Construct the *<word(s) excised>* FJD having JF and FID– supplement of half the abm– continue FI and make Ide– dIe. Lastly join eg then will de, ge, and Te be the three required lines.[7]

You may perceive that the question admits of as many solutions as there are points in the law. There is a nice little crambo[8] arising out of the above solution which may be put to Paul's Neroical annihilator – namely– 'required the locus of the vertices of all the triangles that will fulfill the conditions of the problem'. This is a stickler of my own, though marked with inverted comas; and to disclose to you the secrets, the locus is a hyperbola. A very big word an exterminator.

Your faithful teacher | John Conwill

RI MS JT/1/11/3521
LT Transcript Only

1. *the electric fluid*: Electricity was widely thought to be a subtle, elastic and weightless fluid.
2. *my last letter*: letter missing.
3. *Mr Fenwick's ... his superintendence*: Lieutenant Robert Fenwick's refusal to grant Tyndall leave to visit his parents; see letter 0085.
4. *If action ... the same*: Conwill evokes (in a loose sense) Newton's third law of motion, which states that action and reaction are equal and opposite.
5. *Nero has encountered Paul*: Although the emperor Nero, who reigned from 54 to 68, is said to have persecuted Christians, his role in the trial of the apostle Paul (Acts 25) is not made explicit in the New Testament. According to some later sources Paul was beheaded by the Romans at the instigation of Nero.
6. *a firm supporter*: this may refer to 2 Timothy 4:17, 'Notwithstanding the Lord stood with me, and strengthened me; that by me the preaching might be fully known, and that all the Gentiles might hear: and I was delivered out of the mouth of the lion.' This has

been interpreted as God having rescued Paul from Nero (Eusebius, *An Ecclesiastical History to the Year 324 of the Christian Era* (London: S. Bagster, 1838), p. 60).

7. *Assume ... required lines*: LT's typescript does not permit a viable reconstruction of Conwill's geometrical argument.
8. *crambo*: rhyme.

From James Davy [3] September 1841 0088

York, 3rd Sep. 1841.[1]

My Dear Tyndall,

I received yours[2] this morning and I must acknowledge that averse as I am to writing more than I can help yet your call for an answer is imperative and shall be complied with. So here's to begin I did not think that you held such a contemptable opinion of me as to think I should form an idea of ill or improper conduct on your part from any thing such a man as Linedale[3] might say – it would as the saying is 'go into one ear and out at the other', but I must be just withstanding, tho' Mr Linedale is a great rascal, he said not one word respecting you, good or bad as well as I remember, so let every one get his own as far as Mr Linedale goes, but know respecting who did say, Eh? you want to know who, <u>do you not wish you may get it</u>! Why my friend Tyndall, I never asked who said or spoke disrespectfully of me, but it was couched in such significant language that it was impossible for me to mistake. I might mistake, if I did I am sorry for it and glad to find that such is not the case, and had it been so it would not annoy me much, but to compromise any person is not a trait in my disposition and therefore once for all I shall be silent on the subject of name, and you need not ask is it so and so, my answer will be always the same <u>do you not wish you may get it</u>. Why I am almost laughing at the idea; give up a correspondent. Oh FY!![4] John Tyndall to ask it, I thought you knew better and to drop this subject once for all I would say that I endeavour to act in such a manner as to court the smiles of none or fear the frowns of all and that man is lifeless who is faultless.

I feel thankful to you for your kind expressions respecting me and mine and when you write to Leighlin Bridge give my best respects and also that of my Rib,[5] to Mr Tyndall senior and your mother. Tell John Parker that if he never writes to me until I write to him it will be <u>NEVER</u>. He has a vast stock of imprudence to say what he did on that head and I cannot but think you would do him some service if you gave him a lecture on that head, all I can say he is an ungrateful little varlet[6] and was it not for his father and mother why I do not think his conduct to me should cause me to trouble much about him. My Rib and John[7] send their best respects to you and I beg you will accept the same from

your sincere friend and well wisher | J. Davy

Give my best respects to Mr Evans and I do assure you I should be glad to see you both in this country.

Direct for Mr James Davy Clementhorpe,[8] York. England – leave out R.S. & M,[9] as soldiers are very little thought of here

RI MS JT/1/11/3556
LT Transcript Only

---

1. *3rd Sep. 1841*: The typescript of this letter bears the date '6th Sep. 1841', but since it must predate the next letter (0089), it has been assigned the nominal date of 3 September 1841.
2. *yours*: letter missing.
3. *such a man as Linedale*: Anthony Linedale had absconded from the Ordnance Survey; see letter 0062.
4. *FY!!*: presumably a variant spelling of fie, an exclamation expressing indignant reproach (*OED*).
5. *my Rib*: Davy's wife; Eve having been formed from Adam's rib (Genesis 2:21–2).
6. *varlet*: a person of a low, mean or knavish disposition (*OED*).
7. *John*: Davy's son John Davy.
8. *Clementhorpe*: a village on the outskirts of York, close to York Barracks.
9. *R.S. & M*: Royal Sappers and Miners.

To John Tyndall, Snr 5 September 1841 0089

Kinsale, | Saturday night | Sep. 5th, 1841.

My dear Father

Though you are now in my debt a letter still you see I am willing to place another item to the debt side of your account with me by writing this one. You never sent me an answer to my last letter[1] and I am anxious to know the fate of it as it contained the second halves of two thirty shillings notes. I send you a couple of enclosures[2] now of a different description; one is a bit of a lashing that a friend of mine in Youghal gave to an old priest there. I know the parties well and am acquainted with one or two of the 'forty apostles.'[3] The statement of the writer of the comment about some of them being quartered on the surrounding farmers is strictly true. When I was examining in the parish of Clonpriest[4] last summer twelvemonth, I had the happiness of meeting one of these embryo priests. He was sitting at my host's[5] fire one evening when I returned from my work. I was not long obtaining my usual seat on the hobstone[6] confronting the intended theologue.[7] After taking one or two glances askance at the gentleman and finding that

> Dark ignorance upon his front upreared
> His gloomy standard –

after finding that

> There was no speculation in those eyes
> Which he did glare with.[8] –

my eyes calmly resumed their former position, resting upon the dying embers of the fire. At length the silence was broken by the Monkish gentleman, who questioned me about the Survey. From this the conversation turned on schools. He was loud in the praise of father Foley's college;[9] there fair sciences smiled and the learned lore of antiquity was unfolded. 'I suppose' said I 'that you are intended for the church?' 'For the church or for the bar' was the pompous reply. I could scarcely smother a chuckle as I pictured to myself the vacant phiz[10] of my new acquaintance peering from beneath a barrister's wig – You have read a good deal of science? said I – Yes – You read Euclid[11] of course? Not all – You have then read his first six books – No!!! I asked him a few more questions the result of which proved to me that he hardly knew his multiplication table. Alas thought I you'll make a hopeful hand at turning a gospel sod.[12]

The other enclosure I send you is a part of a letter which I received from Corpl. Davey[13] the other day. You may see by it that I have been made the subject of calumny by the execrable cur Linedale. He was in York looking for employment on the English Survey. His mission proved fruitless, but while there it appears that he has been most industrious in vilifying me. He must have framed his story in a very plausable manner as Corpl. Davey has scarcely breathed a doubt of the accuracy of his statements. I could scarcely believe that human nature could sink to such a degree of superfined rascality – of sublime villany – as Linedale has exhibited. I have stood beside the bed of that wretch when uncared for by the world he lay rotten with disease and cooled his burning tongue, nay further when he was unable to procure for himself a solitary morsel and without the price even of his miserable bed, when in fact he was about to be consigned to the tender mercies of Youghal Main Street,[14] I was the first to come forward and obtain from a chosen few in the office the means of rendering him comfortable until he recovered. This then is my traducer, this the scoundrel who has dared to brand <u>my</u> character with infamy. I have no doubt but he has cast the dark venom of his vile tongue on me in Bagnalstown also. Those who wish to place credence on his foul allegations may do so. Conscious of my own integrity I can smile on the infamous treachery of this monster of a man. I now drop the fellow in the tomb of oblivion shrouded with unutterable contempt.

I have no less than five persons in my debt at present for letters so that I now may rest awhile until some of my correspondents liquidate my claim. How are mother and Emma has that villanious complaint left my mother's cheek yet?[15] If I was near her I'd kiss it away.

Good bye | Your affectionate son | John Tyndall.

Corpl. and Mrs Davey desired to be remembered to you.

RI MS JT/1/10/3239–40
LT Transcript Only

---

1. *my last letter*: letter 0086.
2. *a couple of enclosures*: enclosures missing.
3. *forty apostles*: It had recently been announced that 'about 40 young missionaries are ardently prosecuting their sacred studies at St Mary's' Catholic College, Youghal (*London Standard*, 23 July 1841, p. 2).
4. *the parish of Clonpriest*: about 4 miles south-west of Youghal.
5. *my host's*: host not identified.
6. *hobstone*: a level stone grate on which things may be set to warm (*OED*).
7. *theologue*: theology student (*OED*).
8. *There was ... did glare with*: cf. W. Shakespeare, *Macbeth*, III.iv.100–1.
9. *father Foley's college*:, St Mary's Catholic College at Youghal had been founded by Father John Foley in 1839.
10. *phiz*: face (*OED*).
11. *You read Euclid*: see letter 0017, n. 3.
12. *turning a gospel sod*: working on the Gospels; thus, perhaps, Bible studies.
13. *a letter ... Corpl. Davey*: letter 0088.
14. *the tender ... Main Street*: probably refers to the populace of Youghal, the main street of which was 'more than a mile in length, with numerous streets branching off it' (James Fraser, *A Hand Book for Travellers in Ireland* (Dublin: W. Curry, 1844), p. 236).
15. *has that villanious ... cheek yet?*: The 'villanious complaint' may refer to Tyndall's mother's illness discussed in letter 0086 or possibly to the charge of 'faithlessness' she had levelled against him (see letter 0085).

From John Tyndall, Snr 6 September 1841 0090

Mr John Tyndall | Ordinance Survey Office | Kinsale | Sep. 6th. 1841.

<*text missing*> that the base brutal and bloody whigs has been kicked out of office and Peel at the head of the torys now holding the reigns of Government,[1] and as soon as everything is settled I will use all my exertions to get you a situation and I think it will go hard with me if I am not victorious as I have some weighty metal to stand in the gap. Mr Conwill was here on Saturday evening last and read your letter[2] which was <*text missing*> give up such exercise for every paper I read gives an account of people being drowned by the boat upsetting[3] and though I have not written to you sooner still my mind is very uneasy since I heard of it.

The people here are outwardly peaceable but the inward man I think much agitated. Exclusive dealing[4] is still carried on to a great extent. Very few Romans[5] come into my house and no shoemaker but Lorry Kavanagh.[6] Captain Woodcock the agitator is dead and buried and would not let a papist Priest near him during his illness though Priest McMahon remained in John Cummins house[7] for five days expecting to be called for. But nothing could bend Woodcock, he ordered his sister Cummins's wife[8] to bring his Bible and prayer book. She did so he place them under his head and told her that it was in their contents he believed and that no Priest should meddle with him, that though he joined them in politics he did not in religion. Mrs Groves[9] is very ill with fever and I am afraid but bad hopes of her recovery. We are all well and hope you are well also

I am your affectionate father | John Tyndall

Remember me to Phil.[10]

RI MS JT/1/10/3241
LT Transcript Only

---

1. *whigs ... reigns of Government*: The Whig administration collapsed on 30 August 1841 with the resignation of Lord Melbourne and the new Conservative government under Robert Peel was installed on 3 September.
2. *your letter*: possibly letter 0086.
3. *an account ... boat upsetting*: possibly the sinking on 9 August of the *Erie* on Lake Erie with the loss of 170 lives; reported in *Cork Examiner*, 1 September 1841, p. 2 and other newspapers.
4. *Exclusive dealing*: see letter 0073, n. 10.
5. *Romans*: Roman Catholics.
6. *Lorry Kavanagh*: *Slater's* lists Jeremiah Kavenah as a shoemaker.
7. *Captain Woodcock ... Cummins house*: Captain Woodcock participated in the agitation mentioned in letter 0066 and may have been Thomas Woodcock of Coolnakisha. John Cummins (*c.*1787–1867), farmer, lived at Coolnakisha.
8. *his sister Cummins's wife*: not identified.
9. *Mrs Groves*: not identified.
10. Phil: Phillip Evans.

From Eliza 8 September 1841 0091

Youghal, Sep. 8th. 1841.

My dear Mr Tyndall

I[1] have to return a thousand thanks for your very kind letter[2] which I received on Sunday evening. I need not say it gave me infinite pleasure to hear you were well and had succeeded in refuting the calumny of that hateful Linedale.[3] I was not surprised at what you related of him as I was aware from the first evening of my acquaintance with him of the excessive dislike he had of you – he tried to turn you into the greatest ridicule, besides representing you as most treacherous and insincere. You may be sure that he did not rise very high in my good opinion, when I heard him descanting at such a rate on the character of one whose part in my opinion he should rather have taken. As for your character, it would be well for him if (while in Youghal, even, he could say) his own was half as blameless. I am sure you will hear a little more of his treachery from Mr Roberts, I suspect he was most assiduous in making mischief in that quarter too, I told Mr Ginty and Jim[4] all I heard on that subject a little before they left Youghal. His character is so despicable that, it should (as you have remarked in your letter) 'be buried in oblivion'.

So Kinsale is not improving since I heard last from your friends there, and by way of comfort I must add I fear it will be insufferable in winter. I was sorry to hear Messrs Ginty, Evans and Tidmarsh had nothing better to employ their thoughts than reflecting on by gone days, tell them from me, they ought to be ashamed to own themselves possessed of so little philosophy as to be dejected by so trivial a circumstance as having to live in Kinsale for a few months – it would be time enough to be cast down when the cares of the world should seize on them. I think you are the only sensible person among them in amusing yourself with every trifle that comes in your way. I suppose you study French still and are by this a great proficient. I hope you will soon be ordered to England as I really pity you for being immured in that horrid place.[5]

We are quite gay here, Collins has arrived and his company[6] I believe. They were here when you first came, I was there one evening and was quite pleased with the performance. Have you any sort of amusement in Kinsale? I heard the greater number of the inhabitants were Methodists,[7] so I dare say they would think such a thing as a theatre very wicked. Pilkington is still in Youghal[8] waiting he says for a letter from his family. I don't know how he contrives to live. Miss Aubrey[9] told me he was in Cork last week with the Lieutenant for a recommendation, it seems Mr Whitingham promised him a discharge, but forgot to leave it, he said he was successful in Cork. I have not spoken to him since the week after you left this, indeed I go out very little to walk, he is not a very great friend of yours but as little as Linedale, but at the same time I dont think him so treacherous. I fancy all his dislike proceeds from envy. I will not say why I think so, lest you may think I was learning to flatter, and of all things in this world I consider flattery the meanest propensity. Je pense que c'est aussi votre opinion.[10] I had a melancholy account from Jim of your trip to the Killarney,[11] and Mr Ginty mentioned that he was much amused with the cool manner in which Jim seemed to wait his doom, how did he look? I suppose most interesting. I am sure my dear Mr Tyndall you are at this time quite tired with such a stupid epistle but as you have promised to become a correspondent of mine, I must try and summon up a little more variety when next I write – to tell you the truth I have so little of my time at my own command that I can scarcely write six lines when I am called away, and then I quite forget what I should write. I feel certain with your accustomed good nature you will excuse all deficiencies. The dancing school will open for the winter on Monday next, I was invited to attend but have declined as I fear the society would be rather too refined for me, the same pupils they had last year are all to commence again, what a set they were. My sister unites with me in kindest remembrances to Messrs G E and T[12] and with most sincere wishes for your happiness and welfare believe me

always dear Mr Tyndall | Your sincere friend | Eliza

I quite forgot to say, I hope your head is now quite well and that I will anxiously expect an answer. I have not time to read over this horrid scrawl as Mrs Bagge[13] is calling me to prepare for a drive to Ballynatray[14] and I would much rather stay at home this evening. I will send you a paper this week.

RI MS JT/1/11/3891
LT Transcript Only

1. *Eliza*: Eliza's identity has not been traced but she is probably also the author of letter 0099, which is signed 'A Bessie'. LT added '? Pouchea' in pencil at the end of the letter. Eliza was reasonably well educated and was employed by a Mr and Mrs Bagge. (*Slater's* lists an attorney named Henry Bagge of Dungarvan). She may be the governess mentioned in letter 0082.
2. *your very kind letter*: letter missing.
3. *the calumny of that hateful Linedale*: see letters 0088 and 0089.
4. *Jim*: Phillip Evans.
5. *that horrid place*: contemporary travel books do not portray Kinsale in this way. Hence Eliza's dismissal of Kinsale may arise from her personal bias in favour of Youghal.
6. *Collins has arrived and his company*: not identified, but probably a theatrical company.
7. *the greater number ... Methodists*: There were two Methodist meeting houses in Kinsale.
8. *Pilkington is still in Youghal*: Previously a civil assistant, Macdonald Pilkington had been imprisoned in June 1841 following a fight (see letter 0064). He was later transferred to the English Survey on 1 January 1842.
9. *Miss Aubrey*: not identified.
10. *Je pense ... votre opinion*: I think this is also your opinion (French).
11. *your trip to the Killarney*: Tyndall went with Evans to view the half-submerged wreck of the steamship *Killarney* (see letter 0049, n. 1) at Barry's Head, east of Kinsale, on 8 August 1841. He later made a handwritten narrative of the perilous trip (RI MS JT/8/1/2, pp. 60–3); see also J. Tidmarsh, 'Reminiscences of John Tyndall', *Southern Australian Register*, 24 January 1894, p. 6.
12. *G E and T*: William Ginty, Phillip Evans and John Tidmarsh.
13. *Mrs Bagge*: Eliza's employer.
14. *Ballynatray*: The Ballynatray House estate is about 3 miles north of Youghal on the River Blackwater.

From John Tyndall, Snr 10 September 1841 0092

Mr John Tyndall | Ordnance Survey Office | Kinsale
Leighlin Bridge | Sep. 10th, 1841.

My dear John

I received a second letter from you[1] this moment after it passed through Stradbally in the Queens County. I am glad to perceive from your letter that there is some mistake with regard to the calumny said to be heaped on you by Linedale[2] I went to Bagenalstown with your letter and read that portion of it that contained the supposed slander to the Payne family. They said that though believing him to be a blackguard still his character of you then was noble, he said that the first gentleman in the land need not be ashamed to acknowledge you for his son, and that if any thing bad of you was said by him to Corporal Davey, it was Anne McGee[3] that prompt him to it as you seemed to be opposed to their marriage as indeed all her friends in Bagenalstown were.

That piece you sent me against popery[4] was excellent, he was an hardy brat that wrote it. I have White's reply to the Jesuit Fisher[5] – a discussion that took place before King James the first[6] – it is a first rate work but too large to send you, having cost a pound when reprinted in 1818. I am glad to let you know that Mrs Groves is better and good hopes of her recovery

I am your affectionate father | John Tyndall

I could not read the whole of Corporal Davey's scrawl myself, or indeed make sense of what I did read.

RI MS JT/1/10/3242
LT Transcript Only

1. *a second letter from you*: probably letter 0089.
2. *the calumny ... by Linedale*: see letter 0089.
3. *Anne McGee*: who had married Anthony Linedale on 29 May 1841.
4. *That piece ... against popery*: not identified.
5. *White's reply to the Jesuit Fisher*: Francis White (1563/4–1638), *A Reply to Jesuit Fisher's Answer to Certain Questions Propounded by His Most Gracious Majesty King James the First*, which was first published in 1624. This may have been a reprinting of the 1816 Dublin edition. John Fisher (1569–1641) was an English Jesuit priest.
6. *King James the first*: James VI and I (1566–1625), King of Scotland (as James VI) from 1567 and King of England and Ireland (as James I) from 1603.

To John Tyndall, Snr 16 September 1841 0093

Kinsale | Thursday evening | Sep. 16th, 1841.

My dear Father

The intelligence that your last letter[1] conveyed gratified me much, I'm glad to find that Linedale was not so bad as he was represented to be, however I have other proofs to shew that he is bad enough.

I was looking over an old Sentinel[2] yesterday that Phil[3] got from home, in it I found that the Dean[4] had just returned to Leighlin from Bandon. I intended to go to Bandon a month ago and had I known that the Dean was there I certainly would have put my project in execution; it would have gratified me not a little to have seen him. Bandon is only nine Irish miles[5] from this.

All our field men are idle at present, they have finished their portion of the Survey. They will leave this for Cork on Wednesday next and on the next day will embark for England. Colder our present Superintendent is going along with them. I'm sorry he's going as I never met a fairer man. Another party will start in November. I expect to be kept till about New Year's Day, and if the Island of Saints[6] be not then willing to provide for me, I must only try my fortune among the English beef eaters. 'Next to mouth next to mind' says the proverb – I believe it's so with me at present, as Evans gave me a beef steak for dinner to day and I believe its uppermost in my stomach and therefore nearest my mouth. Evans is our messman[7] here. He provides everything; I dont know what I'm to get for dinner until its laid on the table and not a word of grumbling is allowed. I for my part seldom feel inclined to grumble as I am able to do my duty on whatever comes before me from a herring to a whale!!!

I am going to a drawing school at present after hours. I commenced yesterday evening. We attend three nights in the week I am sure it will be of great service to me should I follow this line of life. There is but another alternative and that is the revenue.[8] I dont care much which my lot falls upon. Should I fail in procuring a situation in the latter, I wont feel much disappointed, as I am pretty good at my present business. There is very little in it a secret to me. I think that I have made as great improvement as any of my comrades for the time I have been on.

Is Mr McGee in Leighlin at present? You never sent me word how he was this long time. How is my uncle Caleb getting on? Do the Rads[9] shew their spleen much towards him? He ought to be very careful for himself, as they would think very little of knocking in a bit of his scull. I have time to say no more as I must be off to stitch my coat – the sleeve of it is torn. Necessity has made a tailor of me. I wish I was near my mother or Emma and I'd accommodate one or the other with a job.

Good bye | Your affectionate son | Jack Snooks.[10]

RI MS JT/1/10/3243
LT Transcript Only

1. *your last letter*: letter 0092.
2. *an old Sentinel*: *CS*, 4 September 1841, p. [3].
3. *Phil*: Phillip Evans.
4. *the Dean*: Dean Richard Barnard.
5. *nine Irish miles*: As 1 Irish mile = 1.27 English (statute) miles (2240, rather than 1760 yards), nine Irish miles would be approximately 11.5 statute miles.
6. *Island of Saints*: Ireland.
7. *messman*: the person responsible for cooking and serving food to a company of people eating together (*OED*).
8. *the revenue*: the Inland Revenue.
9. *Rads*: the Radicals, who might wreak revenge on Caleb Tyndall for having confronted the Repeal agitators and wounded Mary McAssey; see letter 0071, n. 6.
10. *Jack Snooks*: Walter and Jack Snooks were among Tyndall's pseudonyms.

From James Davy 16 September 1841 0094

York 16th Sep. 1841.

My dear friend,

I am sorry that you think so much of the remarks I made to you after what I said in my last[1] I said that I thought I could not mistake the person but it appears I did as I do not think for a moment that you would pen a falsehood I ever had and still hold a much better opinion of Mr Tyndall and I do feel thankful to you and Mr Evans for your kind defence of my conduct when I might be spoken ill of not that I value the opinion of many in the old 5 Division[2] yet permit me to say that I do not think my successors would please every person more than I did, I was very sorry to hear that Hugh Murphy was discharged for dishonesty[3] – this I do believe was a foul charge against him which every action of his life as far as I knew him give the lie to, and I would have considered it my duty if I could have been of any use to him to say that I never knew a man in his place or station conduct himself better or more honestly than Hugh Murphy.

You have been so kind as to tell me that you value my good opinion, for which I thank you, but permit me to say that actions speak louder than words and I would request you to lay aside any bad feeling that you may have formed against Corp. Griffin. I know that he is a curious man in his way of thinking but I cannot for a moment think him a bad man. I do not like to say that he did not write some remarks respecting the Division or not because was I to do so you would form an opinion of some other person. This much I will say that several persons in the Division have wrote to me since I have been in England and every letter has had remarks of what this man said and what another said about me, and as no one person ever considered that I would mention their names, so I think I would be doing an improper act to do so. This I will say, that I do not think you have any guess of the person who drew such a picture that I concluded it was you and nothing but your denial of it would convince me that I was mistaken. Now I hope this will suffice. Believe me when I say I shall at all times be glad to hear from you and should be glad, very glad, to find that those great talents which the giver of all good has blessed you with would be brought to bear on your chief good and that they may never raise in judgment against you – you might be able to do great things remember this, to glorify your God below and find your way to heaven is the Chief End of your existence and I would pray that God might enable you to see what is your duty and what would eventually be your temporal and eternal happiness. Give my best respects to Mr Evans and accept the same from

Your sincere friend and wellwisher | J. Davy

Mrs Davy and John[4] sends their best respects to you.

RI MS JT/1/11/3557
LT Transcript Only

1. *my last*: letter 0088.
2. old 5 Division: While in Ireland James Davey had served in the 5th Division, C District – the division for which Tyndall worked – now renamed the 4th Division.
3. *Hugh Murphy was discharged for dishonesty*: possibly the Thomas Murphy who served in the 5th Division from May 1840 to June 1841, when he was discharged (NAI OS/1/18).
4. *John*: John Davy, James Davy's son.

From John Tyndall, Snr 18 September 1841 0095

Mr John Tyndall | Ordinance Survey Office | Kinsale
Leighlin Bridge | Sep. 18th. 1841.

My dear John

I received a letter from you[1] this morning and am glad to hear that Linedale's conduct was not so bad as was represented and, tho' at the same time not so innocent as to cause an acquittal. If he has said anything against you, blame Anne McGhee, for it is she was angry with you for writing anything against Linedale's being married to her.[2] I am happy to find that your appetite is so good that you can swallow a whale with as great ease as you would a herring. This is stomach extraordinary! Mr Conwill would be glad to have the like at present, but he has not as he lies very ill with a cold and not able to write to you. I am very glad to know that you will take my advice with regard to your boating excursions as such a practice is dangerous. With regard to what the County Surveyor has yearly, as you expressed a wish to know in your letter to Mr Conwill,[3] he has they say a hundred a year and his deputies from fifty to sixty each. If we could get for you a situation in the revenue it would be much better and after some time there will be exertions made for you. At present the Government is only in its infancy[4] – I mean in point of time – for in other respects it is the strongest, O'Connell says, ever formed.

That drawing school you have commenced with will be of great service to you and you do well to let nothing pass that will have a tendency to your improvement. I have now to ask you what did that ill looking head and face represent that you placed on your letter it represents some person with a tremendious big nose. John McGee is in Leighlin, his wife[5] is teaching school. They are badly off. He was at this house last night for a long time speaking to your mother and myself.

Your uncle Caleb is well, but there was a cow stabbed or shot on his land on Thursday last. It did not belong to him, but was sent there to graze by some person in Bagenalstown, as it was on the farm he took from Tom Wall[6] the deed was done. He has great need to mind himself and so has your father. But I think for my part that I will take care of myself as good people are scarce. Patt Moran[7] is pronounced past recovery in a fever and its not expected he will pass this night. We are all well, I will now conclude.

Your affectionate father | John Tyndall

RI MS JT/1/10/3244
LT Transcript Only

---

1. *a letter from you*: letter 0093.
2. *Linedale's being married to her*: Anne M'Gee and Anthony Linedale had married on 29 May 1841.
3. *your letter to Mr Conwill*: letter missing.
4. *the Government ... its infancy*: Robert Peel's Conservative administration had commenced on 30 August 1841.
5. *his wife*: Helen M'Gee.
6. *Tom Wall*: Tom Wall had left Ireland for Tasmania– see letter 0015.
7. *Patt Moran*: Patrick Moran, the father of Patrick Francis Moran (who later became Catholic Archbishop of Sydney and was proclaimed Cardinal in 1885; see letter 0130, n. 2), died 18 September 1841.

To John Tyndall, Snr 23 September 1841 0096

Kinsale, Thursday[1]

My dear Father

As there is a slackness in the work here I deem it no crime to devote a little of that time which in busy times would have been devoted to the queen[2] in writing to you. The state of the office to day is strikingly contrasted with its state yesterday. All was bustle then; all is silence now. Our Surveyors mustered here on yesterday morning and after a final settlement set out for Cork. They will embark for England on tomorrow morning. Their destination is Skipton[3] – a good sized town on the borders of Lancashire and Yorkshire. Another draft will be sent over in November. We are all to remove to Cork then, and a few draftsmen only are to remain to finish up the work. I expect to be among the latter number. They were talking in Mr Fenwick's office the other day about those who were likely to remain. I was named among the number. This I consider rather flattering, as none but the best men in the other divisions were mentioned, and it shewed that they had a pretty good opinion of me when they were pleased to class me among them. I agree with you in thinking that a situation in the Revenue[4] would be the best that could be procured. I think it would be best not to be in any hurry about making an application. It would be best to wait until everything will be settled. However you will be the best judge how and when to act. Wait till you find an opportunity that will suit. I think you will be very apt to be successful; you have the interest of the Captain, the Dean,[5] and Mr Bredin. If you could procure that of Mr John Alexander it would be better than all. Take your time. Proceed slowly and surely, and I think in the end we will succeed. If not, however, I will not be in the least cast down. I can only try my fortune in another way. Often I feel my heart swell with gratitude towards you and my mother for placing me in a position that will enable me to take up almost any situation – I mean by my education. 'tis true I know but little comparatively, but what I do know enables me to make myself generally useful. It would be the highest pleasure of my life to be able in some measure to repay you – to be placed in a position where I might be able to prove to you that I fully appreciate all you have done for me. For myself I care but little but it would be the darling object of my soul to see you independent.

Remember me to my uncle Caleb and all my friends, I believe I have but few foes in Leighlin.

Your affectionate son | John Tyndall

RI MS JT/1/10/3245
LT Transcript Only

---

1. *Thursday*: LT gives postmark as 'Sep. 24th, 1841', which was a Friday.
2. *time ... devoted to the queen*: Tyndall's activities working for the Ordnance Survey, which was initiated by Her Majesty's Government in response to the *Report from the Select Committee on the Survey and Valuation of Ireland* (1824).
3. *Our Surveyors ... Skipton*: Skipton is a market town in the West Riding of Yorkshire; the surveyors sent there included Thomas Charles Higginson; see letter 0097.
4. *the Revenue*: the Inland Revenue; see letter 0093.
5. *the Captain, the Dean*: Captain William Steuart and Dean Richard Barnard.

From Thomas Charles Higginson1 October 1841 0097

Gisburne,[1] Oct. 1st. 1841.

My dear Tyndall

How can I appologize for not keeping my promise I cannot say in your own mind if he could possibly have written he would the only way left me now to regain that esteem in your mind which I have forfeited is by recounting my adventures since we parted, but I am afraid I have not the ability to do so, it would require the genius of a Tyndall. It was with a sorrowful countenance I parted from my dear friends which was diminished nothing in my journey for the day been wet and I muffled up to the eyes I could do nothing but think of my past happiness which I am afraid will be some time in returning. We arrived in Cork after getting a good wetting at 2 o'clock. Made the best of my way to some place where I might rest my limbs and dry my unfortunate self. Oh tis well for you sitting on your stool caring for neither wind or weather. on Thursday[2] I made myself look a little decent and determined to make out a certain person if possible. On my way I met Maryanne Kearney,[3] to my grief I learned that poor

Mrs Foy was gone to be bourne from whence no traveller returnes.[4] Maryanne said a great deal against Foy's good nature to Mrs Foy, she said it was a sorrowful day they met him, but what changes there are in this life – after parting with her my sorrow was changed to joy – guess who I met Miss – Miss D.[5] We walked up Patrick Street, she asked for – I need not tell you who the lad is. I will leave you to guess, you know who sympathised with her parting from Youghal, ay better than I can tell you. After walking and talking for some time we parted to meet no more – we met – we have parted – the pang is now o'er, yet I feel broken-hearted to meet her no more. On Friday morning at 10 o'clock we bade farewell to Cork's own town – the day was beautiful and the sun shone with all its brightness on the shores of our own dear country, making us curse our stars that we had ever left it. When we were going out at the mouth of the harbour, the waves began to rise and of course the vessel. At the first heave Colder threw his accounts ever-board.[6] I braved the billows laughing at them all until we were about 50 miles on our way, when a stream might be seen issuing out of my mouth about three yards long. When that was over I was quite well for some time but all was over with me before night came on. It was a deck passage[7] we had, but any person that wished might go between decks, there was no second cabin, I preferred to stay on deck. I sat between three or four women with an umbrella over our heads all night, which was beautiful, scarcely as much wind as would fill the canvass. We arrived in Liverpool on Saturday at half past three in the evening the view of Liverpool coming up the harbour is splendid. After getting my dinner I walked through the town seeking some friends I have there, which happily I found. I was shown all the places which was worth seeing – at night the markets are beautiful, regular promenades, but I must return to Cork. The first night I was in Cork, Ferris[8] and I went to the play to hear Miss Romer and Mr Wilson singing[9] , they are first rate. On Sunday I spent the day out, dined at 1 o'clock and then went to see every place in the town. I was made a present of some sketches of the principle places in the town. I went to church in the evening; while praying very devoutly with my eyes fixed on a nice girl I was disturbed by the noise of the steam coach[10] going under me, there is a tunnel from the station in Lime Street to Edge Hill one mile and half long.[11] On Monday at 2 o'clock we started from Liverpool to Manchester[12] 30 miles through a beautiful country the fare is 2s. 9d. I had time to go see my uncle in Manchester he gave me Comstock's Natural Philosophy[13] The same day we started for Hibden Bridge[14] the fare is 2s. 3d. About half way we went through a tunnel 3 miles long[15] I thought I would never get out of it it is a curious thing that I scarcely saw any red fields all green which makes it very pleasant for us. Hibden Bridge is a nice village, I slept in it one night in a very good inn which would be called an hotel in Ireland, and all I was charged was 3 pence for my bed. Every house is neat and clean. On Tuesday I went to Skipton, 21 miles, through Keighly which is half way the country all round was very hilly,

Lieut. James[16] has taken an office in Skipton I am changed from your Division to Lieut. Stace's, Moran and Early[17] is with me. I applied to get back to my own Division, but could not, as Mr Stace's was so small. I think Colder will do some thing for me yet, Skipton is a nice town – the Earl of Thanet has a beautiful castle at the end of the town. The town is full of Inns which are the only places for a person to stop in. It is also very cheap, I could get my diet there for 9s. 6d. per week Lieut. James has his own party in Skipton, Mr Fenwick is in Grassington, 11 miles north of Skipton, and we are 11 miles west – the country about this town is very thick with trees. There is no word about the big pay yet. I think it will begin from the first of this month. I have no more to say about the journey. I hope Tidmarsh did not injure any of the Yankees opposite,[18] or that Ginty did not hurt their bellies with his stick. Poor things how they flutter. Now Tyndall take my advice and don't go too far with the joke with a certain person down Street, you know I heard enough about it before I left you. Remember me affectionately to Ginty, Jim, Tid,[19] Cuddy, Mr Chadwick the last not least George[20] tell him I will write to him in a day or two

Your ever attached friend | Harold[21]

RI MS JT/1/11/3730
LT Transcript Only

---

1. *Gisburne*: The village of Gisburn (Lancashire) is 11 miles west of Skipton (North Yorkshire).
2. *Thursday*: 25 September 1841.
3. *Maryanne Kearney*: not identified.
4. *Mrs Foy ... returnes*: death of the wife of Thomas M. Foy.
5. *Miss D*: not identified.
6. *Colder threw his accounts everboard*: Alexander Calder was sea-sick.
7. *deck passage*: overnight passage without cabin accommodation.
8. *Ferris*: Thomas Ferris was a civil assistant who undertook surveyor working in the 5th Division of C District.
9. *Miss Romer and Mr Wilson singing*: The renowned soprano Emma Romer (1814–68) and Wilson – probably the Scottish singer John Wilson (1800–49) – were appearing at the Theatre Royal, Cork (*Cork Examiner*, 24 September 1841, p. [3]).
10. *steam coach*: Although a number of steam-powered coaches had been developed in the early nineteenth century, Higginson is referring to a railway train.
11. *a tunnel ... half long*: a railway tunnel on the Liverpool and Manchester Railway opened in 1836 in which carriages, after being uncoupled from their engines, were lowered by gravity to Lime Street station in central Liverpool and raised by a rope from a stationary engine to Edge Hill station in the southeast of the city.
12. *from Liverpool to Manchester*: on the Liverpool and Manchester Railway, opened in 1830 as the world's first passenger line.
13. *Comstock's Natural Philosophy*: *A System of Natural Philosophy: in which the Elements of that Science are Familiarly Explained and Adapted to the Comprehension of Young Persons* by the American physician John Lee Comstock (1789–1858). This was probably the

revised edition published in London by Scott, Webster, and Geary in 1839.

14. *Hibden Bridge*: Hebden Bridge in the West Riding of Yorkshire is about 22 miles north-east of Manchester. Higginson would have travelled from Manchester to Hebden Bridge on the Manchester and Leeds Railway, which had begun with a line from Manchester to Normanton in 1836, and opened the next section, from Normanton to Hebden Bridge in October 1840, only a year before Higginson's journey.
15. *a tunnel 3 miles long*: The Summit Tunnel near Littleborough and Todmorden on the Manchester and Leeds railway was 3080 yards long – about a mile and two-thirds.
16. *Lieut. James*: Henry James (1803–77) was 1st Lieutenant in the Royal Engineers in charge of the 1st Division, C District. With the support of his influential father-in-law Major-General Edward Matson, James became Superintendent of the Ordnance Survey in 1854 and was knighted in 1860.
17. *from your Division ... Early*: Higginson was working under William Crawley Stace (1806–68), who was in charge of 6th Division, C District. Having joined the army in 1827, Stace worked for the Ordnance Survey in Ireland from 16 May 1836 to 17 January 1842, was based principally in Mallow, County Cork, and later saw service in Yorkshire, Gibraltar, the Cape of Good Hope and St Helena. After he left the Survey Major Waters ordered his successor (Lieutenant Charles Wilkinson) to re-level 'a great portion' of Stace's work (NAI OS/1/17–19). Thomas Moran and John Early had both previously worked in Ireland as civil assistants with surveying teams in the 5th Division, C District, but had been transferred to the English Survey.
18. *the Yankees opposite*: not identified.
19. *Jim, Tid*: Phillip Evans and John Tidmarsh.
20. *George*: probably George Holmes.
21. *Harold*: a nickname of Thomas Higginson's.

From John Conwill 5 October 1841 0098

October 5th 1841.

Had I a poet's, painter's skill
Or a strong creative will
I'd write in polished lays
Of Tyndall and of old Kinsale
Of Leighlin's sons and Donerale[1]
Well known in modern days.

'My dear Tyndall'

I have once more scared away the king of terrors[2] after having laboured hard during fourteen or fifteen day's severe illness. As I am very incapable of writing a long letter I must now turn my thoughts another way and send you something on grammar and a few selections from two letters written by Tony O'Neill's son[3] on the Repeal of the Union.[4] You may judge for yourself on the grammatical acumen displayed by the classical dolt of Ln.[5] On grammar – 'While over all hangs the rich purple eve', 'As conscious of its being her last farewell',[6] – In what case is 'farewell' in this sentence?[7]

So sung
'The glorious train ascending: he through heaven,
That opened wide her blazing portals, led
To God's eternal house direct the way.'[8]
What does 'led' govern in this and on what does 'way' depend?
'The free fair homes of England!
Long, long in hut and hall
May hearts of native proof be rear'd
To guard each hallow'd wall.'[9]

In what case are fair homes & <*word(s) missing*>. Now to O'Neill's letter[10] – 'On reading a few days back &c' 'But for an Irishman, for a Catholic, too, to endure and support such principles, would be to cast an indelible stigma upon the most grateful, the most generous, and most desirous, too, of Ireland's future bliss.' Mind this: 'The writer asks what would be justice to Ireland. Fellow-countrymen mind the answer which it gives – shudder at the thoughts of it. Ponder it, tell it to your &c, and when you have it instilled &c, teach them to remember it.'

'To be silent on the present most important period that has ever been read in the annals of Irish History, and to regret little which has been long and often ardently sought by Ireland's sons, to be imprinted on the Irish character, & having now fixed that distinguishing mark upon every true Irish man's heart, let its indelible stamp pass on to the future men of Ireland.' 'To refuse to give utterance to the throes which swell my heart when I touch at all the subject of repeal, would be to choke the thrill which pervades the veins and hearts of the Irish people.' Here is young O'Neill of classic notoriety for you!

It appears I made a mistake in defending Paul[11] – however it may be corrected thus:– Assume any abc erect gh gd and to join dh construct the [yld] having ly e dh & [yld] supplement of half the cmb; the cbm being the supplement of the assumed abc,[12] I presume to day that this will satisfy you as to Paul's defence.

Not being able to write more | I remain your ever sincere teacher | J. Conwill

RI MS JT/1/11/3522
LT Transcript Only

---

1. *Donerale*: Doneraile, a town in the north of County Cork.
2. *the king of terrors*: a synonym for death in Job 18:14.
3. *Tony O'Neill's son*: The name 'Tony O'Neill' is probably incorrect since in letter 0101, the author is named as the son of Terence O'Neill, the proprietor of the Swan Inn, Leighlin Bridge. The son's name is not known.
4. *Repeal of the Union*: see letter 0051, n. 9.
5. *the classical dolt of Ln*: the son of Tony/Terence O'Neill; Ln is abbreviation for Leighlin.
6. *'While over ... last farewell'*: Neither of these statements is a true sentence. The initial words 'While' and 'As' normally function as conjunctions in a subordinate clause but here no main clause exists.

7. *In what ... this sentence?*: Although the word 'farewell' functions as the complement of the subject 'its', the addition of the word 'of' raises the possibility that the whole participle clause might be considered to be in the genitive case.
8. *So sung ... the way*: J. Milton, *Paradise Lost* (1667), VII.573–6.
9. *The free fair homes ... hallow'd wall*: F. Hemans, 'The Homes of England' (1828), 25–8.
10. *O'Neill's letter*: not identified.
11. *a mistake in defending Paul*: see letter 0087.
12. *Assume any ... assumed abc*: LT's typescript does not permit a viable reconstruction of Conwill's geometrical argument. Cf. text of letter 0087.

From A. Bessie 5 October 1841 0099

Youghal, Tuesday night| Oct. 5th. 1841.[1]

My dear Mr Tyndall

I could scarcely refrain from laughing at the idea of your expecting a lecture from me for writing a letter on the Sabbath day.[2] If you were serious I fear you have a much better opinion of my religious principles than is deserved. I agree with you in thinking it would be rather a hard case, circumstanced as you are, if writing a few friendly lines could be accounted a breach of the Sabbath, in that case you have only acted as I would have done myself, therefore I cannot condemn you.

The next point we have to settle is on the subject of correspondence, I must tell you that no arguments you could adduce make me change my opinion of my being the obliged person, even the strongest quotation from 'Jack and the Bean Stalk'[3] would fail of convincing me to the contrary.

Now on the subject of formality, I will adopt your own method, and lay both Albert and Walter aside,[4] as I told you in my last I consider your own name preferable to either, but at the same time do not for an instant think that my friendship for you is, or ever can be in the slightest degree lessened, those are the only two subjects on which we will differ, And now My Dear Mr T. accept my warmest thanks for the very kind assurance you have given me in your last of never tiring of my dull correspondence, and believe me the feeling is mutual, for I will weary of all I value most on earth when I do of you. Before I proceed further I must enquire how your cold gets on, and caution you to be very careful of yourself, remember that colds if neglected lead to many diseases so if you are not better take my advice and apply to a physician. I speak from experience, I got a severe cold at the dancing school last winter and the consequence was that I was for the entire summer suffering more than I could describe, so now let me entreat of you not to trust to your own skill if you are not by this quite recovered, and when you write again which I am wicked and unreasonable enough to hope may be on Sunday next don't forget to say how you are, as I will be miserable until I know. Now for a little news.

On Friday night last,[5] a set of fellows broke into the shops of Judge and O'Brien,[6] both grocers in the main street, and made off with every thing they could lay hands on and I dare say would have gone on through the town with their depredations but were surprised by Dominick P. Ronayne,[7] who was returning from a dinner party, when they made off with themselves. The next day three of them were taken on suspicion, and one of them turned Queen's evidence.[8] I suppose they will be transported.[9] Old Mr Gardner,[10] the attorney, died on Sunday night of spotted fever. He has left a large family to whom I much fear he will be a serious loss.

I must now tell you how I spent Sunday, I went to Templemichael church[11] with the Bagges,[12] it was a delightful day, we left this at ten, arrived there at eleven and walked about Mr Armstrong's gardens[13] till church time; from that we went after prayers to Ballinatray[14] where the Sunday school is held, while Mr B. was examining the children there, Mrs Bagge and I went to see the gardens, I never was more delighted, the flower garden is immense and beautifully laid out, there indeed it may truly be said that

> The tints of the earth and the hues of the sky
> In colour though varied in beauty may vie.[15]

A more enchanting scene need scarcely be imagined. I had no idea of seeing more than a few autumnal flowers, but instead of that it was rich in all the glowing beauty of summer. It seemed a perfect paradise. Hope spread her balmy pinions[16] over me and it was with reluctance I quitted that fairy scene – nothing was wanting to give it effect, the black water runs at the foot of the garden, and to its edge was wooded with the most beautiful trees and shrubs. It would serve to remind one of that world to which we are all hastening, and yet so frail are our natures we cannot deny ourselves the perishable pleasures of this fleeting life even to obtain for ourselves an eternity of bliss. I never felt more happy than when I remained in that lovely spot, how much I wished for

> One to listen and reply to thoughts
> With which my heart beat high[17]

I know you will laugh at such a description, but I say to <u>thee</u> just what suggests itself to my own wild fancy. I am sure you will not show my letter to any person, but when you have read, burn them. I hope Jim[18] is getting the better of his malady, pray present my kindest remembrances to him and Edwin[19] I would write to him to night but have a headache

And ever my dear Mr Tyndall | believe me yours | sincerely attached | A. Bessie

Ecrivez bientot,[20] and excuse incoherences, my head pains me so that I could not read this over, nor would I write to night to any person but yourself.

RI MS JT/1/11/3892
LT Transcript Only

1. *A. Bessie*: LT suggests that 'A. Bessie' is probably the same person who signed herself 'Eliza', see above, letter 0091. The content and style of both letters support this suggestion. The name Bessie is a diminutive of Elizabeth.
2. *writing a letter on the Sabbath day*: Those who observed the Sabbath (the Lord's Day) strictly would refrain from any work including letter-writing.
3. *'Jack and the Bean Stalk'*: a popular fable, several versions of which were published in the early nineteenth century.
4. *lay both Albert and Walter aside*: Tyndall's used the pseudonym Walter Snooks and apparently also Albert.
5. *On Friday night last*: report in *Cork Examiner*, 8 October 1841, p. [1].
6. *the shops of Judge and O'Brien*: *Slater's* lists John Judge & Co. and John Francis O'Brien, both grocers of South Main Street, Youghal.
7. *Dominick P. Ronayne*: from a prominent Youghal dissenting family.
8. *turned Queen's evidence*: when a participant in, or accomplice to a crime gives evidence for the prosecution.
9. *transported*: to one of the penal colonies in Australia.
10. *Old Mr Gardner*: William Gardiner, a solicitor, died 3 October 1841 aged 67.
11. *Templemichael church*: a Church of Ireland parish church in County Waterford built in 1823.
12. *the Bagges*: Mr and Mrs Bagge; see letter 0091, n. 1. 'A. Bessie' appears to have been employed by Mr and Mrs Bagge, possibly as a governess.
13. *Mr Armstrong's gardens*: probably the gardens of Templemichael House in County Waterford owned by Captain Thomas Armstrong.
14. *Ballinatray*: an estate in the parish of Templemichael in County Waterford.
15. *The tints ... may vie*: Lord Byron, *The Bride of Abydos* (1813), I.i.11–12.
16. *pinions*: wings (*OED*).
17. *One to listen ... beat high*: Byron, *The Bride of Abydos*, I.iii.30–1.
18. *Jim*: Phillip Evans.
19. *Edwin*: not identified.
20. *Ecrivez bientot*: write soon (French).

To John Tyndall, Snr 7 October 1841 0100

Kinsale, Thursday[1] night.

My dear Father,

I think it is high time to break the ice by which each of our pens has been this good while surrounded. I have waited long in vain for something in the shape of a letter from you, but silence has bound your pen it appears. I hope on the receipt of this you will burst the shackles and not suffer yourself to be again manacled. I know you will tell me you have very little time. Look at the way I am placed and I'm sure you'll never say a word about want of time. Here I am working from 7 to 5 o'clock[2] each day without intermission; then the moment I despatch my dinner

I have to tackle at my drawing or otherwise bear a lecture on the foolishness of idleness. Being determined to make the most of my time I work late every night, you will surely then confess that you have much more time than I have. Besides that, I have no less than four or five who expect a letter from me now and then. I have devoted this night solely to the purpose of writing. I began a letter to day in the office, but did not like to take time to finish it, so you see that your son is rather a conscientious fellow. Why you live in strange times in Leighlin now. I'm sure you have an abundance of news and yet you are very sparing of it. I got an English paper (The Age) the other day from a friend of mine in Youghal and in it I saw the name of a Mr Tyndall mentioned in connexion with that of Doctor Morris on account of the latter's refusing to dress Bergin's wounds.[3] Now I ask you is it not too bad for me poor pilgarlic[4] to have to wait until it finds its way into the columns of an English newspaper for the account of such an occurrence as this. See what it is to be obliged to live at the World's End.[5]

Great changes have taken place among us lately. Mr Fenwick has been ordered off the Survey. He has been promoted to Captain and is going out on foreign service.[6] We dont know who we are to get in his place. Some think we will be removed to Cork under the eye of Major Waters; this is the general opinion, but there is no certainty in the matter.

The English Survey is I believe a bad speculation. Our men have arrived there and there is not a word of raising their pay.[7] Provisions are much dearer there than they are in Ireland so that the Surveyors are very tightly pinched to live in it at all.[8] I wish I could lay my hand on the shelf behind your back as you stand in the shop. I think I'd be disposed to break the 8th commandment[9] and steal a pair of shoes from you. My stock is almost run out. However I'll keep them patched up for a while longer. Have you shifted your lodging yet? Did you see Jos. Wright.[10] Are my mother and Emma well? Give my love to both

Good night | Your affectionate son | John Tyndall.

RI MS JT/1/10/3246
LT Transcript Only

1. *Thursday*: LT gives postmark as 'Oct. 8th, 1841', which was a Friday.
2. *from 7 to 5 o'clock*: from 7am to 5pm.
3. *I got an English ... wounds*: William Bergin, a Catholic and a tenant of Colonel Henry Bruen, was severely injured in Leighlin Bridge during a violent assault by a group of men. The dispensing apothecary, Mr Morris, refused to treat his injuries. 'Mr. Tyndal, who was on the spot, remonstrated with him for his unfeeling conduct, and immediately sent for Dr. Roche, who was promptly in attendance, and rendered every assistance to the poor man' (*Age*, 26 September 1841, p. 7; reprinted in *CS*, 18 September 1841, p. [3]). Dr Benjamin Roche of Bagenalstown is listed in *Slater's* as a physician or a surgeon. The *Age and Argus* was a conservative weekly newspaper founded in 1825 and costing 6*d*.
4. *pilgarlic*: a pitiable, lowly, or foolish person (*OED*).

5. *the World's End*: see letter 0072, n. 1.
6. *Mr Fenwick ... foreign service*: Letter 0097 indicates that Captain Robert Fenwick had been posted to Skipton or Grassington, Yorkshire, but he was subsequently sent to Bermuda where he would die of yellow fever in August 1843; see letter 0226, n. 12.
7. *not a word of raising their pay*: Civilian assistants on the Irish Survey thought that they would be far better paid when they joined the English Survey.
8. *Provisions are ... at all*: In fact Tyndall would tell his father soon after arriving in England: 'Market rates much the same, perhaps a little dearer than in Cork, so that I can manage to do pretty well'; see letter 0164.
9. *the 8th commandment*: 'Thou shalt not steal' (Exodus 20:15).
10. *Jos. Wright*: see letter 0083, n. 14.

To the Editor of the *Carlow Sentinel* 12 October 1841 0101

Kinsale Oct 12/41[1]

Sir,

In casting my eye over a Sentinel a few days ago,[2] the account of a repeal meeting held in Leighlin Bridge met my view – I was astonished to find some of the persons mentioned there figuring as heroes of the platform – Timothy Hennessy Esq[r3] holds a prominent place in the catalogue of patriotic worthies – I find also from another source that the son of Mr Terence O'Neill[4] of the Swan hotel Leighlin Bridge had thought proper to obtrude the produce of his tiny pen on public notice – I have read portions of two letters written by him on the subject of Repeal![5] Now this is overbearing – to see fellows with just as much brains as gateposts aspiring so high is a downright outrage on common sense – I could not let them pass without administering a slight castigation – Should it be deemed worthy of a place in the columns of the Sentinel[6] it may be of some service to one or other of these puny champions.

I remain Sir – | your most obed[t] humble serv[t] | John Tyndall

The Eds of the Carlow Sentinel

I have reasons for not wishing my name to be mentioned in connexion with this nor even the quarter from which it came Should it gain insertion you will be so kind as to suppress both and merely place to it the letters – W.S.[7]

What sounds are these which strong and clear
Strike full upon my ravished ear?
'Tis freedom – bursting from the night
Of ages, sheds her glorious light
O'er Erin's[8] undulating plains
Her cloud-capped hills and mouldering fanes!
List![9] Carlow list! while young ONeill
Vociferated – Repeal! – Repeal!
Hark! rising to the shrilling cry,
The voice of Leighlin answers nigh.
See! as the swelling notes ascend
The poplars on the Bawnogue[10] bend!

And echo wafts it far and wide;
Along the Barrow's[11] placid tide.
And hark! along the eastern line,
The 'Captain's'[12] geese in cackle join
In council sworn – every one
To help their brother of the Swan*
With tail erect in wild amaze,
Each hungry ass in concert brays!
And had they Baalam's gift[13] you'd hear
Them give their fellow ass a cheer
Oh! who can con the glorious theme
Nor glow with all a patriot's flame?
Strike! Erin Strike! the happy lyre,
Your Sons have caught the gen'rous fire;
And Tory Bruen waxes pale
Before the doughty young O'Neill;
And Tory tyrants now confess
This youth the lustre of the press.[14] –
And Tim[15] honoured in modern Story,
(Sure youthful hearts will pant for glory)
From young ambition's gleaming pyre
Has snatched the title of 'Esquire'!
A man of stalwart consequence,
Though some assert a lackbrained fool,
The deadly foe of common sense
The wise man's scoff – the villain's tool
A bubble-full of empty pride,
The filth of agitation's foam,
That drops into oblivion's tide
'Unwept – unhonoured – and unknown'.[16]
Oh! is it then such */pulling/* things
That Erin's dignity maintains?
Ah no! – each apish essay flings
A deeper tint upon her stains
Presumptuous Jackanapes! – resign
Your tiny pens to other hands
Fulfill your destiny's design
Go mete the tape and measure drams!
Nor once again vain fools aspire
Beyond the counter's genial noise,
In scorn my glowing thoughts expire
Goodbye ye patriotic boys! –

W. S.
RI MS JT/8/2/1/1–2

---

* the Swan hotel

1. This letter contains a rough copy of the poem subsequently published in the *Carlow Sentinel.* The editor, not named here, was Thomas Harris Carroll.
2. *a Sentinel a few days ago*: 'The Repeal Meeting at Leighlin', *CS*, 25 September 1841, p. [3].
3. *Timothy Hennessy* Esq[r]; not identified.
4. *son of Mr Terence O'Neill*: see letter 0098, n. 3.
5. *two letters ... Repeal*: see letter 0098.
6. *Should it be ... the Sentinel*: published as 'The Leighlin "Orators" – or, the Late Repeal Meeting', *CS*, 16 October 1841, p. [3]. The published version contains a number of minor changes, principally added punctuation.
7. W. S.: Walter Snooks, a pseudonym of Tyndall's.
8. *Erin's*: romantic name for Ireland.
9. *List!*: listen.
10. *Bawnogue*: probably Bawnogue or Bownoge, near Baltinglass, about 12 miles north-west of Carlow.
11. *Barrow's*: the River Barrow, which flows through County Carlow.
12. *Captain's'*: probably Captain Thomas Woodcock, mentioned in letters 0066 and 0090.
13. *Baalam's gift*: the gift of prophesy (Numbers 22:1–3).
14. *This youth the lustre of the press*: the two published letters by the son of Terence O'Neill.
15. *Tim*: either Timothy Hennessy or Terence O'Neill's son, both mentioned earlier in this letter.
16. *'Unwept – unhonoured – and unknown'*: slight misquotation of 'Unwept, unhonoured, and unsung' in W. Scott, *The Lay of the Last Minstrel* (1805), VI.i.16.

From John Tyndall, Snr 13 October 1841 0102

Mr John Tyndall | Ordnance Survey Office | Kinsale
Leighlin Bridge | Oct. 13th. 1841.

My dear John

I am sure you are nearly worn out of patience expecting something in the shape of a letter from me I have been very much unsettled since the 29th of September in consequence of removing to the house your uncle Will lived in.[1] This is the chief cause of my not writing to you ere this. You say we live in strange times at present in Leighlin Bridge. You say what's true, as all the bad feeling that existed during the Election is still in full force, and exclusive dealing the order of the day. You sent 'the Age' newspaper to me and in it a short paragraph taken from the Carlow Sentinel[2] about Bill Bergan being beat in this town. Such also is the truth, and severely, and every one that's seen or heard of the transaction says it was a miracle I was not murdered myself as I was with him a little before the severe assault was given. Now to give you the particulars of that base act. I was standing at your uncle Bill's door speaking to him and some other person that now I forget. Bergan passed by. At the time we took no notice of him, nor he of us; but as soon as he reached Tom Hughes's corner[3] the yell of 'black sheep'[4] was set up. As soon as I heard it, I walked down the street after him, thinking that by my presence I might be a check on any person I knew; which I evidently was, for no person but a stranger made use of it as I passed along the street. When he reached the top of the bridge he was surrounded by a mob of strange reapers. When I saw him thus situated I walked quick up to him and advised the reapers to let him pass through the town peaceably, that if they done any thing to him they would be the sufferers. I might as well be whistling jigs to a mile-stone[5] as speaking to them; and when I saw the storm rising I advised Bergan to return with me. He refused to do so. I then thought it would not be well done to shrink from him at this stage of the proceedings. I proceeded onwards, thinking the agitation would subside and during my passage to the corner of the street leading to old Leighlin I received two blows with the fist and a very severe kick and at the same time was surrounded by four or five hundred villians with reaping hooks in their hands and did not know one of them. When I saw that I could neither protect myself nor the unfortunate man I was with (though up to this I received more injury than he did) I told him to step with his back to the wall of Mary Hanlon's corner[6] till I'd bring the police. So I ran over the bridge amidst the most noisy yells and groans, and when I came to the Barracks[7] they were not there. So I hastened back through the gang I passed before, without receiving the least insult. When I came up to Bergan he was weltering[8] in his gore. I then found that the police were on duty at Mr Faulkner's, a magistrate, who lived where Mr Wynne lived,[9] I then ran towards the place but met them where Mr Watson lived.[10] We then met Willy Tyndall your uncle who was in the crowd during the whole transaction but dare not come near me for if he did us both would be killed. He said he knew one of the leaders in this tragedy. We then went in search of them, and during our absence the Woodhouses[11] with some other persons brought Bergan over the bridge to have his wounds dressed by Doctor Morris[12] – who savage like refused to do so. There was a trial on him at the dispensary the other day, but he was acquitted, as there was a point of law in his favour: that was he acted in his private capacity not as apothecary for the dispensary. When will you come home? If you dont like the Survey dont stop with them. I am very tired John and must stop, but we are all well here.

I am your affectionate father | John Tyndall

I forgot to tell you that when Morris refused I went for Doctor Roach[13] who was dining at his mother's. He left his dinner and came with me, it was Caleb was alluded to in 'the Age' as he heard Morris refuse the Woodhouses to dress Bergan.

RI MS JT/1/10/3247
LT Transcript Only

---

1. *removing ... lived in*: William Tyndall appears to have moved to Carlow. According to *Griffith's Valuation* (County Carlow, p. 179), John Tyndall, Snr, occupied 22 Bridge Street, a property valued at £1 5*s*. 0*d*, his landlord being William Stewart.

2. *You sent ... Carlow Sentinel*: see letter 0100, n. 3.
3. *Tom Hughes's corner*: Thomas Hughes lived at the corner of Chapel Street (now Saint Laserian's Street) and Poe's Hill, which is a link between Chapel Street and Milford Street. William Bergan's pursuers came from County Kilkenny, to the west of Leighlin Bridge.
4. *'black sheep'*: a disreputable or unsatisfactory member (of a family, etc.); a bad character (*OED*); but in this context probably means traitor to the cause (of the repeal of the union between Ireland and Britain).
5. *whistling jigs to a mile-stone*: 'whistling jigs to a milestone, in hopes of making it dance' is an Irish saying meaning that the other persons were unresponsive. See, for example, Countess of Blessington, *Grace Cassidy; or, the Repealers. A Novel*, 2nd edn, 3 vols (London: Richard Bentley, 1834), vol. 2, p. 75.
6. *Mary Hanlon's corner*: A Margaret Hanlon lived at the corner where Milford Street and Chapel Street meet, Old Leighlin.
7. *the Barracks*: probably the Constabulary on Carlow Street, Tyndall's father having crossed the bridge.
8. *weltering*: the action of turning or twisting the body about (*OED*).
9. *where Mr Wynne lived*: presumably when Lieutenant George Wynne had been stationed in Leighlin Bridge he lived in Rathvinden Cottage, the home of the magistrate Henry Faulkner.
10. *where Mr Watson lived*: Thomas Watson lived on Milford Street, Old Leighlin, facing the River Barrow.
11. *the Woodhouses*: not identified.
12. *Doctor Morris*: described as 'a dispensing apothecary' in *CS*, 18 September 1841, p. [3].
13. *Doctor Roach*: see letter 0074, n. 22.

To John Tyndall, Snr 19 October 1841 0103

Kinsale, Oct. 19th, 1841.

My dear Father

A more welcome sight than your last letter[1] did not greet my eyes these six and thirty years!![2] My mind was on the rack of anxiety on account of being so long without hearing from you. I was quite certain that all was not right; and being informed that the fever was raging in the vicinity of Leighlin[3] added not a little to my disquiet. But all my gloomy anticipations were hushed into oblivion when your well known handwriting met my view. Hurrah! I cried – and a standing leap accompanied the whoop – he is still able to write! I hope you are comfortably situated in your new house[4] I think the change will be for the better, the other was both too small and too far removed from the thoroughfare. Where has Bill pitched his tent?[5]

The account you gave of Bergan's affair was quite new to me I did not imagine that you had any thing at all to do with it. You behaved bravely, I must say. You may laugh at me, but youd hardly imagine the long proud stretch I gave my neck after reading the account of the transaction to our fellows. You see a son can plume himself on his father's courage.

We are still in uncertainty here as to the result of the movements that have recently taken place. We are to get a new officer, but he has not yet been seen. A Captain Rimington[6] who had charge of a division in Dungarvan has been fixed upon, tho' some imagine that Lieut Beatty[7] is to take charge. This latter was stated in the newspapers.[8] However one thing is certain that we are to remain in Kinsale. I am not sorry for this, as I would not like to be interrupted in my drawing, and, besides that, moving about is expensive.

You may recollect hearing of Foy's being reduced from the rank of Corporal for disobedience of orders in marrying.[9] He got a very nice wife, for her he sacrificed his hopes of promotion in the corps, a blight was cast upon the poor fellow's prospects, and now the being for whom he gave up all is a tenant of the lonely tomb. Mrs Foy died of consumption and Tom had not the poor gratification of closing her eyes or of hearing her last farewell.

I'm beginning to fear that Carlow will hardly <u>ever</u> be the same social place it was when I left it. I see that some of the Leighlin boys[10] have grasped the pen and rung changes on the subject of <u>repeal</u>. Did you see last Saturday's Sentinel?[11] and, if so, did e'er a little rhyme of satire meet your view? I hear that a fellow signing himself as W.S.[12] gave some of the young patriots a dressing.

You have asked me when I'll go home. It is impossible for me to answer this question situated as I am at present. One thing you may rest assured of, and that is that I'll go <u>as soon as ever I can</u>. I am, if possible, more anxious to see you all than you are to see me. In a week or two perhaps when everything is settled I will be able to answer that question. You tell me to come home if I am tired of the Survey. I am not at all tired of it. Many a better man is not half so well off as I am on the Survey. I am therefore content with my present lot, and if providence does not think proper to place me in a better situation I shall remain content. Give my love to my mother and Emma.

Your affectionate son, John Tyndall

RI MS JT/1/10/3248
LT Transcript Only

---

1. *your last letter*: letter 0102.
2. *these six and thirty years!!*: as Tyndall was then about 20 years old, this overestimate was for rhetorical effect.
3. *the fever ... Leighlin*: 'Spread of Fever', *CS*, 2 October 1841, p. [3].
4. *your new house*: see letter 0102, n. 1.
5. *Where has Bill pitched his tent?*: William Tyndall had moved to Carlow.
6. *A Captain Rimington*: Thomas Hosmer Rimington (1801–87), a 2nd Captain in the Royal Engineers, had been in charge of A District and since July 1840 had been stationed at Dungarvan, Co. Waterford. However, he appears not to have taken up the position in Kinsale but left the Survey on 21 November 1841 (NAI OS/1/16–18).
7. *Lieut Beatty*: According to a report in the *Cork Examiner* of 13 October 1841, p. [1],

Lieutenant Andrew Beatty took over the command of 4th Division, C District, from Lieutenant Robert Fenwick.

8. *This latter ... the newspapers*: for example, *Freeman's Journal*, 11 October 1841, p. 1.
9. *Foy's being ... in marrying*: Because he had been absent without leave and married without permission, on 15 December 1840 Thomas M. Foy was demoted to the rank of Private, transferred to the 4th Division and his pay reduced from 1*s*. 6*d*. to 6*d*.
10. *some of the Leighlin boys*: the son of Terence O'Neill having had two letters on the subject of repeal published; see letter 0098.
11. *last Saturday's Sentinel*: *CS*, 16 October 1841, p. [3]; see letter 0101.
12. *a fellow signing himself as W.S.*: Walter Snooks, Tyndall's own pseudonym; for the 'little rhyme of satire', see letter 0101.

From John Tyndall, Snr 24 October 1841 0104

Mr John Tyndall | Ordnance Survey Office | Kinsale
Leighlin Bridge | Oct. 24th, 1841

My dear John

I received your letter[1] after passing through the Queens County and part of the County Kildare, and am glad to hear that the one you received from me caused you so much pleasure as to make you jump, buck-like, when you saw my well known hand in Kinsale. I have now to accuse you with a very great sin – that of humbugging your father with regard to the transaction with Bergan. You said that your neck stretched crane-like when you read of the courage I displayed on that occasion. At all events, to throw joking over board, I did show some pluck on that occasion, and if the police had been at home at the time I would have had some of the ringleaders apprehended. I forgot to tell you that after I left Bergan in search of the police the mob asked who I was. They were told my name was Tyndall, a brother to the man who shot the woman.[2] They said if they knew that I would be prevented from seeking for police – that is be murdered. They told lies though, as I passed them in a few minutes after without receiving the slightest injury Ally Kinsela, who first called 'Black Sheep',[3] and a man by the name of R. How, who was seen striking Bergan with a stone, are returned to the assizes to abide their trial for the offence.[4] I am happy to find that you are contented with your present situation. It's an old saying Don't throw out the dirty water until the clean comes in,[5] however if you did not like the Survey I am still able to support you bad as the times are and as extensive as exclusive dealing is carried on. Its thought the Colonel[6] wont come home immediately; be that as it may, I intend going amongst my friends shortly to see what plan is best to be adopted with regard to the getting of a better situation for you than the one you have at present. If I am beaten in my efforts so was Bonaparte.[7] Patt Moran is dead and buried[8] and Bill Tyndall living in Carlow. We are all well.

I am your affectionate father | John Tyndall

RI MS JT/1/10/3249
LT Transcript Only

1. *your letter*: letter 103.
2. *the man who shot the woman*: Caleb Tyndall who wounded Mary McAssey; see letters 0071, 0073, 0075 and 0078.
3. *'Black Sheep'*: see letter 0102, n. 4.
4. *their trial for the offence*: 'William Tyndal sworn – I saw a great crowd in Leighlin-bridge on the evening in question; I saw Bergin passing down the street, and heard a woman named Alley Kinsella, say "there is a Black Sheep" – when the crowd commenced hooting and shouting, and followed him towards the bridge; I was afraid they would murder him, as there were about 200 men there; ... I saw the defendant, James Roe, take up a large stone and burying it in Bergin's head who, on receiving the blow, fell heavily near the wall, where he attempted to defend himself against his assailants'. Report of the trial of Alley Kinsella and James Row at Bagenalstown Petty Sessions in *CS*, 25 September, p. [3].
5. *Don't throw ... comes in*: a proverb first recorded in the fifteenth century.
6. *the Colonel*: Colonel Henry Bruen.
7. *Bonaparte*: Napoleon Bonaparte (1769–1821) was defeated at the Battle of Waterloo in 1815.
8. *Patt Moran is dead and buried*: see letter 0095, n. 7.

To John Tyndall, Snr 26 October 1841 0105

Kinsale, | Oct. 26th, 1841.

My dear Father

I sit down to acknowledge the receipt of yours of the 24th.[1] Whether was it – you or my letter[2] – that 'passed thro' the Queen's County and County Kildare?' I suppose it was the latter, as letters are often sent round the world for sport. You bring forward a charge against me in your last letter to which I plead Not guilty. You say that I was guilty of a great sin in humbugging you. I assure you this was not the case. What I stated was literally true. Why you had a most narrow escape from the Kilkenny savages.[3] And what were the lies you spoke of in your last letter? I admit the justice of the proverb you quoted.[4] There is no fear of my giving up what I have at present until I have a better provision made for me. You forgot to answer my question respecting the bit of poetry in the Carlow Sentinel. Did you see any thing from W.S. in Saturday week's paper?[5] If you see next Saturday's paper[6] be so good as to say whether this W.S. has put any thing in it or not – I have particular reason for wishing to know this. I have nothing more to say. I am just after despatching an enormous dinner – a regular beggar's dish[7] – a curious compound of lumpers,[8] onions, and lots of fat mutton, the relics of Sunday's meal.

I remain your affectionate son | John Tyndall.

RI MS JT/1/10/3250
LT Transcript Only

---

1. *yours of the 24th*: letter 0104.
2. *my letter*: letter 0103.
3. *the Kilkenny savages*: the Catholic mob from County Kilkenny that attacked William Bergin, whom Tyndall's father helped to save.
4. *the proverb you quoted*: 'Don't throw out the dirty water until the clean comes in'.
5. *in Saturday week's paper*: Tyndall's first poem (see letter 0101) appeared in *CS*, 16 October 1841, p. [3]. He published it over the signature W. S., the abbreviation of his pseudonym Walter Snooks.
6. *next Saturday's paper*: Tyndall's poem 'Carlow' (see letter 106) was published in *CS*, 30 October 1841, p.[3].
7. *beggar's dish*: The 'beggar's dish' was 'a regular and proper part of his [the mendicant's] professional accoutrements' and was used by 'itinerant professed beggars and ... wandering tribes of gypsies' to collect offerings. However, according to the article 'Beggar's Dish' in *Chambers's Edinburgh Journal* (25 February 1832, p. 27) it had by the 1830s largely gone out of fashion.
8. *lumpers*: a large, coarse but nutritious type of potato widely grown in Ireland. The failure of this potato crop a few years later resulted in the terrible Irish Potato Famine.

To the Editor of the *Carlow Sentinel* 26 October 1841 0106

Kinsale Oct 26th '41[1]

Sir,

Encouraged by the success of a former experiment[2] I once more obtrude myself on your notice – my present performance speaks for itself – further preface would be needless – should this attempt meet your approbation pray be so kind as to suppress my name as before

I remain Sir &c | John Tyndall

The Ed. of the Carlow Sentinel

Carlow[3]

Sweet spot! where first th' imperfect accent hung
In lingering lispings on my infant tongue
Where young imagination first took flight
And roam'd unshackled thro' the realms of light,
When bright and happy life's perspective shone
And golden vistas opened farther on
Each dancing stream that woos the sun's bright rays
Each wee wild flower that blossoms on thy braes[4]
Each shrub that sheds its fragrance in the dell
Around my soul has cast a hallowed spell –
Sweet blissful spot! where Barrow[5] fair and free
Rolls liquid chrystal to the distant sea

Fall oft when morning tinged the orient sky
And heav'ns calm azure spread itself on high
I've trod thy banks in summer beauty drest
T'inhale the breeze that freshened o'er thy breath
And when from Zenith[6] fall the scorching ray
In reckless frolic gambolled with thy spray
Ere sanguine strife thy hollows had bestained
Or friendship mourned her Sacred Courts profaned
Her glimmering incense cheered each passing gloom
The heart its altar – union its perfume
Oh then approving heaven smiled to see
Congenial bosoms throb in sympathy
Alas how changed, these Halcyon days[7] are o'er
And drooping Carlow must their loss deplore
Now agitation whelms her like a flood
And noisome vermin* gloat upon her blood
Where peace sat smiling gloomy terror reigns
And Ate[8] revels o'er thy lovely plains
Discords black banner now each Zephyr fills
And darkly hovering o'er thy sunny hills
Each cloudlet big with desolation hangs
While reeking 'neath O'Connell's vampyre fangs[9]
Prostrate you lie – each burning tear that's shed
Calls down damnation on his guilty head
What tho' the fiat[10] linger on the tongue
of Justice – tho' the hair-held sword[11] has hung
So long above thee waiting the decree
of slumbering vengeance to descend on thee
Tho' Titan-like[12] you raise your impious hand
And brandish in high heav'ns face your brand
Thy day arrives – the widow's withering ban[13]
Shall burst in thunders o'er thee – bloody man.
Nor lost to fame in ages yet to come
Fell infamy still brood upon thy tongue
And execution grim shall mark the spot
Where low you lie – detested not forgot

RI MS JT/8/2/1/3–4

---

1. *Kinsale Oct 26th '41*: This letter contains a rough copy of the poem subsequently published in the *Carlow Sentinel*. The editor, not named here, was Thomas Harris Carroll.
2. *success of a former experiment*: see letter 0101 and W. S., 'The Repeal Meeting at Leighlin', *CS*, 25 September 1841, p. [3].
3. Carlow: published in *CS*, 30 October 1841, p. [3]. The published version contains a number of minor variations, principally added punctuation.

* Bugs &c.

4. *braes*: steep banks bounding a river valley (*OED*).
5. *Barrow*: The 120 mile long River Barrow flows through Carlow and discharges into the sea at Waterford.
6. *Zenith*: the highest point of the celestial sphere as viewed from any particular place; the upper pole of the horizon (*OED*).
7. *halcyon days*: a period of calm weather, anciently believed to occur about the winter solstice when the halcyon bird, a species of kingfisher, was breeding (*OED*).
8. *Ate*: In Greek mythology Atë is the goddess of mischief, delusion, ruin, and folly.
9. *O'Connell's vampyre fangs*: Tyndall charges Daniel O'Connell with responsibility for the divisive agitation during the recent election.
10. *fiat*: an authoritative sanction, an authorization (*OED*).
11. *the hair-held sword*: Tyndall conflates that symbolic Sword of Justice with the sword that hung over the fourth-century BC courtier Damocles, suspended above his head by just a single horse hair.
12. *Titan-like*: The Titans were the first pantheon of Greek gods, who, despite their great strength, were overthrown by the younger Olympian gods.
13. *the widow's withering ban*: In 1815 Daniel O'Connell had killed John D'Esterre, a member of the Dublin Corporation, in a duel. His widow Jane Lucretia D'Esterre (1797–1868) angrily refused O'Connell's remorseful offer to share his income with her, although she did accept an annuity to support her daughter. A ban is an imprecation of a curse, an execration or malediction expressing anger (*OED*).

From Thomas Charles Higginson 27 October 1841 0107

Skipton,[1] Oct. 27th 1841.

My dear Tyndall

Rejoice – rejoice – if you have any love for me rejoice I am once more in the office. I had a letter written and was just sending it off when I received orders to go into the office – you see what it is to have a friend in Court.[2] I wrote to Colder to have me changed from the Division I was in to the old one,[3] but what was my astonishment when I was ordered into the office. We petitioned for an increase of pay but received answer that if we liked to resign we might. Yesterday the surveyors got an increase, that is those that did not petition, those that had not 12s a week were increased to it, but those that were over it might remain, so those that petitioned – any that had not 1s.9d per day – were increased to it, so that your humble servant got nothing.[4] But I dont care now that I am in the office. I was perfectly tired of that horrid Division.[5] I am lodging in an inn at 8s.6d per week Colder desires me to tell you he would have written to you but he has not a moment to himself, and has been ill ever since he came here. I am his secretary. I am much obliged to you for answering my letter so quick, yours was the first I received, I have had the pleasure of seeing Roberts's marriage in the Warder.[6] It cut a great swell no less than John George Roberts Esq.[7] I hope he has done it to better himself. Blaty, a brother of Blaty[8] that is with the major[9] is going to the Tower of London to be employed there – a good exchange. You must be thankful that I have written so much as bad as it is for I am almost mad with these horrid teeth of mine. Remember me to Ginty, Jim, Tid[10] and all the office fellows

Your ever devoted friend | Harold[11]

P.S. I hope Ginty didn't hurt the Yankee Bellies[12] since.

RI MS JT/1/11/3731
LT Transcript Only

---

1. *Skipton*: see letter 0096, n. 3.
2. *a friend in Court*: a person in a position to use his or her influence on one's behalf (*OED*).
3. *to the old one*: the 5th Division (that had become the 4th Division before this letter was written), C District, in which Higginson had worked with Tyndall in Ireland.
4. *your humble servant got nothing*: Since August 1840 Higginson had been earning 1*s.* 10*d.* per day, that is 11*s.* a week.
5. *that horrid Division*: the 1st Division was located at Skipton.
6. *the Warder*: the *Dublin Warder*, a Protestant anti-Catholic weekly newspaper owned by Joseph Sheridan le Fanu.
7. *John George Roberts Esq*: Roberts had been discharged from the Irish Survey on 9 January 1841 for 'idleness' (NAI OS/1/18). The *Dublin Evening Mail* of 11 October 1841 carried the announcement of the marriage on 6 October at Aghada Church of John George Roberts, 'of Kieucon [?], county of Westmeath ... to Esther Armstrong, of Corkbeg, county of Cork' (p. 3).
8. *Blaty, a brother of Blaty*: probably Lieutenant Andrew Beatty of the Royal Engineers and one of his brothers. The Map Office was located in the Tower of London, but had subsequently to be moved due to a fire on the night of 30 October 1841.
9. *the major*: Major Marcus Antonius Waters.
10. *Jim, Tid*: Phillip Evans and John Tidmarsh.
11. *Harold*: a nickname of Thomas Higginson's.
12. *Yankee Bellies*: see letter 0097, n. 18.

From John Tyndall, Snr 30 October 1841 0108

Leighlin Bridge | Oct. 30th. 1841.

My dear John

I received your letter of the 26th instant[1] and in it a second tirade of humbugging. You ask me whether it was myself or your letter that passed through the Counties Kildare and Queens. You know well my meaning, therefore the crime you are guilty of will not be forgiven without sincere repentance. You ask me why I did not answer the question about the poetry that appeared in the Carlow Sentinel of Saturday week.[2] I forgot to do so, not having the letter with me at the time, as Mr Conwill took it in a mistake a few nights before, and he a little tipsy, but sent it back the next day. But to come to the point, I did see the poetry, and admired it because it was exceedingly well done. And Emma read it also. And witness her sagacity, when after reading it she said it was you composed it. I had very strong suspicions that you were the person, as Tim Hennessey Esqr. and young O'Neill were mentioned. But I have full confirmation that you were the author from Bill Tyndall who was here from Carlow last night. As he was passing Mr Malcomson's house in Carlow the other day, Mr Malcomson[3] called him in and asked him had he any one belonging to him in Kinsale. He told him he had a nephew. He then asked him your age He told him about 18 years.[4] He then said that you were uncommonly talented and desired Bill to write to you and let you know that he would be glad to receive communications from you and that his paper should be at your command.

I am just after reading this day's Sentinel and in it a piece headed original poetry Carlow,[5] which I think is your production also. Its beautifully done. I would be glad to see such productions from you often, because it will not only improve your mind but also bring you into notice. It fills my heart with gladness to find I have a son able to perform the like and though you were often not pleased with my surly looks still I did not neglect the main point – I mean with regard to your education and the only thing that prevented you from going through the degrees of college was my poverty, however *<rest of letter missing>*

RI MS JT/1/10/3251
LT Transcript Only

---

1. *your letter of the 26th instant*: letter 0105.
2. *the poetry ... Saturday week*: 'The Leighlin "Orators" – or, the Late Repeal Meeting' published in *CS*, 16 October 1841, p. [3]; see letter 0101.
3. *Mr Malcomson*: Henry Malcomson (*c.* 1796–1843), proprietor and publisher of the *Carlow Sentinel.*
4. *about 18 years*: As Tyndall's birth certificate is missing, most accounts give his year of birth as 1820, which would make him 21 at the time of this letter. This estimate by his

uncle William suggests that he may have been younger. However, his uncle would not have seen Tyndall for about 18 months.

5. *a piece headed original poetry Carlow*: *CS*, 30 October 1841, p. [3]; see letter 0106.

To John Tyndall, Snr [3] November 1841 0109

Kinsale, Wednesday night. | Nov. 1841.[1]

My dear Father

I now sit down to acknowledge the receipt of yours of I dont know what date.[2] Your last letter surprised me not a little. I thought W.S.[3] was likely to remain in obscurity but I find he has been dragged from his lurking place, a connexion having been traced between him and a denizen of the World's End.[4]

A very trifling circumstance first induced me to trust any of my productions to the mercy of the press Mr Conwill sent me in a letter a couple of extracts from O'Neill's letter.[5] With amazement I scanned the bombastic stuff over and over. I then thought of composing a bit of a rhyme to stick at the commencement of my next letter to Mr Conwill. Well I set about it, and when finished the thought struck me that it would have a much better effect in the columns of the Sentinel. I hardly dared to think however that they would consider it worthy of insertion. I packed it off, and was agreeably surprised to see in the Sentinel[6] for the following week my little satire. The success of this experiment induced me to try my hand again, and the success of my second attempt far exceeded my most sanguine expectations. I received the paper which contained the poetry fresh from the Sentinel office, and the communication you made respecting what Mr Malcomson[7] said will I think induce me to make a third attempt.

I wonder do the young patriots know the person[8] they may thank for the whipping they got. I hope not, I would not like the thing to become too public.

I suppose Mr Conwill has told you the reason of my using the letters W.S. Myself and Roberts were coming from Bagnalstown one evening and this side of the old elder bush which stands half way between Bagnalstown and Leighlin we christened ourselves – I was named Walter Snooks and he Sam Weller.[9] We never called one another by any other name, and the morning Roberts and I parted[10] – and we parted on bad terms – the last words were 'good bye Walter' and 'good bye Sam'. This is the cause of my using the initials W.S. I may here remark that Roberts has got married,[11] and in the list of marriages in the Cork and Dublin papers the name of John George Roberts Esq stood prominent. Send me word how everything stands in Leighlin at present. Did my satire cause any sensation among the <u>repealers</u>! I suppose they cursed the author of it. However they are a parcel of gentlemen I care very little about and, as far their abilities, tho' some of them may have taken out their degrees for aught I know, I'm not at all disposed to quail before them. A poor fellow like me has no reputation to lose, and if by any means the beam of talent should lean on my side, why the literary fame of my opponents is damned for ever. So that I may gain something in the conflict but can lose nothing. This is a drowsy epistle It's very late and I am half asleep writing it

Your affectionate son | Jack | alias | Snookes

RI MS JT/1/10/3252
LT Transcript Only

---

1. *Nov. 1841*: LT has added 'probly 4', the 4th being a Thursday. As the letter is headed 'Wednesday night', the likely date is 3 November.
2. *yours of ... what date*: letter 0108.
3. *W.S.*: Tyndall's pseudonym Walter Snooks.
4. *the World's End*: see letter 72, n. 1.
5. *a couple ... O'Neill's letter*: see letter 0098.
6. *in the Sentinel*: *CS*, 16 October 1841, p. [3]; see letter 0101.
7. *the communication ... Mr Malcomson*: see letter 108, n. 3.
8. *do the young patriots know the person*: Tyndall himself; see reference to 'young patriots' in letter 0103.
9. *I was named ... Sam Weller*: The name Walter Snooks (Tyndall) does not appear in the fiction of the period, whereas Sam Weller (John Roberts) was the witty, worldly cockney who served as Mr Pickwick's valet in Charles Dickens' popular novel *The Pickwick Papers* (1836–7).
10. *the morning Roberts and I parted*: Roberts was discharged from the Survey on 9 January 1841.
11. *Roberts has got married*: see letter 107, n. 7.

To the Editor of the *Carlow Sentinel* [8] November 1841 0110

Kinsale Monday Nov[1] – 1841

Sir

A voice from the South[2] again salutes your ear – the kindness you have evinced in giving my puny productions a place in your paper demands my warmest thanks – it emboldens me to throw myself a third time on your mercy in the hope that my present performance will be equally successful with my former ones

John Tyndall

The Ed. of the Carlow Sentinel

I thankfully acknowledge the receipt of last Saturday week's Sentinel – the kindness was quite unexpected | NB. Decoy as usual[3] –

The Testimonial[4]
Hark the voice of empire calls
Forward to the shrine of fame
on its cloud crown'd capitals

Emblazon Bruen's name

Lo the answering signal brand
Flashes on Mononia's[5] rills
Gleaming bright from strand to strand
Sheds its light on Ullin's hills[6]

As turns the Moslem to the shrine
When the last tinges of the sun
In all their golden glory shine
Above the distant horizon
Each grateful eye is burned on thee*
Hibernia's new Thermopylae[7]

The soaring condor plumes his wing
on Chimborazo's lofty peak[8]
And hears the mountain echoes ring
In dread explosion far beneath
Amid the elemental war
The spirit of the tempest rides
And flashing from his cloudy car
Red lightnings hiss along the mountain's sides
Unmoved – unruffled and serene
The tenant of the crag looks down upon the scene

So stood brave Bruen undisturb'd he viewed
The scowling cloud of agitation lour[9]
With steady eye – his lofty brow unmoved
He calmly waited the impending shower
And when at length the demon of the storm
Let loose the thunders from his red right hand
The dauntless chief on freedom's pinions[10] borne
Unfurled her flag and drew her flaming brand
And mantling on his cheek the patriots glow
He hurled defiance at his gorgon[11] foe

And oft to battle for the right
He led his trusty men
And oft was worsted in the fight
Yet Bruce-like[12] fought again

And conquered too – the wild hurrah
Has reached the distant sky
And echo from her mountain hold
Has answered cry for cry
---

* Carlow

Tho' rent by many an adverse breeze
Upon the battle plain
His glorious banner freely waves
Unsullied by a stain

While surpliced fiends[13] – Hell's viper spawn
Before the standard bow
Lerne[14] twines the laurel wreath
To bind her champions brow.

Now the voice of empire calls
Forward to the shrine of flame
On its cloud crowned capitals
Emblazon Bruen's name[15]

RI MS JT/8/2/1/5–6

---

1. *Monday Nov*: This letter contains a rough copy of the poem subsequently published in the *Carlow Sentinel*. The editor, not named here, was Thomas Harris Carroll. As the poem was published on 13 November, this letter was probably written on 8 November 1841.
2. *from the South*: from County Cork.
3. *Decoy as usual*: Tyndall requests that the poem appear over the initials 'W. S.', Walter Snooks being a pseudonym of Tyndall's.
4. The Testimonial: Published in *CS*, 13 November 1841, p. [3]. The published version contains a number of minor changes, principally added punctuation.
5. *Mononia's*: Mononia is the Province of Munster, comprising Counties Clare, Cork, Kerry, Limerick, Tipperary and Waterford. The principal source for 'Mononia' was the poem 'Remember the Glories of Brien the Brave' in Thomas Moore's *Irish Melodies* (1821).
6. *Ullin's hills*: 'And Ullin's hills be silent as the grave' (G. Harvey, trans., *Ossian's Fingal; an Ancient Epic Poem* (London: Valpy, 1814), II.xv.8).
7. *Hibernia's new Thermopylae*: Colonel Henry Bruen's electoral success at Carlow in 1841 is here compared with the famous battle at Thermopylae in 480 BC, when the Greek forces repelled a far larger army of Persian soldiers. Hibernia is the Classical Latin name for Ireland.
8. *Chimborazo's lofty peak*: one of the highest mountains in the Andes in South America.
9. *lour*: variant spelling of lower.
10. *pinions*: wings (*OED*).
11. *gorgon*: In Greek mythology the Gorgons were three sisters whose hair was made from venomous snakes and who turned those who looked at them into stone.
12. *Bruce-like*: The great Scottish warrior king, Robert the Bruce (1274–1329), defeated a far larger English army at the Battle of Bannockburn in 1314. He later invaded Ireland to assist the Irish struggle against England.
13. *surpliced fiends*: wearing the loose vestment of white linen worn over a cassock by Roman Catholic priests.
14. *Lerne*: possibly the Lernaean Hydra that, in Greek mythology, was a many-headed serpent.
15. *Emblazon Bruen's name*: Colonel Henry Bruen's electoral success; see n. 8.

From John Tyndall, Snr 14 November 1841 0111

Mr John Tyndall | Ordnance Survey Office | Kinsale
Leighlin Bridge | Nov. 14th, 1841.

My dear John

After a long silence I now sit down to scribble over something in the shape of a letter to you. I would have written to you ere this but was waiting for the accomplishment of an event that was said would take place on yesterday and that was no less than the marriage of your cousin Emma Tyndall with a young man named Young[1] now living at Kilgraney[2] – a man holding a very fine farm of land under Lord Beresford,[3] and in very good circumstances. It has not taken place yet, but I think will some day this week. The way this attachment was formed was during Caleb's stay in the jail. This man being brother to the turn-key[4] paid him many visits when he met with Emma, who often went there to see her father, by which means he fell dotingly in love with her – or with Caleb's hundred pounds which he is to give her as a fortune. I think her much too young to be married to Young, but of this they are the best judges.

You asked me to send you word about every thing connected with this place. All I have to say is that, if the annals of infamy was searched over, so infamous a den as the den of tigers that inhabits the town you were born in could not be found. For my part all they can do against me they are doing. The consequence is I am selling I may say nothing. But my trade is better than it ever was, which I hope will counteract in some measure their villiany. Witness your pet John Mooney of bigoted notoriety; I think he was not inside my door since the night Caleb fired the shot,[5] and when Mrs Steuart attacked him for such conduct he told her it was the dearness of my leather that caused him to act so. But my leather was not too dear till the morning after the shot was fired. So much for Mooney's mean subterfuge. All the bad spirit that attended the Election still remains here in the inward man, but the outward man is peaceable, because they must be so – and give them no thanks – as we have a Government now[6] that will see the laws executed with justice and impartiality. I will now conclude with letting you know that I read in yesterday's Sentinel another beautiful production from you headed 'original poetry for the Carlow Sentinel' 'the Bruen Testimonial,'[7] and your Uncle Willy told me this evening that you were the author of it, but I knew that before he told me so. We are all well.

I am your affectionate father | John Tyndall

Will you be home at Christmas?

RI MS JT/1/10/3253
LT Transcript Only

1. *a young man named Young*: George Young, who married Emma Tyndall, daughter of Caleb, on 22 November 1841 at Dunleckney (Irish Church Records).
2. *Kilgraney*: Kilgraney is about 3 miles south of Bagenalstown.
3. *a man ... Lord Beresford*: Young's land and buildings were assessed at a net annual value of £65 (*Griffith's Valuation* (County Carlow, p. 244)). On Henry de la Poer Beresford, see letter 0071, n. 5.
4. *the turn-key*: James Young, the jailer at Carlow jail.
5. *the night Caleb fired the shot*: 27–8 June 1841.
6. *we have a Government now*: The Conservatives, under Robert Peel, had been returned to power on 30 August 1841, replacing Lord Melbourne's Whig administration, which had lasted 6 years.
7. *yesterday's Sentinel ... Bruen Testimonial*: *CS*, 13 November 1841, p. [3]; see letter 0110.

To John Tyndall, Snr 19 November 1841 0112

Kinsale, | Friday night[1]

My dear Father,

Tho' it is now pretty late, I consider this a favourable opportunity for writing to you. If I am a little drowsy, I have one comfort and that is I am undisturbed by the fly that was tormenting me a few minutes ago. In one corner of the room stands Evans, scarcely able to drag his jacket off his back with downright laziness, and ever and anon stretching his mouth to an amazing capacity as yawn after yawn gurgles in his throat. In another corner stands Tidmarsh in his shirt scratching himself, this he considers as one of the greatest luxuries on earth – while in the middle sits your son with his back to the dying embers of what was once a good fire scribbling away as fast as his pen can go

The old proverb says that miracles will never cease.[2] This is verified every day; the news of Tom Magwire's[3] turning protestant or Dan O'Connell[4] conservative could not have surprised me more than the intelligence of my cousin Emma's intended marriage. Egad[5] she intends to make early work of it! Be that as it may, she has my best wishes for her welfare. My prayer is that every day may add to her store of happiness. I don't know Mr Young,[6] but I'm told that he is a fine young fellow, and holds a capital farm.

I'm sorry to hear that the seeds of strife have not yet been rooted out of Leighlin. When I see John Mooney I'll talk to him in the language of a poet.

You have asked me will I go home at Christmas I received your letter in the evening of some day this week. Well, in the morning of that day I was speaking to Sergeant Carey – our present Superintendant – about getting a month's leave. He told me that he thought I could get it. So that I think you may expect to see me demolishing a whole goose or something like it in Leighlin on Christmas day. No exertion shall be wanting on my part to bring this about, and I have every reason to expect that my wish will be obtained. I have a good notion of becoming teatotaler,[7] as I fear my Leighlin friends will be inclined to make me tipsy thro' pure goodnature. I'm extremely sorry that I am not at home for Emma's wedding. Och if I was there, I'd caper 'the humours of Glin'[8] in prime style. I made an attempt once at going through the figure of <u>eight</u>.[9] Mind it was a first attempt. Judge then of your son's amazing talent at this business. I went within one of the number – I made a perfect <u>nine</u> of it. Was not that a good offer

You spoke of my last production in your letter. If you had not spoken of it I would not have known that it got inserted, as I have not got the paper. If you could by any means get it for me I would be thankful to you. Ask Billy Tyndall[10] could he make it out. I know if he can he'll forage it out for me. How did he know that the poetry was mine? Mind I dont put half the heading to these pieces that you see to them. I merely put: 'The Testimonial' as a heading. The words 'Original Poetry for the Carlow Sentinel' were put in by the editor himself.[11] I'm glad to find this, as it shews that they don't consider it unworthy of their notice. I'll have another piece to send next week. My pen refuses to do any more for me. I am monstrous sleepy.

good night | Your affectionate son | Jack

Tell my mother that I eat a most enormous mess[12] of bacon and cabbage for dinner to day.

When will parliament meet? When will the Col.[13] be home? Do you think he'll be at home at Christmas?

I was talking to John O'Neill the other night about different situations. He says one in the bank of Ireland would be first rate after a little time. Mind I dont mention this on account of any anxiety I feel to leave my present situation, only to give you a greater scope should you be speaking to any of the gentlemen. To fill this however I should have good bail.

RI MS JT/1/10/3254–5
LT Transcript Only

---

1. *Friday night*: LT gives postmark as 'Nov. 20th, 1841', which was a Saturday.
2. *miracles will never cease*: a proverb first recorded in the eighteenth century, but drawing on Genesis 8:22.
3. *Tom Maguire's*: Father Thomas Maguire.
4. *Dan O'Connell*: Daniel O'Connell.
5. *Egad*: a softened oath originally derived from A God! (*OED*).
6. *Mr Young*: see letter 0111, n. 1.
7. *teatotaler*: variant spelling of teetotaler.
8. *'the humours of Glin'*: 'The Humours of Glin' ('Cuilinn Uí Chaoimh'), a popular Irish jig.
9. *the figure of* <u>eight</u>: a (principally Scottish) country dance in which each dancer follows a path like the figure 8.
10. *Billy Tyndall*: his uncle, William Tyndall.

11. *my last production ... editor himself*: *CS*, 13 November 1841, p. [3]; see letter 0110. The *Carlow Sentinel*'s editor was Thomas Carroll.
12. *mess*: a serving of food (*OED*).
13. *the Col.*: Colonel Henry Bruen.

To John Tyndall, Snr 28 November 1841 0113

Kinsale, Sunday[1] morning

My dear Father

In the postscript to your last letter[2] you requested me to write to you immediately, this injunction I should have attended to had I not been waiting for the result of an application which I had made respecting my getting leave at Christmas. We were all bustle last week, preparing for the reception of Major Waters, our District Superintendent, and Col. Colby the Superintendent of the Ordnance Survey (I should have put his name first, but no matter). We received intelligence from Cork that they would certainly be here on either Wednesday or Thursday last. I had a written application placed in the hands of Sergt. Carey who has charge here. In it I requested him to use his influence with Major Waters on my behalf. He promised to lay it before the Major when he came, but we were disappointed in our expectation, neither Col. nor Major has made his appearance here. Be that as it may, I intend forwarding my application to Cork this week. I have every reason to hope that it will be granted.

My little family was on the point of being divided. Last week we were near losing poor Evans. Start not – I don't mean to say he was going to kick the bucket, but he was put down as one of a party who are going to England on the 11th of next month. His name was forwarded to Major Waters along with several others. I believe the Major objected to sending draughtsmen away and Sergeant Carey told Evans the other day that 'the probability was that he would not go' I am extremely glad of this. Phil[3] and I have weathered the gale for a long time together and it would cost us a pang to part. Phil however is not idle. He is at present in the field looking out for something more substantial than the Survey. So you tell me the die is cast and Emma Tyndall married.[4] I would have given a good deal to be present on the occasion. It would have been glorious to see me stuck to some <u>auld</u> horse's tail lashing away with might and main.[5] I hope however to see her happy at Christmas.

A fortnight ago Cuddy and I had it laid out to get leave together he to go to Tyrrell's Pass[6] and I to Leighlin. We intended to take the packet from Cork and go by Dublin.[7] On calculation I found this to be the cheapest way of travelling. But unfortunately our schemes were all abortive. Cuddy forms one of the party which proceeds to England, so that in the event of my getting leave I intend to coach it, instead of sailing to Dublin and thence by the flyboat[8] &c. home.

You shewed your prudence by not going to the wedding when you apprehended such fatal circumstances;[9] but I know how I'd act myself had I been in your place. I received the Sentinel you sent me for which I thank both you and my uncle Caleb

Your affectionate son | John Tyndall.

RI MS JT/1/10/3256
LT Transcript Only

1. *Sunday*: LT gives postmark as 'Nov. 28th, 1841'.
2. *your last letter*: letter missing.
3. *Phil*: Phillip Evans.
4. *Emma Tyndall married*: see letter 0111.
5. *might and main*: utmost or greatest possible power or strength (*OED*).
6. *Tyrrell's Pass*: in County Westmeath.
7. *take the ... by Dublin*: probably on the Cork to Dublin route operated by the St. George Steam Packet Company.
8. *flyboat*: a flat iron canal boat that could achieve speeds of up to 10 miles per hour.
9. *such fatal circumstances*: the reasons why Tyndall's father did not attend Emma Tyndall's wedding were presumably set out in the missing letter.

To John Tyndall, Snr 7 December 1841 0114

Tuesday night | 1841.[1]

My dear Father

I was making a little account book for you this night when I received your letter.[2] I was anxiously expecting one from you for some days. I had written part of a letter to Mr Conwill, and in it I was determined to request him to tell you to write to me when lo! the appearance of your letter rendered all unnecessary. You may see by my employment that there is not much doubt of my going home at Christmas. I wrote to Mr Little[3] the coach agent in Fermoy this morning requesting him to let me know the best mode of conveyance from this, and to give me some information respecting the fares. I'm sure he'll tell me all.[4] So you tell me that Mr Steuart and Mr Bredin have been kind enough to approve of my attempts at poetry, this is a cause of great satisfaction to me, as I know them to be persons of good taste and judgment in such matters. I heard Mrs Steuart's taste praised long before I left home. I hope in my future attempts I shall continue to deserve their approbation. I sent off a long piece this morning – the subject is 'Landlord and Tenant'[5] – it's blank verse. Tell me what piece you mean by the last. I sent one on the registry[6] and I don't know whether it gained insertion or not as I have never seen the paper. It begins with these words:– | 'Child of the North' &c.[7]

If you could procure the paper for me that contains this I would feel very grateful, I send you the piece you requested – take care of it and send it back to me, but tell me what do you want with it? You did not tell me this and I am puzzled to find it out. Mr Conwill told me every way he was circumstanced in his last letter,[8] you no doubt will think it odd to see your direction on a letter to him, but it was to save the penny[9] and I had his written first and therefore could not enclose it in this small sheet. I am extremely glad to find that things are coming to a settlement in Leighlin. Where is Jim Walsh now? I'll have to be drawing on you for something to cover my feet when I go home, my old boots will hardly carry me to Leighlin they are like a patch work quilt. Now I beg of you not to let your next letter be so long coming as your last, write soon and always believe me

My Mother's son | John Tyndall.

RI MS JT/1/10/3258
LT Transcript Only

---

1. *Tuesday night | 1841*: internal evidence suggests Tuesday 7 December 1841.
2. *your letter*: letter missing.
3. *I wrote to Mr Little*: letter missing.
4. *the best mode ... tell me all*: Deborah McAssey conveys some of this information after staying with James Little; see letter 0125.
5. *'Landlord and Tenant'*: *CS*, 11 December 1841, p. [3] and 18 December 1841, p. [3].
6. *on the registry*: After 6 January 1841 letters handed into a post office could be registered (i.e. entered into a register) and a supplementary payment made in order to increase the safety of delivery.
7. *'Child of the North' &c*: 'The Battle of the Constitution is to be Found at the Registry', *CS*, 27 November 1841, p. [3].
8. *his last letter*: letter missing.
9. *save the penny*: Since the advent of the Penny Post in 1840 the cost of sending a letter was 1*d*., a sum that Tyndall had saved by enclosing this letter to his father inside the presumably larger letter to John Conwill.

From Phillip Evans 15 December 1841 0115

Tuesday night, Dec. 15th 1841.[1] | PRIVATE

My dear Jack

Here goes to give you a sketch of Cuddy's and my proceeding since we parted.[2] You know we parted but not for ever, sorrow take this bad ink. I say, Cuddy, ring the bell and ask Mrs Morgan[3] for other ink Mrs M. Oh yes, Sir would you like red ink? Evans, yes that will do. Well Jack my boy, as I was saying, we left your place any way and I was hardly out of sight when I commenced pouring out a volume of execrations upon Hunter[4] for not lending me his rubber[5] as the rain commenced to descend pretty fast, however it was now too late to think of this I made up my mind for a fine ducking, but indeed I must say that we escaped very well. I pass over the road without meeting anything attractive save a fox hunt, which I admired very much, I have no time to say any more about this, so I landed at the post office. Here we were informed that the coach would not pass for four hours – this was indeed very annoying. how to pass away this long time was a mystery, we strolled over to the village a ¼ of a mile from the office and after a few minutes hesitation bolted into a public house with our God save-all-here,[6] planted ourselves down at the fire, and got something to knock the cold out of us. We then came to the conclusion of taking a walk to pass away the time, so away we started on the road to Cork, but faith our walk was soon interrupted and our five link[7] paces shortened so much so that we had to come to the right about face, the rain commenced to pour upon us in torrents, so back to the nine-mile house[8] we were forced to march. In we landed to the place we had so recently left here we had to content ourselves till 5 ½ o'clock, now we start for our resting place where we got about 7 o'clock, left our boxes in the coach office, then out with my sketch which I found very accurate. Tell Mr Davy that I found the M. Chapel ½ a link out of position[9] and that he made a mistake in the number of arches in the bridge, for which I must get him checked ½d a day. Well Sir from it I found out Mrs Barrey's,[10] without making the least inquiry, tumbled (dont think when I say tumbled that I was drunk) in here. I saw Mr Todd, Mr Martin[11] being out, who came in a few minutes after. Took tea and after some chat got a bed, determined to be up at 8 o'clock in the morning. Here I cannot pass without telling you that I never experienced such unbounded kindness and friendship in all my life as from Mr Martin and Mr Todds. It is quite impossible for me to describe their kind feelings for us. Well Jack, we slept very well and was at office before the time this morning, was introduced to the Sergt.[12] the first thing was to sign our names on a roll as being present, that this is done there is no checking I might say. In a few minutes what was brought to myself but a content sheet of upwards of 6000[13] to transfer and plot the fair plan of it. ho! ho! says I, see what I have come to at last. You know I didn't like the idea of plotting[14] at all; however there was no use in complaining, so to work I set, got some triangles transferred[15] and a few lines plotted to day. Cuddy is at the Registery in a different room, nothing only accuracy and take lots of time – no matter about time, have the work correct is the word. Mr Martin is in the room with me and if his kindness last night could be surpassed he done it to day in the office. In fact every one of them are very kind. Tell Mr Collins I delivered his notes and that I am in the room with Turner. I saw one fair plan to day done by a young hand I have my own notions about the drawing but John dont let any one know this, not even Ginty or Tid[16] for I fear it might come here I could say much more but Cuddy is annoying me, however I will say a little more. We are now in our lodgings very comfortable a fine big woman – the first word was that we should call her nothing but Mother. Jack that's the woman for you boy. O! fine steak for dinner plenty of onions and pickled cabbage for dinner, but Sir you must know that this dinner was got ready in 5 minutes. A fine dancing room, in which Jim's Bessy and Hunter's Sally[17] could figure in first style in a set of lancers,[18] yes boy, and at Christmas we will have a dance Cuddy is blasting my soul and yours too for keeping him out of bed, but man I cant tell you half enough. our Bed room is in first order, Bandon is a splendid town, oh Tyndall if you were here to night when the two bands were playing, I think you would apply to come here. lots of girls fine heifers[19] here boy O! I will be introduced or acquainted here shortly. I will be making up a letter for you again, you answer this. I am afraid you will never make this out, Cuddy wont let me say any more

Good night | alls well | Jim[20]

Write soon Sunday boy O!

Jack my boy come immediately to pertake in the smiles of heaven born angels I and Jim is nearly heart broken, for why, we are at a loss how to badger the one half of the try <*word(s) missing*>

Yours | Sam.[21]

RI MS JT/1/11/3567
LT Transcript Only

---

1. *Tuesday night, Dec. 15th 1841*: LT notes that the letter sent was from Bandon, a town on the River Bandon, about 7 miles north-west of Kinsale.
2. *since we parted*: Phillip Evans and Martin Cuddy had recently been transferred to the 2nd Division, C District under the command of Lieutenant Charles Wilkinson (see letter 0144, n. 10) and were stationed at Bandon, County Cork.
3. *Mrs Morgan*: At Bandon *Slater's* lists Robert Morgan (boot and shoe maker), Mary Ann Morgan (milliner) and William Morgan (worsted manufacturer), any of whom may have supplemented their incomes by taking in lodgers.
4. *Hunter*: The name is given by LT as 'Gunter', but it is most likely to be either Richard Hunter or his brother William Hunter, both of whom were civil assistants working with the 4th Division, C District.
5. *his rubber*: a rubberized waterproof coat, manufactured using a process invented by the Scottish chemist Charles Macintosh.
6. *God save-all-here*: a common Irish salutation on entering a house.
7. *link*: see n. 9; notably Evans even estimates the distance of a casual walk in the units of measurement used by the Ordnance Survey.
8. *the nine-mile house*: probably a safe house; may relate to the Ordnance Survey's use of a base line of 8 miles.
9. *the M. Chapel ½ ... position*: a link is a unit of length of 7.92 inches. There are 100 links to the chain and 8000 links (80 chains) to a statute mile. Corporal James Davy's error was therefore 3.96 inches (10.1 cm). As the distance measured would have been considerable, an error of half a link would be negligible. Thus Evans was complementing Davey on the accuracy of his work. 'M. Chapel' has not been identified.
10. *Mrs Barrey's*: not identified; oddly, Evans suggested that his landlady was called Mrs

Morgan at the beginning of the letter, so this may be an error on his part.

11. *Mr Todd, Mr Martin*: John Todd, a civil assistant working in the 2nd Division, C District, under the command of Lieutenant Charles Wilkinson. Possibly John Martin, who was a civil assistant in the 5th (now 3rd) Division.
12. *the Sergt.*: not identified.
13. *content sheet of upwards of 6000*: probably a list of more than 6,000 locations.
14. *plotting*: transferring information from field books.
15. *some triangles transferred*: triangles drawn to represent the distances on the ground produced by the method of triangulation.
16. *Tid*: John Tidmarsh.
17. *Jim's Bessy and Hunter's Sally*: Phillip Evans's lady friend Bessie and Sally, the friend of William or Richard Hunter; Bessie and Sally not further identified.
18. *a set of lancers*: a group of quadrilles danced by eight or sixteen couples.
19. *heifers*: derogatory slang for women or girls, comparing them to young cows that have not yet calved (*OED*).
20. *Jim*: Phillip Evans.
21. *Sam*: John Roberts was given the nickname 'Sam Weller'; see letter 0109. However, Roberts is not known to have been in Bandon at the time this letter was written. LT has suggested that 'Sam' was Martin Cuddy.

To John Tyndall, Snr 17 December 1841 0116

Kinsale, Thursday night. Dec. 17, 1841.

My dear Father

I believe I was born for disappointment. All my long cherished hopes relative to going home at Christmas are now blasted. This day gave the final stroke to my expectations. I was fed up with the hope of getting away without the least hesitation and until yesterday I had not the least reason to believe that my hope was illfounded. This makes my disappointment the more severe. Yesterday morning the division Clerk[1] told me that Sergt Carey had received a letter from Major Waters in which he stated that in the event of any man applying for leave at Christmas he would not grant it. I then went in to Carey to know the truth of the matter myself. He confirmed all I had heard before. I represented the way I was circumstanced – told him I had everything arranged, that I had business which concerned me deeply to transact, and that any disappointment might be of lasting injury to me. In fact I said all that could be said on the subject. He then told me that if I wrote a letter to him requesting a fortnight's leave he would forward it to Major Waters. I thanked him for even this small favour. Came home and wrote the letter. Well, this morning I think the Sappers, that is the noncommissioned officers, had a consultation about the matter as one of them came in and told me that Serjeant Carey did not wish to forward my application, that he would much prefer my deferring it until the County Cork would be finished, and moreover that there was not the slightest chance of success even if the application was sent as he would have to let the Major know that I could not be spared. 'Well' said I, 'he can do nothing more than refuse me, so I would wish to have the application sent'. Carey then got it for his perusal, the result was that he said it would not do at all. Now I had shewed the thing to men of ability and experience beforehand, and their opinion was that it could not be better suited to the purpose it was designed for. But no matter for that, I should remodel it according to the views of my masters. This would couch the thing in such terms as would preclude any possibility of success, the very point on which I depended most being expunged from my original request. After a moments consideration I tore the thing up and threw it in the fire. This is the fruit of my being held under the Superintendence of such fellows. No matter – grumbling would do me more harm than good so I must pocket my resentment.

Major Waters says that every man must be out of Ireland by the 31st of March;[2] this is the reason of my detention as the work is in a very backward state. But if my application had been sent off when I first made it I'm quite sure I would not have been refused. I may bite my lip in vexation that's all the good it will do me.

The motive, next to that of seeing you, which made me anxious to get home was the hope that we might make interest enough to get me a situation. This being now out of the question, everything devolves on yourself. I hope you wont let the opportunity when the Colonel[3] is at home pass unimproved. If you do, such another may never present itself. He can do wonders and I am sure you have a good many friends who have great influence with him, You know the two situations which I look to. Good bye

Your affectionate son | John

RI MS JT/1/10/3259
LT Transcript only

1. *the division Clerk*: not identified.
2. *out of Ireland ... March*: The Survey of Ireland was nearing completion and many of the personnel had either moved to England or were completing their tasks in Ireland. Tyndall did not leave Ireland until early August 1842.
3. *the Colonel*: Colonel Henry Bruen.

To John Tyndall, Snr 30 December 1841 0117

Kinsale, Wednesday night | Dec. 30th, 1841.

My dear Father

I believe I promised you once that I would write to you weekly[1] no matter whether you wrote to me or not. Well here goes to redeem that pledge as I believe it's more than a week since I last wrote. I have received nothing in the shape of a letter from you in the mean time, so that this will place you two in my debt.[2] The hope that you are waiting until you have some good news to communicate enables me to wait with patience.

I spent my Christmas very comfortably here. Tom Foy, a lad named Holmes and I dined together. Our establishment has undergone some alterations since Evans's departure.[3] Since I came to Kinsale I had Ginty lodging with me besides Evans and Tidmarsh. Foy in the mean time was lodging in the barracks. Well an order came that all the Sappers were to leave barracks to make room for some troops that were coming from Cork. Foy therefore had to seek lodgings elsewhere. He has taken a room in the same house with me[4] and we all mess together. Holmes is living with him – he's a nice quiet lad – and as far as Foy himself goes a better messmate I would not ask. He is very much changed since his imprisonment. The death of his wife[5] has caused the poor fellow to become quite religious. So that he possesses all the good qualities that he possessed in Leighlin unattended by that wildness and love of sport which characterised him there; and when I speak of Foy's good qualities I must say that they are many. I think however that we are not doomed to be long together; we will not be so in Kinsale at all events. We have got orders to prepare to be off to Cork in three weeks at the utmost. Sergeant Carey thinks we'll go at the end of next week. Major Water's confidential man[6] was here yesterday, inspecting the office, and thro' him the order came, so that I must soon be packing up my traps for the grand parade.

I was in Bandon[7] on Sunday week, seeing Evans and Cuddy. They were very comfortably situated at the time, but since that I hear they have exchanged their lodgings for others still more comfortable. Bandon is a fine town, far very far superior to Kinsale. It's lighted with gas.[8] It's a good smart walk from this to it. The roads on Sunday week were very heavy on account of it's having snowed all day without intermission, so that I had a heavy jaunt trudging home on Sunday night

But I was near forgetting to tell you how I spent my Christmas day. I was at preaching in the morning at 6 o'clock then after breakfast got a gun and went out along with a friend of mine named Chadwick. We walked all day here and there and could scarcely get any thing. I had serious notions of wreaking my vengeance on an unfortunate little wren that sprung up before me now and again. In the evening we found ourselves a great way from home with empty bellies and empty pockets. We faced the road however, and before we got half way my companion began to wish for a cold potato or even a few skins he often said he would be glad to make his Christmas dinner on what he'd pick off a hat full of skins. I thought of the little child with the dirty nose that you robbed at Ballytarsna,[9] and wished that fortune might throw such another in my way. My wishes were fruitless. We trudged on, getting more and more hungry. Hunger will break thro' a stone wall, so that you must not be surprised when I tell you that I hit on the expedient of begging my way home. Accordingly I bowled up to the first cabin I met with a smiling – 'Sews merra widh'[10] I asked for something to eat, even a cold potato, but without success, and to the shame of hospitality be it told that I was refused in this manner in no less than a dozen places along the road. We at length hit on a place where we got two potatoes apiece and a wee bit of mutton and they enabled us to get home. Maybe I wouldn't cut into the goose then!

I would feel thankful if you'd send me the paper containing the piece 'Child of the North' &c.[11] and if you be done with my first piece,[12] send it to me also.

I sent you the Times and Waterford Mail,[13] did you get them?

I have not heard from Mr Conwill this long time. I cannot devine the reason of this. Ask him when you see him what have I done to merit this silence.

How are my mother and Emma? Give my love to both. I hope my mother is not fretting on account of my not being let home. I felt the disappointment as keenly as any one, but when I see things irrevocably fixed I find there is no use in repining. Instead of being dejected at such little mishaps, I for my part ought to feel thankful for what I enjoy – good health and strength. I have now been nearly three years on the Survey, and during that time I have never lost a single day thro' sickness. So that I hope my mother when she considers this will not let light matters trouble her mind.

Good bye | Your affectionate son | John.

RI MS JT/1/10/3260–1
LT Transcript Only

---

1. *I promised ... weekly*: see letters 0054 and 0062.
2. *two in my debt*: including letter 0116.
3. *Evans's departure*: see letter 0115, n. 2.
4. *the same house with me*: see letter 0072.
5. *his imprisonment ... his wife*: see letters 0097 and 0103.
6. *Major Water's confidential man*: not identified.
7. *Bandon*: see letter 0115, n. 1.
8. *It's lighted with gas*: It is not known when the Bandon was illuminated by gaslight or when the Bandon Gas Company was formed. The Dublin Gas-Light Company had been formed *c.* 1821.
9. *the little child ... at Ballytarsna*: this appears to relate to a notable event in Tyndall family history; Ballytarsna is about 6 miles south of Carlow.
10. *'Sews merra widh'*: probably Dia is Muire dhuit, a common Gaelic greeting meaning, literally, God and Mary be with you, but generally understood as God bless you or God save you.
11. *'Child of the North' &c.*: 'The Battle of the Constitution is to be Found at the Registry', *CS*, 27 November 1841, p. [3].
12. *my first piece*: W. S., 'The Repeal Meeting at Leighlin', *CS*, 25 September 1841, p. [3]; see letter 0101.
13. *the Times and Waterford Mail*: Tyndall probably means *The Times* of London, which circulated widely in Ireland at this time; the *Waterford Mail* was a twice-weekly conservative Protestant newspaper founded in 1823.

1842

To John Tyndall, Snr 16 January 1842 0118

Robroy Hotel,[1] Cork. | Jan. 16th, 1842.

My dear Father

After a weary week of incessant tossing I have at length found time to sit down and write to you. You may see by casting your eyes to the right hand corner[2] that I have left Kinsale and am at present a denizen of 'Cork's awn town' tho' this can scarcely be considered as strictly true as my present quarters are not within the precincts of the City.[3]

Last week we were on the tiptoe of expectation thinking that every post that came had the order for our removal.[4] The order however did not come until Saturday night.[5] On Monday we broke up and a lot of us purchased our liberty[6] until 12 o'clock on the following Thursday. It was specified that Wednesday was to be the day on which we should flit. Well I packed up on Tuesday having previously determined to walk to Cork, a distance of about 16 miles. Tuesday you may recollect was a very fine day and we had sanguine expectations of having a pleasant walk on the day following. But all my hopes as to this matter were crushed the moment I opened my eyes at about 4 o'clock in the morning. The rain was pelting most unmercifully against my bedroom window, while the wind moaned piteously thro' the crannies of the houses. I however got up and breakfasted early, consulted the wind's eye and found it enveloped in clouds; in fact there was not the least sign of it ceasing. So that I had to be content until 12, as I was told that the day would then clear up. Well 12 came and no cessation of rain; so did one, two, three. I then gave up all hopes and sat by the fireside until bedtime, then I started to hammock. Wakened at 4 o'clock on the following morning[7] and heard the rain still pelting. It cleared up at 6 o'clock, so I breakfasted and started along with four or five more at 8 o'clock. The roads were very heavy and by the time that we got to the barrack hill not one of us was in much humour for jumping a five barred gate. We got a little refreshment and then were obliged to unpack and drag the tables, stools &c. in to the barrack. We did not get away until dark and that was a bad time to look for lodgings. We (that is Foy and Holmes, a friend of mine named Chadwick and myself) set out on the search. We rambled a great way but without success. We then stepped into an eating house and got each a snack and finally determined to sleep there that night. Well bed time came and we were ushered into our dormitory. The look of the place was not very prepossessing. A broken skylight was all the window it could boast of thro' which the rain trickled down the walls. The beds seemed something worse of the wear, and the idea that they might contain bugs compelled me to give more than one involuntary shrug of my shoulders. Well, we stripped and Chadwick went to bed. I then thought of looking about me. I examined the bed clothes raised and found them very different in colour from snow. I went farther, I raised the tick and there to my no small disgust I found a foolish bug foraging about. I suppose he only formed the vanguard of a host of them. I call him foolish because if he had kept under cover he might have got his maw[8] filled before morning. The effect that this discovery had on poor Chadwick was amazing. An electric shock could not produce the like. He sprung from his lair like a mountain cat. We all dressed again, settled for what we had got and sallied forth once more. We got a good bed at the Munster Inn.[9] Before we left Kinsale the four I have mentioned had determined on stopping together; we found however that it was difficult to get accommodation for four and thought it judicious to separate. Chadwick and I have pitched our tent at an inn which stands nearly opposite to the barrack gate, and as the office is situated in the barrack it is very convenient. Some of our lads have to walk nearly a mile every morning from their lodgings in the City up a tremendous hill.[10] Our board and lodging will stand in about 9/- a week here. Our board consists of two meals daily. No tea at night. In fact this is a luxury I have dispensed with these 18 months so that I don't care about it. I could make this epistle twice as long, but I have three or four more to write and so must cut this short.

I suppose you have seen Mr Alexander, and in your next letter will be able to tell me the result of your conference with him. I am as stout as an ox and have a capital edge for my dinner which I expect to see appearing every minute. Give my love to my mother and Emma

Good bye | Your affectionate son | John

Direct your letter to me – Ordnance Survey Office Military Barracks Cork.

RI MS JT/1/10/3262–3
LT Transcript Only

---

1. *Robroy Hotel*: the Rob Roy & Commercial Inn, King's Square, Cork (*Slater's*).
2. *the right hand corner*: where the sender's address would have appeared on the original letter.
3. *present quarters ... the City*: Cork Barracks and the Rob Roy Inn were both just outside the City limits

4. *our removal*: from Kinsale.
5. *Saturday night*: 8 January 1841.
6. *purchased our liberty*: were given leave of absence; the civil assistants may have been required to pay a premium.
7. *the following morning*: Thursday 13 January 1841.
8. *maw*: the stomach of an animal (*OED*).
9. *the Munster Inn*: probably the Munster Hotel, Coburg Street, Cork.
10. *up a tremendous hill*: Cork Barracks are located at the top of a steep hill to the east of the city.

To John Tyndall, Snr 26 January 1842 0119

Strand Road, Cork. Wednesday[1]

My dear Father

I am beginning to think that the post offices are dealing unfairly with one or the other of us. I wrote to you on Sunday week[2] and I think it highly improbable that you should delay until now without writing to me, this induces me to enquire if it be the post office I have to blame for being so long without hearing from you. When I last wrote to you I was quartered at the Robroy hotel.[3] This my letter stated. I now stop on the Strand Road[4] having removed from the Robroy of Friday last. My mother I'm sure will be glad to hear that my hostess is a quaker. Her name is McCarthy.[5] I hardly thought there was a reformer of the name to be found.[6] I am comfortably situated. These 'Verilies'[7] have everything very nice and regular, its a quiet place, removed upwards of half a mile from the din of the City. I am still employed colouring the City of Cork. I am making great improvement, I was very fortunate in being transferred from Kinsale, the style of drawing adopted for the City of Cork is quite different from our usual method. It is a very useful style, and I have now a good opportunity of making myself master of it. Ask my mother did she ever hear of Sunday's well.[8]J. How, 1841–3), vol. 1, p. 47). If she did she knows of the ground that I am employed at. Little I thought when she used to be telling me of the Grand Parade and the dyke[9] that it would be my fortune to be rambling thro' each, yet so it is.

We had a dreadful storm here last night.[10] It continued all day unabated. The barrack got a great stripping. The slates were actually flying like feathers thro' the air. It was rumoured to day that three women were killed in the meat market by the falling of tiles. I cant vouch for the truth of the report. When I was coming out of the barrack gate this evening I cast my eyes on the Robroy hotel. There were two splendid bow windows in front the lower one has every pane in it shivered to pieces and there was not a vestige of the upper window to be seen at all. Not a sash remained. The house was nearly unroofed. Several others in the vicinity fared little better. The Garrison hotel[11] had even the leading of the roof torn away, and some houses lost not only their slates but their rafters also. I cant say what appearance the city presents as I have not been there since the storm abated. Our little domicile fared well. I believe it did not lose a slate. Its very well sheltered. I received a paper last night from Emma Young.[12] This name sounds odd to me. I was near writing it Tyndall. It was the Sentinel I received. I dont see W.S. figuring there now the fellow has become dumb. They expect to have finished the Irish Survey at the end of March. Surely I'll get away to see you then.

Your affectionate son | John

Direct your letter to Cpt.[13] Carberry McCarthy, Strand Road Cork, for me. Have you seen Mr Alexander?

RI MS JT/1/10/3264
LT Transcript Only

---

1. *Wednesday*: LT gives postmark as 'Jan.27th, 1842', which was a Thursday.
2. *I wrote ... Sunday week*: letter 0118.
3. *the Robroy hotel*: see letter 0118, n. 1.
4. *the Strand Road*: Strand Road is to the north-east of the City of Cork.
5. *My mother ... McCarthy*: Tyndall's mother (Sarah, née McAssey) was descended from the Malone family of Ballybrommell and had included Quakers among her great grandparents. The only direct evidence that connects Sarah with the Quakers is a letter of hers in which she mentioned that she had been taken by her father to a Quaker meeting (N. McMillan and M. Nevin, 'Tyndall of Leighlin', *Carloviana*, 27 (1978–9), pp. 22–7). No person named McCarthy has been found in the records of the Cork Quaker meeting. Although Tyndall describes Mrs McCarthy as a Quaker, it seems likely that she (like Tyndall's mother) was not a practising Quaker but was descended from Quakers.
6. *a reformer ... be found*: McCarthy is usually an Irish Catholic name, and thus most likely the name of a nationalist and supporter of the repeal of the Union with Britain.
7. *'Verilies'*: truthful people; Quakers are committed to dealing honestly and, more broadly, to living in accordance with their religious principles
8. *Sunday's well*: part of Cork situated in the north-west of the city. Mr & Mrs Samuel Carter Hall described it as providing 'a magnificent view of the river and of the landscape for many miles around it. "Sunday's Well" derives its name from one of those sacred fountains ... where devotees assemble, at particular periods, under the belief that the water is blessed and cures all disorders'(*Ireland: Its Scenery, Character, &c.*, 2 vols (London:
9. *Grand Parade and the dyke*: The Grand Parade in Cork was wide and lined with fashionable houses; the Mardyke (the shortened name being 'dyke') was a tree-lined promenade beside a canal.
10. a *dreadful storm here last night*: on 25–6 January ('Awful Hurricane and Loss of Life', *Cork Examiner*, 26 January 1842, p. [2]).
11. *Garrison hotel*: the Garrison Hotel was opposite the Barracks in Cork.
12. *Emma Young*: Tyndall's cousin Emma (née Tyndall) who had recently married George Young; see letter 0111.
13. *Cpt.*: There is no evidence that Carberry McCarthy was an army Captain. 'Cpt.' may be a mistranscription of 'Cpl', although no Corporal of that name has been identified. It is possible that he was captain of a ship, but that has not been corroborated. A contemporary reference suggests he may have been a clerk (*Cork Examiner*, 5 August 1842, p. [2]).

From John Tyndall, Snr 1 February 1842 0120

Mr Carberry McCarthy | Strand Road, Cork | for John Tyndall, | Leighlin Bridge | Tuesday night Feb, 1. 1842.

My dear John

I think you are anxiously expecting something in the shape of a letter from me and though you think there is blame to be laid to the charge of the post office be assured that there is not as I received your two last letters.[1] I was waiting for Mr John Alexander to come from Dublin where he has been this long time. The day I went to Milford[2] he was not there, but I saw Mr Charles[3] and read your letter for him. He said he had heard of you and would be glad, if in his power, to serve you. But he was afraid it was not; as England was filled with engineers; that the man to whom he is bound is often without employment, that you would have better chance in Dublin, but that as soon as he would go over to London – which will be next week – he will speak to his master for you.

I was in Milford this day when he repeated the same words to me and asked me did I hear from you since. I also saw Mr John[4] who was at the time very busy. I told him my business. He said he would be glad to have it in his power to serve a protestant man, especially such a one as you were, and requested of me to leave your letter with him till the next day I go to Milford. Its useless to speak to Mr Steuart on this occasion, as he is using all the interest he has to get the eldest of the Rev Mr Hickey's sons,[5] who is very badly off, some situation and wrote to Colonel Bruen in London on the subject, who told him that at present nothing could be done for him. They would be glad to get a situation for him of sixty or seventy pounds a year. You may guess from this the low ebb Mr Hickey's family has come to, and the great difficulty of getting any situation. Keep this a secret, as they would not wish to have it known. I have not spoken to the Dean[6] yet, but will shortly, and will let you know his answer. The Colonel[7] was at home for a few days, as your uncle Caleb was with him on Saturday last and expects to be appointed High Constable at the Assizes;[8] and only he went off on Monday though the day being tremendious wet I would have got letters from the Dean and Mrs Steuart to him. We had a dreadful storm here a few days ago,[9] five or six thatched houses were blown down, windows broken, and slate houses stripped. We were obliged to keep the shutters of the shop window up all day else the house would be blown down. I left a letter this day with Mr Conwill as I was going to Milford. I think it was from Evans, as the Bandon post mark was on the letter. When I hear of anything worth letting you know you shall hear from me. We are all well | Your affectionate father | John Tyndall

The bug was dreadful. Why is W.S.[10] dumb?

RI MS JT/1/10/3265
LT Transcript Only

1. *your two last letters*: letters 0118 and 0119.
2. *Milford*: the Alexander family's estate on the River Barrow 3 miles north of Leighlin Bridge.
3. *Mr Charles*: Charles Leslie Alexander (1820–88), who later entered the Church, was one of the sons of John Alexander.
4. *Mr John*: John Alexander (1802–85), was another son of the above. He had studied at Trinity College, Dublin, served as High Sheriff of Carlow (1824) and in 1853 was elected Tory MP for Carlow Borough.
5. *Rev Mr Hickey's sons*: possibly a son of the Reverend James Hickey.
6. *the Dean*: Dean Richard Barnard.
7. *The Colonel*: Colonel Henry Bruen.
8. *appointed High Constable at the Assizes*: Caleb Tyndall was 'appointed High Constable for the Barony of Idrone west'; see letter 0130.
9. *dreadful storm here a few days ago*: on 25–6 January.
10. *W.S.*: Walter Snooks, one of Tyndall's pseudonyms.

From Deborah McAssey 4 February 1842 0121

Fermoy[1] | Feb. 4th 1842.

My dear John

I am sure you will be surprised to find my letter dated from this. I came down here yesterday with Miss James[2] who was coming to see her sister Mrs Little[3] and will stop here until Monday morning next when we return home. I would be very glad to see you if you could come down. If not write by return of post I would have written to you long ago only I heard that you had left Kinsale and consequently I did not know where to direct. I know you will say I am never without an excuse but I am not the only girl that knows how to make one. We are still thinking of going to America. I expect Ballybromwell[4] will not belong to my father against[5] I reach home. We expect to leave Ireland in April. I hope that you will come to the Co.[6] Carlow before then – If you cannot see me here send me a long letter and a great deal of news and don't forget to let me know how you leave all your little girls. You need not fear my telling Rachel.[7] I had great sport coming down here, we travelled on a jaunt car[8] and were obliged to stop one night in a village 7 miles at this side of Clonmell.[9] There was great preparations there for a patron[10] the next day the patron was to be for making up matches coming up to Shroff.[11] I must have been very unfortunate in being there a day too soon or I might have a chance of a Co. Tiperary Irishman. There is no help for misfortune. I saw Wm. Chapman[12] coming through Clonmell. He was very glad to see me and made me promise to stop with him a day on my return. He and his wife look uncommonly well and have a nice place. He had £100 a year from Mr. Hughes.[13] His employer died about one hour before I reached Clonmell I suppose the business will be carried on by his sons.[14] I am anxious to hear how

Wm., will be circumstanced. Pardon great haste. Let me either see you or hear from you before I hear this. Direct to me at Mr Little's

I am ever your | affectionate cousin | Deb. McAssey

RI MS JT/1/11/3759
LT Transcript Only

1. *Fermoy*: a town in County Cork, about 20 miles north of Cork.
2. *Miss James*: not identified.
3. *Mrs Little*: Sarah Little (née James), who had married the Fermoy coach agent James Little on 22 October 1840 (Irish Church Records).
4. *Ballybromwell*: a village in the parish of Fenagh, County Carlow.
5. *against*: before (*OED*).
6. *Co.*: County.
7. *Rachel*: probably Rachel McAssey (1822–1908), Deborah's younger sister.
8. *jaunt car*: a light two-wheeled carriage pulled by a single horse.
9. *Clonmell*: Clonmel is a market town in County Tipperary.
10. *patron*: a feast day of a patron saint, especially in Ireland (*OED*); it was presumably the patron for Saint Brigid, on 1 February.
11. *Shroff*: a dialect word, seemingly specific to Tipperary, for Shrovetide, the three days before the beginning of Lent (see [O. J. Burke], 'The Bridegroom of Barna', *Blackwood's Edinburgh Magazine*, 48 (1840), pp. 680–704, on p. 699). Shrove Tuesday, which in 1842 fell on 8 February, was a traditional day for Irish weddings because of the prohibition on celebrations during Lent; see letter 0047.
12. *Wm. Chapman*: possibly a relation of Marianne Chapman; see letter 0033, n. 6.
13. *Mr. Hughes*: probably Thomas W. Hughes, owner of a mill on the quay in Clonmel.
14. *the business ... his sons*: In 1846 there were corn merchant businesses in Clonmel called Thomas Hughes and Son, Bridge Street, and Thomas and James Hughes, Quay Street (*Slater's*).

To John Tyndall, Snr 5 February 1842 0122

Cork, Saturday night

My dear Father

Your long expected letter[1] came at length to hand on I believe Wednesday night. It appears that you had good reason for your delay, namely on account of Mr John Alexander's absence. Mr Charles's[2] account of England is not a very good one. However the world is wide, and I'm not afraid to face it. I have at present a competence[3] and no doubt as to its permanence. To it therefore I shall stick until something good presents itself. You may have remarked in the course of my letters a tone in some very different from that of others. From some you may have gathered that I am rather impatient, while from others you might conclude that I am perfectly satisfied with my present situation. The fact is this: no letter of mine was penned thro' a spirit of impatience except that which I wrote at the time of Mulligan's removal,[4] and for this I was sorry. It is not discontent that induces me to write, and urge on you the necessity of exertion. No, I am not at all discontented with my present situation, but I have an eye to futurity and I know the value of time. This makes me dread the idea of wasting it. You have a better opportunity of knowing when an application should be made than I have, and you therefore should not let my letters urge you to make it at an improper time. I would therefore say, use your own judgement in this matter. Mr Alexander's interest is powerful,[5] and tho' at present it may not have scope to work in, still I am sure that at a future day it will be found availing. It gives me great satisfaction therefore to find him on my side.

Tidmarsh's father, who is at present in Leeds, sent him advertisement the other day taken from the Leeds Mercury[6] – It stated that there was a person wanted who knew everything respecting the Surveying, Mapping and computing of estates and Townships. Mr Tidmarsh desired the young fellow,[7] in the event of his not being able to fill the situation, to speak to me about it. I accordingly applied but have got no answer as yet. I have no expectation of being successful. This however I dont much care about as from the nature of the advertisement I dont think the gaining of the situation would compensate for the loss of my present one.

I got a letter from Debby yesterday.[8] She is at present in Fermoy at Little's[9]

Give my love to my mother and Emma I'll write to them shortly.

Direct your letter as usual, I mean as you directed your last.

Your affectionate son | John.

RI MS JT/1/10/3266
LT Transcript Only

---

1. *Your long expected letter*: letter 0120.
2. *Mr Charles's*: Charles Leslie Alexander; see letter 0120, n. 3.
3. *competence*: a sufficiency of means for living comfortably (*OED*).
4. *that which ... Mulligan's removal*: letter 0036.
5. *Mr Alexander's interest is powerful*: John Alexander was a very successful and highly-respected mill-owner with a large estate at Milford.
6. *advertisement ... Leeds Mercury*: 'To LAND SURVEYORS.–WANTED, in a Land Surveyor's Office, in Yorkshire, a PERSON of good Character and sober Habits, who has been accustomed to the Surveying and Planning of Townships and Estates. None need apply who are not fully competent to Survey, Plan, and Cast the same with accuracy and expedition. It is not necessary that he should be able to make finished Plans and Field Books. Apply, by Letter, addressed to No. 9, Mercury Office, Leeds, stating qualification and salary requested' (*Leeds Mercury*, 22 January 1842, p. 5).
7. *the young fellow*: John Tidmarsh.
8. *a letter from Debby yesterday*: letter 0121, from Deborah McAssey.
9. *She is ... at Little's*: see letter 0125.

From John Conwill 12 February 1842 0123

February 12th 1842.

'My dear Tyndall'

I return you my most ample thanks for the splendid eulogium conferred by you on me in your last luminous epistle.[1] I assure you that the crow mentioned in the fable, never felt more nettled at the insiduous flattery of sly reynard,[2] than I did at the beautiful line of demarkation drawn by you, between Hennessy's capabilities and mine. Imagine not that I deem your praises insiduous: by no means. I always look upon any thing flowing from your pen as the emanation of a pure and unprecedented mind – I question not your sincerity towards me and I unequivocally proclaim that whenever I become the object of your encomiums, the most unbounded enthusiasm seizes my imagination, and that the galvanic battery[3] with three hundred plates in operation could not add more to rouse my spirits at any time they may become dormant, than the vivid flashes of Tyndall's pen. You are a strong reasoner on the human mind, and when I contemplate the view which you have taken of frail man, so far as Hennessy is concerned, there cannot be a question about the generality of your predictions. You have said that 'time might teach him sense,' and you have asked 'when could he rival me at old Diophantus'. Now hearken – The other day a note was presented to me from Hennessy praying that I would condescend to remain in the school on last Saturday till two o'clock that he might communicate many strange occurrencies to me together with exculpating himself for his unparalleled ingratitude towards me. I remained, according to his importunities, and you may be certain that he approached me in the most crouching manner. I looked on the fellow with disdain and pity. What did he want? To be serious, he wanted the solutions to the Diophantine problems in Bonnycastle's Algebra.[4] I have refused him for some time in order that he might take a lesson from experience and hence learn to appreciate those who could 'stand by him when in deep distress.' I also apprised him that he should reward me with plenty of the ryno[5] if I should yield to solve the questions – this he promised to do.

Your mother feels much elated at, and pregnant with, the hope of seeing you shortly. You know March is fast approaching – she likewise forms very sanguine hopes that the spirit will move your hostess to preserve you from the damnable bugs of Cork.[6] I am sure the one that was about to attack you, intended to ape the frog which thought to swell itself to the size of an ox.[7] Why, the nasty thing, if it did not attack you so soon it might, in some part of the night, take a few good pieces out of your hind quarter.

Wishing you all kinds of felicity | I remain your faithful teacher | Conwill

RI MS JT 1/11/3523
LT Transcript Only

---

1. *your last luminous epistle*: letter missing.

2. *sly reynard*: a cunning anthropomorphous fox in European folklore, who, in a story from the ancient Greek collection *Aesop's Fables*, flatters a crow into singing, thereby dropping a piece of cheese from its beak.
3. *galvanic battery*: an electrochemical cell, named after the Italian physician Luigi Galvani (1737–98), which produces electrical energy from chemical reactions within it.
4. *Bonnycastle's Algebra*: Either *Introduction to Algebra* (London: J. Johnson, 1782) or *A Treatise on Algebra in Theory and Practice*, 2 vols (London: J. Johnson, 1813) by John Bonnycastle (1751–1821), Professor of Mathematics at the Royal Military Academy in Woolwich. On p. 132 of the former work Bonnycastle observed: '*Diophantine problems* are those that relate to the finding of square and cube numbers, &c. and are such as are generally capable of a great variety of answers'. On Diophantus, see letters 0054, n. 6 and 0060, n. 10.
5. *ryno*: money (OED).
6. *the damnable bugs of Cork*: see letter 0118.
7. *frog which ... an ox*: In another story from *Aesop's Fables*, a conceited frog attempts to inflate itself to the same dimensions as an ox, but bursts in the process.

To John Conwill [17 February 1842] 0124

Thursday night

My dear Sir

The thanks which you have awarded me in your last letter[1] are quite unmerited. The statement of my opinion as to the difference existing between you and Hennessy was merely a tribute due to your abilities; justice and truth demanded it, and shall Tyndall refrain from giving his teacher his well won meed when thus circumstanced. No! I would assert it in the teeth of Hennessy himself and those of any other boord popinjay,[2] nor would I shrink to declare the same from the mountain fastnessess of the ridge[3] where Mickey Byrne's[4] eloquence wakes the echoes of the hill to Newtown Chapel[5] whose walls have often rung in answering peals to the vociferations of Professor Murphy.[6] Aye if occasion required I'd assert your superiority amid the threatening scowls of the sons of science in Carlow College.[7] I would not cower before the mighty O'neill,[8] nor should the classic Hennessy call an additional streak of paleness to my cheek!

'Thrice is he armed who hath his quarrel just'.[9] What I have said then, being merely your due, it was an act of <u>duty</u> not of <u>merit</u> on my part to state my opinion.

So Mick[10] has at length found the want of a friend, I told you time would teach him sense, it is not until a man is deprived of a benefit that he is able to appreciate its value, and it was not till Hennessy found himself on his own hands, unsupported, that he was able to form a true estimate of your assistance. Never mind, each day will waft away its own cloud of vanity, and observation and experience will shake from Mick's mind the dusty tinsel which he may have borrowed from the great fellows in the Metropolis.

I am now going to trouble you. I as well as Hennessy feel the want of a helping hand. God forbid that my opinion of myself should ever mount so high as to induce me to think myself independent of others; each day's experience teaches me the folly of such a notion and I firmly hope that time, instead of effacing the idea, shall like a stream in its bed imprint it the deeper. But you are asking by this time what I want. Well I want to know how a table of natural sines[11] is constructed– I know everything about the projection of them– now by having the radius BA and the angle ABC, how am I to find the line AC?[12]

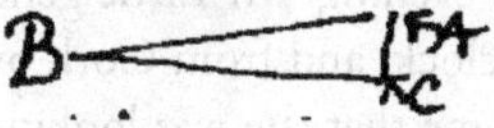
[13]

it will refresh my memory. I wish I was near you till you'd give me a good raking on Log^{s}.[14] Do you think I'll understand the algebraic formation of them? I'm improving myself greatly in my office business, my drawing of the city of Cork is thought a good deal about. Remember me to your mother.

good bye | Your affectionate pupil | John Tyndall

RI MS JT 1/11/3529
LT Transcript Only

1. *your last letter*: letter 0123.
2. *boord popinjay*: an ill-mannered dandy or foppish person (*OED*).
3. *the ridge*: a townland area on the western side in the Slieve Margy range of hills, opposite Newtown and at the start of the Castlecomer Plateau.
4. *Mickey Byrne's*: possibly Denis Byrne, also known as Monk Byrne; see letter 0066, n. 8.
5. *Newtown Chapel*: St Patrick's Church, Newtown, County Carlow.
6. *Professor Murphy*: see letter 0069, n. 9.
7. *Carlow College*: founded in 1782 by James Keefe and Daniel Delany, both Roman Catholic Bishops, following the relaxation of penal laws restricting Catholic education. The College was accredited by the University of London in 1840.
8. *O'neill*: possibly Terence O'Neill.
9. *'Thrice is ... quarrel just'*: W. Shakespeare, *Henry VI* part 2, III.ii.233.
10. *Mick*: Michael Hennessy.
11. *table of natural sines*: For a right-angle triangle the sine of an angle is the ratio of the length of the side opposite that angle divided by the length of the hypotenuse. A table of natural sines gives the values of the sines for each minute (and perhaps second) of the quadrant.
12. *how am ... line AC*: It is not clear which angle (if any) is a right angle or whether BAC = BCA.
13. *Gibson*: probably *The Theory and Practice of Surveying* (New York: Evert Duyckinck, 1821) by the British-American mathematician Robert Gibson (d. *c.* 1761), newly arranged with a new set of accurate mathematical tables by James Ryan. Pp. 73–81 contain a table in which the 'natural sines are exhibited to every degree and minute of the quadrant' (p. 71).
14. *Log^{s}*: Logarithms.

From Deborah McAssey 24 February 1842 0125

Ballybromell[1] | Feb. 24th 1842.

My dear John

Mr Little was handed your letter[2] on Sunday the 4th, while we were at dinner at a Mr Smyth's[3] he opened it and read part of it, after which he excused himself to me saying he thought it was for him. Up to that time I was expecting to see you. The mail coach starts from Cork every evening[4] Mr Little generally is the driver from Fermoy[5] to Cork on the one o'clock, and from Cork on the six o'clock coach, in the evening. Mrs Little[6] told me that she was looking out for you Christmas eve – she expected to be able to give you a good fat tumbler before the coach you would come on would leave the office in Fermoy. You say that you will see me before we would leave home I think if ever you come to the Co.[7] Carlow you will be likely to find some of the family in the old spot. I was very anxious to get home from Fermoy, thinking that the place would be sold before I could get back. Mr and Mrs Little thought to keep me until Easter but I would not consent. I got home on the beginning of last week. The place was still unsold. There was a good many coming to look at it. On Friday last Mr Robert Watson[8] came and bought it for himself and his brother –John Leckey Watson[9] – wrote the agreement and took it home to Killconner[10] to his father.[11] On Saturday morning Mr Robert came to let my father know that his father would not give the money. So the sale ended there. Perhaps it is all for good. Mr Faulkner was here on Monday last and seemed glad that my father had given up the thought of going and said that he would do any thing in his power for him. We got a letter from Tom[12] last week. They are well. Dan and Bess[13] has got a daughter. Sam[14] says that he will go to America this season. I think all the boys will go one after another.[15] Tom promised in his letter to come home next fall. All our family are well. The travelling did me a great deal of good, we had fine weather and a very pleasant visit. But travelling through so many markets without any person bidding for Miss James[16] or me, I think we may give up all hope of ever being sold however we have no reason to complain when the Fermoy boys did not send us to Nenagh[17] Shroff Tuesday[18] night. Let me know in your next how you escaped and what court you summond the bug to that insulted you on the night you arrived in Cork.[19]

Yours affectionately | D. McAssey

RI MS JT 1/11/3760
LT Transcript Only

1. *Ballybromell*: see letter 0121, n. 4.
2. *your letter*: letter missing.

3. *Mr Smyth's*: not identified.
4. *The mail … every evening*: The Cork-Dublin Night Mail Coach left the coach office at 12 Dawson Street in Cork each evening, stopping at Fermoy, Clogheen, Cloumel, Kilkenny, Royal Oak, Leighlin Bridge, Carlow, Castledermot, Kilcullen, and Nass, before arriving in Dublin the following night (*The Traveller's New Guide Through Ireland* (Dublin: John Cumming, 1819), p. 554).
5. *Fermoy*: see letter 0008, n. 5.
6. *Mrs Little*: see letter 0121, n. 3.
7. *Co.*: County.
8. *Mr Robert Watson*: son of John Watson and brother of John Leckey Watson; see nn. 9 and 11.
9. *John Leckey Watson*: Watson was elected a life member of the Royal Dublin Society in 1834, and listed as a Magistrate for Carlow in *The Dublin Almanac, and General Register of Ireland* for 1847. In 1860 he was the High Sheriff of Carlow (Subscribers List for *Ierne: or, Anecdotes and Incidents During a Life Chiefly in Ireland* (Dublin: W. Curry, 1861), p. xvi). He was a Justice of the Peace for 1867–9, but was deleted from the Royal Dublin Society's membership list in 1870. The Watson estate in Fenagh consisted of 687 acres in 1878.
10. *Killconner*: a village in the parish of Fenagh, County Carlow.
11. *his father*: probably the John Watson Esq. recorded in the Tithe Applotment Book for Killconner in 1826.
12. *Tom*: not identified.
13. *Dan and Bess*: probably Daniel McAssey (b. 1806), Deborah's younger brother, and his wife Elizabeth.
14. *Sam*: probably Samuel McAssey (b. 1815), Deborah's younger brother.
15. *I think … after another*: Deborah had five brothers, and by April 1847 the entire McAssey family, including their parents John and Mary, had emigrated to America, settling in Wisconsin.
16. *Miss James*: sister of Sarah Little; see letter 0121.
17. *Nenagh*: a market town in Tipperary.
18. *Shroff Tuesday*: Shrove Tuesday; see letter 0121, n. 11.
19. *the bug … in Cork*: see letter 0118.

From George Wynne 24 February 1842 0126

To Mr John Tyndall Junior, | Wm Steuart Esqr., | Steuart's Lodge, | Leighlin Bridge, | Ireland. Zanthe, Greece.[1] | 24th February 1842.

Sir,

In reply to your letter reminding me of the promise[2] I give you a Testimonial[3] of your character and abilities whenever it might be of service to you, I have great pleasure in stating

That I have the very highest opinion of your integrity and good conduct; during the time you were in my Division of the Ordnance Survey of Ireland[4] you exhibited the greatest diligence and attention to the duties you were employed on, and became in a very short time a superior Draughtsman and Surveyor.

From my general knowledge of your character I feel assured that you will never discredit the high character I feel myself justified in giving of you and that you will honourably acquit yourself to your future employers whoever they may be.

I am, Sir, | Your most obedient servant | GEO. WYNNE

RI MS JT/5/17a/41
LT Transcript Only

---

1. *Zanthe, Greece*: an island (now called Zakynthos or Zante) in the Ionian Sea, west of the Greek mainland. The island had been made a British protectorate at the Congress of Vienna in 1815, and Wynne spent four years there with the 97th Regiment of Foot.
2. *your letter ... the promise*: letter missing.
3. *a Testimonial*: Wynne had initially declined to provide Tyndall with a testimonial (see letter 0011), but had apparently now changed his mind.
4. *my Division ... of Ireland*: the 5th Division, C District.

From Martin Cuddy 8 March 1842 0127

Tuesday night | March 8th 1842.

Tyndall, excuse me for not writing to you for some time past but I assure you I had not one minute to spare for nearly the last fortnight

I'm bothered with whisky I'm bothered with <u>love</u>. I suppose that Jim gave you a scrap of my proceedings.[1] Well then Jack I will give you another <u>scrap</u>. To <u>begin</u> from the first onset, you know my <u>cap</u> – it were the cause of the row, good luck to it, for it went first, and I followed after. Miss Charlotte Shine had it for three days and she sent me word I would not get it till I went for it myself – I would not go, but I sent the following verses:

> Beauteous maiden, pray give me that
> Which shelters me, from spring's rude blast, my cap
> For when the thunder's storm fierce doth clap
> The veil it falls, mine ears to wrap
> None other comforter have I –
> Save <u>him</u>, who rules heavens arch, the sky
> And oh! may that great being bless,
> And protect thy loveliness.
>
> (do not laugh at my composition).

Well Jack the day after, she received the above, she sent for me, it was night, I met her in a dark hall in her own house, oh! Jack tongue cannot describe our meeting. Behold your poor Blackthorn[2] closeted with the grandest young lady in Bandon. I was so very <u>shy</u>, what a bull[3] I made, why Jack I was that daring devil to kiss her, and oh! how sweet the taste, it is on my lips yet. I see her every night. She is a lovely creature. Somewhat about Cupid's[4] size but not so bulky. Now Jack your advice. What am I to do, she has forty or fifty pounds a year. I do not know what ready cash. She is no less a personage than a relative to Lord Bandon.[5] At all events I think I will crook her, but dang it, the time is so short, watch the Cork Constitution[6] you may see in it:– Married by the Rev – Mr M________ Cuddy, Sappers, Clk.[7] to the Honble.[8] Miss Charlotte Shine of Bandon. Dear Jack if this business happens, I will be well off, and you too, for dang me but you will be my brotherinlaw that's if you like my fair sister. I have a great deal to inform you of, twenty letters would not contain all. Cupid is standing to me well, he sometimes receives a message for me, blow me, but he must have one of my sisters too, we all three will live together what blowouts we will have, but all that I have informed you of be buried within your breast. I now bid you farewell for a while, answer this, let me know all particulars: remember me to Ginty, Tidmarsh and Payne.

I am yours | Cuddy

RI MS JT 1/11/3754
LT Transcript Only

---

1. *I suppose ... my proceedings*: Tyndall had not yet received Phillip Evans's letter of 7 March 1843 as it was misdirected; see letter 0129.
2. *your poor Blackthorn*: Cuddy's poetic pseudonym.
3. *bull*: blunder (*OED*).
4. *Cupid's*: Phillip Evans.
5. *Lord Bandon*: James Bernard, 2nd Earl of Bandon (1785–1856).
6. *Cork Constitution*: a twice-weekly Protestant newspaper costing 6*d*.
7. *Clk.*: Clerk.
8. *Honble.*: Honourable.

From Maria Payne 12 March [1842] 0128

Bagnalstown | March 12th

My dear John,

I am afraid you will often have cause to reprimand me though I assure you that you are not forgotten I think too highly of your friendship to forfeit it lightly. Circumstances prevented me answering your letter[1] sooner not want of inclination. You have given a long glowing description of the ladies of the South,[2] I am sure if they could have read it they would feel highly indebted to you and greatly flattered by your praises. I am surprised if you are quite heartwhole[3] – in days past you were wont to say that you were a stoic[4] is that the case still? I own I doubt it. You too that possesses such a romantic sentimental disposition. Surely you can not be insensible to so much beauty as you have described. You accuse me of blarny,[5] if I possess any I think I must have derived it from you, as I assure you if I had the least tendency to vanity your extravagant praises as to my singing would have turned my head. However I have to return you many thanks and I hope the day is not long distant when I shall have the pleasure of personally thanking you. Emma[6] also feels indebted and desires me to present her kindest love to you. In conclusion I would just say that you are anxiously expected in the sweet little town of B–[7] and also that I am

Your sincere friend | Maria Payne.

P.S. your cousin's address is | Mrs George Young[8] | Shygaff | Bagnalstown

Write soon and give me a description of Cork I know when you like that you can shine in that way – excuse this scrawl as I am very busy.

M. Payne.

RI MS JT/1/11/3842
LT Transcript Only

1. *your letter*: letter missing.
2. *the ladies of the South*: presumably of County Cork.
3. *heartwhole*: not emotionally attached; not in love (*OED*).
4. *a stoic*: an exponent of the ancient Greek philosophy established by Zeno of Citium in the third century BC, which enjoined the acceptance of natural happenings and avoidance of destructive emotions.
5. *blarny*: variant spelling of blarney, smoothly flattering or cajoling talk (*OED*).
6. *Emma*: Emma Young (née Tyndall), Tyndall's cousin.
7. *B–*: Bagenalstown.
8. *Mrs George Young*: Tyndall's cousin Emma, the daughter of Caleb Tyndall, had married George Young on 22 November 1841; see letter 0111.

From Phillip Evans [13 March] 1842 0129

Bandon | Sunday night 1842

My dear John

By this time you have set me down as being very ungrateful or negligent. Well I don't blame you as I <u>now</u> see you have not got my <u>answer</u> to your last,[1] but to commence my story, look at the <u>date</u> of the one which accompanies this,[2] read it through paying particular attention to that part which refers to B. –n.[3] Well you have read it, you see that it was written on <u>last Monday night</u>[4] (now a week gone by) on the following morning, Tuesday, as I was passing through the shop, gave it (enclosed in an envelope) to the shop girl directing her to have it sent (by the boy) immediately to the post, well Sir every day since I was expecting to get an answer from you, but no such thing. B. –n was also wondering at your silence as he wrote on Tuesday[5] and referred you to my letter that went the day before <u>as we thought</u>. About Thursday he managed one way or another to speak a few words to her,[6] she said she was surprised at a letter which came from him she could not understand its contents at all, and that it commenced with 'John'.

He told me and we did not bestow a thought upon it, as we sent on each letter, imagined that it was some one that sent it on a joke to her. Well at this time it was out of her power to give it but promised to do so as soon as possible. Friday he could not see her but by appointment she got up at 7 o'clock Saturday morning and they had a regular chat, in this conversation the hard word passed and he has won her consent to ask for her, even now he did not get the letter which began with John, but Sir yesterday evening at 9 o'clock he got it enclosed in a 3 corner note, brought it home and on our opening it what was before our eyes only the selfsame note which accompanies this, the selfsame note that I enclosed and gave to be sent to the post for you. Yes John there was my hand writing before me. O! John it would be impossible for me to think of hinting to you how I felt, suffice it to say that I would not believe until I had gazed well upon it I then did not know where I was or what I was doing, I am afraid to do any thing about it as it might injure —.[7] I think it lay in the till and that another girl who is a cousin to the hostess and who I firmly believe is capable of doing any thing, came to the till, broke the seal, read it, and then enclosed it to Miss C —.[8] In all the course of your life did you ever hear of such an action, and the worse of it is that we are afraid until a future day to say anything about it, no one knows anything about their courtship only those who saw this letter, and myself and one sister, I am sure from what was laid before her eyes in my letter, that she must be one in a million to speak to him at all, he did not see her since he got it until about an hour ago. She told him not to make himself the least uneasy about it as she knew it must have come from an enemy of his or something to this effect. She has directed him to go to Mr Gilmore (Methodist preacher)[9] and ask him to acquaint her father[10] about it, she thinks her father will make enquiries about his character &c &c &c. Herself has told him he must leave the Sappers stiff chat boy O! He is just after reading a letter for his father going by this night's post acquainting him about it he seems to be quite determined upon it provided that things turn out to his satisfaction. You will without delay send me the enclosed which accompanies this back again as I will want it. John the thought of all this is not out of my head yet I am very badly able to write at all however I will say a few words about myself now. The men who are to leave this division for England have been selected and their names sent to the Major.[11] I am not one of them, I am to remain in Cork. This I am sure gives both of us great pleasure. I will soon be by your hip, what is this I have said, will I have to go to Cork to get by your hip, no! I rather think not, I would sooner say that you will have to come to Bandon and get by my hip in room of B.–n for he says he will not do it if you don't come – faith I think you are to be the grooms-man.[12] So boy O! I think you will see Bandon and me before I see you or Cork. Many a wish he gave to have you here this evening when he was writing as he says you are a good general and adviser. Will you let me know did Ginty get an answer to the one I sent him last, I know that I wrote to him with very little delay, but likely he never got it. Through the whole was it not a good job that Evans was not mentioned

Write immediately | Yours ever | J.

Bandon | Monday night | March 7th 1842.

My dear John

Hark! What sounds are those which salute my ear, is it the wind roving through the streets, or the rain smiting against my window. No, no, it is more cheerful sounds than these, it is this, the glad tidings that though we were parted for a time, yet, we will meet again. Yes John in defiance to those who were the cause of our separation we will again see each other once more as you have already said it is only when friends are away from each other that they can see the good and pleasure of each others society. Time has passed away very slowly since I came here and more so than ever since the information you gave me about our going to Cork. O! on the other hand I imagine this month very very long I am wishing daily for the hour to arrive when our hands shall be locked together, but then the bright anticipations which are before me (should they not be of long duration) cheers me on that I will feel the time revolving away.

It is possible that I will be one of those who will remain in Cork as Mr W.[13] did not intend sending me away on the 18th (which order has been cancelled) I have done a little drawing which pleased him.

There is a great quantity of heights to be inserted, two of St. Stace's[14] men were at them. He told Turner who had the charge of the drawing that he thought I would do them well, being a neat draughtsman, and also that he did not approve of how the others done them. I have done some he told me he couldn't find any fault in indeed said he they are very well done. I merely mention this to show what grounds I have for supposing that it may happen that I will remain in Cork and I believe the heights will be last on hand. I will do every thing in my power to stay (that is if you are stopping too). Well Jack you are pretty well off in town as usual getting 'acquainted' every day with the fair ones – get on my boy and prosper. May every day bring more pleasure to you this is the fervent wish of Jim[15] – I can't say prayer for you know such a thing does not trouble me often. You ask me how is Martin[16] getting on. O! then I will tell you boy o! he is doing well if he can accomplish his design, courting a Miss Shine who is a member of a most respectable family in town, I hear from good authority that her mother[17] is related to Lord Bandon.[18] There are now 4 unmarried sisters and I believe for an undoubted fact that each have £40 per year don't mistake me here when I say 40. He has sent letters and *<word(s) missing>* and received others in reply yes and more than this has had a chat and more than this too – what? a kiss, she is about my size (small enough) and very fair every other way you take her, we went to preaching last night at her request and from stolen looks and other movements (mind the old chat was there) I believe she is inclined for the job, and faith I don't joke when I say that Martin will if possible take her by fair or foul means for the 40. By the bye Jack dont be surprised if you hear of a runaway match. I must now tell you about a little conversation he and I had in bed the other night. Arah Jim I would like to be asking Jack Tyndall to the wedding he would to take wrinkles out of that little pot belly of yours. Jim to Martin: It wont be your fault I am sure if you dont <*word(s) missing*> the 40. Mary Jane[19] is here still I met her on the green on Friday evening accompanied by Miss Montjoy[20] – we had a walk and some chat. I am able from good authority to congratulate you on your being <*word(s) missing*> all their fathers drawing the big plans. Yours is the best of any done – this I understand is allowed even by the incomprehensibles. More power to your elbow Jack my boy. Is there any news from our home I have not heard any I am at a loss not getting the Sentinel[21] for the last six weeks. My uncle has to send it to a lodger who is from home. I believe if it should be my happy lot to remain in Cork we wont fall out about the messmanship.[22] We are all very anxious here to know if Major Waters will be on the English Survey.[23] Can you give this information in your next? Martin says only he is so extremely busy with his darling he would write but promised not to delay much longer.

Good night | Yours ever | Jim

RI MS JT 1/11/3569
RI MS JT 1/11/3568
LT Transcript Only

---

1. *your last*: letter missing.
2. *the one which accompanies this*: see the second letter below, dated 7 March 1843.
3. *B. –n*: Martin Cuddy. LT's handwritten insertion proposing that this means 'Bob Martin' (i.e. Robert Martin) is erroneous as Martin was never stationed in Bandon and had already been transferred to the English Ordnance Survey in August 1841 (NAI OS/2/18). Cuddy, on the other hand, went to Bandon in May 1841 as part of the Irish Survey's 2nd Division, C District, and stayed there until he was transferred to the English Survey on 8 April 1842 (NAI OS/1/18–19). B. —n therefore probably stands for Blackthorn, Cuddy's poetic pseudonym; see letter 0127, n. 2.
4. last Monday night: 7 March 1842.
5. *he wrote on Tuesday*: letter 0127.
6. *her*: Charlotte Shine.
7. —: presumably Cuddy.
8. *Miss C—*: Charlotte Shine.
9. *Mr Gilmore (Methodist preacher)*: There are no records of a Methodist preacher of this name in County Cork at this time, so this is probably Reverend John Benjamin Gillman (1800–88), a Methodist preacher ordained in 1823 and appointed to the Cork circuit in 1840, before being transferred to Dublin in 1843.
10. *her father*: presumably William Hunter Shine of Bridge Street, Bandon, who was the town's Postmaster (*Slater's*).
11. *the Major*: Major Marcus Antonius Waters.

12. *grooms-man*: a young man acting as friend or attendant on the bridegroom at a marriage, either alone (as 'best man') or as one of a company (*OED*).
13. *Mr W.*: Major Waters.
14. *St. Stace's*: see letter 0097, n. 17. William Crawley Stace was in fact a Lieutenant at this time, rather than a Sergeant.
15. *Jim*: Evans means himself.
16. *Martin*: Martin Cuddy.
17. *her mother*: Eliza Shine of Bridge Street, Bandon.
18. *Lord Bandon*: see letter 0127, n. 5.
19. *Mary Jane*: not identified.
20. *Miss Montjoy*: possibly the daughter of Thomas Montjoy of North Main Street, Bandon.
21. *Sentinel*: the *Carlow Sentinel*.
22. *messmanship*: the responsibility for cooking and serving food to a company of people eating together (*OED*). When they were stationed in Kinsale in September 1841 Tyndall told his father 'Evans is our messman here. He provides everything; I dont know what I'm to get for dinner until its laid on the table and not a word of grumbling is allowed'; see letter 0093.
23. *if Major ... English Survey*: Waters did not join the English Ordnance Survey. Instead, after he left the Irish Survey in November 1842 he was sent to the West Indies; see letter 0179, n. 3.

From John Tyndall, Snr 16 March 1842 0130

Mr Carberry McCarthy | Strand Road, Cork | for John Tyndall | Leighlin Bridge | March 16th, 1842.

My dear John

I am sure you are worn out of patience, daily expecting something in the shape of a letter from me. The fact of the matter is I had nothing worth your notice to communicate, for the idea of the Corn, the Flour and the Provision Bills[1] becoming the law of the land has turned every thing in this country in the shape of trade or business upside down. One portion of the people in extacy thinking that abundance are in store for them while another portion thinks that misery and want will be the consequence. In the midst of all this confusion it is useless for me to use any interest I may have to forward yours. But when the proper time comes and everything is settled I will do all in my power to forward your interest, but at present I can do nothing. But John about your coming home, when will such an event take place? I have at present a letter on the dresser from you to Mr Conwill. He will be here to-night at 8 o'clock – as he is teaching young Moran[2] – when he will receive it. Mrs Murray complains much of William[3] that he has not written to her these six months, and would be glad to know from you where he is, and how he is getting on. Your uncle Caleb has been appointed High Constable for the Barony of Idrone west and Mathew Griffith for east.[4] I would not bother you with those lines but that I thought you might be uneasy. I will now conclude with letting you know we are all well

I am your affectionate father | John Tyndall

A rumour has come to town this moment that Patt Monahan has arrived in Carlow to abide his trial for the murder of Margaret Murphy. I think its truth, the Assizes is going on.[5] This is the first day its sitting.

RI MS JT 1/10/3267
LT Transcript Only

---

1. *the Corn ... Provision Bills*: Throughout February and March 1842 the House of Commons was debating Robert Peel's Bill to Amend the Laws for the Importation of Corn, which proposed implementing a reduced sliding scale in the duty on imported corn, bringing it down to only 1*s.* when the price of domestic wheat fell to 73*s.* per quarter. The Bill was passed on 29 April 1842. Another Bill to prevent flour being substituted for foreign wheat secured in bonding warehouses, put forward by the MP William Hutt, was also debated in March.
2. *young Moran*: probably Patrick Francis Moran (1830–1911), son of Patrick Moran (d. 1841; see letters 0095 and 0104), who left Leighlin Bridge later in 1842 to study at the Pontifical Irish College in Rome, before emigrating to Australia where he became Roman Catholic Archbishop of Sydney and was proclaimed the first Australian Cardinal in 1885.
3. *William*: William Murray.
4. *uncle Caleb ... for east*: Idrone in County Carlow was divided into two distinct Baronies in 1802 under the provisions of an act passed in 1799. Martin Nevin and Norman McMillan have suggested of Caleb Tyndall's appointment, 'This was his reward for services to the landlord Bruen', ('John Tyndall of Leighlin', *Carloviana*, 2 (1978/9), pp. 22–7, on p. 24).
5. *A rumour ... going on*: The rumour seems not to have been true, as this case was not heard at the Spring Assizes, nor at the Summer Assizes in July (see 'Carlow Spring Assizes', *CS*, 19 March 1842, pp. [2–3]; and 'Carlow Assizes', *CS*, 26 March 1842, pp. [2]).

From Phillip Evans — 17 March 1842 — 0131

Bandon | Wednesday night March 17th 1842.

My dear John

I was not long breaking open your letter[1] last night as I was rather uneasy till I should hear from you. Your enclosure surprised me as I knew not how or what way you came by it. I thought more bad work had happened,[2] but if I judge right from what you say I sent it through mistake. I must through some bull[3] have left the terrible one on my table and I suppose the folks of the house got it and sent it to – I cant account for it any other way however enough of this.

John knowing your anxiety about how certain things stand (the <*word(s) missing*> you thief) I loose very little time in giving you some information, you may recollect that I mentioned about their[4] going to walk after prayers on Sunday – they did for about an hour and I can tell you that this was great honour. Well Sir, as usual we went to preaching and was surprised a little to see the whole family looking now and then at B–[5] she too would look and hang down her head, B– could not see – that night at all, but at 7½ next morning went into the shop and to his astonishment found the poor thing in a flood of tears O says she, did you hear any thing from Ellen[6] who is a messenger for them, no says B– O! I am afraid you will have to give it up, my father[7] was told about the walk yesterday &c &c &c

At this part of the proceedings the girl opened the parlour door and said her father was coming down stairs. Oh no says Miss, he is not dressed yet. O indeed he is Miss, and will be down in an instant. So my poor B– had to fly. – Tuesday (yesterday) passed away without a breath of any kind of news untill about 8 o'clock, when Jim[8] went out and met Ellen who told me that a sister told the father how C.[9] had been walking with Srs.[10] on Sunday, he got enraged, got a whip and welted[11] her and another who accompanied her, I picked a great deal out of this so much so that a R–y[12] would be quite acceptable yes with a heart and a half. This day had passed without any news, but in about an hour we expect to hear how things are going on. I forgot to mention that she told him she was at her F–[13] disposal he could call her off to ls.[14] we do not know nor have we any way to know whether that is the case or not but this I know that great confidence can be placed on her authority B is fully determined not to do anything rashly, has not the slightest notion of any such thing except things are settled to his utmost satisfaction, and I entertain great doubts about this, the time is so very short. I have now given you in detail as well as I could of the whole proceedings up to this moment I hope they are satisfactory, I have now to congratulate you all on your recommendations. I am laying out 6d for you boyO! Yesterday the return, which was about the men going to England, was sent back from the Major;[15] he was surprised that more were not named for embarkcation and directed the list to be revised and more sent. Well this news put me in a queer way, I immediately set myself for crossing waters but I have again escaped for the present. Should I continue this way I will have a month at least with you, but I am afraid it will be returned again, tho' we added 8 to the former one. I am glad the Sentinel[16] afforded you so much pleasure. I am just after reading 8 columns of the Southern Reporter taken up with a speech of Sir Robert Peel's,[17] by the bye he is a great fellow. I think I will be able to send it to you tomorrow, it is well worth your reading. I say are you and Ginty on good terms, if so tell him I'll break his big head when I go to Cork for delaying so long in answering my letter. And now I'll bid you good night the time for news has arrived

Yours | Jim

RI MS JT 1/11/3570
LT Transcript Only

1. *your letter*: letter missing.
2. *more bad work had happened*: see letter 0129.
3. *bull*: blunder (*OED*).
4. *their*: Martin Cuddy and Charlotte Shine.
5. *B–*: probably Blackthorn, the poetic pseudonym of Martin Cuddy; see letter 0129, n. 3.
6. *Ellen*: not identified.
7. *my father*: see letter 0129, n. 10.
8. *Jim*: Evans means himself.
9. *C.*: Charlotte Shine.
10. *Srs.*: Sappers.
11. *welted*: beaten, thrashed (*OED*).
12. *R–y*: possibly Runaway.
13. *F–*: Father's.
14. *ls.*: not identified, but presumably a place name.
15. *Major*: Major Marcus Antonius Waters.
16. *Sentinel*: the *Carlow Sentinel*.
17. *Southern Reporter ... Robert Peel's*: 'Imperial Parliament', *Southern Reporter and Cork Commercial Courier*, 15 March 1842, pp. [2–3]. The speech was that made by Robert Peel in the House of Commons on 11 March in which he sought to 'develope the views of Her Majesty's Government with respect to the financial and commercial policy of the country' (p. [2]). In this speech, Peel announced the reintroduction of income tax at 7*d.* in the pound. Ireland, however, was excluded because 'in the state of society in Ireland there is something peculiar, which makes the devising of machinery for its [i.e. income tax] collection matter of grave consideration' (p. [2]). Instead the Irish duty on spirits was raised from 2*s.* 8*d.* per gallon to 7*s.* 10*d.* The column featuring the part of the speech dealing with Ireland was printed in a larger font size than the other seven columns.

From John Chadwick 22 March 1842 0132

Kinsale | March 22nd 1842.

Dear Tyndall

A thousand apologies for not sooner fulfilling my promise to you of writing, but indeed my fault is so great that I almost think that the best way I can manage is to leave a *[word illeg]* and let your forgiving disposition imagine the gnawing of my reproving conscience and by judging of the effects of the still small voice be inclined to deal leniently with me and temper the wind to the shorn lamb or ram whichever you please. Seriously speaking I have not had a spare moment hardly since I came home which, joined with my usual supineness in writing nearly steeped my memory in the lethe[1] of forgetfulness to-wards you till I was awakened from my stupor by considering what idea must the ci-devant[2] lodger have of his old crony, so verily the spirit moved me at last to perpetrate this scribble. I was laid up with the <u>hoith</u>[3] of sickness for 8 or 9 days after I returned and ever since my convalescence I have been busy to the utmost extent no survey humbug but do downright matter of fact work, and it is only within this last day or two that I have had any time and almost the first use I make of it is to write to you. I had to get the shop all altered painted and papered and to clean out the dust and accumulations of centuries. I was almost beginning to think that the place was like the stables of Argus.[4] Had I not left Cork at the time I did, I would most assuredly have got a heavy fever and as it was it was only by taking every remedy that I escaped it. My greatest torture was my throat, for I could not swallow for nearly a week. I now enjoy good health and find that the state of life to which it has pleased God to call me agrees with me much better than my old sapping and mining, and I like the business well enough. Tell Latimer that is hardly worth writing to him as I expect to see him so soon. As to yourself I fully expect that you will come with him according to promise, so do not fail to come for I am longing to see you. Though Mathews is in the town I have hardly had time to see much of him. Remember me to all friends too numerous to mention. I will settle about Baker's scale[5] when you come here on Good Friday. Remember me to Mr and Mrs Carberry.[6] Concluding

I am faithfully yours | John Chadwick.

RI MS JT 1/11/3494
LT Transcript Only

---

1. *lethe*: one of the five rivers of Hades in Greek mythology. The dead drank its waters to experience oblivion and forget their mortal lives.
2. *ci-devant*: former, especially an erstwhile noble following the French Revolution (*OED*).
3. *hoith*: not identified.
4. *stables of Argus*: presumably a mistaken reference to the stables of King Augeas, which, in Greek mythology, contained more than one thousand immortal cattle and had never been cleaned before the task was set for Hercules as the fifth of his penitential labours. Argus was a hundred-eyed giant.
5. *Baker's scale*: presumably a scale belonging to John Baker, a Private in the 13th Company of the Royal Sappers and Miners, who worked in the 3rd and 4th Divisions, C District of the Irish Ordnance Survey, before being transferred to the English Survey in May 1842 (NAI OS/1/16–19). The scale in question is likely to be a Vernier Scale; see letter 0150, n. 24.
6. *Mr and Mrs Carberry*: Carberry McCarthy and his wife.

From Thomas Charles Higginson 23 March 1842 0133

Skipton | 23rd March 1842.

My dear Tyndall,

I received yours[1] to day and am going to turn the tables on yourself in your own words 'has Higginson done anything to offend you?' but I suppose it is such a long time since you received my last[2] that you forgot whether you had received it or not before Geo. Holmes left Cork for the Re-levelling.[3] I wrote to you and him the same day and received two letters from him since but you have never condescended to answer the one I sent you so I don't know what to say you see the ill nature does not lie on my side. Ginty I suppose says the same thing that he did not receive my letter in Kinsale long before he left it. I thought I occupied a small portion of his affection alas but that thought has long since vanished.[4] Indeed I suppose the vanities of Cork have monopolized all your affections so that an old faithful friend is entirely forgotten I could lay twice as much on your illnature which you try to turn upon me as the easiest way of apologizing. I think I have now done enough to eradicate the false impression that has arisen in your mind. I suppose you will be over here on the 1st. of April but I'm afraid I won't have the happiness of seeing you as I think the divisions will be formed before that time. Capt. Tucker was in the office to day, he said that one of the 4 divisions thats here, that is the field party, is to go to Manchester in a few days. Lancashire is to be surveyed 1st and then go north so that I think you wont come further than Manchester which you need not be sorry for as this is a nasty place.[5] I am expecting to be going off the survey in about a month or six weeks[6] but no matter where I go to I shall always feel it the greatest happiness to correspond with my dear Tyndall. The work here is going on very quick one of this division (1st)[7] got a situation of a 100 per year with an attorney this week that is a good beginning. Recommendations are going on very slowly I have got now only 13/6.[8] Write to me as soon as you have a leisure moment. Dont remember me to Ginty but do to Foy and Chadwick

Your ever faithful friend | T. C. Harold[9]

RI MS JT 1/11/3732
LT Transcript Only

---

1. *yours*: letter missing.
2. *my last*: probably letter 0107.
3. *the Re-levelling*: the correction and reestablishment, by means of spirit levels, of the elevation of the height reference points used on the Irish Ordnance Survey.
4. *I thought ... since vanished*: Perhaps prompted by Tyndall telling him of this sentiment, William Ginty wrote to Higginson a day later on 24 March 1843 (RI MS JT 1/11/3598).
5. *this is a nasty place*: Higginson's opinion had changed since his arrival in October 1841,

when he told Tyndall 'Skipton is a nice town'; see letter 0097.

6. *I am ... six weeks*: Higginson subsequently enlisted in the Royal Artillery on 10 April 1842; see letter 0182.
7. *this division (1st)*: the 1st Division of the English Ordnance Survey.
8. *a situation ... only 13/6*: £100 per annum corresponds to approximately £1 18*s*. (38*s*.) per week, about three times Higginson's own pay of 13*s*. 6*d*. a week.
9. *T. C. Harold*: Harold was a nickname of Higginson's.

To John Tyndall, Snr 24 March 1842 0134

Strand Road, Cork. | Thursday night

My dear Father

I received both your letters[1] and indeed both were highly acceptable to me, the first because it was long and anxiously waited for, and the second on account of its containing a document which I quite agree with you in calling precious.[2] I feel extremely thankful to Mr Wynne. I was promised a good testimonial by Lt. Whittingham, but the thing must have escaped his remembrance, as he went away without leaving me a single line. This therefore makes Mr Wynne's testimonial the more valuable. I would not give it at all events for a five pound note. It shall always be my aim to render myself worthy of the good opinion of my employer, and I flatter myself that no person with whom I am at present connected would be disposed to say any thing of me at variance with what Mr Wynne has testified. I received a letter to night from Mr Murray requesting some information respecting William.[3] He left Ireland with the first draft that went to Yorkshire,[4] and I am unable to say where he is at present. I'll make every enquiry to morrow relative to his affair, and if I dont get a satisfactory account I'll write to England. You may tell Mr Murray this.

I remain | Your affectionate son | John.

RI MS JT 1/10/3268
LT Transcript Only

1. *both your letters*: letters missing, as neither of the letters described correspond with letter 0130.
2. *a document ... calling precious*: letter 0126.
3. *William*: William Murray.
4. *the first draft ... to Yorkshire*: for the Ordnance Survey of that county, possibly leaving Ireland in October 1840, when Murray's work for the 5th Division, C District of the Irish Survey ceased.

From John Conwill 26 March 1842 0135

March 26th 1842.

'My dear Tyndall'

There is something so picturesque, so sublime, so poetic in your varied mode of letter writing that the reader must exclaim: these productions are emanations of a person whose mind is stored with an inexhaustable treasure of historical and natural abilities. I am often wrapped up in silent admiration at the vivacity displayed in your multifarious correspondences. You comment very largely on the socialists;[1] really I think them queer fellows so let them tug away at their oars.

You seem to think little in some degree of the writings of Rousseau, Hume and Payne,[2] but at the same time say give you the bold Infidel[3] who never crouches to servility. Well, though I am no advocate for infidelity, I firmly believe the works of those men tend greatly to expand the mind. I have my own opinion, you have yours about the matter. We will not quarrel, as, comparatively speaking, I know nothing when brought into competition with any individual possessing one particle of brains.

I am now convinced that you have received an answer to your query about what I said in my last letter[4] about myself knowing nothing comparatively speaking. Nor do I. What am I in the scientific world when compared to Newton, Simpson, Lagrange, Laplace, Landon, Bonnycastle &c.[5] – nothing – nothing – nothing. In fact I deem myself below zero when compared with many of my contemporaries in a literary point of view. Pardon this hasty scrawl. Spare your old teacher who would burn the candlestick and all to relieve you out of difficulties about natural Sines.[6] – Do not burn his whiskers with your cutting sarcasms – for Heaven's sake spare him.

Your faithful teacher | J. Conwill.

RI MS JT 1/11/3524
LT Transcript Only

---

1. *the socialists*: Thinkers who were identified as socialists in the early 1840s generally expounded views of the ideal arrangements for social harmony and equity. The most notable of these utopian socialists were Henri de Saint-Simon, Charles Fourier, and, in Britain, Robert Owen.
2. *Rousseau, Hume and Payne*: Jean-Jacques Rousseau (1712–78), French philosopher who stressed the importance of sentiment and subjectivity, and whose political writings, especially *The Social Contract* (1762), helped inspire the French Revolution; David Hume (1711–76), Scottish Enlightenment philosopher who advocated empiricism and took a notably sceptical approach to religion; Thomas Paine (1737–1809), English political theorist who, after emigrating to America in 1774, participated in the country's struggle for independence from Britain, before moving to Paris and publishing *The Rights of Man* (1791), which defended the French Revolution. In *The Age of Reason* (1793–4), Paine

rejected Christianity in favour of deism.

3. *the bold Infidel*: possibly Robert Owen (1771–1858), Welsh social reformer and advocate of utopian socialism whose *A New View Of Society* (1813) insisted on the importance of education for the cultivation of individual morality and the alleviation of social problems. Three years before Conwill's letter, Owen, who rejected conventional Christianity, had complained 'I have been called–an atheist, an infidel, an unbeliever' (*Report of the Discussion between Robert Owen, Esq., and the Rev. Wm. Legg* (London: Simpkin, Marshall, 1839), p. 20). According to Elbert Hubbard, Tyndall went to Dublin in 1847 to hear Owen lecture, and 'sought Owen out at his hotel, and they talked, talked till three o'clock in the morning' (*Little Journeys to Homes of Great Scientists*, 2 vols (East Aurora, NY: The Roycrofters, 1905), vol. 2, p. 67). Later in the same year, Tyndall joined Queenwood College, which had been founded by Owen in 1841 as Harmony Hall, although by this time his opinion of Owen's socialism had become more sceptical.
4. *an answer ... last letter*: see letters 0123 and 0124.
5. *Simpson ... Bonnycastle &c.*: Thomas Simpson (1710–61), English mathematician and Regius Professor of Mathematics at the University of St Andrews; Joseph-Louis Lagrange (1736–1813), Italian-born mathematician and astronomer who spent most of his career in Prussia and France, becoming Professor of Analysis at the École Polytechnique in Paris following the French Revolution; Pierre-Simon Laplace (1749–1827), French mathematician whose *Celestial Mechanics* (1799–1825) was central to the development of mathematical astronomy and statistics. Revered as the French Newton, Laplace had considerable political influence during the regimes of Napoleon and the restored Bourbon monarchy; John Landen (1719–90), English mathematician elected a Fellow of the Royal Society in 1766; on Bonnycastle, see letter 0123, n. 4.
6. *natural Sines*: see letter 0124, n. 11.

From Phillip Evans 30 March 1842 0136

Bandon | March 30th 1842.

My dear John

<*words missing*> I received from Ginty), however tho' I was satisfied about my letter going to hand safe, I was determined to rouse you from your slumber by again putting pen and paper to work this night.

Breathe not the name of forgiveness to <u>Phil</u> I know you now too well to think that you should require such a thing from me but my dear John enough of this you <*word(s) missing*> number of men going to England, but after all I have escaped, I am to remain in Cork until our work will be completed. Perhaps we will not be separated again, if I was sure of this what a source of comfort it would be to me. However we will be together for some time any way. John your information about Roberts surprised me. I cannot find language to express the horror which I feel at his base and brutal conduct O! shame Sam Roberts[1] how will you, you can never, atone for this savage conduct. I do assure you John, though I do not know the unfortunate young woman[2] yet I feel very much for her meeting with such a depraved character. Martin[3] is still getting more and more into his darling's[4] favour, but it is no go, she is afraid to apprize her father[5] about it from what he said after her walking the Sunday which I spoke to you about in a former letter.[6] It is wonderful in the extreme the love she has for him and after all the many risks she runs to get speaking to him, she would do any thing he might suggest, no matter what that might be, but then the father might cut her off. The only consolation left her is that a year or two passes quickly and it may then be in her power not only to be the commandant of herself but also of her purse, she is from experience the most faithful creature ever I saw. Well John they turn out queer dresses when ever they go out, it would dazzle your eyes to see them on Sunday last, but their dresses is not the same as the most <*word(s) missing*> has met with great applause.

Good night, my dear John | Yours Phil.

P.S. A newspaper came here addressed to you it bore the well known stick[7] of Harry Wright alias Barlow,[8] which was confirmed by his broad seal he wrote a few words on it expressing a wish for you to write, but it was so old I think it useless to send however I'll bring it to you, he went to my father, and requested me to write, this I have done. I think he will join again.

RI MS JT 1/11/3571
LT Transcript Only

1. *Sam Roberts*: John Roberts, whose Dickensian nickname was Sam Weller. Roberts had presumably abandoned the wife whom he married in October 1841; see letter 0107, n. 7.
2. *the unfortunate young woman*: Esther Roberts (née Armstrong).
3. *Martin*: Martin Cuddy.
4. *his darling's*: Charlotte Shine.
5. *her father*: see letter 0129, n. 10.
6. *a former letter*: letter 0131.
7. *stick*: presumably handwriting, although no such usage of stick is recorded in the *OED*.
8. *Harry Wright alias Barlow*: probably Gustavus Hamilton Wright, who had left the Irish Ordnance Survey during the autumn of 1839, returning by April 1840, and then leaving again between August and October 1841. There is no record of him joining a further time as Evans speculates.

To John Tyndall, Snr 3 April 1842 0137

Strand Road, Cork | Sunday morning

My dear Father

I suppose you intend to impose on me the task of writing two letters to your one. This is any thing but fairplay, as I am sure you have far more time than I have. I am just after breakfasting, and an awful mess[1] I can demolish. I'll have an addition to my family on Wednesday. Phil[2] is coming from Bandon, he will remain with me till the party starts for England, and then I start home; about the 1st of May will be the time.[3]

Your affectionate son | John.

RI MS JT 1/10/3269
LT Transcript Only

---

1. *mess*: a serving of food (*OED*).
2. *Phil*: Phillip Evans.
3. *till the party ... the time*: In fact Evans did not travel to England until 20 May 1842, along with William Ginty, George Holmes, Richard Hunter and William Latimer; see letter 0144.

To John Tyndall, Snr 8 April 1842 0138

Cork | Friday night

My dear Father

I suppose you have made a vow against writing to me, this is now my third letter without having received one in return.[1] I would not write this were it not to communicate something respecting my expected furlough.[2] I wrote and applied for leave the other day, it was presented to Major Waters. I asked leave for a month, but he has even exceeded my request, he has given me from May 1st to June 10th, nearly six weeks. I suppose you'll all be tired of me before that time is expired. You may however expect to see me toddling into Leighlin about May day, after my two years tramp. It will be two years the 1st of May next since I first entered the town of Youghal[3] as a Citizen of the world. Give the accompanying parcel to Mr Murray. I saw about one hundred and fifty of our fellows set off today in the steamer for England among the number were Cuddy and Joe[4] and Tom Payne. Mrs Payne remains in Cork.

Your affectionate son | John Tyndall.

RI MS JT 1/10/3270
LT Transcript Only

---

1. *my third letter ... in return*: see letters 0134 and 0137.
2. *furlough*: leave of absence (*OED*).
3. *two years ... of Youghal*: In fact, Tyndall's arrival was probably on 3 May 1840; see letter 0001, n. 2.
4. *Joe*: Joseph Payne.

From John Conwill 9 April 1842 0139

April 9th 1842.

'My dear Tyndall'

I believe you will coincide with me in saying that good letter writing is a sure indication of an expanded mind, and therefore I am justified in my remarks on your graphic mode of narration;[1] for as a man of no small notoriety[2] said:–

Of all the arts in which the learned excel
The first in rank is that of writing well.[3]

As to 'Hume and his colleagues',[4] I leave every individual that pleases to indulge in his own ideas about their having extended their notions of revealed religion[5] too far; but this much I can say, though I am no advocate of theirs, that I often found their arguments too stubborn for pulpit orators and Altar declaimers. However, humble as I am, I think I could upset many of their far-fetched objections to Revealed Religion. I am sure they would appear nought in your view. I now fondly anticipate that ere long, you will become a star of the first magnitude in the Mathematical world, as you are now traversing through the regions of geometry, Algebra, and Mechanics, to the difficulties in Tyson[6] – <*words missing*>

Give my best regards to Evans as I understand he has been removed to Cork.[7]

Yours truly | J. Conwill.

RI MS JT 1/11/3525
LT Transcript Only

---

1. *my remarks ... of narration*: see letter 0135.
2. *a man of no small notoriety*: John Sheffield, 1st Duke of Buckingham and Normanby (1648–1721).
3. *Of all the arts ...writing well*: paraphrase of the opening lines of Sheffield's *Essay on Poetry* (1682), 'Of Things in which Mankind does most excell, | Nature's chief Master-piece is writing well', 1–2.
4. *'Hume and his colleagues'*: see letter 0135, n. 2.
5. *their notions of revealed religion*: In Christianity, revealed religion is founded principally on the revelations of God to mankind in the Bible. The Scottish Enlightenment philosopher David Hume contended, in *An Enquiry Concerning Human Understanding* (1748), that as miracles were a violation of the laws of nature, there was not sufficient basis in the testimony of the Bible to believe they had occurred.
6. *Tyson*: Edwin Colman Tyson (b. *c.* 1796), Fellow of Catherine Hall, Cambridge, and Master of the Royal Mathematical School at Christ's Hospital. Tyson's principal mathematical publication was *Key to Bonnycastle's Guide to Arithmetic* (London: J. F. Dove, 1829); he also edited, revised and enlarged several of John Bonnycastle's textbooks on algebra and geometry. On Bonnycastle, see letter 0123, n. 4.
7. *I understand ... to Cork*: Phillip Evans had been sent from Bandon to Cork ahead of his journey to Liverpool to join the English Ordnance Survey; see letter 0137.

From John Tyndall, Snr 10 April 1842 0140

Mr Carberry McCarthy | Strand Road, Cork | for John Tyndall
Leighlin Bridge | April 10th, 1842.

My dear John

I received your third letter[1] this morning, and am happy to hear that you will be at home by the first of May. You accuse me with making a vow never to write to you again, be assured such is not the case. I had nothing worth your notice to communicate I hope I wont have cause to regret the neglect of your interest, as I have (I am afraid) let one opportunity pass without endeavouring to advance your temporal welfare. The Colonel[2] was at home for a few days, and being ignorant of the proper mode of acting I let him set off without making any application to him on your behalf, but I hope as the old saying is, 'all is not lost that's in danger',[3] and that a day may come when my application will be made with good effect. You still hold a situation under the Government and continue to do so as long as you think proper, and if at any time you want to withdraw from them you have a home and of course a hearty welcome to return to. A couple of such letters as that of Mr Wynne's[4] will do a great deal to advance you, and I think you will always merit such praise. I would be glad to know what mode of travelling you will adopt, as if you come on the night coach as Phil Evans did we might remain up for you. Your mother says she will make use of everything that comes in her way to make herself fat by the time she sees you. Emma sent you a paper though you did not acknowledge the receipt of it.

I am your affectionate father | John Tyndall.

RI MS JT 1/10/3271
LT Transcript Only

1. *your third letter*: letter 0138.
2. *Colonel*: Colonel Henry Bruen.
3. *'all is not lost that's in danger'*: a proverb first recorded in Randle Cotgrave's *A Dictionarie of the French and English Tongues* (1611).
4. *such letters ... Mr Wynne's*: letter 0126.

From John Tyndall, Snr 20 April 1842 0141

Mr Carberry McCarthy | Strand Road, Cork | for John Tyndall | Leighlin Bridge | April 20th, 1842.

My dear John

I received your letter[1] this morning and in it a lecture on my niggardly method of writing now I think you dont act fair with me in this particular, and that for the best of all reasons the first is I am not half nor quarter as good a pensman as you are and therefore not so well inclined to sit down to write, and secondly my diction is a century behind yours. Now couple these reasons with the fact that I had nothing that I considered worth your notice to communicate and then I think if you are not very unreasonable you will acquit me of any seeming neglect in writing to you. I read that portion of your letter to Anne Moony[2] about her sons illness and sent her to Mrs Steuart with it. Mrs Steuart desired her to let me know that she would keep your letter till she'd see me herself and in the mean time prepare the medicine for John Moony which will not be ready for a couple of days and then will have it sent by post for him to Cork.

Your description of Madlin Brook[3] was very handsome and I think it will be a long time till you catch any more pinkeens[4] in its straim.[5] Mr Bredin was here this day and I let him know I received a letter from you this morning I told him you were listening to Bishop Tonson[6] in Christ Church[7] Cork on Sunday last and the way you described his eloquence. He shook his head and said that John, meaning you, would be a great man yet. I told him that if such an event took place it should be by your own exertions. He said there would be exertions made for you but at present the Whigs had every situation filled by their own creatures.[8] We buried Robert McDonnell[9] on Monday evening last in Nurney[10] after a long and painful illness and left a large and helpless family in great distress. You have not told me the mode of travelling you intend to adopt, if you take my advice you'll come on the Mail[11] it will be here at half past 12 or one o'clock.

Send me word has your feet changed much since you left home that I may have a pair of shoes made for you. We are all well

I am your affectionate father | John Tyndall

Write soon that I may have the shoes for you.

RI MS JT 1/10/3272
LT Transcript Only

1. *your letter*: letter missing.
2. *Anne Moony*: mother of John Mooney.
3. *Madlin Brook*: The River Madlin runs next to the town of Farranatreney in County Carlow.
4. *pinkeens*: minnows (*OED*).
5. *straim*: variant spelling of stream.
6. *Bishop Tonson*: Ludlow Tonson (1784–1861), 3rd Baron Riversdale and, from 1839, Bishop of Killaloe and Clonfert.
7. *Christ Church*: Protestant Church on South Main Street in Cork, built in 1725.
8. *at present ... own creatures*: From 1835 Ireland was administered by the Whig triumvirate of the Earl of Mulgrave (Lord Lieutenant), Viscount Morpeth (Chief Secretary) and Thomas Drummond (Undersecretary), and 'By 1840–1 it has been estimated that one-third of the key legal and executive positions in Ireland were held by "anti-Tories"' (A. Jackson, *Ireland 1798–1998: War, Peace and Beyond*, 2nd edn (Chichester: Wiley Blackwell, 2010), p. 45). Tyndall's father had exulted when the 'bloody whigs' were 'kicked out of office' in August 1841, and assured his son that now he could 'use all my exertions to get you a situation' (see letter 0090), although clearly it took much longer for Whig

influence to wane in Ireland.
9. *Robert McDonnell*: not identified.
10. *Nurney*: see letter 0056, n. 14.
11. *the Mail*: see letter 0125, n. 4.

To John Tyndall, Snr 20 April 1842 0142

Wednesday night, 12 o'clock.

My dear Father

I was reading a letter from Mr Conwill[1] a few minutes ago and before I had it half finished, I took up the pen determining to write to you, tho' you are at present a letter in my debt. I am after returning from hearing a lecture on the invocation of Saints and Angels and the worship of the Virgin Mary in Christ's Church.[2] The occupant of the pulpit was a most able fellow his name is Bleakley,[3] he did his business cleverly – the church was crowded. The words of his text were 'Whatsoever ye shall ask the father in my name he will give it you –'[4] therefrom deducing the fact that intermediate intercession[5] were unnecessary, were I not sleepy I'd give you some of his arguments. I am as well as usual and that is as well as I can be, how are my mother and Emma. I think there's no fear of yourself, you are pretty hardy. Good night

Your affectionate son | John Tyndall.

RI MS JT 1/10/3257
LT Transcript Only

1. *a letter from Mr Conwill*: Unless Tyndall means letter 0139, written several days earlier, this letter from Conwill is missing.
2. *Christ's Church:* see letter 0141, n. 7.
3. *Bleakley*: Reverend John Bleakley of St Patrick Street, Cork, Curate of Christ's Church (*Slater's*).
4. *'Whatsoever ye ... give it you –'*: John 16:23.
5. *intermediate intercession*: The Roman Catholic doctrine of Intercession asserts that, as well as Christ, saints and angels can offer up their own prayers on behalf of men.

From William Ginty 19 May 1842 0143

Friday.

My dear Tyndall,

I got a read of your eight days composition[1] yesterday and in reply I must say it was all that I wished for – or expected it to be, a lively narrative of incidents of travel particularly romantic and ludicrous. What a shocking specimen of Human Depravity that coachman was! How very admirable your perfect composure and resignation to the will of the Omnipotent must have been to eye of a fellow traveller during the impending thunder-storm. (And your left arm – what a narrow escape indeed – really I wonder it was not dissolved!) You will pardon me when I say that I cannot help smiling at the extremities to which you tourists are sometimes driven. I cannot imagine how ever you brought yourself or had nerve enough to buy a pennyworth of that barbarous bread (Lord, how I should like to see a little of it through a microscope) And then you had to eat it! Oh shocking! I would not be surprised if you were laid up with a surfeit now that you have time to consider on the horrid deed. How very much your dear C- - -t- - -a[2] would be affected did she but know the dangers to which you were exposed Poor dear girl she would at all events be very uneasy until your return. Let me congratulate you on the happiness which awaits your arrival. Just fancy yourself some Sunday evening about the 10th of June next standing before the door of Put. St. Chapel[3] with your arms gracefully entwined round that fatal lamp-post and exclaiming with all the fervency and ardour of an enthusiastic lover

'Here's the post she loved so much
Where I was often planted'!!!

fancy with what delight you will hear the theme of the sacred preacher dying into echoes

'By Beauty mellowed o'er the pavement sweep'

fancy the creaking of the door bursting on your ravished ears as it opens to emit the fair form of your charmer – fancy – oh! happiness unparalleled – pleasure indescribable joy unutterable – the grasp of that silk gloved hand and the – the – the cool 'How d'ye do'. 'Happy to see you' – 'Quite well I hope'

'The just reward of so much love'
'But let that pass' -----

Yes! 'This survey is a failure'
List[4] ye Tyndall now to me
Quondam[5] mess-mate – no repealer[6] –
I'll reveal myself to thee
First there is our embarkation
On the 20th of May
Then we leave our native nation
Friends and country – far away
Then again the Progress Journals[7]
Of which no tidings yet are come
The devil sweep his hotest funnells.
With the man who stopt the fun.
The Tuesday previous to our sailing
Dunn[8] and Foy and some few more

Will sail for Dublin, without failing,
T'enlist in Larcoms fire-proof core[9]
'Professor',[10] 'Phil',[11] Oneill and others
Amongst which are Doig and son[12]
(How cruel 'tis to sunder brothers)
These must go with chains[13] to run
But ah, the worst now follows after,
Which makes their misery more intense
Surely 'tis a sad disaster
These must work for eighteen pence
This cursed order – damn its writer!
Came from England t'other day
By the way of healing plaster
To expectants of more pay.

Enough of this – The following gents will be your colleagues on and after the 10th of June. – Sinnett, Foy Scally,[14] Holland,[15] Latimer, McGowan,[16] Eivers, Bloomfield, Wm. Hunter, Davy Doig !!! One of the Todd's[17] and all the D.O.[18] chaps. Evans is beginning to look sea-sick already. Poor little devil he wont be the weight of a lark when he lands. Nothing will do Tid[19] but a dog – blow'd if I wont ship him out for the Isle of Man as I go by. We finished the job and got the R.M.D.[20] I will write again as soon as I get settled among the John Bulls[21] once more. Jack and Jim[22] and I and Dick H.[23] and Kelly[24] are qualified as plotters and tolerable draughtsmen. That saves our bacon. Did you take the rust off your chain yet. We have not yet heard our destination. I perfectly agree with you with regard to Carroll. Before I read your comment I had formed that opinion, if I dont mistake he was drunk while writing it the similies that he used in the description of his dress &c smells very much of the effusion of a 'Whiskified soul'. I rather think that our few bobs[25] is like McNab's Church of Rome[26] 'They are gone to the bats and to the moles'[27] – he will be certainly a grace to a Reading Desk. I should not wonder if he would sometimes fall asleep and wake shouting 'What's Trumps'.

If there are extremes in human wretchedness and if the unfortunate afflicted is to be pitied – Mrs Payne is. She is lying on a sick bed with none to attend to her most trivial wants but her unfortunate starving children. She seems to be a playtoy in – the hands of the demon of misery – a poor afflicted stranger without I may say a friend to even enquire after her. Giblin[28] is in hospital and what will it be after the 20th. The miserable pittance which Joe[29] sends or is able to send her is not near enough to sustain nature even in health. Heavens Tyndall has this woman no friends – is she altogether an outcast from humanity and the human. There was some talk in the office this morning of doing something for her but I am afraid it will be a failure. Survey men are not very purse-proud previous to a long journey. I really pity her from my soul, but that does her no good. She was foolish that she did not go home long ago.

I'm glad to hear Murray is better, if he was here he would surely have to go chain[30] in England for 9 bobs[31] a week. I suppose you will write again before we embark. Give my kindest remembrances to your sister Emma (I forgot you have but one) but I suppose she completely forgets even my name

Yours ever faithfully | William Ginty

David Bates is to meet us in Liverpool I told him to look out for the Ocean bound from Cork[32] with a load of pigs and sappers clerks!!!

I hear the city of Hamburgh is completely burnt thro' accident[33] attended with a great loss of life and property.

Friday evening

Dear Tyndall I'll just open this to let you know something more about Mrs Payne. Sinnet came up to our room to day and asked for Payne's address and says 'I'll write to him to day and if they dont send their mother some money to take her home or keep her from starvation I'll lug the whole before Major Waters'

What she will do I dont know. She must go into the Poor House or starve[34] and that immediately

Yours &c. | Ginty.

RI MS JT 1/11/3578
LT Transcript Only

---

1. *your eight days composition*: letter missing.
2. *C- - -t- - -a*: Christina Tidmarsh.
3. *Put. St. Chapel*: Ginty's abbreviation does not correspond to any street names in Cork at this time. He may have meant Princes Street, where there were two Protestant chapels (*Slater's*).
4. *List*: listen (*OED*).
5. *Quondam*: former holder of an office or position (*OED*).
6. *no repealer*: opposed to the Repeal Association, which campaigned for the repeal of the Act of Union and restoration of the Irish Parliament.
7. *Progress Journals*: presumably the 'Journal of Progress and Weekly Report'; see letter 0038, n. 3.
8. *Dunn*: John Dunne, a civil assistant who worked in the 4th Division, C District of the Irish Ordnance Survey. He had joined the Survey in May 1838, and was removed to Dublin in May 1842, after which he no longer appears in the records (NAI OS/1/17–19).
9. *Larcoms fire-proof core*: Thomas Aiskew Larcom (1801–79) was a 2nd Lieutenant in the Royal Engineers who joined the Irish Survey in 1826. Two years later, Colonel Thomas Colby appointed Larcom as his assistant in the central organization of the Survey at Mountjoy House, Phoenix Park, Dublin, where Colby, in 1825, had constructed a fire-proof store for maps and documents.
10. *'Professor'*: not identified.
11. *'Phil'*: possibly Phillip Evans.
12. *Doig and son*: David Doig, Snr and David Doig, Jnr, both civil assistants in the 4th Di-

vision, C District of the Irish Survey. Doig, Jnr had joined the Survey in June 1837, and was the higher paid of the two. His father joined the Survey in January 1839. They were transferred to England in, respectively, May (Snr) and August (Jnr) 1842 (NAI OS/1/18–19).

13. *chains*: surveying chains of 66 feet in length; see n. 30.
14. *Scally*: Alexander Scally, a civil assistant who joined the Irish Survey in December 1827 and worked in the 1st Division, C District (NAI OS/1/16–19).
15. *Holland*: John Holland, a civil assistant in the 1st Division, C District. He joined the Irish Survey in September 1836, and was transferred to England in August 1842 (NAI OS/1/17–19).
16. *McGowan*: John McGowan, a civil assistant who worked in the 4th and 5th Divisions, C District. He joined the Irish Survey in November 1835, and was transferred to England in August 1842 (NAI OS/1/17–19).
17. *Todd's*: possibly John Todd; see letter 0115, n. 11.
18. *D.O.*: District Office.
19. *Tid*: John Tidmarsh.
20. *R.M.D.*: not identified.
21. *John Bulls*: the English. John Bull, who was created by John Arbuthnot in the pamphlet *Law is a Bottomless Pit* (1712), was an archetype of the bluff, freeborn Englishman.
22. *Jack and Jim*: John Tidmarsh and Phillip Evans.
23. *Dick H.*: Richard Hunter.
24. *Kelly*: probably George Kelly, a civil assistant in the 4th Division, C District. He joined the Irish Survey at the beginning of 1838, and was transferred to England in May 1842 (NAI OS/1/16–19).
25. *bobs*: bunches or clusters of leaves, flowers or fruit (*OED*); Ginty seems to be using it figuratively to describe his and Tyndall's collections of poetry.
26. *McNab's Church of Rome*: not identified.
27. *'They are gone ... the moles'*: adaptation of Isaiah 2:20.
28. *Giblin*: possibly John Giblin, a civil assistant in the 5th Division, C District. He joined the Division in *c.* June 1840, but at a rate of pay (1*s.* 8*d.*) that suggests he may have been on the Irish Survey before and was returning. He was transferred to the English Survey on 5 August 1842 (NAI OS/1/17–19).
29. *Joe*: Joseph Payne.
30. *chain*: the method of land surveying in which linear measurements of distance are made using a chain of 66 feet in length, with the area to be surveyed divided into an outline and the framework comprising triangles.
31. *bobs*: shillings.
32. *Ocean bound from Cork*: PS *Ocean*, a paddle steamer built for the St George Steam Packet Company and launched in 1836.
33. *the city ... thro' accident*: Between 4 and 8 May 1842 about a quarter of the old city of Hamburg was destroyed by fire, with 51 people killed and the town hall and three churches destroyed. The conflagration started when a small fire in the Slödings market reached a warehouse containing spirits.
34. *She must ... or starve*: The Irish Poor Law Act of 1838 established a workhouse in each Poor Law Union on a similar basis to the new Poor Law introduced in England in 1834, with the exception that in Ireland no out-relief was available in any form.

From William Ginty 29 May 1842 0144

Sunday, 29th May 1842.

My dear Tyndall

When you proceed a few lines further you will not be surprised at my delay in writing. We sailed from Cork on the 20th as appointed – the passage was a short one (28 hours) and so calm that there was none (among the masculine gender) sick but Tom Mathews and Freeborne[1] – as for Tom I think he would get sick on an artificial lake or pond. I will say nothing concerning the scene from Cork to Cove but this – you have still to behold the most beautiful prospect that ever eyes dwelt with extacy on (barring the great places abroad intirely); again, if when coming you are lucky enough to have fine weather – the Welsh Alps will excite and draw forth no small quantum of your admiration from the picturesque and romantic – again, if the weather is calm you will not mistake the direction in which Liverpool lies ere you are within 30 leagues of it. Over it a dark cloud of smoke eternally hangs whose density will lead you at first to imagine that it is one of 'Scotia's lofty mountains'[2] mingling with the clouds in the far perspective – Again, when you approach it nearer – I tremble for you! poor nervous young man! I would advise you to open your eyes by degrees, stooping under the bulwark of the steamer so as to prevent the entire prospect from bursting before your eyes instantaneously. So sure as you act otherwise (recollect I warn you!), as sure will you lose the little wit you have and become altogether teetootally non compos mentis.[3] A thought has just struck me that, by giving you some idea of what you may expect, it perhaps might diminish the impending calamity in some degree – Prepare yourself – brace your nerves, ahem! Now just imagine every cabbage stalk from where you stand (query – sit?), into Carlow and 6 miles round you transubstantiated into masts for ships – every kettle (with or without a pipe) in the Barony into a steamer – the shank of every old Dhudeen[4] in the Province into the chimney of a factory – every hole in my face into a dock – all the sparrows that you have seen for the last 7 years into Policemen, Excisemen, &c., &c., &c. and every corn-creak in Whyte's toy shop in Cork into a windmill. – these, with the assistance of a fertile imagination and a naggin of Tomkin's best, may enable you to form a distant idea of 'Liverpool' alias confusion ad infinitum.

But let that pass – the doom's prepared
Tyndall the sight awaits thee and then thy fright's reward.

After landing we went and got a good dinner and then back again to the dock to wait for further orders. Then Jim[5] and I left Holmes, Tid,[6] Hunter, W. Latimer and Mathews there to receive orders 'for them, for us, and for the whole' – crew – while we went in quest of a place to lay our weary heads and more wearied bones. Jim and I being sole dictators with power absolute, were not long 'closing the bargain' – 6d per head per night – excellent beds and the best of accomodation at Pace's, 84 Porter Street[7] (see enclosed sketch)[8] whereunto thou doest well to attend shouldst thou be coming this way, 'thigum'.[9] To make a long story a short one, Capt. Tucker didn't come near us until noon on Monday. After selecting and arranging for about an hour, he commenced thus: – 'Freeborne, Ginty, Hunter, Latimer, Tidmarsh, you will stop in Liverpool and report yourselves to Lt. Hamly 75 Great Georges St. (his office) – Squad the 2nd. Evans – Malone – and 3 or 4 more of Wilkinson's[10] surveyors – to proceed to Preston next morning, and all the rest to Manchester'. We went up to George's Street immediately and to our unutterable dismay soon learned that we were to go to the Field! Hunter, Tid and Freeborne to Prescot,[11] about 8 miles from here; Latimer and I to remain here. May be this cruel separation didn't make us hang our lips like so many motherless foals. David Bates and his brother George[12] (to you a Saxon) came to meet us from St. Helens[13] on Saturday evening and remained with us until Sunday evening, very kind of them I must say. I intend going to see Dick[14] and Tid next Saturday evening. Here we are then, chaining[15] away like blacks. I will commence to survey Tuesday or Wednesday. It set very severe on me the first two days, but now I care nothing for it. We were up at 4 o'c. the first morning and away before breakfast, 6 miles, to survey the strand at low water (Lancashire), where we got up to our knees in mud – a good introduction! I have not heard from any of the lads yet. After I finish this I have four more to write, so you will excuse its many blunders. Until further orders my address is – at Mrs Clarkson, 120 Front Portland St. (off Vauxhall Road) Liverpool. This is my lodgings. Now concerning the things which are to come I know little and care less – one thing I know that this is a queerish place. Oh! if you saw the c.a.s.[16] that are surveyors here. Picture to yourself the most rugged mountaineers that the wilds of Connemara[17] could produce possessed of all the ferocity of hyenas – square this, and you may form an imperfect idea of these nondescripts. One of them has but the left hand and they say he keeps the best book in the lot. I hope Murray is getting better – he was lucky enough not to come here for it would kill an ass – I have my eye on a half dozen of nice coffins for ourselves. We saw the Great Western[18] just sailing for New York as we came in to port – 4 masts and awful big paddle boxes. They are building another here longer.[19] I must now reluctantly conclude

Adios, mon cher ami[20] | Yours ever faithfully | William Ginty

Write soon!!!!!!!!!!!!!!!

You will find a great many sharks tendering their services to conduct you to nice lodgings – beware oh beware, Jim and I were soft enough to go with one of them, but we soon found out our mistake. There are two houses close by together in Porter Street, dont forget the name or No., or you will go to the wrong one. However it is very likely that Capt. Tucker, ahem! and I will be at the Dock to receive you. You can fare well and cheap in Paces if you go the right way about it and it is a most respectable House – mind, when you order dinner dont order dinner for so many – or you will be salted,[21] order what you think will do.

RI MS JT 1/11/3579
LT Transcript Only

---

1. *Tom Mathews and Freeborne*: Thomas Matthews, a civil assistant who worked in the 3rd and 4th Divisions, C District of the Irish Ordnance Survey. He joined the Survey in December 1837, and was transferred to the English Survey in May 1842. James Freeborne, a civil assistant who worked in the 1st Division, C District of the Irish Survey. He joined the Survey in April 1839, and was transferred to England in May 1842 (NAI OS/1/16–19).
2. *'Scotia's lofty mountains'*: A. M'Kenzie, 'Anecdote of a Highlander', in *Poems and Songs on Different Subjects* (Belfast: Alexander Mackay, 1810), p. 112.
3. non compos mentis: not of sound mind (Latin).
4. *Dhudeen*: short clay tobacco pipe.
5. *Jim*: Phillip Evans.
6. *Tid*: John Tidmarsh.
7. *Porter Street*: a small side street only a few minutes' walk from Clarence Dock, where Ginty had landed.
8. *(see enclosed sketch)*: sketch missing.
9. 'thigum': I understand (Gaelic).
10. *Wilkinson's*: Lieutenant Charles Edmund Wilkinson (1807–70) of the Royal Engineers, who commanded the 2nd Division, C District of the Irish Survey. He later rose to Major-General, and in 1860 spent three months as the acting Governor of Ceylon.
11. *Prescot*: Prescott is a town in Lancashire about 8 miles east of Liverpool.
12. *his brother George*: George Bates, a civil assistant in the 2nd Division, C District of the Irish Survey. He was dismissed at some point during the first half of 1838, and unsuccessfully attempted to claim outstanding pay from the Survey in July 1838, before leaving for England with his brother David (NAI OS/2/16).
13. *St. Helens*: a town in Lancashire about 14 miles east of Liverpool.
14. *Dick*: Richard Hunter.
15. *chaining*: see letter 0143, n. 30.
16. *c.a.s.*: civil assistants.
17. *Connemara*: a region in the west of County Galway containing the Twelve Bens mountain range.
18. *Great Western*: SS *Great Western*, a paddlewheel steamship designed by Isambard Kingdom Brunel and built for the Great Western Steamship Company. Launched in 1838, the SS *Great Western* served as a passenger ship on Atlantic crossings, between Bristol or Liverpool and New York, for eight years.
19. *They are building another here longer*: Ginty presumably means the SS *Great Britain*, which, at its launch in 1843, was the longest passenger ship in the world, measuring 98 metres, 12 metres longer than its sister ship the SS *Great Western*. Like that earlier ship, however, it was built in Bristol not Liverpool.
20. *Adios, mon cher ami*: farewell, my dear friend (Spanish / French).
21. *salted*: defrauded by paying for something that has been made to appear more expensive than it actually is (*OED*).

To Laurence Eivers May 1842 0145

<*beginning of letter missing*>

On Thursday last being a holiday I started at 5 o'clock for Gowran[1] accompanied by a young fellow – a relation of Burchell's[2] I arrived in Gowran shortly after 7, called at Lady Dover's gate[3] and found that the person who had the hydraulic ram[4] in charge had not yet made his appearance – I found the house – rapped him out of bed and was conducted by him to the object of my search – what you told me concerning this engine excited my curiosity and I had half a dozen conjectures formed concerning the affair before I saw it. The moment it met my view I must confess I felt disappointed, it was so utterly unlike any of my preconceived ideas, so I had to crush all my airy conjectures and bring my faculties to bear on the reality. There is something sympathetic between the head and hand when a fellow's imagination is on the stretch – any one located in my immediate vicinity as I gazed on the hydraulic ram might behold a practical illustration of what I mean. I walked and gazed and stood and gazed and according as my ideas approached the climax of confusion the more frequently my claws and cranium came into contact. One comfort I had amidst the chaos of my brain, the engine keeper promised to explain all to me when I had satisfied myself looking at it. In a short time I requested him to do so, Well, Sir, he placed a hand on each ball, gave two or three ominous '<u>hems</u>' and commenced in the strain of a Connemaran[5] Cicero[6] the forensic swell was singularly blended with the Connaught[7] brogue he hiccupped forth words the pronunciation of which would strangle an ordinary man, in such a manner that if I were to be hanged for it I could not prevent the idea of Corney Delany's[8] wrinkles from obtruding itself on my fancy, you may judge from this that the harangue was in no degree calculated to dispel the obscurity of my mind, it required the exercise of all my gravity to keep my risible muscles in proper subjection, the fact of it is when the man had finished I endeavoured very quietly to forget all he had said. By degrees the mysteries of the apparatus began to unfold themselves and half an hour's investigation gave me a tolerable idea of the principle on which it worked. I shall now send you my solution of the matter: –[9]

L is a metal tube about 4½ inches in diameter one end H communicates with a cistern which is always kept full of water, this end is 10 feet higher than the foot of the tube E – H is a stopper which turns on a hinge, there is a cord attached to it by means of which the end of the tube can be opened and shut at pleasure. B is a spherical valve of highly polished metal, which exactly fits into a round aperture, the only communication between the tube and Ball A – To prevent this spherical valve being dashed aside there is a bridle placed over it within about an inch of its surface. G is a pipe which communicates with the ball nearly on a level with the valve B. Y is a valve adjusted so that it will move up and down at certain times by the action of the water in the tube – when the stopper H is raised the water flows through the tube L and were there no obstacles would continue to flow until it would reach its own level X in the tube G, but it must be remembered that the ball A is full of air, when the water enters the ball through the valve B it prevents all communication between the air in the ball and that outside, hence, according as the water presses upward the air becomes condensed in the top of the ball, a reaction therefore takes place and the water is in its turn pushed back, the valve B descends and the water is impelled with great force through the valve G. It is plain that even if both the valves A and B were dispensed with, the aperture which X fills being stopped and that which B fills being left open, the same would take place to a certain extent, the air would be condensed and the water impelled on the principle of the force pump[10] – when however the water flows through L the valve Y is raised thereby stopping its aperture, the moment the water has attained its utmost height the valve descends thus allowing a quantity of the water to escape which flows into the trough M. This destroys a portion of the upward pressure and allows the air within the ball to exercise an expansive force much superior to that which it would exercise had the valves Y and B been dispensed with.

There is a small cube of brass at a through which an aperture sufficient to admit a small sewing needle runs to the interior of the tube. I put a small pin into it and on my taking it out again a thread of water was ejected with great force.

This engine propels water to the height of 87 feet.

J. Tyndall

RI MS JT 1/11/3526–7
LT Transcript Only

1. *Gowran*: a town in County Kilkenny.
2. *Burchell's*: possibly David Burchell, a cabinet maker from Bagenalstown, County Carlow, whose son was baptised in February 1841 (Irish Church Records).
3. *Lady Dover's gate*: Lady Georgiana Howard (1804–60), widow of George Agar-Ellis, 1st Baron Dover, who had died in 1833. The family seat was Gowran Castle.
4. *hydraulic ram*: a cyclic water pump invented in France in the mid-1790s that uses hydropower to pump water to higher elevations than the original source. James Fraser's *A Hand Book for Travellers in Ireland* (Dublin: William Curry, 1844) notes that the 'demesne of Gowran ... may be said to be in a transition state ... But even now, in the incipient state of the improvements ... evidences of what is intended may be seen ... from the supply of spring water to the inhabitants, brought up from the lower levels by means of the hydraulic ram' (p. 162).
5. *Connemaran*: an inhabitant of Connemara; see letter 0144, n. 17.
6. *Cicero*: Marcus Tullius Cicero (106–43BC), celebrated Roman orator and prose stylist.
7. *Connaught*: an ancient province in the west of Ireland.
8. *Corney Delany's*: a character in Charles Lever's novel *Jack Hinton, The Guardsman* (1844).

9. *my solution of the matter*: The structure and operation of hydraulic rams was a topical issue at this time, with, for instance, Thomas Ewbank's *A Descriptive and Historical Account of Hydraulic and Other Machines for Raising Water* (London: Tilt & Bogue, 1842) offering a detailed account of the workings of various hydraulic mechanisms (pp. 367–72). The diagram accompanying this letter has not been located.
10. *principle of the force pump*: Force pumps use a plunger to force liquid upwards and expel it under pressure.

From William Wright 11 June 1842 0146

Cincinati June 11th 1842.

My oldest and most valued friend

I have lately received your address[1] from my sister Rebecca and after a lapse of 18 months or more again take up my pen to address you and now that I have done it I hardly know how to commence there are so many thoughts of Bygone days crowding on my memory that I am at a loss to know what to write. I need hardly add that you Jack are always connected with those happy recollections. Oh for the days of our boyhood, why did they pass so quickly or why did we not value them more at the moment scarce had we time to sip their sweets when they vanished and left no trace but memory. Even that is a sweet solace (as in the words of that beautiful opera Sonnambula[2] which often recur to me and make me think of old times)

> Still so gently o'er my senses stealing
> Memory still brings back the feeling
> Spite of all my cares revealing
> I love them dearly, dearly do I love them still[3]

Many a changing and exciting scene have I passed through since last I saw you and often did I think of you and wish you were beside me. When watching our gallant ship cutting its way through the mighty deep or gliding through the noble Hudson[4] or standing by Niagara's mighty waters I often thought how much more they would have interested me had you been also a spectator. It seems to me that you are the only person of my early years whose sympathy and feelings flow in the same channel as my own and none who will understand them as well, feebly as I have attempted to portray them in words. Although there are some of my old companions with whom I spent more time and had more intercourse yet your absence is the only one I feel for and whose presence if ever I behold it will be most welcome. I hope it will not be long until I do see you though I should never wish to see you in America, at least until your Father and Mother would be released from all their troubles[5] (& may that day be far distant is the sincere prayer of your friend). If ever you do come to this land bring with you no hopes of speedily realizing a fortune or that it is much easier acquired than at home, if you do you will be disappointed. For I must acknowledge it has been my case. Not that I was very sanguine or had very high notions of this country before I arrived in it, or that it is hard to satisfy me. I have no doubt if I had settled about Carlow it might have been better for me. Trade is much worse here at present than I have ever seen it in the old country, we have hardly anything in the shape of cash, any business that is done is by trading one article of <*word illeg*> for another I do not at all regret coming out for I have seen and learned a good deal since I did come whether I profit by it or no. Again if you do think of coming out here you will have to lay aside all love and veneration for all old England's laws and constitution. If you do not you will often be an unwilling listener to words that grate upon your ears. I admit that Americans have accomplished a mighty work and that there is a good deal in their character to admire. But they entirely deface it by their self conceit vain boasting and determined hostility to England the very author of our greatness – for it is English gold and English hands that has constructed their vast railroads and canals through their unlimited forests and make them teem with life and plenty.[6] And for the payment of which the honour of their several States is pledged but which I believe never will be redeemed.[7] For it is a bad sign of it when the Governors of their States starts and is elected on the repudiation ticket or in other words pledged to resist the payment of their State debts both principal and interest. And then they openly and continually express themselves that they would be glad of engaging in a war with England. If it was only to show how well they could thrash her. For as they remark they thrashed her when in their infancy and now that they are arriving at the years of maturity they would be much better able to do it; forgetting that at the time they did engage with her last[8] England had almost the whole world against her and not one ally. It is amusing to see some of the remarks of the press on the subject and how easily they think it would be for an American army to pitch their tents in the Regents Park and their officers make Head Quarters of the Tower.[9] They must truly imagine that the spirits of our Pictons and Nelsons[10] and our cock of the walk Wellington[11] are all vanished from British hearts. I calculate they would be a wee bit mistaken, For in the prophetic words of the immortal Shakespear[12]

> England never did, nor never shall,
> Lie at the proud foot of a conqueror
> But when it first did help to wound itself
> Come the three corners of the world in arms
> And we shall shock them Naught shall make us rue
> If England to itself do prove but true[13]

You see I give Shakespere credit for being a prophet as well as poet at least in these few lines. Now Jack I must lay by my martial spirit and confine myself to a little domestic affair. I would not have alluded to the above subject but that I am often compelled to be a listener to the most scurrilous remarks on my native land (For I never think of drawing a line of distinction between England and Ireland), so I am obliged to let off a little steam occasionally and as I know you will have a little forbearance with me I have puffed off all the extra that has been accumulating for some months. In short if you wish to know the real American character read Mrs Trollop's Domestic Manners of the Americans,[14] it is the nearest the mark of any publication I have seen on the subject. I have again struck off my intended track but I will strive and keep on it as well as I can. I wish you were a little closer to me and I would have a small job for you to do for me, I mean to be a sponsor for my little son he is a very fine little fellow. Now Sarah[15] desired me to mention that it was not her which prevented me from writing to you but that she often desired me to do it I am sorry to be obliged to confess that such was the case. Not that I wanted inclination to do it, but that I put it off from day to day expecting to have something extraordinary to mention until I found I let it go too far, and then I began to expect a line from you and that you would think my letter had miscarried, (I am determined to be candid with you). But now that I see no chance of first receiving a letter from you, I have broke the ice and if the rent again closes it must be your fault this time for I am sorry to say it has been my fault more than once. I hope you forgive me for it. And although the wide Atlantic rolls between us we will again converse about the mossy banks of our own dear Barrow[16] upon whose smooth surface we have spent many a cheerful hour. And through its small tributaries we have often chased the spangled trout and swiftly gliding pike making the old Moat[17] resound again with our loud husas when we would seize our victim and bear it in triumph to the shore I do believe if I was standing this moment upon the bank of Maudlin stream[18] and you along side of me I would at once roll up my sleves tuck up my trowsers and in right away and enjoy the chase of a small trout as much as ever I did. And if you did not follow me, I would brand you with such a name as always brings the blush into the face of a true hearted Irishman. If you are in company with any of my old associates please to present them my kindest regards hoping that all of them enjoy health and happiness. And that it may be your lot your Father's, Mother's, sister's and relations is the constant wish and prayer of your humble servant. You will please remember me to them all. If you receive this please to let my father know that you have heard from me and that we are all in good health here. Along with this letter I send you some newspapers. I hope it will not be long until I receive a letter from you and you must give me the whole full and true account of all your affairs especially with the ladies you know I made a confidant of you and turn about it only fair play so do not forget it. I was very near forgetting to mention a subject which will surprise you a little I have only been out 4 days fishing and shooting since I came here. Not that I could not make time, but the few days I was out I got heartily sick of it. You are aware I suppose that there is no hinderance to your shooting where you choose so that its not from want of room, but the game is so very scarce you might hunt a whole day without meeting anything worth shooting. Quail are the only birds at all plenty Out in the Back Woods[19] there are turkey but there is so much cover for them that it is hard to get at them there is very little water fowl to be met with. But up near the Sucker[20] in the North part of this State they are very plentiful. On my road here from N. York[21] I stopped a few days with the Rainsfords.[22] They live on the banks of the Niagara river about 2 miles from Lake Erie out of which the Niagara flows, and about 20 miles from the falls. You can hear the noise distinctly at their door when the wind blows in the direction from the falls. On the river I have seen some thousands of water fowls of different sorts but it is very hard to get within gun shot of them. George told me that some seasons of the year the river does be completely covered over with them. On the whole shooting is as good in the old Ireland if one had a little more liberty. I believe it is the very hinderance that makes it sweet And as for fishing it is much more pleasant, the fish are not so plenty and the season of the year for fishing the sun is so hot that you cannot possibly stand out. Now Jack I must bid you good bye, it is want of room which compels me now to do it. So farewell my old friend and believe me I remain

Yours always | William Wright.

You will direct your letters to Richard Haughton Main Street Cincinati for W. Wright and I will be sure to get them. I am doing business for the above named individual and receive a weekly stipend of 10 Dollars which is pretty fair considering the present state of trade.[23] Woolen and cotton fabrics are worth double as much as at home but provisions are very cheap so that a person having more salary here the outlay is so much more than at home you arrive at the same conclusion that there is a great deal to spare I expect to get business for myself as soon as trade takes a turn for the better once more

Farewell | W. Wright

Send me a newspaper. I shall commence and send you some weekly. Remember me particularly to Joseph Payne and W. Murray the gardener. If you know anything of my old Beau A.S.[24] let me have it.

Yours | W.W.

RI MS JT/1/5/1800–3
LT Transcript Only

---

1. *your address*: LT's handwritten insertion suggests this letter was addressed to 'Carberry McCarthy | Strand Road | Cork' and then 'forwarded to Preston'.
2. *Sonnambula*: an opera semiseria by Vincenzo Bellini and Felice Romani, first performed in Milan in 1831, which relates the romantic misadventures of the sleepwalking Amina amidst a pastoral setting in Switzerland.
3. *Still so gently ... love them still*: from a solo sung by Amina's lover Elvino in act II, scene II

of *La Sonnambula*.

4. *the noble Hudson*: the River Hudson, which flows through New York State.
5. *your Father ... their troubles*: Wright is perhaps thinking of Psalms 25.22, 'Redeem Israel, O God, | out of all his troubles!', and means when Tyndall's parents are no longer alive.
6. *it is English ... and plenty*: In the 1830s American capital projects were financed by the selling of bonds on British money markets, where Anglo-American banking houses offered cheap and plentiful credit, largely because the bonds were guaranteed by state governments.
7. *And for the ... be redeemed*: In 1837 the American financial system collapsed, creating a recession that made it increasingly difficult for states to pay back their debts to British banks. Between 1841 and 1842 eight states and one territory defaulted on their interest payments, and there was much agitation in favour of repudiating the loans altogether.
8. *the time ... her last*: presumably the War of 1812, when Britain was also engaged in a global conflict with Napoleonic France and its only ally Austria had just been defeated.
9. *the Tower*: the Tower of London.
10. *Pictons and Nelsons*: Thomas Picton (1758–1815), the most senior officer of the British Army to die at the Battle of Waterloo; Horatio Nelson, 1st Viscount Nelson (1758–1805), the flag officer of the Royal Navy who died at the Battle of Trafalgar.
11. *Wellington*: Arthur Wellesley, 1st Duke of Wellington (1769–1852), the commander of allied forces at the Battle of Waterloo in 1815.
12. *the immortal Shakespear*: William Shakespeare (*c.* 1564–1616).
13. *England never ... prove but true*: Shakespeare, *King John*, V.vii.112–18, with Wright substituting prove for rest in the final line.
14. *Mrs Trollop's ... the Americans*: a travel book by Frances Milton Trollope (1779–1863) published in 1832 by Whittaker, Treacher & Co. in both London and New York. In the book, Trollope details her time living in Cincinnati, and is highly critical of American society.
15. *Sarah*: Sarah Maria Wright (née Neale).
16. *our own dear Barrow*: the River Barrow, which passes through Leighlin Bridge.
17. *the old Moat*: the Moat of Ballyknockan, the only remaining part of the Dinn Righ, the ancient seat of the Kings of Carlow, south of Leighlin Bridge near the west bank of the River Barrow.
18. *Maudlin stream*: a small tributary of the River Barrow on the southern outskirts of Leighlin Bridge.
19. *Back Woods*: wild, uncleared forest-land, especially that of North America; also a remote and sparsely inhabited region (*OED*).
20. *the Sucker*: possibly the Sucker Lake in Indiana, the neighbouring state to Ohio.
21. *N. York*: New York.
22. *the Rainsfords*: not identified.
23. *the present state of trade*: see n. 7.
24. *my old Beau A.S.*: not identified.

To John Tyndall, Snr 12 June 1842 0147

Cork | June 12th, 1842.

My dear Father

This is to let you know that I arrived safe and sound in Cork on Friday at half past 3 o'clock, having cut my way gallantly thro' volumes of dust which the small breeze with which we were favoured scattered abroad in the most liberal profusion. I was as white as if the contents of half a dozen of the Milford conductors[1] were heaved on me. I dashed into Carberry's,[2] however, ran up stairs and pushed in the door of my little room without any ceremony. There to my no small surprise an old lady was seated, and strange to say seemed quite at ease. She appeared in fact as if she thought that she had a right to sit there. Upon enquiry, I found that she had taken the place during my absence. I contrived however to wash myself and change my shirt. Mrs McCarthy told me that the house was engaged but at the same time kindly said that she would contrive to accommodate me until I could procure lodgings. Having finished my scrubbing I started out to see some of my old comrades. I was just in time to bid Foy farewell. He started for Liverpool at 5 o'clock on the evening in question. Well, travelling has arrived at a pitch of great perfection. On Friday morning I might be seen watching the rising sun at the hole in the wall on Irish road, on the evening of the same day I might be seen on board the 'Ocean' steamer,[3] helping Foy to get his luggage packed. Having seen Foy off I started in quest of lodgings. I was not long in procuring them. I am snugly situated in Pine Street, where two of our lads have been living this good while. We three now form a mess, so that I am not at all lonesome. I'm just going down to see Mrs Payne so I must conclude. Tell my mother that I'm as snug as a thrustle.[4]

Your affectionate son | John Tyndall

Direct your letters to me at Mrs Cotters No 6 Pine Street

RI MS JT 1/10/3273
LT Transcript Only

---

1. *Milford conductors*: The town of Milford in County Carlow was famous for the large flour mill established by John Alexander in 1790. Mr and Mrs Samuel Carter Hall, in their *Ireland, Its Scenery, Character, &c.*, observe that 'Mr. Alexander carries on his trade in corn at eight different places in Carlow and the adjoining counties, from whence it is transported to Milford, to be converted into flour, and thence distributed through the country or exported to the English markets' (3 vols (London: Jeremiah How, 1841–3), vol. 1, p. 406). By conductors, Tyndall seems to mean vessels for the storage of flour, although this usage is unusual and appears to have confused even his own father; see letter 0149.
2. *Carberry's*: the house in Strand Road, Cork, where Tyndall had been lodging with Carberry McCarthy and his wife.
3. *the 'Ocean' steamer*: see letter 0143, n. 32.

4. *thrustle*: throstle, or song thrust (*OED*).

From John Chadwick 26 June 1842 0148

Kinsale | June 26th 1842.

My much esteemed and very reverend Tyndall

I received your bagatelle[1] (I need hardly say only for fashion sake) with a degree of pleasure bordering on rapture indeed the expectant ardour of a nymph in girlhood's rosy joy when looking forth from her bower for the individual with the bags who bears her first clandestine billet-doux[2] on the wings of a lazy letter carrier could never be equal to mine when I saw our Mercury[3] come into my establishment and hand me an epistolary morsel directed in your peculiarly neat hand to my unworthy self though, apropos, bloody end to any one who would so style me except my own pen in its meek humility. I tore it open though it was worthy of a better fate than disembowelling, nevertheless with all the gusto of a medical tyro I performed the Cesarean operation[4] and devoured the contents with all the ferocity of an ogre, though the contents related to nothing but a hot journey and an episode about an old faggot of a woman who has dared to usurp a den where sappers held their orgies,[5] for which crime she deserve to be tried by a packed jury of parallel rulers and sentenced to have six inch traces pricked off on her buttocks until they were as tattooed as a New Zealander's frontispiece.[6] Yet still even if you had penned a letter to me full of nothing at all, it wont have brought up old recollections from the gulfs of memory bearing an instant impress to the heart and recalling past scenes when we dived into the same dish of sprats and were co-partners on a featherbed. I have been long expecting a letter from you, indeed I was almost guilty of thinking you had forgotten me, but the receipt of yours broke like a beam of sunshine upon the cloudy sameness of my Kinsale existence, warming and irradiating it for the time. Indeed it is gratifying to one who has retired from the busy world like me to find that at least one old associate holds me in a snug place in his memory – all I want here to make me quite content (I will not say happy for that is not of earth saith the preacher), [7] is the companionship of a congenial spirit like yours. That is all the loss I sustained in leaving the survey[8] it is a pity that you could not find some line of life that would be parallel to mine, but the paths of life in Kinsale are all too lowly for a soaring spirit like yours, and the tide in the affairs of men spoken of by our immortal bard,[9] ramifies itself into a number of petty channels in this town which, were they collected into one tideway, would be but sufficient to keep one poor devil afloat. Nevertheless, to continue the figure, if you would venture in my bark for a season and favour me with your promised visit, you would be welcome to remain while there was a grain of meal in the kist,[10] that is if you would promise to be provident and not oblige me to tell you that the 'hough's in the pot' as the border raider's wives were wont to say when there was nothing left of the cow but the shinbone, as a signal for them to make a fresh haul.[11] I cannot say when I may go to Cork I was planning to go to Liverpool in July, but fear I cannot leave this to go such a distance. Write to me and let me know where you intend to reside in Cork also when you can come down to me. Write soon – The Survey officials are now reduced to a select few. I suppose you have not many companions. I may surprise you with a visit ere long but could not name the time. All friends here are well. Mrs C.[12] desires to be mentioned to you as also Miss Donaclifte. My young un is growing long and round and shouts as hard for her meals as if she was 4 years old instead of 4 months,[13] she is a remarkable little baste.[14] Everything goes on here in the same monotonous manner, for Kinsale is as unchangeable as the laws of the Sweedes and Prussians.[15] Remember me to the companions of my former devilments, that is to such of them as remain on the green isle. I hope that you will carve 'forget me not' on my ci-devant[16] table and stool before they go across the seas. Nothing gave me greater pleasure than to hear of Tom Foy's promotion.[17] If I go to Liverpool I will have an opportunity of seeing some old messmates there. I must now conclude wishing you lots of prosperity and heaps of happiness and remain

your steadfast crony that was | John Chadwick | Adieu

I hope that craw fish[18] of a Sinnett behaves decent to you I hear he is the only one of the lords anointed left among you.

RI MS JT 1/11/3495
LT Transcript Only

---

1. *your bagatelle*: letter missing.
2. *billet-doux*: love letter (French).
3. *Mercury*: messenger, a person who brings news (*OED*).
4. *Cesarean operation*: a surgical procedure in which a baby is cut out of the mother's body. The term derives from the Roman Imperial law, or Lex Caesarea, that required babies to be removed from mothers who died in childbirth.
5. *an episode ... their orgies*: see letter 0147.
6. *tattooed as a New Zealander's frontispiece*: The Maoris, the indigenous Polynesian people who settled in New Zealand, practice a form of body art known as moko. As they regard the head as the most sacred part of the body, elaborate tattoos covering the face are a marker of high social status in Maori culture.
7. *that is not ... the preacher*: possibly an allusion to Ecclesiastes 6.1–12.
8. *leaving the survey*: Chadwick resigned from the Irish Ordnance Survey on 26 February 1842 (NAI OS/1/19).
9. *the tide ... immortal bard*: 'There is a tide in the affairs of men | Which, taken at the flood, leads on to fortune', *Julius Cesar*, IV.iii.218–19. The immortal bard is of course William Shakespeare.
10. kist: a large chest (*OED*).
11. *'hough's in the pot' ... fresh haul*: The explanatory notes that Walter Scott appended to his

poem *The Lay of the Last Minstrel* (London: Longman, Hurst, Rees, and Orme, 1805) include the following passage: '"A saying is recorded of a mother to her son (which is now become proverbial), *Ride Rowlie, hough's i' the pot*; that is, the last piece of beef was in the pot, and therefore it was high time for him to go and fetch more". *Introduction to the History of Cumberland*' (p. 321).

12. *Mrs C.*: Chadwick's wife, née Fudger (see letter 0177), possibly the daughter of either Harris or William Nicholas Fudger, who were both boot and shoe makers on Main Street, Kinsale (*Slater's*).
13. *My young un ... 4 months*: The birth of Chadwick's daughter seems to coincide with his resignation from the survey exactly four months earlier; see n. 8.
14. *baste*: beast (*OED*).
15. *unchangeable as ...* Prussians: not identified.
16. *ci-devant*: former, especially an erstwhile noble following the French Revolution (*OED*).
17. *Tom Foy's promotion*: Thomas M. Foy had been promoted from Private to Lance Corporal in the Corps of Royal Sappers and Miners on 27 May 1842 (NAI OS/1/19).
18. *craw fish*: variant spelling of crayfish, a freshwater crustacean.

To John Tyndall, Snr 28 June 1842 0149

Cork, Tuesday morning, 6 o'clock

My dear Father

I have been this hour back exploring the mysteries of a quadratic equation.[1] It has proved a gordian knot[2] to me and being heartily sick of striving to unravel it, I think it will afford me some degree of relaxation to write to you

It's a good while since I received your letter, you should have had one from me ere this had I not written to Mr Conwill and in his letter[3] requested him to read it for you.

You appear anxious to prevent my mistaking the evening you ment when you spoke of my arrival in Cork, I see the Queen's County[4] trip has not yet quitted your mind. You imagine I have committed a great bull[5] in speaking of the Milford flour conductors.[6] I shall now try and relieve myself from the weight of your argument. Here are your words 'I would be glad to know which was it the metal conductors or the flour they contained which you drew your figure from, you say you were as white with dust as if the contents of a half a dozen of the Milford conductors were heaved on you. Now the conductors are metal and how metal could cause you to be white or dirty I am at a loss to conceive' !!!!!! Bravissimo![7] This is worthy of the opponent of Lynap.[8] Did I speak of the conductors being heaved on me? No. I spoke of the contents of the conductors, alias what the conductors contain; alias flour. Look to a former part of the page and you will see the word CONTENTS staring you in the face. Who made the bull now? Go now to my mother and let her console you after this drubbing. I hope you wont get this before breakfast as I would not wish you to lose your appetite which I am afraid the thrashing I have given you would infallibly distroy. You spoke of my being affected at Foy's going away,[9] and also of my being sorry for leaving my old lodgings; why if such things would grieve me I'd be wrapt in eternal sorrow. I have learned to treat such things with the most sublime indifference. I went down to see Mrs Payne last night on entering her room judge of my astonishment at seeing Maria[10] there. What in the name of wonder can have induced her to take this step. She speaks of going to America, in fact I was down at the emigration office last night enquiring about vessels. She knows her own business best.

With respect to my being introduced to Mr Carroll, its a thing that would gratify me much. But to go to him depending on the merits of any of my rhymes for a favourable reception is a thing I could not think of. I know the merit of my productions as well as any one, and I feel convinced that it is not such as would justify me in imagining that on their account I was worthy of the notice of any literary man.[11] This is my candid conviction. You ask why I did not waken you the morning I came away. 'twould be a sin, you were revelling in the very luxury of sleep, there was no fear of my forgetting my shirt. I dont care much about Burns.[12] I got a letter from Ginty the other day,[13] the drollery is inimitable. Give my love to my mother and Emma. I am very comfortable in my present lodgings.

I remain | Your affectionate son | John

RI MS JT 1/10/3274
LT Transcript Only

---

1. *quadratic equation*: an equation having the form $ax^2 + bx + c = 0$, where x represents an unknown, and a, b and c are constants with a not equal to 0.
2. *gordian knot*: the unyielding knot by which the ox-cart of the Phyrgian king Gordias was tied to a post in his capital city Gordium. When Alexander the Great visited the city in 333BC the only way he could untie the legendary knot was to cut it with his sword.
3. *his letter*: letter missing.
4. *Queen's County*: the County immediately north-west of County Carlow, now called County Laois.
5. *bull*: a self-contradictory proposition (*OED*).
6. *Milford flour conductors*: see letter 0147, n. 1.
7. *Bravissimo!*: most excellent (Italian).
8. *Lynap*: not identified.
9. *Foy's going away*: Thomas M. Foy had been transferred to the English Ordnance Survey on 10 June 1842; see letter 0147.
10. *Maria*: Maria Payne.
11. *the notice ... literary man*: Thomas Harris Carroll was the editor of the *Carlow Sentinel.*
12. *Burns*: presumably the Scottish dialect poet Robert Burns (1759–96).
13. *I got ... other day*: possibly letter 0143 or 0144.

From William Ginty [July] 1842 0150

Liverpool | Sunday morning.

My dear Tyndall,

I received your letter[1] about 10 minutes ago while sitting at breakfast. If you can recollect the contents of my last epistle[2] the word Crosby[3] may be familiar to you as I said it was probable I would be sent there when I commenced surveying. 'My fears have boded all too right'.[4] There I was sent on last Tuesday week fully equiped with a chain and a man to carry it,[5] a pole and flag a spud (or juvenile spade), book and cover, and as many copies of the act of Parliament[6] as would thatch a house. In the printed form before the Extract (from the Act), there is an article headed thus:– 'I Henry Tucker Captain R.E.,[7] and officer acting under the orders of the Master General[8] and Honble.[9] Board of Ordnance, appoint you as one of my assistants &c. &c.'. Now Sir, owing to the gross ignorance and selfishness of the inhabitants, I have to shake – not one – but a whole dozen of them at the owner of every wheat field; and, as few of them can read, I generally read for them, thus:–

'I Henry Tucker Captain R.E., being an officer acting under the orders of the Master General and Honorable Board of Ordnance, (Here I stop short; another word and away with the Captaincy.) I look into the fellow's face with as much Fenwickism[10] as I can scrape up, and then I turn to the Extract and read the penalty for him. I served 2 or 3 this trick already, for I draw this inference or at least come to this conclusion – that if Captain Tucker's name would have any effect on their beer-imbued brains, the effect would be more powerful if they could be led to imagine that they were confronting Captain Tucker in propriâ personâ.[11] I have the chainman as well discipled for the stratagem as need be. When I have performed my part, I walk off with 'all that strut en militaire'[12] to a respectable distance, while he endeavours to bring them to a compromise lest 'the Captain' should enforce the penalty. One old farmer was so fully convinced as to my consequence &c. that he wanted me to come over to the village near my work to have a glass of wine. What a pity it is I'm a teetotaler! Jack[13] and Dick[14] had a very respectable watch-maker waiting on them at their lodgings in Prescott[15] to see if they could employ him. Jack's behaviour on this occasion was on a piece with his conduct in general, which I highly commend. They took the fellow into their antechamber and told him they would be with him presently – up stairs they went and got two of the largest books they had and came down and placed themselves in the two farthest extremities of the room from the anxious expectant. After a while the books were opened and Jack's first question was did he know fluctions[16] well. The poor fellow, knowing as much about even the word as he did about astrology, was completely confounded; by a great deal of exertion he mustered up nerve enough to stammer out a solitary No! a few more like this and they would be under the necessity of calling in their medical adviser. Humanity and the fellow's countenance alone induced them to refrain. Thus ended the exam. The poor watchmaker left them convinced of his own insignificance and the rashness of the step he was about to take.

William Latimer and I were out seeing them last Sunday week. Prescott is 8 miles from here. Crosby 7. William was out with me last Sunday, I am with him this Sunday, and the other two lads were to be in here last night but they have disappointed us, for which I shall talk to them 'in the language of a poet'.[17] Alas, poor Evans, he is alone in Preston. The poor fellow seems to be disconsolate and, entering deeply into the spirit of the matter, he says with Tom Moore[18] 'Twas ever thus &c'[19] As for Holmes he seems to be if anything worse; he says he is 'estranged from those he would have for ever nigh' but as this gent talks a great deal about some fair charmer beyond the broad waters of the Atlantic I dont wonder at his discomposure. The only thing I regret is the separation from Tid[20] and Dick, it is a sorry thing to imagine ourselves in the same division[21] and still apart. That's what I call 'Disunited tho' not Separated' the plea of the cousin was not able to cozen out the desired effect – How did you hear that pray? I commenced surveying Wednesday week and I have finished the first Δ[22] My work averages 40 acres or 130 chains per day since I commenced, and I have given entire satisfaction to the Corpl. in charge.[23] He is I think a right honest man, all the men speak well of him. I have to come to Liverpool every Saturday for my cash. William seen Foy last Sunday evening when he returned from Crosby. I am very sorry I did not see him, particularly as I hear he has got back the V[24] Lt. Hamley sent him to Manchester.

Now just mind what I'm going to say: you, and every one of you, will be scattered to the four winds of heaven 24 hours after you land in Liverpool. Some of ye may and will get into the office or offices – for there are some of ye fit for nothing else. When ye come ye will be served just as we were (and, mind, this is not a supposition). Your tables[25] &c will be put into the stores, and Captain Tucker will tell ye off in half dozens and away with ye to different parts, and most likely all to the Field for they have nothing to do in the office for they have no points[26] observed. In this office, which contains 40 men, they are plotting the work on cartridge paper from the reduced distances; and by that they are able to tell the lines that require remeasurement[27] and they are getting those done so that when they do get the points there will be no delay. So now any of ye that has sense will have a good pair of strong shoes of Irish manufacture to prepare for the worst. There are some of ye (between ourselves) that I suppose will go to work with dancing shoes and white silk stockings. It is more than probable that you will find us four waiting for you at the Clarence Dock.[28] My best respects and kindest remembrances to Messrs Hunter, Bloomfield, McGowan,[29] Eivers, Latimer &c &c &c. And to all those that are at Mrs Cotter's particularly Mr Robinett.[30] –

Tell me, did not Evan's brother and Mr Young[31] that accompanied him, go to live there? They agreed with her before I left that.

I would'nt doubt Willy Hunter for a bit of mischief. So he told you that I used to be kissing Miss Cotter[32] That infant. She is not ripe enough for me, he told you this to take the suspicion off himself. 'I could a tale unfold'.[33] Ask Mary[34] for me how is her Globe.[35] Tell Bloomfield that as yet I had no opportunity of making any zeal manifest.

We are all most beautifully sun-burnt if my face or at least the skin of it was analyzed it would make raw sienna.[36] Dick is the brownest of the lot and its a toss up between Jack and I. William[37] has escaped best – for he pulls his cap over his eyes. Straw hats would be of better service in going thro' hedges and briers.

RI MS JT 1/11/3583
LT Transcript Only

---

1. *your letter*: letter missing.
2. *my last epistle*: letter missing.
3. *Crosby*: a town in Lancashire about 6 miles north of Liverpool.
4. *'My fears ... too right'*: slight misquotation of the lines ' ...my hourly fears, | My dreams have boded all too right' in T. Moore, *Lalla Rookh: An Oriental Romance* (1817), III.i.275–76.
5. *a chain ... carry it*: see letter 0143, n. 30.
6. *the act of Parliament*: the Ordnance Survey Act of 1841, which granted surveyors the right to enter 'any land, ground or heritages of any person or persons whomsoever, for the purpose of making and carrying out any survey' (1841 c. 30 (Regnal. 4 & 5 Vict) section 2).
7. <u>R.E.</u>: Royal Engineers.
8. *Master General*: George Murray, who was Master-General of the Ordnance from 1834 to 1835 and between 1841 and 1846.
9. *Honble.*: Honourable.
10. *Fenwickism*: presumably the manner of Captain Robert Fenwick.
11. *propriâ personâ*: in person (Latin).
12. *'all that strut en militaire'*: Lord Byron, 'Farewell to Malta' (1811), 20.
13. *Jack*: John Tidmarsh.
14. *Dick*: Richard Hunter.
15. *Prescott*: a town in Lancashire about 8 miles east of Liverpool.
16. *fluctions*: a variant spelling of fluxions, the term that Isaac Newton used for differential calculus.
17. *'in the language of a poet'*: not identified.
18. *Tom Moore*: Thomas Moore (1779–1852), an Irish poet and friend of Byron.
19. *'Twas ever thus &c'*: likely an allusion to the lines 'Oh! ever thus, from childhood's hour, | I've seen my fondest hopes decay' from Moore's *Lalla Rookh*, III.i.280–81.
20. *Tid*: John Tidmarsh.
21. *the same division*: the 5th Division of the English Ordnance Survey.
22. Δ: presumably triangulation.
23. *the Corpl. in charge*: probably Corporal Charles Sheperd (b. 1804) of the 14th Company

of the Royal Sappers and Miners, who was second in command to Lieutenant William George Hamley in the 5th Division in Liverpool.

24. *V*: possibly a Vernier Scale, a device used for making accurate measurements invented in 1631 by the French mathematician Pierre Vernier (1580–1637).
25. *tables*: plane tables, devices that afforded a solid and level surface on which to make field drawings.
26. *points*: trigonometrical points, fixed surveying stations.
27. *reduced distances ... remeasurement*: probably related to the change from the 2 inches to 1 mile scale in which some parts of Lancashire were originally surveyed, to the standard 6 inch scale.
28. *Clarence Dock*: dock on the River Mersey, part of the Port of Liverpool. The dock was opened in 1830 as a facility for steamships, and named in honour of the Duke of Clarence (1765–1837).
29. *McGowan*: see letter 0143, n. 16.
30. *Mr Robinett*: presumably a lodger at Mrs Cotter's lodging house in Pine Street, Cork, although he was not involved in the Irish Ordnance Survey.
31. *Evan's brother and Mr Young*: not identified.
32. *Miss Cotter*: presumably the daughter of Mrs Cotter.
33. *'I could a tale unfold'*: W. Shakespeare, *Hamlet*, I.v.15.
34. *Mary*: not identified.
35. *her Globe*: not identified.
36. *raw sienna*: a yellow-brown colour pigment that was produced in the Italian city of Sienna during the Renaissance; see letter 0154, n. 2.
37. *William*: presumably William Latimer.

From William Ginty 11 July 1842 0151

Great Crosby,[1] Sunday evening

Dear Tyndall

I received your invocation[2] to the sanctified shades of Dunran[3] and Mickey F.[4] in company with five others on last Wednesday morning – One from Cupid[5] – one from 'the lodger'[6] one from 'cousin Jack'[7] (as he styles himself) one from David Bates and one from Ginty's father. When I got a glimps of these and recognized the well known 'sticks',[8] to use a hacknied expression, my feelings were more easily imagined than described. I grinned, I laughed, I scratched my head and danced one of Aubrey's favourites[9] in my extacy – all around me became a 'passing dream'[10] until I devoured their precious contents. As I have nothing else to write about I'll just give you a synopsis of the impression that each and every one of them made upon my mind then and there, Commencing with your invaluable self – Oh ever venerated shades of the Kinsale 'Inadequates', offer up your orisons[11] in behalf of a brother – I dont well know how to comment on yours. I am very proud to hear that my probono publico[12] affords so much amusement – it was not treating me fairly to show off my bladderings before a room full of critics, but then I suppose they knew Ginty better than to pay too much attention to his

Nonsense precipitate like running lead
That slipp'd thro' creaks and zizzags of the head[13]

Devil a many laughed at yours, for this simple reason there was none seen it, more was the pity. The chainman[14] I had with me that day did not know what to make of me. When I would take a fit of laughing he would laugh too, not at your letter, though it was the primary cause – but at me because he imagined I was either a fool or non compos mentis.[15] Just as we were both enjoying it, an old woman was passing by the road and she poor credulous being imagined we were laughing at herself and so she commenced an harangue that astonished my ignorance and made me laugh myself nearly into fits; two men that worked at a hay cock nearly over us, seeing the sport we had, thought it was or must be something good, and so over they came and seeing nothing more laughable in the company than the old damsel they commenced laughing also at her; this drew forth her 'invectives long and loud' so that I thought a retreat advisable. So you see what a piece of mischief your letter caused.

Now for Jim,[16] this 8th wonder of the world seems to be in a bad state – he has dipped deep into the pathetic – he says you and I are all that has remembered him – as for Holmes and Tid[17] they are long since out of his books. In his own words he is 'alone uncared for and forgotten' I have advised him to treat the matter as philosophically as he can, and to turn misanthropist – by acting thus – when he imagines himself 'alone' to console himself by repeating the adage which says – 'better alone than in bad company'[18] and when he thinks he is uncared for and forgotten to amuse himself by singing Burn's ditty:—

I care for nobody no not I
Since nobody cares for me[19]

But I'm afraid he wont relish this much – for I think nature never designed Jim for a Timon.[20]

No. 3 Tidmarsh – I have no correspondent that enters so fully into a detailed account of affairs in general as this gent. One of the leading particulars of his inimitable epistle is 'that he had stirabout[21] for breakfast for a change' This is what I call a change with a vengeance, and in my opinion another change would be very advisable as soon as possible. Changes are all very fine as long as there is no changing for the worse. He learned all this quackery in McGrath's.[22] You were always too fond of trying stomachic experiments. I think this piece of valuable news was a 'slip', for he does not go much further before he informs me 'by way of a soft'ner' that they are to have a dumpling for dinner. Now this latter point I take to be an 'invention' for this simple reason that if he had anything good to tell he would tell that first, and that the certainty of having a dumpling for dinner in the wilds of Lancashire would make him soon forget that he had Burgoo[23] for breakfast. However, he informs me that Dick[24] and he will be in Liverpool on the 30th without fail. I imagine that every hour is a day until that wished for time arrives. It would seem that fate has allotted it for the gathering together of the scattered tribe of Erin[25] and that for one night at least we will be 'one fold and one shepherd'[26] So all that remains for you to do is to keep a sharp look out for 4 fellows with faces approaching to the brunette – or the very antipodes of your own; and mind,

> Mislike them not for their complexions
> The shadowy livery of a burnished sun[27]

No. 4 Holmes – after wasting half a sheet of letter paper in a practical treatise on inaccessible distances,[28] he winds up with an invocation to me and all good people to pray for the preservation of the Corkonian for in his own dry witty style he says he is afraid that the Field work will soil his boots and 'may be his socks' I think he will come off very safe if it doesn't go something higher even to soil his shirt – what do you think? Now do not forget to send the exact time of your embarkcation – and if you have a piece of a hard pencil in your pocket in Liverpool I'll think it very hard if you dont give it to me. William Latimer desires to be remd.,[29] and would thank you if you will tell Geo.[30] to write to him this week. My best respects and kindest remembrances to Mrs Cotter and Mr Robbinett[31] and to the inamorata[32] of Willy Hunter. I hope that sweet dream of your slumbers in Geo. St.[33] is well. Tyndall – false fickle man, do you forget Kinsale and 'the lady of the raven plume'[34] Oh! Oh! Oh! The heart of man is deceitful above all things. Concerning that piece of gospel that you were enquiring about, I heard it for a fact. Holmes tells me that they will get pay when sick by obtaining a doctor's certificate.[35]

I have no more news nor room.

so goodbye | Yours very sincerely | Ginty.

RI MS JT 1/11/3581
LT Transcript Only

---

1. *Great Crosby*: a district of the town of Crosby; see letter 0150, n. 3.
2. *your invocation*: letter missing.
3. *Dunran*: a famously picturesque glen in County Wicklow.
4. *Mickey F.*: not identified.
5. *Cupid*: Phillip Evans.
6. *'the lodger'*: not identified.
7. *'cousin Jack'*: probably John (Jack) Tidmarsh.
8. *'sticks'*: presumably handwriting, although no such usage of stick is recorded in the *OED*.
9. *one of Aubrey's favourites*: possibly John Aubrey (1626–97), whose *An Idea of the Education of Young Gentlemen* (1684) proposed that dancing be included in the educational curriculum, or, more bathetically, the Miss Aubrey mentioned in letter 0091.

10. *'passing dream'*: Although the expression was used frequently in nineteenth-century poetry, Ginty's literary predilections suggest he is quoting it from Thomas Moore's *The Loves of the Angels: An Eastern Romance* (1823), I.329.
11. *orisons*: prayers (*OED*).
12. *probono publico*: for the public good (Latin).
13. *'Nonsense precipitate ... the head'*: A. Pope, *The Dunciad* (1728), I.123–4.
14. *chainman*: an Ordnance Survey employee who carries the surveying chain of 66 feet in length, and helps in stretching it between ranging rods; see letter 0143, n. 30.
15. *non compos mentis*: not of sound mind (Latin).
16. *Jim*: Phillip Evans.
17. *Tid*: John Tidmarsh.
18. *'better alone than in bad company'*: a familiar proverb dating back to at least the fifteenth century.
19. *Burn's ditty ... cares for me*: The lines are not in fact by the Scottish dialect poet Robert Burns (see letter 0149, n. 12), but from an air in Isaac Bickerstaffe's comic opera *Love in a Village* (1762), V.viii.7–8.
20. *Timon*: Timon of Phlius (*c.* 320–*c.* 230BC), Greek philosopher and satirist who maintained a sceptical outlook on life and human motivation.
21. *stirabout*: porridge made by stirring oatmeal in boiling water or milk (*OED*).
22. *McGrath's*: the lodging house in Cork Lane, Youghal, where Ginty lived with Tyndall and Tidmarsh in 1840 and 1841. The latter later recalled of it: 'our host–a sharp, keen, intelligent Roman Catholic– ... delighted to take a part in the fray [of religious disputation]. His name was McGrath. We christened the house "McGrath's Hotel", and for years after it was known by this distinction' (J. Tidmarsh, 'Reminiscences of John Tyndall', *South Australian Register*, 24 January 1894, p. 6).
23. *Burgoo*: thick oatmeal gruel or porridge (*OED*).
24. *Dick*: Richard Hunter.
25. *tribe of Erin*: romantic name for the Irish.
26. *'one fold and one shepherd'*: John 10:16.
27. *'Mislike them ... burnished sun'*: adaptation of W. Shakespeare, *The Merchant of Venice*, II.i.1–2.
28. *inaccessible distances*: where impassable obstructions, including thickets, buildings, rivers and lakes, interfere with direct continuous measurement of a line, requiring the surveyor to take a measurement of an equal or proportional accessible distance, or, if this is not practical, to calculate the distance by computation from a measured diagram.
29. *remd.*: remembered.
30. *Geo.*: George Latimer.
31. *Mr Robbinett*: see letter 0150, n. 30.
32. *inamorata*: female lover or sweetheart (*OED*).
33. *that sweet ... in Geo. St.*: presumably the 'never forgotten ever present Peri of 99 Great Georges Street Cork, who has thrown the once adored "lady of the raven plume" into the drear and distant perspective' that Ginty told Thomas Higginson of in a letter from 24 March 1843 (RI MS JT 1/TYP/11/3598).
34. *'the lady of the raven plume'*: Ellen from Kinsale; see letter 0186, n. 6.
35. *get pay ... doctor's certificate*: There is no record suggesting that George Holmes's report was correct. The removal of pay during illness was one of the principal grievances of the civil assistants on the English Survey, especially as the responsibility for supporting sick colleagues fell upon the civil assistants themselves. In Preston Tyndall helped in 'organis-

ing a sort of Friendly Society and Sick Club' to provide 'succour to the members when sickness or any other exigency may render that succour needful' (LT, 'Biography', vol. 1, pp. 69–70).

From John Conwill 23 July 1842 0152

<beginning of letter missing>

Let AB[1] = 480000 miles and Cq or qD – the side of a right angled triangle opposite to an angle of 5° 20',[2] the inclination of the moon's orbit to that of the earth, the hypotenuse[3] of the right angled triangle be 240000 miles.

I considered the earth's path to be surrounded by a glass tube of an oval shape its diameter being AB, CD.

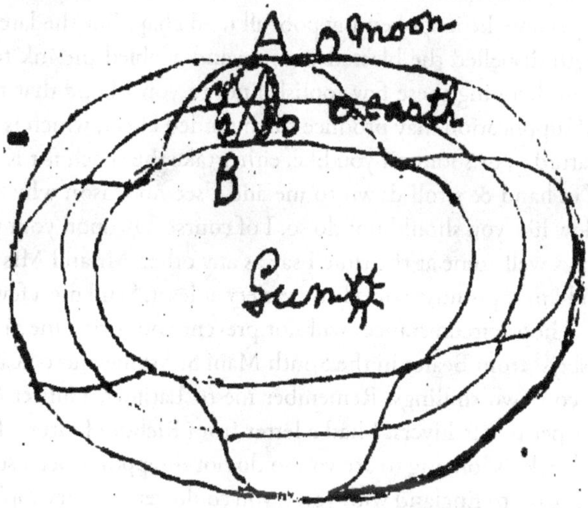

Could the moon leave a trace behind and that the tube specified could be brought into action a very lengthy spiral would be seen traced on its surface, which line would be the moon's path around the sun.

July 23rd. 1842. | J. Conwill.

RI MS JT 1/11/3526

1. *AB*: the diameter of the moon's orbit.
2. *an angle of 5° 20'*: 'the degree of inclination of the moon's axis to her orbit ... is 84° 51' 11"; or it is 5° 8' 49" out of the perpendicular' (*Library of Useful Knowledge: Natural Philosophy III* (London: Baldwin & Craddock, 1834), p. 83).
3. *hypotenuse*: the side of a right-angled triangle which subtends, or is opposite to, the right angle (*OED*).

From John Chadwick 27 July 1842 0153

Kinsale | July 27th 1842.

Dear Tyndall

I suppose you have concluded by this time that the blackguard in Kinsale has quite let you slip out of his cullender memory or that he has been taking two throws from the stream of lethe[1] but such has not been the case. You may lay the blame upon that foul fiend called procrastination, who is always so ready to step in between your deeds and intentions and by suggesting excuses throws a murky cloud upon the clear day of friendship, and it has been through the aforesaid hazy atmosphere that I have been looking towards you this some time past, to my shame be it spoken my poor ill used chap. But the late dry weather has at length dispelled the blasted vapours and enabled the ink to run free in my pen – in dictating these few foolish lines to you I hope that the following prayer and supplication may produce the intended effect, which is that you will on next Saturday, or sooner if you like, either take the coach for Kinsale or take your staff in hand & stroll down to me and I see no reason why a strong agile young fellow like you should not do so. I of course rely upon your promise, and you might as well come at the time I say as any other. Mr and Mrs Abbott[2] will be here and I may promise you a little variety at least. Send me a few lines before you come. I hope circumstances will not prevent you. Bring me ¼ gross of split studs for stays[3] from Beales in the South Main St.[4] somewhere near the church. They will cost two shillings. Remember me to Latimer, Hunter &c. Give the enclosed paper to Mr Eivers. I had a letter from Richd. Hunter a few days ago. Miss Chadwick[5] is longing to see you so do not disappoint her I suppose it will be soon hey over to England with you if you could get the fairy cap[6] it won't save your bowels or disorderafication. I have no news. Matters go on here without much excitement or variety but I expect you will enliven me a bit if you come

In conclusion | I remain very faithfully yours | J. Chadwick.

RI MS JT 1/11/3496
LT Transcript Only

---

1. *stream of lethe*: see letter 0132, n. 1.
2. *Mr and Mrs Abbott*: not identified.
3. *stays*: corsets (*OED*).
4. *Beales in the South Main St.*: Mary Ann Beale kept a haberdashery shop at 92 South Main Street, Cork (*Slater's*).
5. *Miss Chadwick*: presumably Chadwick's baby daughter; see letter 0148.
6. *fairy cap*: not identified.

To John Tyndall, Snr [27] July 1842 0154

Cork, | Wednesday evening

My dear Father

Having returned from the office, dined heartily on hashed mutton and new potatoes! and perpetrated a most comfortable shaving operation to the no small detriment of my whiskers! I now sit down to gratify you in giving you a short epistle as you say that you would prefer a short one to yourself before a long one to any body else.

I am glad to find that you were prepared to acknowledge the force of my arguments respecting the Milford conductors[1] – I floored you completely. Things are going on as usual here. I have three or four methodists preaching to me daily, very nice fellows indeed and I think consistently religious. We are expected to sail on the 29th. There will be three or four lads waiting for us in Liverpool – Ginty among the number – to conduct us to nice lodgings. Ginty has been employed in the field since he went there, he tells me that if the skin of his face was analyzed it would make raw sienna,[2] a colour resembling yellow soap. We heard to day that the Survey of London[3] is about to be commenced immediately. If this be the case it's probable that we who have been employed at the city of Cork will be drafted thither. I cant vouch for the accuracy of the report. With respect to Mrs Payne it is quite unnecessary to advise me not to suffer myself to be humbugged.[4] Stern necessity compels me to look sharp. A fellow on the eve of a long journey wants every shilling he can muster. Mrs Payne knows this, at all events she has not applied to me since I came here. She gets sewing to do from a man named Booth,[5] I dare say you know him, his father is a Carlow man. This, together with what Joe[6] sends her, ought by good management to support her well. I am sorry to hear of Mr Holden's death, he was a man for whom I had a great regard.

Give my love to my mother and Emma good bye

Your affectionate son | John.

I weighed myself two days ago and found that I had increased half a stone since I came to Cork – I mean from Kinsale, I'm thriving. | J.T.

Let me know how John Murray is when you write next | J.T.

RI MS JT 1/10/3275
LT Transcript Only

---

1. *Milford conductors*: see letter 0147, n. 1.
2. *he tells me ... raw sienna*: see letter 0150, n. 36.
3. *Survey of London*: A Survey of London at a scale of 5 feet to 1 mile, to be paid for by the Exchequer, was proposed in early 1842, presumably to assist with much-needed improvements to the city's infrastructure. Colonel Thomas Colby estimated that the cost of

surveying and publication, to be carried out by the Royal Engineers, would be £104,000. However, the Treasury, keen to keep a tight rein on public finances, rejected the proposal. The plan was revived in 1848, following a request from the Metropolitan Sanitary Commission in November 1847, and the passing of the Public Health Act earlier in 1848. The Survey of London, superintended by William Yolland and Lewis Alexander Hall, was concluded in 1850.

4. *humbugged*: hoaxed or cajoled into doing something (*OED*).
5. *Booth*: not identified.
6. *Joe*: Joseph Payne.

From William Ginty [31] July 1842 0155

Sunday evening

Mon cher[1] Tyndall,

Yours of the 24th[2] I received just in time to apprize me of your non-appearance on the shores of Great Britain on yesterday[3] 'Heaven and Earth but this is wondrous strange'[4] Tyndall has forever disclaimed 'the lady of the raven plume'[5] Ah! me! Time was and it was not thus – Oh no! but she has been replaced by another dark haired beauty. That is the reason if I opine rightly. But alas for Tyndall the scene and hour will soon be past, heavens grant him fortitude to withstand the shock of next Thursday.[6] That fatal eventful – ever-to-be-remembered, never-to-be-forgotten Thursday. Tyndall I feel for you. Fancy has led me to lift the sombre curtain that shrouds futurity – A few days on in the perspective I behold the fatal day – a gloomy cloud seems to overhang its very dawn, as the day advances it becomes thicker and more dismal – night comes and it is intolerable. Out of this gloom I see a form issuing wan and dejected (miserable man! thou surely art the sufferer) slowly and mournfully he saunters along on his sad sad errand – he has reached - - - - - here I must stop, all is confusion, description fails me, even I am seized by a tremour – A pause – I hear cries the most pitious – Another – A door opens, the same sorrowful figure issues from its portals – slowly it closes after him – Closed and for ever – he turns to take a last fond look, then retraces his trembling tottering footsteps hark he sings a favourite air, but oh how that tone is altered, how pathetic is the strain, I can just catch the words as they fall from his quivering lips – They are as follows:-

No more shall Tyndall in concert sing
  A lay[7] in a Southern clime[8]
Nor strike the knocker as the bell he'd ring
Neath the window of 99[9]

This was the lay of his wretched self
  Tho' not in his merrier hours
As sadly he strayed from his last farewell
Neath the 'lamp post's' iron towers

Oh where are the songs of that happy place
  That were sung Cupid's javelins to bear
Of that angel now there's not left a trace
  To tell she was lovely and fair
(here he placed his hand over his heart)

The grief grows grim o'er this aching spot
  Where once grew hope's brightest flowers
Land of my Charmer thou'lt ne'er be forgot
  While a vestige remains of her powers.

If you do embark on Friday the 5th this will reach you just in time to tell you that if you want your travelling bag in L.Pool[10] dont put it in the hold in Cork for as sure as you do so sure will you not get it until Monday morning. For unless they are in early on Saturday they wont have time to unload and on Sunday they dare not nor would not unload nor even uncover the hold Bracken[11] told me last night that Jack[12] and Dick[13] have asked a day's leave (Monday) we will also have it. I was at the docks on yesterday when the Ocean[14] came in. Jack said in the last letter I got from them that he expected to meet his father in L.Pool. I am very sorry indeed for Miss T- - -s[15] illness I hope for her sake as well as for yours that she be all right soon. Remember me to Mrs Cotter and to all enquiring after my nabs.[16] I rather think you wont have a great task to perform!!! Dont forget No 86 Porter St.[17]

Ere I conclude this short essay
Please list with patience to a lay
Such as your reverence gave me ere
I left the land that nurst my fair
That fair for whom amongst the rest
You claimed a portion of my breast
Despite my oft repeated cries
On all the gods to prove them lies
(Oh! Heavens if they would claim their parts
Where would I find enough of hearts
A heart-less villain then I'd bend
At beauty's shrine till time would end)
But fate has ruled it for the best
My heart lies buried in My breast
While yours – Oh Tyndall! – I would stop
Did I not know you'll bear this shock
This 'tale of woe' may thee prepare
To bear a shock still more severe
Yes! yours is gone and in thy breast
Drear sorrow stalks a mournful guest
'A few short hours or days' at most
And thou wilt count the bitter cost
Of Love in vain – of beauty fled

Of bliss then talked of as the dead
'Affections trampled' – 'hopes destroyed'
Thy love-fraught breast an aching void
No 'Charmer' now to look a smile
And sooth thy amrous breast the while
No – all are gone – love – hope – and joy
Go take thy last fond look – poor – boy!
Air 'The light Guitar'[18]
Oh leave the gay and beauteous one
  Her smiles and eyes so bright
And roam towards other lands unknown
  Bid her a long good night
Perhaps when there thou'lt yet find one
  And have another 'Bar'
But should'st thou not – thou'rt not undone
  Just think on her afar
Unknown sweet lips with tender tales
  May yet cheer up thy gloom
May kindle in thy breast a flame
  To quench but in the tomb
But if their tales should make thee sad
  Should aught their beauty mar
Just think of love in other times
  And sigh for one afar
So fare-thee-well until the time
  When wafted o'er the billowy brine
To England's shore 'a stranger land'
  I'll grasp thy longed for welcomed hand
While you with manly accents say
  At meeting on that wished for day
'"A long farewell to love I gave"'
'My grief I buried in the wave'

And prove to Ginty there and then
That Tyndall 'is himself again'

RI MS JT 1/11/3582
LT Transcript Only

1. *Mon cher*: my dear (French).
2. *Yours of the 24th*: letter missing.
3. *your non-appearance ... yesterday*: Tyndall had initially expected to begin the two-day journey from Cork to Liverpool on 29 July; see letter 0154.
4. *'Heaven and ... wondrous strange'*: conflation of W. Shakespeare, *Hamlet*, I.v.166 and 168.
5. *'the lady of the raven plume'*: Ellen from Kinsale; see letter 0151, n. 34.
6. *next Thursday*: 4 August 1842.
7. *lay*: a short lyric or narrative poem intended to be sung (*OED*).

8. *a Southern clime*: that of Cork in the south of Ireland.
9. *99*: 99 Great Georges Street, Cork, the address of Peri; see letter 0151, n. 33.
10. *L.Pool*: Liverpool.
11. *Bracken*: Lance Corporal John Bracken (b. 1804/5) of the 14th Company of the Royal Sappers and Miners, who had superintended the field parties of the 3rd Division, B District of the Irish Ordnance Survey (NAI OS/1/16–18), and now worked in the 5th Division of the English Survey in Liverpool.
12. *Jack*: John Tidmarsh.
13. *Dick*: Richard Hunter.
14. *the Ocean*: see letter 0143, n. 32.
15. *Miss T- - -s*: Miss Tidmarsh, probably Christina.
16. *my nabs*: myself (*OED*).
17. *Dont forget No 86 Porter St*: presumably a reminder not to go to the wrong lodging house. Ginty had recommended Pace's at 84 Porter Street, with 'excellent beds and the best of accomodation' for '6d per head per night', but warned that there were 'two houses close by together in Porter Street, dont forget the name or No., or you will go to the wrong one'; see letter 0144.
18. *Air* 'The light *Guitar*': a musical arrangement first published by Tulley & Co. in 1835.

To John Tyndall, Snr 1 August 1842 0156

Cork, | Monday

My dear Father

Our doom having been sealed and our sailing warrant signed I hasten to say that we start from this on Friday next at 3 o'clock.

We were to sail on Friday last,[19] but owing to the unfinished state of the work we were detained a week. We expect a pleasant trip, as the weather is likely to continue fine, some of our fellows are sick in anticipation. For my part, I dont think my stomach will get a single turn. Its too well toned. As I have a good deal of business on hand I must cut this short. The next time my mother sees me she may expect a little of the English twang on my tongue. Give my love to her and Emma. I suppose she thinks a trip to England a great thing. For my part I care no more about it than a trip to Carlow.

I cant say where we'll be stationed, as we have been applied for by every officer in England. A Capt in Skipton[20] applied for four of us. I was among the number. They imagine I suppose that they may expect to find passable draftsmen among us.

Pardon this hurried scrawl

I remain | Your affectionate son | John

I'll write to you when I get settled in England. You need not write until you hear from me again.

RI MS JT 1/10/3276
LT Transcript Only

19. *We were to sail on Friday last*: 29 July 1842; see letter 0154.
20. *A Capt in Skipton*: Tyndall possibly means Lieutenant Henry James (see letter 0097, n. 16), although he was not made a Captain in the Royal Engineers until 1846. Many surveyors from the C District of the Irish Ordnance Survey, including Thomas Charles Higginson, had already been sent to Skipton, a market town in the West Riding of Yorkshire, in September 1841; see letter 0096.

From John Chadwick 3 August 1842 0157

Kinsale | August 3rd,

Dear Tyndall

I send the books according to promise. Give the letter to my sister Mrs Harrison[1] when you go out of the Clarence Dock[2] gate. Any one will shew you Mr Harrisons,[3] corner of William Street opposite the upper dock gate of the dock. You and Latimer can go there but go decent (offence apart) as they set a great opinion on appearances in Liverpool. You may also see my elder brother – and my fat comfortable-looking sister. I have no news. I got home quite safe and was glad that I had not stopped for the reason you know. Please God we may meet again at the furthest next summer. Till when all good angels guard you! and may the spirit of good luck lead you into pleasant quarters and comfortable quibberies[4] and Lord send that the roast beef and pudding of old England[5] may put some lining on your skeleton. Keep clear of the exvertuous[6] females of Liverpool and don't be deluded but live wholesomely and chastely until I see you again. Kiss Latimer for me and kiss the quay wall and kiss Miss Tidmarsh who lives on it. Tell them for me that I will call on them when next in Cork. And now my poor old comrade farewell, adieu, goodbye, good luck to you, caed mille falthagh[7] &c. Amen. Yours while memory sparkles

J. Chadwick – late C.A.[8]

Do not forget the World's End[9] whose occupant will be looking into futurity for a letter from you. My younger brother[10] is worth cultivation if you have time for it, you may be fixed in Liverpool. I send a few lines to Dick Hunter.

RI MS JT 1/11/3498
LT Transcript Only

1. *Mrs Harrison*: On 14 July 1844 Tyndall recorded in his Journal that Chadwick's 'sister Mrs Harrison to whom he gave me a letter of introduction when I first went to Liverpool ...is dead' (vol. 13a, p. 47).
2. *Clarence Dock*: see letter 0150, n. 28.
3. *Mr Harrisons*: not identified.
4. *quibberies*: not identified.
5. *the roast beef... old England*: As in the patriotic ballad 'The Roast Beef of Old England',

composed by Henry Fielding for *The Grub-Street Opera* (1731), the dish symbolized English national prosperity and power.
6. *exvertuous*: presumably unvirtuous.
7. *caed mille falthagh*: céad míle fáiltea, Gaelic for a hundred thousand welcomes.
8. *C.A.*: Civil Assistant.
9. *the World's End*: see letter 0072, n. 1.
10. *My younger brother*: possibly Richard Chadwick, who had joined the Irish Ordnance Survey in July 1839 before resigning in November (NAI OS/1/16).

To Mr. Burchell [6] August 1842 0158

<*beginning of letter missing*>[1]

The noon of night saw me pacing the deck of the vessel, gazing at times on the Tusker revolving light,[2] sometimes it presented the appearance of a mere spark, then as it grew gradually larger it shed a lurid glare around; anon, bursting forth into a full blaze of brilliancy it cast its beacon flare far and wide.

High above stretched the waveless sea of heaven studded with its lustrous systems, the rays from the thousand stars pierced the welkin[3] & while gazing on the spangled empyrean[4] I could not help bursting forth in the sentiment and language of Byron;

> Who ever gazed upon them shining
> And turned to earth without repining
> Nor wished for wings to flee away
> And mix with their eternal ray?[5]

Here I stand with pencil in hand, leaning on the gunwall[6] of the Ocean steamer;[7] the mighty Atlantic urging his boisterous progeny – the crested billows – to wage war against the sides of the oak Leviathan[8] which bears me on. On the West nought comes within my ken[9] save the billowy waste, here and there bearing a snowy sail which 'walks the waters like a thing of life',[10] far on the East rise the steep sides of old Snowdon.[11] How many reflections does it call forth! I had read of Llewellyn[12] when a child – where is now the clarion of the Chieftain which once aroused the sleeping echoes of the hills? where the many 'mingled cries' which once rung through its thunder torn crags? Silent all silent! Yet Snowdon is unchanged, it still frowns as in days of yore on the peasant's cot which lies buried in its shadow, while the ephemeral day of those who were once lords of the hill has long since closed for ever.

RI MS JT 1/11/3527–8
LT Transcript Only

---

1. *Mr. Burchell*: see letter 0145, n. 2; Tyndall had travelled to Gowran with a 'relation of Burchell's' in May 1842.

2. *Tusker revolving light*: The lighthouse on Tusker (or Tuskar) Rock, off the cost of Wexford, employed a revolving catoptric light with three faces, two white and one red. The optic was supplied by G. Robinson of London, and the light was first used in June 1815.
3. *welkin*: sky (*OED*).
4. *empyrean*: the highest heaven; in ancient cosmology, the sphere of the pure element of fire; in Christian use, the abode of God and the angels (*OED*).
5. *Byron ... eternal ray?*: George Gordon, Lord Byron (1788–1824), *The Siege of Corinth* (1816), xi.248–51.
6. *gunwall*: variant spelling of gunwale, the upper edge of a ship's side (*OED*).
7. *the Ocean steamer*: see letter 0143, n. 32.
8. *Leviathan*: sea monster in Job 41:1–34.
9. *ken*: range of sight or vision (*OED*).
10. *'walks ... thing of life'*: Byron, *The Corsair* (1814), iii.93.
11. *old Snowdon*: the highest British mountain outside of Scotland, located in Gwynedd in north Wales.
12. *Llewellyn*: Llywelyn ap Iorwerth (*c.* 1172–1240), Prince of Gwynedd and defender of the Welsh against incursions from England.

From William Ginty [6 August] 1842 0159

Here[1] I am since 12 o'clock waiting for you fellows[2] to come in (now 2)[3] I told you I would be back shortly – also that I would come here about noon – but I suppose you are not so feminine as to think a shake of the paw essentially requisite under present circumstances – yet I suppose it would have been the last – perhaps for ever.

Well I wish you prosperity with a sincerity of no common kind – I will write to you shortly. I'll write – I'll write while I have a hand to guide the pen or an eye to see the paper.

Good Bye | Yours ever faithfully | Ginty

Taylor is here waiting also – and off I must go – or it will be too dark. 12 miles[4] no laughing matter!

'Well I will dream &c.'[5]

I may be back on Monday,[6] if so it will be early

RI MS JT 1/11/3580
LT Transcript Only

1. *Here*: presumably Clarence Dock in Liverpool; see letter 0150, n. 28. LT added a handwritten insertion to this letter stating 'Picture of Clarence Dock &c. accompanies this'.
2. *you fellows*: Tyndall travelled to England with Robert Baskin, David Doig, Jnr (see letter 0143, n. 12), John Holland (see letter 0143, n. 15), William Hunter, George Latimer, John McGowan (see letter 0143, n. 16), John O'Neill, John Todd (see letter 0115, n. 11) and another civil assistant called William Henehan.
3. *since 12 o'clock ... (now 2)*: Tyndall's ship, PS *Ocean*, did not arrive until 8pm; see letter 0160. This was presumably a late arrival as when Thomas Charles Higginson sailed

the same route from Cork to Liverpool in September 1841, he told Tyndall that he 'arrived in Liverpool on Saturday at half past three in the evening [i.e. afternoon]'; see letter 0097.

4. *12 miles*: Ginty presumably means the round trip from and to Crosby; see letter 0150, n. 3.
5. *'Well I will dream &c.'*: 'Well – I will dream that we may meet again', Lord Byron, *Childe Harold's Pilgrimage* (1812–18), II.ix.5.
6. *I may be back on Monday*: Tyndall's account of his arrival in letters 0161 and 0162 suggests that he was met by Ginty, and William Latimer, as soon as he arrived, so Ginty must have changed his plans, perhaps after meeting Latimer.

To John Tyndall, Snr 7 August 1842 0160

Liverpool | Sunday

My dear Father

I hasten to transmit to you the news of our safe arrival in Liverpool after a beautiful passage. We started from Cork at 3 o'clock on Friday[1] and were safely landed here on yester evening at 8 o'clock. I was no way sick during the passage, tho' some around me were puking with might and main.[2] I cant say where I'll be sent. Tomorrow will decide. We are to meet our officer at 12 o'clock tomorrow. I have not time to give you a description of the incidents of the passage. I write this fearing you might feel some anxiety on account of my being thrown on the tender mercies of the broad Atlantic. Give my love to my mother and Emma

Good bye | Your affectionate son | John.

RI MS JT 1/10/3277
LT Transcript Only

1. *Friday*: 5 August 1842.
2. *might and main*: utmost or greatest possible power or strength (*OED*).

To John Tyndall, Snr 10 August 1842 0161

Preston, Tuesday night | Aug. 10th, 1842.[1]

My dear Father

Having borne the tossing of the sea, the noise of the seaport, and the whirlwind motion of the steam coach,[2] with the fortitude and patience of a Kerry pony,[3] I now sit comfortably down to a comfortable little table situated in a most comfortable little room and in the most comfortable manner possible commence a letter to you. I have no less I'm sure than a dozen expectants on my list, but your claim to an epistle supersedes all others therefore you are the first that I shall write to.

After the bustle which took place previous to our departure from Cork, I felt happy when, relieved from care, I found myself pacing the deck of the ocean steamer[4] as she swept majestically down the Lee. We passed Cove[5] on our way and soon found ourselves at the mouth of the harbour shaded by the frown of Carlisle and Camden forts,[6] the former standing on the left the latter on the right. Here the sea was pretty rough, it having blown freshly in the former part of the day. Before I started I could not help picturing to myself a dozen unfortunates retching their very bowels out in unison with the lurching vessel. Among the group I in imagination placed myself, and indeed, thought I, if my fancy picture the thing truly, I'll certainly cut a most interesting figure The picture however had nothing in reality to corroborate it. I got over without a single qualm. We are quite unable to appreciate the comforts of life until we have been deprived of them. I proved this on Friday night. I would think as much hay as a donkey would take in his mouth a luxury. I paced the deck of the vessel nearly the whole night, sometimes I'd throw myself on a coil of ropes and lie for half an hour when the fresh cold blast would rouse me up again to keep my blood in circulation by walking. Things however went on comparatively well. We had a fine starlight night. I amused myself sometimes for an hour together stargazing. On the next day we came in sight of the coast of Wales, along which we bore. We could see the Snowdon hills rising far to the right. At 8 o'clock that evening we arrived in that modern Babel[7] Liverpool, met Ginty on the quay, and were conducted by him to comfortable lodgings. On yesterday[8] at 11 o'clock we received orders to be at the railway station at ¼ before 5 o'clock. We were there, and started for Preston.[9] Tell my mother that she can hardly imagine the delight I experienced in travelling steam coach rate. It's extremely pleasant. We met Evans at the railway terminus.[10] He had a good dinner prepared for us, after which I went in quest of lodgings – there not being room in his lodgings for myself and two more who were determined not to separate. I shall only add that we procured very comfortable ones. In my next I'll go into further detail.

Good bye | Your affectionate son | John.

Remember me to Mr Conwill.

RI MS JT 1/10/3278
LT Transcript Only

1. *Tuesday night | Aug. 10th, 1842*: Tyndall seems to have written the wrong date, or to be writing after midnight, as the Tuesday was 9 August.
2. *steam coach*: see letter 0097, n. 10.
3. *the fortitude ... Kerry pony*: The mountain breed of Kerry Bog Pony is traditionally used for hauling peat from the bogs of south-west Ireland, and has a reputation for hardiness combined with a gentle temperament.
4. *the ocean steamer*: see letter 0143, n. 32.

5. *Cove*: see letter 0068, n. 3.
6. *Carlisle and Camden forts*: twin forts at the narrowest point of the entrance to Cork Harbour, both were built in the early 1780s during the American War of Independence.
7. *Babel*: the name used in the Hebrew Bible for the Mesopotamian city of Babylon; also the location of the hubristic tower, in Genesis 11:1–9, that prompts God to confound the languages of men so that they could no longer communicate with each other and strive for things that are beyond their reach.
8. *yesterday*: Monday 8 August 1842.
9. *started for Preston*: As the direct route on the Liverpool, Ormskirk & Preston Railway did not open until 1846, Tyndall would presumably have made the journey from Liverpool to Preston in two stages. Firstly on the Liverpool and Manchester Railway, opened in 1830 as the world's first passenger line, and then on to Preston on the North Union Railway, opened in 1838.
10. *the railway terminus*: There were two railway stations in Preston at this time, the North Union Station, opened in 1838, and the Maxwell House Station, opened in 1840. With the latter serving the Lancaster and Preston Junction Railway, the route of Tyndall's journey suggests he arrived at the North Union Station.

To John B. Edgeworth[?] August 1842 0162

<*beginning of letter missing*>

...that ever grinned in the Pope's belly.[1] I'll not be a beholder of my woe. Acting on this resolution, I closed my peepers, but vain are human resolutions – sure angels have bartered heaven for woman's eyes and is it a matter of surprise that Tyndall found his air-founded resolutions vanish like a vapour when, as the steamer paused at Passage,[2] he was told that half a dozen of the said eyes twinkled on the quay. From this out I possessed fortitude enough to enjoy the scenery as I passed down the harbour. Cove[3] was quickly passed. Shortly after we were shaded by the frisure of the two forts which guard the harbour's mouth,[4] then Sir we emerged into the world of waves and like a bubble on a stream we danced upon the billows. Here and there might be seen a face writhing in grim contortions ... Holland[5] was the first to give way, he leaned over the bulwark,[6] the picture of despair ... as ghostly as a church yard deserter.

After having amused myself sufficiently at Holland's expense, my mind took a more sentimental turn – my eyes turned westward to behold the glories of the setting sun in peaceful splendour, he dipped beneath the horizon, when the mists of evening closed the view and shut this land of the west from my gaze for ever and ever. The moon &c ... but a truce to this awhile. Latimer and Baskin shortly after their embarkation – being desirous of pampering the flesh as most of the latter day saints are – agreed for a berth each, for which they paid 8 shillings. Well Sir, when night closed round and we unfortunate pilgarlicks[7] were seeking a spot here and there on which to snatch a doze, the two gentlemen alluded to retired very comfortably to their warm beds between the hours of 12 and 1. However, as I was pacing the deck to keep my blood in circulation, to my no small astonishment I beheld by the starlight the woebegone countenance of Baskin before me, the sublimity of misery hung in wrinkles round his eyeballs. What! said I, is it you I see? what brought you here? Oh, hiccupped he, I met more company than I agreed for. I remained as long as I could, and that was until the bug-bites rose like small potatoes on my arms. To corroborate his assertions, he placed my hand upon his wrist which soon convinced me of the prodigious effect of the bite of a bug – I must say that an hour's sleep on Saturday morning would have been the very acme of felicity. I endeavoured to find a place to stretch myself and was not long in fixing on a spot in the bow[8] of the vessel under the ropes. Here I lay, Sir, on the hard deck with a plank beneath my head for a pillow and was in the act of perpetrating one of the most comfortable slumbers that ever crept over a mortal's eyes, when suddenly the sound, aye and the substance too, of many waters struck on my ear from the confounded waterpipe of a sailor who was cleaning the vessel. On account of the darkness of the corner I had chosen, the fellow did not see me. I dare say however that whether or not the result would have been the same. At half past ten we got in sight of Holyhead[9] – from this we kept in sight of the Welsh Coast as far as it extended. At 8 o'clock we landed on Clarence quay[10] – were met by Ginty and Latimer and conducted by them to lodgings at Pace's[11] – 84 Porter Street[12] – It would be a difficult thing to convey an idea of Liverpool, it is a perfect Babel,[13] the rumbling of waggons, the gee wo of the waggoner, the clang of the tram cars and the cry of the fishwoman all keep up such an eternal din that a fellow should be furnished with brazen organs to be able to bear it all. I rambled through the best part of the city on Sunday. On Monday[14] at <u>11</u> o'clock we got orders to be at the railway station Lime Street[15] at half past 4 o'clock. After some mischances, such as forgetting my bag and giving a fellow half a crown to return for it, we at length mustered at the proper time, in proper order, at the proper place <*rest of letter missing*>

RI MS JT/1/11/3926
LT Transcript Only

1. *in the Pope's belly*: The papal stomach was a familiar trope in the rhetoric of Irish anti-Catholicism, with a mid-nineteenth-century toast of the Protestant Orange Order declaring of those who refused to drink it: 'May his soul be in the pope's belly' (H. Newland, *The Erne: Its Legends and Its Fly-fishing* (London: Chapman and Hall, 1851), p. 45).
2. *Passage*: Passage West, a port town on the west bank of Cork Harbour.
3. *Cove*: see letter 0068, n. 3.
4. *the two forts ... harbour's mouth*: Carlisle and Camden forts; see letter 0161, n. 6.
5. *Holland*: see letter 0143, n. 15.
6. *bulwark*: the raised woodwork running along the sides of a vessel above the level of the deck (*OED*).
7. *pilgarlicks*: pitiable, lowly, or foolish persons (*OED*).
8. *bow*: the fore-end of a ship or boat (*OED*).

9. *Holyhead*: a port town in Anglesey, an island off the north coast of Wales.
10. *Clarence quay*: see letter 0150, n. 28.
11. *Pace's*: the lodging house at which William Ginty had found 'excellent beds and the best of accomodation' for '6d per head per night'; see letter 0144.
12. *Porter Street*: see letter 0144, n. 7.
13. *Babel*: see letter 0161, n. 7.
14. *Monday*: 8 August 1842.
15. *the railway station Lime Street*: the terminus of the Liverpool and Manchester Railway, opened in 1836, although construction was not completed until the following year.

From John B. Edgeworth 15 August 1842 0163

4 Mount Verdon Terrace[1] | August 15th 1842

My dear Tyndall

Many thanks for your very acceptable letter.[2] I would not have asked you to write to me if I did not intend answering you indeed it would be a very bad return for your great kindness to me whilst in the office here. I still continue to go to the barracks and have finished three of the plans on the trace. Mr Bloomfield very kindly gives me every assistance, I assure you your letter was not the least too long I am only afraid that you must have been tired writing it, for I judge from myself that I never would be able to surmount such a long epistle – I am very glad you were not sick going across the channel as when one is not ill nothing is more delightful than a sea voyage particularly when there is coasting. remember me most kindly to Mr Latimer and all who think it worth while asking for me indeed I shall not easily forget the kindness I experienced from you all in Cork We have had very hot weather since you left – Major Waters is leaving Passage[3] and is coming up to lodgings in Cork Mr Moor[4] or as Mr Bloomfield calls him 'your little old crowl'[5] is in the same room as us and of course we have sometimes a word about the 'Darbies'[6] – The Sawyers who threw the vitriol at Mr Wilson[7] were shipped off for Dublin in the Eagle steamer great crowds were assembled and the prisoners were obliged to be escorted from the jail to the quay by a party of the 10th hussars[8] and some police however there was no violence committed A few nights ago the police were in great commotion about Wilson having stabbed a man – The facts were as follows – there was a crowd passing by the saw mill and it being a wet day a recruit belonging to some cavalry regiment went to take shelter in the mill when Wilson came down and told the man to leave the place he refused and a struggle ensued during which I hear the man called him a 'bloody informer' and Wilson took out a dagger and struck the recruit who was in plain clothes the man was not dangerously wounded I believe he is quite recovered now[9] If you think it worth while I hope you will write to me soon and you may as well leave out the Mr to my name you see I have not used it to yours which I hope you wont think too great a liberty – I am sure you must think plotting exceedingly dry work. I daresay you would like field work better[10] do you expect to be sent to the field if you know Mr Eiver's direction will you send it to me as I should like to write to him at some future time

I must now bid you farewell | and believe me | sincerely and truly yours | J.B. Edgeworth

RI MS JT 1/11/3558
LT Transcript Only

---

1. *Mount Verdon Terrace*: in Cork, on the north side of the River Lee.
2. *your very acceptable letter*: possibly letter 0162.
3. *Passage*: see letter 0162, n. 2.
4. *Mr Moor*: probably James Moore, a civil assistant who joined the Irish Ordnance Survey in May 1834 and worked in the Examination Office for C District, located in Cork (NAI OS/1/17–19).
5. *crowl*: a stunted or frail person (*OED*).
6. *'Darbies'*: not identified, but darbies is a slang term for handcuffs (*OED*).
7. *The Sawyers ... Mr Wilson*: On 13 August 1842 the *Spectator* reported, 'At Cork Assizes, on Friday, Daniel Lonergan, John Drew, Daniel Sullivan, and Timothy Herlihy, were tried for an aggravated assault on Mr. John Wilson, the proprietor of some patent saw-mills; upon whom they cast a quantity of oil of vitriol, by which one of his eyes was burned out of the socket, and the other was seriously injured. The workmen of the firm to which he belonged were discontented because they had introduced steam-power into their mills, which diminished the demand for labour; and the prisoners entered into a combination to put the steam saw-mills down. On the 1st March, Mr. Wilson was waylaid, and the vitriol was thrown in his face. The proof rested principally on the evidence of an approver, and of Mr. Wilson, who himself identified the four men. They were found guilty, and sentenced to transportation for life' ('Ireland', p. 779).
8. *10th hussars*: 10th Royal Hussars, a cavalry regiment of the British Army.
9. *Wilson having stabbed ... recovered now*: On 16 August 1842 the *Morning Post* reported, 'at Cork, a riot arose in consequence of a report that the four sawyers just convicted of maiming Mr. Wilson, were about to be transmitted for transportation to Dublin. A mob collected about Mr. W.'s mills, and a young man named Jowell, a recruit, drawn amongst them by mere curiosity, being thrust forward by some persons towards Mr. Wilson, a struggle took place between them, when Mr. Wilson drew a dirk and stabbed the young man under the ribs and severely wounded him. Great excitement was caused by the circumstance. The young man, however, is not dangerously hurt' ('Ireland', p. [2]).
10. *think plotting ... work better*: Under Colonel Thomas Colby's superintendence of the Ordnance Survey, the activities of plotting (transferring information from field books, usually done in an office) and surveying (conducted in the field) were kept separate, and no surveyor was allowed to plot his own work.

To John Tyndall, Snr [18 or 25] August 1842 0164

18 Oxford Street, Preston | Thursday evening

My dear Father

I should have written to you since I received your last[1] ere this – better late than never. You ask me what the Englishmen were seeking when they turned out. The thing was confined to those working in the factories who were puffed into insurrection by the harangues of some Chartist delegates.[2] The number of such persons in this part of the country is enormous. Five or six hundred men and women might be seen at a time issuing from a single factory. And you could scarcely find a spot in Lancashire from which you could not see multitudes of chimneys; they are not confined to the towns. Between this and Liverpool the country, which is very beautiful, is thickly studded with shafts. You ask me am I in any danger – not the least indeed – I could pass very quietly thro' ten thousand Chartists and be taken for a brother. The country however is quite peaceable at present. The thing was checked in its bud, the people saw that decisive measures were about to be taken – this quelled the insurrectionary spirit.[3] You ask me how I like England. I was never more comfortably situated in my life. Market rates much the same, perhaps a little dearer than in Cork, so that I can manage to do pretty well.[4] As for my pay being increased,[5] there is no talk of it. I was completely humbugged with respect to pay. Before I left Youghal I was promised an increase; being drafted to Kinsale the officer there knew nothing about me, and so I had to remain six months without any hope of an increase. Shortly before I went home I was recommended. My recommendation was with Col Colby while I was at home, tho' I believe I never told you so. The thing however was smothered, I never heard a word about it since. I am now cast on the tender mercies of another stranger – Capt Tucker – a man always famous for his aversion to increasing the men's pay, he has the Superintendence of the Survey of England[6] and indeed it could not fall into worse hands – worse I mean for those employed on it. He imagines no doubt that he is working in an economical manner, but dearer work was never done than his will be. No man will exert himself under such a man, and the consequence is that he is not getting <u>half enough</u> of work done. The expense attendant on the correction of wrong work – work done by inferior hands – would remunerate men of ability. We are not allowed however to question the policy of our rulers. I spoke of being humbugged[7] with respect to pay on account of my being shifted so often. I hope however this will ultimately be of service to me. As I often told you my improvement overbalances my pecuniary loss, and tho' this may be of no service to me at present, it <u>may</u> at some future period. Seldom do I feel discontented, and when I do the hope of better times cheers my gloom. True there are some men very far above me in pay, and – I hope I may say it without being considered egoistic – I <u>do</u> say it without vanity – not a whit above me in other respects. That however does not make me repine. I have several of the men who were stationed in <*rest of letter missing*>

RI MS JT 1/10/3279
LT Transcript Only

---

1. *your last*: letter missing.
2. *puffed into ... Chartist delegates*: On 12 August 1842 a mass meeting of *c.* 3,000 workers called a strike in Preston's cotton factories. The *Preston Chronicle* reported that 'their objects were stated to be the attainment of "the people's charter", and of "a fair day's wage for a fair day's work"' ('Serious Disturbances by Turn-outs and Chartists', 13 August 1842, p. [2]). The People's Charter of 1838, whose proponents became known as Chartists, demanded democratic reform of the political system, but had been rejected by Parliament.
3. *The thing was ... insurrectionary spirit*: On 13 August 1842 the strikers marched on the centre of Preston in order to confront workers who were breaking the strike. They were met by police and soldiers at Lune Street outside the Corn Exchange. The *Preston Chronicle* reported that 'an attempt was made to reason with the parties, and they were informed that if they did not disperse, and cease their riotous conduct, orders would be given to fire upon them. The Riot Act was read, and the police having been beaten back, the order to "Fire" was given, and several were wounded' ('Preston Riots', 13 August 1842, p. [3]). Four of the strikers were killed.
4. *Market rates ... pretty well*: Tyndall had earlier worried, as he told his father in October 1841, that the 'English Survey is I believe a bad speculation ... Provisions are much dearer there than they are in Ireland so that the Surveyors are very tightly pinched to live in it at all'; see letter 0100.
5. *my pay being increased*: Tyndall's pay had been increased to 2*s.* 10*d.* in either March or April 1841, but he had received no increases since then.
6. *he has the Superintendence ... England*: Captain Henry Tucker was in fact only formally in charge of the 2nd and 3rd Divisions of the English Ordnance Survey, which he commanded from the Divisional Office in Wakefield, although he does seem to have overseen much of the work in the other three Divisions. It was Colonel Thomas Colby who, until 1847, remained Superintendent of the Surveys for both Ireland and England.
7. *humbugged*: imposed upon, hoaxed, deluded (*OED*).

From Christina Tidmarsh 26 August 1842 0165

Cork | Aug. 26th, 1842

Dear Mr Tyndall

I would have answered your letter[1] before but not knowing exactly where to address – however now I send it under cover to John[2] knowing it would be the surest way. He must be greatly improved since he left Ireland – I hope you will succeed in your exertion to get him with you. Father is not come home yet – when he wrote last he was preparing to visit the Giant's Causeway[3] with a few friends as he was in that county. You had a good escape from the bugs[4] I should have laughed heartily at Mr Baskin and his companion had I seen them I suppose you were greatly amused with all you saw since your arrival. Were Mr Ginty's similes of Liverpool[5] very exact? I hope you had many an occasion since to break the resolution you formed of leading a retired life – should you live as retired as you intended if ever I have the pleasure of seeing you I shall expect to behold a hermit.

May sends her warmest regards to you – and desires me to say she goes to her duty every Sunday evening, but is neither a methodist nor a Bigot. We often laugh heartily when we think of Mr Hunter's hat.[6] Grandmama and Mama desire to be remembered to you

Yours sincerely | Christina Tidmarsh

RI MS JT 1/11/3867
LT Transcript Only

1. *your letter*: letter missing.
2. *send it under cover to John*: to her brother John Tidmarsh.
3. *Giant's Causeway*: a famous landmark in County Antrim formed of interlocking basalt columns.
4. *a good escape from the bugs*: see letter 0118.
5. *Mr Ginty's similes of Liverpool*: see letter 0144.
6. *Mr Hunter's hat*: possibly relating to an incident with William Hunter's hat on the sea voyage to England.

From John Conwill 3 September 1842 0166

<beginning of letter missing>

The velocity of bodies descending in this manner is quite different from that of bodies falling through vacuum to the earth – in the latter case the body receives no impulse in any direction save that which goes through the earth's centre, while in the former case the moment the body enters the earth it is pulled back by a force due to the distance between it and the earth's surface (vide[1] A) hence we draw two queries:–

1.To determine the velocity of a body at any point between the earth's surface and her centre supposing it could pass in vacuo[2] to that centre without taking away attraction or extension

2.Will the momentum of the body be completely destroyed when it arrives at the earth's centre? If not, how far will it move beyond that centre?

- - - - - - - - - - - -

- - Now to your queries, and 1st with regard to light, heat &c. We must premise the following observation – If a substance of any density be spread over a surface of any given dimensions, when that substance is spread on a surface of twice, three times, four times &c that of the given one, its density will be only one half, one third, one fourth &c of its first density.

Let 'a' be a luminous body whence rays of light issue on the surfaces 'nrmo' 'SRQT' which are similar but one surface being one fourth the other,[3]

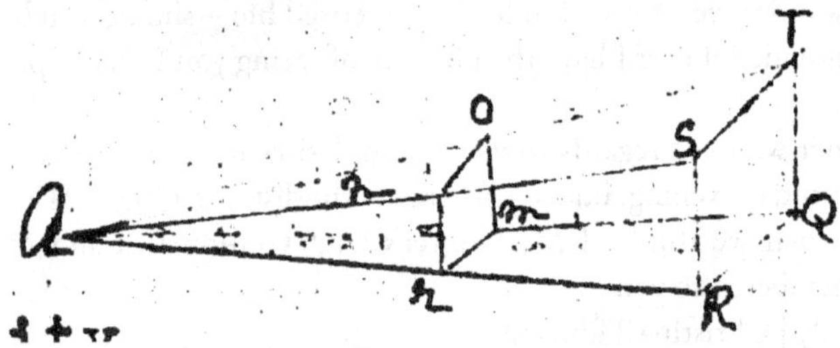

From what has been premised the density of light on 'or' is four times its density on 'TR', that is to say

surface or : surface TR :: dens[ity of light] on 'TR' : dens[ity of light] on 'or'

— Again, from the similar [triangle]s Arm, ARQ & ANR, ASR we have

Ar : rm :: AR : RQ

Ar : rn :: AR : RS

Multiplying these proportions

Ar^2 : rm x rn :: AR^2 : RQ x RS – Alternately

Ar^2 : AR^2 :: or : TR :: dens[ity of light] on TR : dens[ity of light] on or

Q E D.[4]

RI MS JT 1/11/3526

1. *vide*: see (Latin), used to direct the reader's attention to a specific point in a text.
2. *in vacuo*: in a vacuum or empty space (Latin).
3. *one surface being one fourth the other*: this is the inverse square law, which applies to the propagation of light and heat.
4. *Q E D*: abbreviation of quod erat demonstrandum, Latin for which was demonstrated.

To John Conwill September 1842 0167

<beginning of letter missing>

Your solution of my query is faultless, I cannot however say this with respect to your solution of the one which related to bodies falling upon the earth[1] – That gravitation exists at the earth's centre I allow, but that it exists there exclusively I deny. attraction is an infusion for a share of which each particular particle of matter can claim.

Let the earth be a perfect sphere, a body descending thro' it, say at 'A' is pulled downwards, if I may so speak, with a force due to the distance 'ab' – it is pulled upwards by a force due to the distance 'ac' and it descends with a force equal to the difference of these (abstracting the idea of its preacquired momentum) until it comes to the earth's centre 'O' where the difference is nothing and there it will remain.

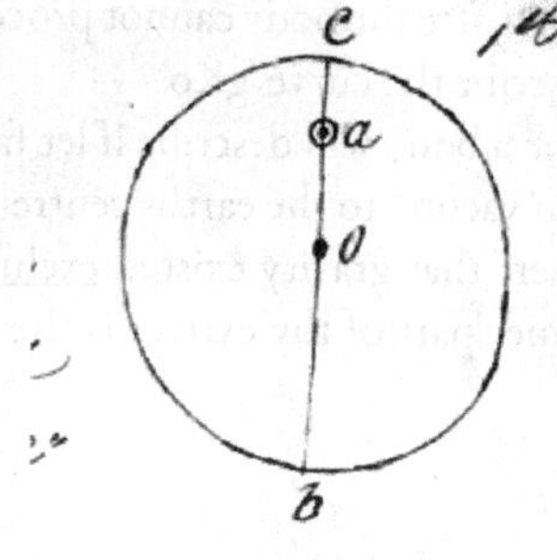

You have virtually admitted this in the query which you attached to your last letter*[2] – for, if attraction existed exclusively at the earth's centre there would be no difficulty in conceiving that a body would wend its way to the spot without turning to the right or left –

A ******A. Let 'a' be the N pole[3] of the earth. if a body be let drop from it – abstracting the idea of density from the earth, but letting attraction and extension remain as at present – it will pass on in a straight line to the earths centre – it is evident that an equilibrium

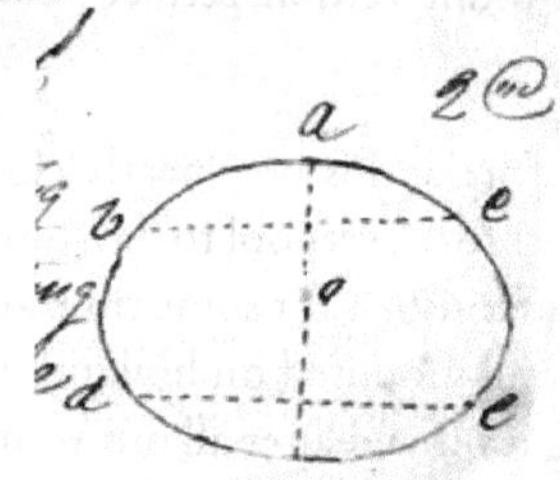

of attraction exists at each side of the body (supposing the earth to be of uniform density) that is, as it is as much attracted towards 'b' as towards 'c' – towards 'd' as towards 'e', it is evident I say that it will not deviate from a straight course downwards to the centre 'o'

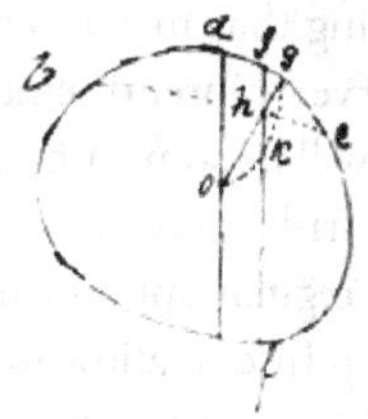

But let us remove it from the pole to the latitude mentioned – say to 'g' – the case is now altered, let us assume a point 'h' in the straight line from 'g' to 'o' and let us conceive the body to be situated there and enquire how it is affected by gravity – draw the parallel 'lhf' – it is evident now that the face of attraction at 'b' is greater than at 'c' because the greater number of attracting particles exist at that side, and if we assume any other two corresponding points we shall have a like result, therefo' I conclude that the attractive force at the side 'b' over balances that on the side 'c' – hence the body cannot proceed in a straight line to the earths centre but will describe the curve 'gKo' - - -

*To determine the line a body will describe if let fall from 54° 22' N latitude supposing it could pass in vacuo'[4] to the earths centre –

– But C. did not assert that gravity existed exclusively at the earths centre the argument in the former part of thy extract is the result of a misconception of my own

J. Tyndall

RI MS JT 2/1/164–3

1. *your solution of the one ... the earth*: see letter 0166.
2. *your last letter*: letter 0166.
3. *N pole*: North Pole.
4. *in vacuo*: in a vacuum or empty space (Latin).

From John B. Edgeworth 17 September 1842 0168

4 Mount Verdon Terrace | Cork. | September 17th, 1842

My dear Tyndall

I believe the Survey of Ireland is very nearly closed[1] Mr Bloomfield Messrs Scally,[2] Moor[3] &c. &c. start for Liverpool to day, only Major Waters and Taylor[4] remain My trace[5] had been finished for some time and I commenced pricking it off[6] on a very nice sheet well mounted on holland[7] by Mr. Bloomfield but alas! I was too long in pricking it off and after all our trouble and after pencilling in a good part of it I found out that it had been shifted[8] and all was gone for nought – however I got another sheet which Mr Bloomfield very kindly mounted for me and he and I set to work and pricked it off in half a day and I am now busy pencilling it in – I must now beg your pardon for being so very prosy in my account of my misfortune and taking up your time reading such stuff – However here goes to tell you something that may interest you more, I am trying to get appointed to the English Survey I intend remaining about a year on it to learn mapping &c. I dare say there will be some difficulty in getting an appointment as all the men who were on the Irish survey have not yet been employed upon the English but if I cannot get a regular appointment I must only offer myself as a volunteer which they cannot refuse to allow me – and which will suit me nearly as well I sincerely hope that I may get to wherever you are do not mention what I have told you to anyone If Major Waters were appointed to superintendance of London[9] there would be no difficulty as he could appoint me himself at once. If ever my dear Tyndall I can be of any use to you in any way you have only to send me word and I shall only be too happy to serve you in any way I can. If ever you require any interest either with the Heads of the Ordnance department[10] or in any other quarter where my friends may have any power I shall be glad to be able to make any return for your attention to me in Cork I hope you will soon let me hear from you and with best kindness and remembrance to Mr Latimer (though he is a methody)[11] believe me

sincerely and faithfully yours | J. B. Edgeworth

RI MS JT 1/11/3559
LT Transcript Only

---

1. *the Survey ... nearly closed*: In fact the Survey would not be formally concluded until 1847, after Parliament had appointed a select committee in August 1846 to assess the work that was still necessary for its completion. However, most of the surveying work in the Districts was completed by autumn 1842, with Survey personnel only remaining in Dublin, Kilkenny and Wexford.
2. *Scally*: see letter 0143, n. 14.
3. *Moor*: see letter 0163, n. 4.
4. *Taylor*: probably Edward Taylor, a civil assistant who worked in the District Office of the C District of the Irish Survey, in Cork, where Major Marcus Antonius Waters was based. He joined the Survey in May 1837, and seems to have left it by January 1843 (NAI OS/1/17).
5. *trace*: a tracing sheet that allows topographic details to be transferred from one map to another.
6. *pricking it off*: using a fine needle to transfer plotting points from an original map to a new one underneath it, often via a tracing sheet.
7. *holland*: laid paper with a smooth and a rough side.
8. *shifted*: where the original and new maps have not been placed exactly together, so that the details traced from one to the other do not appear in the same places.
9. *If Major Waters ... of London*: see letter 0154, n. 3.
10. *Ordnance department*: presumably the head office of the Survey of Ireland at Mountjoy House, Phoenix Park, Dublin.
11. methody: Methodist (*OED*).

From William Ginty 3 October 1842 0169

Crosby Monday evening

Mon Cher[1] Tyndall

Here goes to discharge a debt (but not the poetical one). I commenced one for you in Ormskirk[2] on Saturday, but I had to leave it there ½ written and this I expect will not be posted these two days or so. I met an old friend in Ormskirk, (at least he came here first and then there to see me) – Burrowes[3] of immortal memory – The terror of Griffin – the dread of Collins – the torment of Marshall[4] – the annoyer of Davy – and the delight of all lovers of genuine wit fun and frolic. He is now one of the 'Tarry tribe' as fine and as noble a looking tar[5] as I ever beheld — whiskered to the eyes and belted round the waist; he is after returning from North America, where (in his own words) 'I've scoured like a reindeer over the wide extended plains of the prairiee; I've traversed the shady banks of the mighty St. Lawrence;[6] yes, and I've knelt, by Heavens, with admiration before the sublime cataracts of Mont Moratz[7] and Niagara' – He seems to like his present occupation right well, he intends stopping 18 months more and then (oh foolish Burrowes!) getting married in Wales (where he is going now to visit his charmer). Oh Tyndall dont you feel devilish curious at the mention of 'his charmer' (ah miséricorde!)[8] He bought 3 or 4 wedges of splendid cavendish.[9] But now to other matters – you know him not. If you did you would not regret the acquaintance.

I went to Ormskirk last Thursday, to an Inspection of stores (by the way I intend giving you a verbatim et literaturm[10] account of the proceedings on the last side of this letter, if my furious imagination does not occupy too much room ere then). Hamley has left word with Sheppard[11] that we five ex-office men are to be taken in-to the office[12] previous to the cold weather which he considers will be next month. Oh' Crickey! This I'm afraid will put a stop to my Prestonic[13] anticipations – alias hopes. Oh unfortunate Ginty – there to be installed amongst a lot of strangers, during secula secularum![14] I must curse a while now, and smoke away the infernal blue devils of hateful disappointment. Oh! infernal disappointment, excruciating tormentor of my soul – agonizing anticipations of the future! Perdition seize thee, fate! chance! or what ever thou art. Why treat a poor devil's expectations so cruelly – But enough of this. I must nerve myself into philosophy, grin at fate, laugh at chance and steer steadily along towards the final goal of all earthly peregrinations. I must rout all the loved ideas – put them all to the left counter march in double quick time, shouting 'forward!' with the lungs of a stentor[15] – away! away! away!

'Gone glimmering thro' the dreams of things that were.
I thought to stand where beauty smiled
  My Friend, beside thee yet
But no! the thoughts are now exiled
  God! that we ne'er had met!
I thought to share with one and all
  Sweet friendship's tête a tête
    To smile and hear a brother's call
'Boys don't yet separate!'

I thought – oh heavens! – and now I think
  Such thoughts are at an end.
On me at least, no more on me
  Shall such loved thoughts descend.

I wish ye all a friend can wish
  All fortunes favourite flowers

Your hearts the resting place of bliss
Your souls of strength the towers

Should time your memories rend
The present may the past recall
To 'mind ye of a Friend.

I must leave the inspection affair a clean side and a new day. I'm in too pathetic a mood to write any more. Them five verses are the extemporaneous essence of Ginty's soul. I will expect to hear from you shortly. I will now say good night and good bye for I suppose I wont have room on the other side to sign myself as usual

Yours ever truly | W. Ginty

I had a letter from Dick.[16] He is sporting himself at salmon fishing and says he will send me a 'pinkeen'[17] in the next letter. Are you not at the completion of the Tithe Survey?[18] And do you expect to be sent to Skipton[19] – I heard so through Wm. Latimer. Remember me to Jim[20] – he owes me one. Holmes got 4d. What did that old single-handed dotard mean by such a circular! We wont get leave to <u>think</u> next, without being guilty of a crime

W,G,

I enclose a note from Jack[21] — I will go to live with him at the end of the week.

THE INSPECTION

A Farce! in two acts

By the author of 'The Moonlight Shore'![22]

Dramatis Personae:

Hamley - - - a Lieut. of Engineers
Sheppard - - - The Corpl., in Charge
Bracken[23] - - - The Corpl. out of Charge
Ginty - - - - -One of the Squad
Labourers lookerson &c &c

Scene — a highway — with a Δ[24] in the perspective (apropos)

Act 1st, <u>The Stores</u>.

<u>Bracken</u> (transferring the stores to Sheppard) 'Wm. <u>Geenty</u>!' – 'Here' – <u>Hamley</u> (interrupting him) 'You're a draftsman are you not?' – 'Yes Sir', (aside) 'a kind of one'. <u>Hamley</u> 'Which do you like best surveying or drawing?' <u>Ginty</u> 'Drawing Sir, I'm hardly able to stand the fatigue'. <u>Hamley</u> – (in a loud tone of astonishment) 'The Devil! <u>you</u> not able to stand it– you seem to thrive on it then'. <u>Ginty</u> (with a blunt emphasis and audible grin) 'Oh do I; <u>I'm very sure I don't</u>' (a general giggle and move). <u>Hamley</u> 'Why are you not?' <u>Ginty</u>:– 'I've

never been used to such awful walking' – Hamley (looking pale and in a tone approaching the pathetic) 'I only wish I could bear as much' – Ginty:– 'I'm willing at all events Sir to do my best' Hamley 'very right very right' Exeunt.

Act the 2nd, The Books

Hamley (taking my book): – 'You learned to survey since you came to England?' Ginty 'I learned the practical part Sir – chaining.'[25] Hamley 'Hum' (turning over to some awful big figures in a diagram) Well! them are woppers' – Ginty 'So much the better for the office men Sir' – Hamley 'Ah, oh, yes, I cant say there is any particular objection' Ginty 'No Sir! there cant be!' – Hamley (turning another page and seeing 'total 167') – 'That's a good day's work' (another, 114) 'That's passable.' – Bracken: 'He is a very excellent steady and diligent workman Sir' – Hamley 'Oh! oh!' – Ginty 'My work Sir for the last two months has been very very difficult, all demesned.'[26] – Hamley 'Not here abouts, eh?' – Ginty 'NO Sir, Crosby, and Ince Halls'[27] – Hamley 'Oh Ince Hall – I hope you will make a clever job of that Ginty – (a nod).

Finis.[28]

RI MS JT 1/11/3584
LT Transcript Only

1. *Mon Cher*: my dear (French).
2. *Ormskirk*: a market town in Lancashire situated between Liverpool and Preston.
3. *Burrowes*: Robert Burrows, a civil assistant in the 5th Division, C District of the Irish Ordnance Survey who left the Survey in April 1839 at the time that Tyndall joined (NAI OS/1/15–16).
4. *Marshall*: James Marshall, a Private in the 13th Company of the Royal Sappers and Miners who worked in the 5th Division, C District of the Irish Survey. He had joined the Survey in *c.* 1832, and was transferred to the English Survey in January 1842 (NAI OS/1/16–19).
5. *tar*: sailor (*OED*), hence the 'Tarry tribe' is a generic name for sailors.
6. *the mighty St. Lawrence*: a river in North America that connects the Great Lakes with the Atlantic Ocean.
7. *Mont Moratz*: not identified, but presumably one of the Laurentian Mountains in Quebec, Canada.
8. *miséricorde*: mercy (French).
9. *cavendish*: cured sweet tobacco (*OED*).
10. *verbatim et literaturm*: word-for-word and letter-for-letter (Latin).
11. *Sheppard*: see letter 0150, n. 23.
12. *the office*: the Divisional Office of the 5th Division of the English Ordnance Survey at 33 Windsor Street, Toxteth Park, Liverpool.
13. *Prestonic*: relating to Preston, where Tyndall was stationed.
14. *secula secularum*: a century of centuries (Latin).
15. *stentor*: herald of the Greek armies during the Trojan War in Homer's *The Iliad*.
16. *Dick*: Richard Hunter.
17. *'pinkeen'*: a minnow (*OED*).

18. *Tithe Survey*: The Tithe Survey of England and Wales was undertaken in accordance with the Act for the Commutation of Tithes in 1836 (1836 c 71 (Regnal. 6 & 7 Will 4)). The Survey produced maps of landholdings from which the new tithe rent charge, based on average prices of wheat, barley and oats, could be calculated. The Act made individual parishes responsible for paying for the production of the maps.
19. *Skipton*: see letter 0096, n. 3.
20. *Jim*: Phillip Evans.
21. *a note from Jack*: John Tidmarsh; the enclosed note may be 'The Inspection'.
22. *'The Moonlight Shore'*: not identified, but possibly a poetic production by Tidmarsh.
23. *Bracken*: see letter 0155, n. 11.
24. Δ: a trigonometrical point, a fixed surveying station.
25. *chaining*: see letter 0143, n. 30.
26. *demesned*: private estates (*OED*).
27. *Crosby, and Ince Halls*: Crosby Hall in Little Crosby, Lancashire, was built in the sixteenth century, with major alterations in the 1780s, as the family seat of the Roman Catholic Blundell family; the neoclassical Ince Hall in Ince Blundell, Lancashire, was built in 1720 as the family seat of the Blundell family (no relation to the Blundells of Little Crosby).
28. *Finis*: the end (French).

From William Ginty [October] 1842 0170

What direful thoughts of sickening gloom,
Around me gathers, this evening noon!(Question)
Enough to quench, the little ray of joy(splendid
That grew with me, ere I was a boy(harmony
Like waves that roll, with thundering, tempest tost (from a
And dash their fury, on the sea beat coast (fiddle
With mad'ning force, and mountains of white foam
Engulphing man, into one vengeful doom
A miserable attempt at stealing anothers laurels!
Like tyranny – rose that demon power
(A.F.R) Griff[1] – dark as midnight hour
With nostrils gaping!!! See behold they come
With fiendish strides, and hell'sh brimstone fume
To suffocate devour and dethrone
All poor surveyors, and hunt them home
For why the millions, T.R.[2] lays is growing slack
Those beggars I'll hunt – Griff your knapsack
To hold the surplus of the Government Stock,
Tis little enough. Me they will shortly dockhooray!
Of all my vestures maps and title deeds(more
For robbing the surveyors of their fees.(music!
But what for that, I to myself must look (Oh
I'm almost blind youth hath me forsook, (Lord!
My ears grow cold, a naccous[3] juice doth flow
From out the poers[4]: variant spelling of pores. – oh! misery sorrow woe
To be deprived of life oh! God what shall I do

For satan claims his own, I cannot look to you
See the fiend comes, while blasted spirits attend
To waft my soul to H - -[5] most wretched is my end
So ends the the life of one, a blasted race did run
A robbing thieving knave. A father's 'cursed son'.
Philantrophist[6]

So far 'first rate'. The real prose of it is that a conglomeration of thoughts assembled at some certain (quere uncertain) period of the day and poured a bucketfull of dirty water on a ray of his before he was born. W.G.

These last 8 defy all common sense! The 'Scotch Reviewers'[7] would throw down their scourging quills in despair and set about writing an ode to the memory of 'damned nonsense' It excels all I ever seen or imagined in the way of literary hotch potch! Certainly it is a philantrophical effusion with a heavy vengeance! In the letter speaking of it he[8] says I 'must not publish it' – as the subject is rather a 'domestic one' This is the first letter I had from poor 'Blackthorn'[9] since I came to England so that it is quite evident that it is his own high opinion of 'the poem' that induced him to write. I'll keep him to it! until I get a few more from him. Byron[10] could not afford as much amusement!

(A.F.R.)

Quere. Tucker or Tom Reynolds of Carlow notoriety.[11]

First he devours all poor surveyors and then hunts them home! For why the millions &c. &c. – !!!– –

Mark! Some dirty stuff is flowing out of his ears!

I received your letter[12] Friday evening as I came in from patroling the woods of 'Crosby Hall'.[13] Tyndall I am surveying a most beautiful place – a gentleman's demesne[14] – it took me two days last week to produce a line of 6000[15] I had to cut my road through wood for one half and to crawl under shrubs and flowers and apple trees and over ice houses garden walls the other. I can just get as much apples, peaches, currants resberries, strawberries, gooseberries, melans,[16] cucumbers, &c &c &c. as I like to eat or pocket. I will be in the demesne for a month yet all I have to amuse me is a pack of hounds that are in a kennel ready to devour me if I put a foot within the precints of their stinking domicile. The hares and rabbits go about in flocks, seriously when we make our appearance at one end of the lawn they just scour away like a flock of sheep from a dog. The pheasants and partridges are innumerable I counted 17 partridges in a cluster ten minutes before I got your letter. I have nabbed one hare already. The chainman I have has pinned two. I killed him with a tip of the pole – we can positively trip them with the chain.[17]

Yours &c. | Ginty.

RI MS JT 1/11/3589–90
LT Transcript Only

1. *(A.F.R) Griff*: possibly a comical allusion to acting Lance Corporal David Griffin, whose subsequent demotion to Private Ginty saw as a deserved vengeance; see letter 0199. The abbreviation (A.F.R.) is used again later in the letter, but its meaning is unclear (although the A may stand for Acting).
2. *T.R.*: possibly either Captain Henry Tucker or Thomas Reynolds; see Ginty's exegetical comments on the poem and n. 11.
3. naccous: not identified.
4. *poers*
5. *H*- -: Hell.
6. Philantrophist: Rather than a misspelling, this poetic pseudonym seems to combine the word philanthropist, indicating a purveyor of practical benevolence, with the Greek term for one who loves to overthrow or ruin, anantrepo (ἀνατρεπω). Ginty's response to the poem suggests he recognized this.
7. *'Scotch Reviewers'*: censorious critics of poetry in journals such as the *Edinburgh Review* and *Blackwood's Edinburgh Magazine*, as satirized by Lord Byron in his poem *English Bards and Scotch Reviewers* (1809).
8. *he*: presumably Martin Cuddy; see n. 9.
9. *poor 'Blackthorn'*: Cuddy's poetic pseudonym; see letter 0127, n. 2.
10. *Byron*: George Gordon, Lord Byron (1788–1824).
11. *Tom Reynolds of Carlow notoriety*: Beyond the Captain Thomas V. Reynolds who had worked on trigonometrical surveys for the Board of Ordnance in the 1790s, no other Tom Reynolds was involved in the Ordnance Survey at this time, so this is possibly the Thomas Reynolds who was a leading member of Daniel O'Connell's Repeal Association; see letter 0066, n. 4. In June 1841, Reynolds, a City Marshal in Dublin, 'went down to the County Carlow to canvass the electors' in the bitterly fought election, and 'remained several weeks', which may account for his apparent notoriety (W. J. O'N. Daunt, *Personal Recollections of Daniel O'Connell*, 2 vols (London: Chapman and Hall, 1848), vol. 1, p. 292).
12. *your letter*: letter missing.
13. *'Crosby Hall'*: see letter 0169, n. 27.
14. *demesne*: private estate (*OED*).
15. *a line of 6000*: presumably a chain line of 6000 feet, made by a chainman and a follower stretching a chain of 66 feet in length between ranging rods and then successively repeating the action; see letter 0143, n. 30.
16. *resberries ... melans*: variant spellings of raspberries and melons.
17. *chainman ... with the chain*: see n. 15.

To John Tyndall, Snr [October] 1842 0171

Preston, | Monday morning 5½ o'clock.

My dear Father

I believe there is an old promise of mine on record to the intent that you should hear from me once a week.[1] Well I believe the specified time has elapsed since last I wrote, so here goes to redeem my promise, tho' I was expecting to hear from you ere this. In the first place my health is as usual. It would be impossible to speak stronger in its commendation. You know my usual state of health – Sound in body and limb, and if I get a fair run, can jump the height of my chin. I'll let you know now how I employ myself. Our office[2] hours are from 6½ to 5 o'clock. They were from 6 to 5½ but were changed on account of the darkness of the weather. We come to breakfast at 10 minutes past 8, and return at 9; come to dinner at 1 and return at 2. So that our hours are nicely divided. When I come out in the evening I have a cup of coffee or perhaps soluble cocoa, resembling very much a medicinal dose in appearance. The moment it's swallowed I am off to school. Dont start, I'm learning French. A man named Malone, a brother of one of our fellows, has set up here lately, he's a classical teacher.[3] He devoted two hours every evening to me alone. His terms are £1 a quarter at 6/8 a month. Its dear, but then his whole time being devoted to me I cant complain. I fear however that he is not altogether as competent as I could wish, I have resorted to a plan, which I often pursued before, to test his knowledge. I raise a difficulty now and then and as I am rather unwilling to take mere assertion from any man we often have a bit of an argument. He usually endeavours to baffle me with Latin analogies, but I see thro' him and keep him to the point. My knowledge of English enables me to give him some dry rubs. To sum up all, he may do very well for one month. I'll remain with him no longer, as I shall then be able to get on myself. It is not for the purpose of being able to call things by duplicate names that I am learning French. I have very little ambition in this way. It is because the writings of many clever Engineers are locked up from me on account of being written in that language. I use French then merely as the key of a treasury, not as the treasury itself. When I want to write letters I get up at 5 o'clock in the morning. 5½ is my usual hour, then I have an hour to spare for reading &c. Direct your next to No 11 Butler Street,[4] as I intend quitting my present lodgings[5] on Saturday night.

Good bye, my love to mother and Emma | Your affectionate son | John

I owe you some money which I hope I'll be able to pay shortly, that sentence sounds rather queer to me. I suppose it will affect you in much the same manner. I know however you'r a kind creditor, you'll not be hard on | J.T.

RI MS JT 1/10/3280
LT Transcript Only

---

1. *an old promise … once a week*: see letters 0054 and 0062.
2. *office*: the Divisional Office of the English Ordnance Survey's 1st Division, located at Preston and used for producing plans of Lancashire and Yorkshire.
3. *A man named … a classical teacher*: An advertisement in the *Preston Chronicle* on 24 December 1842 announced that 'Mr. Malone, F.C.D.' had taken over the 'Select Classical and Commercial Academy' in Chapel Walks, Preston on 26 September 1842. The advertisement claimed that 'Mr. M.' had 'gone through a full academical course of edu-

cation, and had considerable experience in the tuition and management of youth', and 'Though a short time in Preston, his system of education had been sufficiently tested and satisfactorily proved by the improvement of the pupils who attend his school'. It added: 'For respectability of character and general competency, Mr. M. can refer to the Fellows of the Irish University, and many Families of the first rank in that country' ('Advertisements & Notices', p. [2]). He was presumably the brother of Denis Malone.

4. *Butler Street*: close to the railway and North Union Station in central Preston.
5. *my present lodgings*: presumably 18 Oxford Street, Preston; see letter 0164.

To John Tyndall, Snr [Autumn] 1842 0172

Tuesday morning

I'm just after demolishing my breakfast, and believe me that 'twas no small one. This is wild morning. The wind is high and the rain pelts most furiously. I have a most interesting prospect at present before me, I gaze out of the window of my room, and the varied beauties of wood and water strike my delighted eyes. Let not the idea of waving foliage however combine with your notion of wood. The wood which I speak of is the piles of red deal and yellow pine which stand in hoary majesty in the timber yard about 100 yards in front of me; and the water! picture not to yourself the unruffled surface of a glassy lake along whose placid bosom the graceful swan glides smoothly. No Sir! it is a pool where the gabbling ducks revel in all the rich exuberance of – mud.

Do you know the address of Mr Ryan the American gent. If you dont perhaps Mr James Thomas[1] does. If possible procure it and send it to me, I want it for a particular purpose. I expect the Sentinel[2] by this morning's post, I sent for it last week. There goes the postman, but no paper, they must have been late in posting it. Mr Eivers, he of whom you heard me speak of often has left the Survey. I hear he has got a situation worth £300 a year.[3]

And now what more have I to say? Verily nothing. This being the case

I subscribe myself | Your affectionate son | John.

RI MS JT 1/10/3286
LT Transcript Only

1. *Mr James Thomas*: see letter 0013, n. 7.
2. *Sentinel*: the *Carlow Sentinel*.
3. *Mr Eivers ... £300 a year*: Lawrence Eivers had possibly left for a job on the Tithe Survey in London, as William Ginty told Tyndall on 23 April 1843: 'Larry Eivers ... has just arrived from the great metropolis. He is walking about here [i.e. Liverpool] ... – to improve his health previous to his removal for the examination of a Township on the Tithe scale about 10 or 12 miles from here'; see letter 0200.

To John Tyndall, Snr 15 October 1842 0173

Preston Oct. 15/42

My dear Father

I received your last welcome letter[1] on Wednesday last. The proposal contained in the latter part of it was worthy of a good father – it was kind – very kind. I have however no idea of assenting to the prudence of such a step – if I should go home and commence Classics with Mr Sheridan[2] all the knowledge I have acquired since I joined the Survey must go for naught, as the application of that knowledge is incompatible with the application of Latin and Greek. Granting even that I was acquainted with these languages what would I have to look to? Nothing without interest. And if interest be of service at any time it is so now. I might indeed soar to the height of authorship, and, to the astonishment of the world, sign the productions of my classics pen from the chill loneliness of a garret room – others have done so before. I have no desire to follow their example. I however give Mr Sheridan credit for the best intentions towards me.

And now to other matters. With respect to your proposed plan, it might be very well under some circumstances but I think not in the present instance. The fact is it would puzzle me to draw up a testimonial which all the persons you have mentioned could consistently sign. Mr Newton knows nothing of my qualifications, Mr Watson[3] knows just as much, the other gentlemen you have mentioned may know a little but not sufficient to warrant any strong testimony on their part. Before laying down my plan of operation I would premise a few remarks – A person seeking for a situation such as I am, must possess not only interest which arises merely out of a desire to serve, he must possess, from proper authority, testimony that he is fully competent to discharge the duties of the situation which he seeks. And I flatter myself that you possess the former in the gentlemen you have named, while Mr Wynne's testimonial[4] supplies the latter. I speak with reference to a particular situation, of which more by and by. My plan then would be this: obtain from each of the gentlemen alluded to a letter expressive of his knowledge of my character and his opinion of my abilities, this – coupled with his request that the Col.[5] would use his interest in my behalf – would I think be pretty strong. These letters will of course be varied by circumstances – the Dean[6] and Captain Steuart for instance can speak more confidently than any else respecting me, their testimony will be the result of personal observation, while Mr Watson's must necessarily be limited to the expression of his desire to serve me on your account. Finally, if these letters, together with Mr Wynne's, were presented to the Col. by Mr Bredin, the latter could verbally, and perhaps more strongly, add his request to the common stock; indeed I dare say Mr B. can form as correct an opinion of my qualifications as any one. He was once kind enough to make me a pupil of his own. This then is my plan of operation, if it meet your approval I hope you will act on it with energy – You may ask what is all this intended to achieve – I'll tell you – the end and aim of it is to procure employment for me in Somerset House London. It is a Map office under the Superintendence of the Tithe Commissioners.[7] Several men who were once on the Survey have procured employment there[8] – Ability to fill the situation and the interest of an influential person are the only things necessary to assure success. With respect to the former, I can bring other testimony than my own assertion, backed by specimens, to prove that I possess it (nothing approaching to vanity pervades my mind while making this statement). As to the latter, it all depends on the result of our application to Col. Bruen.

Wishing that it may have the desired effect | I remain your affectionate son | John

P.S. It is now upwards of two years since Mr Wynne left the Survey[9] – his testimony is founded on his knowledge of me at that time. I had ample means of improving myself since and I availed myself of every opportunity to do so. | J.T.

RI MS JT 1/10/3281–2
LT Transcript Only

1. *your last welcome letter*: letter missing.
2. *Mr Sheridan*: possibly Bernard Sheridan BA of Bagnalstown, who took in boarding and day students (*Slater's*).
3. *Mr Watson*: possibly John Leckey Watson; see letter 0125, n. 9.
4. *Mr Wynne's testimonial*: see letter 0126.
5. *the Col.*: Colonel Henry Bruen.
6. *the Dean*: Dean Richard Barnard.
7. *Tithe Commissioners*: Officials appointed to administer the Tithe Commutation Act of 1836; see letter 0169, n. 18. The Tithe Commissioners were based at Somerset House, on the Strand in central London.
8. *Several men ... employment there*: Tyndall possibly has Lawrence Eivers in mind; see letter 0172.
9. *two years ... the Survey*: Lieutenant George Wynne relinquished command of the 5th Division, C District of the Irish Ordnance Survey at Youghal on 18 September 1840; see letter 0011.

From William Ginty 11 November 1842 0174

Kirkby Lonsdale[1] | Co.[2] Westmoreland | Friday Nov. 11th 1842.

My dear Tyndall

Your exquisite and genuine Irish sentiment[3] I received yesterday evening just as I was returning from work wearied and drenched to the skin 'My eyes!' cried I as I opened its folds 'here's a God-send to amuse me while my coat is drying' – it did amuse me. You touched the happy chord when you said:

The tulip Bill in splendour glows
But wants the perfume of the rose

Tis a good sentiment – and more – tis most admirably applied! I subscribe to the whole of that with all my soul. I too meet with some and could meet with many – but oh! my thoughts are Erin-ward[4] I love bright eyes! they are here! I love the clustering ringlets and the snowy breast! They too are here! But ah! I must go further Westward for the 'faithful feeling heart' without which all is insipid – foul and worthless. Thoughts of Ellen[5] and her numerous stereotypes[6] places an insurmountable barrier between me and them and so to the devil with the whole of them (and they are going to him fast) They're not worth so much scribbling.

With regard to the 'Crimson glow' the 'glorious blush' and the reverse 'Pale haggard and emaciated' all I have to say is that I'm sorry it is not a fact for your own sake.

I will quote the concluding sentences of your letter and then make a comment on the logic of it. 'The streaks of morn appear in the East – the Ghosts are vanishing – I too must away' 'What the devil d'ye mean' 'The Ghosts are vanishing You too must away' – you must imagine me very stupid by this repetition of yours

I 'lit my tapèr'[7] last night about an hour after I received your hint about Windermere[8] and to work I went. You will find 46 lines of my 'bladdering'[9] on the 3rd side. A kind of a preface to the main design. At some short future period you shall have the rest, on these conditions: you must send me 'a summat'[10] in return for this, and as your muse is mounted on a thoro'bred, it will not clip much off your sleep. My poor pegasus is broken winded[11] and so must take his time – particularly going up a hill.

Confound that old bothered Captain,[12] I often offer up a few prayers for his future welfare. What a hard-hearted old wretch he must be. Jack[13] and Dick[14] were here last Sunday – or properly speaking from Saturday evening to Monday evening. We of course felt happy whilst together. We will meet tomorrow (Saturday) evening in a village called Ingleton.[15] Monday morning of course will find us together – its the only pleasure we have. We still form the equilateral triangle in our relative positions. I had a letter from Ormskirk[16] which contained an introduction to a host of fair ones – how Jim[17] would seize the opportunity of getting so numerously 'acquainted'. Holmes got a letter from Higginson with all his adventures, he sent it to me but unfortunately it was lost or mislaid, I begged him to repeat it. I have not seen Sinnett since I came here. We arrange all matters thro' the post. Tell George that William[18]: George and William Latimer. received his letter. Last Saturday we four bought a 'Glengarry'[19] each. The natives dont know what to make of us – we look so very martial. Dont forget the 'Summat'

Yours as usual | Ginty

'Well I will dream that we may meet again
And woo the vision to my vacant breast'[20]

(My best respects to the Miss – Jack is thinking of writing to Miss G.[21] about Jim – sarve him right.) | W,G,

You need not send my letters to Clapham[22] –
Kirkby Lonsdale, | Co. Westmoreland will find me out.

Fast towards the West th'untiring sun
Was hurrying just as I began
To wind my way up thro' the dell
To thy dark summit 'Berkenfell'.[23]
To work went both my hands and feet
And many a briar my nose did greet
The dell once past I stood to view
What I had done and yet to do
Pulled out my pipe and struck a light
When two dear friends hove into sight
Yelept[24] Sublimity and Reason
I bowed and said 'you've come in season'.
'I want advice, so give it now
'You live above? and you below?'
'Yes!' cried the pair, 'we'll calmly chat
'The matter oer', so down we sat.
Then came the struggle – Reasons cause
Was that called 'first of nature's laws'.
With serious accents first he spoke
And said 'this freak may prove no joke'
'Oft will thy wearied limbs bespeak
'A swift retreat adown that brake.
'A Chamoise Goat – a North Wales sheep
'Would shudder at yon craggy steep,
'Beware! beware! with me descend
'I've always been and am your friend'
He ceased – Sublimity sprang to his feet
And pointing upwards – cried in sweet
Voluptuous accents 'come! come! with me
'I'll nerve – thy fragile feeble knee
'Is there a being with a soul
'Who would not dare to reach that goal'
_ 'There is! there is! I will!' I cried
And jumped exulting by his side!
On – on – still onwards – upwards flew!
Poor Ginty – Jack and so would you!
O'er rock – crag – heath alike pell-mell
To thy dark summit – 'Berkenfell'!

_ 'Twas even when I was seated there
The earth seemed mingled with the air
A torrent leaped beneath my feet
And tumbled down a steep retreat
Wraped in a cloud – astride a crag
Propped by my pole – (I lost the flag!!)
'Here let me' cried I 'scan – peruse
'Dame Natures glories' – Hail my muse!
To be continued (?)[25]

RI MS JT 1/11/3585
LT Transcript Only

1. *Kirkby Lonsdale*: a town in Westmoreland (now Cumbria).
2. *Co.*: County.
3. *Your exquisite ... Irish sentiment*: possibly the poem 'To Ginty' composed in October 1842 (RI MS JT 8/2/1/7).
4. *Erin-ward*: towards Ireland, Erin being a romantic name for the country.
5. *Ellen*: from Kinsale.
6. stereotypes: duplicate prints from the original printing plate.
7. *tapèr*: wax candle with a long wick (*OED*).
8. *your hint about Windermere*: It is not known what Tyndall's hint was, but Windermere is the largest lake in England, set amid picturesque scenery in Westmoreland.
9. *'bladdering'*: blathering; talking foolishly or nonsensically (*OED*).
10. *'a summat'*: a something.
11. *your muse ... broken winded*: In Greek mythology the Muses were three goddesses of literary inspiration. They were born as three sacred springs, which burst forth when the winged horse Pegasus touched the ground on Mount Helicon with his hooves. Later, Athena tamed Pegasus and presented him to the Muses.
12. *Captain*: Captain Henry Tucker.
13. *Jack*: John Tidmarsh.
14. *Dick*: Richard Hunter.
15. *Ingleton*: a village in the Yorkshire Dales, close to Lancashire.
16. *Ormskirk*: see letter 0169, n. 2.
17. *Jim*: Phillip Evans.
18. *George that William*
19. *'Glengarry'*: a military cap worn by Scottish and Irish regiments (*OED*).
20. *'Well I will ... vacant breast'*: Lord Byron, *Childe Harold's Pilgrimage* (1812–18), II.ix.5–6.
21. *Miss G.*: not identified.
22. *Clapham*: not identified.
23. *thy dark summit 'Berkenfell'*: this seems to be an imaginary mountain.
24. *Yelept*: an archaic poetic term for friend.
25. *To be continued (?)*: the poem is continued in letter 0176.

From William Ginty 1842 0175

Kirkby Lonsdale | Tuesday evening

My dear Tyndall,

My unfortunate head and ribs are not quite recovered from the shock they received this morning by that rigmarole production of yours.[1] Acrostic the first is what I call genuine Irish sublimity, acrostic the 2nd genuine Tyndallish sentimentality.[2] What a devilish pathetic genius you are! You ought to get a martingale or a crupper on that pegasus of yours[3] for she rides over the fondest hopes of the soul without either christian charity or human forbearance. Its well for you that I possess a benign and benevolent disposition or I could make you writhe under the agonizing lash of merited Justice

Why blast and damn and sink and fire
I hardly can restrain my ire
Nor spout it forth in maddning rhyme
– Was ever mortal since the time
Of Eve's dessert on ripened apples
So cruelly used – my spirit grapples
The very steel within my fist
And yells Revenge! then with a twist
Transfixes it within my grasp
Oh damn me Tyndall I could rasp
Some filings off your very core
You foul and stinking rooting boar ('bore')
You cod me! eh! oh blow me Jack
You'd not tried this if you'd look'd back
To other days when you and I
Alone had 'kenned'[4] your villainy
'Tongue hath not named nor heart conceived'[5]
Eye hath not seen nor ears believed
That man possessed of human soul
Could do the deeds my pen control
Think'st thou thy base deceitful heart
Could brook one gaze on them nor start
In horror and remorse – Yet I
Could lay them all before thine eye
Then wretch! beware thee of that tale
'Twould wax thy cheek more deadly pale –
And mend thy manners every one
You 'white unshrouded skeleton'.[6]

I had a letter from Sinnett last night after I posted Jim's[7] – he gave me corrobarative evidence respecting our transit and says it is probable that I will not be left to finish the remts.[8] even in Kirkby. I say that was a right good decision of the Master General.[9] You must excuse the brevity of this as I have two more to write.

Good night | Yours &c. | Ginty.

RI MS JT 1/11/3588
LT Transcript Only

---

1. *rigmarole production of yours*: letter missing.
2. *Acrostic the first ... Tyndallish sentimentality*: An acrostic is a type of poem in which the initial letter of each line spells out a word. All of the acrostics in Tyndall's extant manuscript poems spell out the names of women, including Christina Tidmarsh, Jane and Miss Hebdon (RI MS JT 8/2/1/7).
3. *get a martingale ... pegasus of yours*: a martingale and a crupper are both tacks used to control horses (*OED*); on Pegasus, see letter 0174, n. 11.
4. *'kenned'*: known (*OED*).
5. *'Tongue hath ... heart conceived'*: slight misquotation of 'Tongue nor heart cannot conceive nor name thee!' from W. Shakespeare, *Macbeth*, II.iii.63.
6. *'white* unshrouded *skeleton'*: 'The Disembodied' (1831), 62, by the Irish poet James Wills.
7. *Jim's*: Phillip Evans.
8. *remts.*: re-measurements.
9. *Master General*: George Murray.

From William Ginty 25 November 1842 0176

Kirkby Lonsdale | Nov. 25th 1842

My dear Tyndall

Enclosed I send you the 'continuation'.[1] I hope it may please you – but I ween[2] not – for it does not please me. Jack[3] took a notion last week to scheme Friday and Saturday[4] and go to the Fair of Ingleton[5] to enjoy the fun with Dick Hunter. They both came over here on Sunday morning Jack went home on that evening Dick the evening after. I've spun this preface merely to tell you that I had a note from Jack on Tuesday saying that Baker[6] had reported him absent from his quarters on Saturday morning, and that Sinnett had written back for 'all the particulars as the charge was a serious one and should go before Lt. Hamley'. Jack's excuse was that his work lay nearer Kirkby than Hornby[7] and so he came in here on Friday night (I of course will swear to that) I hope it wont injure him in getting the two shillings. I think he will get over it with a slight reprimand, if his S–r[8] has not been seen by Baker. I've opened a score[9] for that monkey sapper to be rubbed off[10] when opportunity offers. I hope I may be always kept away from them as I am at present. I just worked 2 hours this week and that within ¼ mile of the town – no sign of compulsion there boy'o. 5 days last week done my fortnight's work. I had a letter from Jim[11] on yesterday. That paragon of cads! for want of other matter he gave me

> A taste of ould times – when 'discussions' were
> in vogue – oh that I was near him just for
> two or 3 hours. I'd make my fist 'acquainted'
> with the softest part of his head.

– Tell me Tyndall – did Bessie[12] answer his letter. I guess not for if she did he would be proud to boast of it. But 'I'll talk of this anon' 'Does Miss Edwards squint?' No! miscreant – dare you trifle so with perfection! Does Claris[13] squint? Does Ellen[14] squint? did Miss Ronayne[15] squint? did the spalpeen[16] Sapper squint? did – did, but enough – Do you recollect when you were 'hailing the last resort' you surely squinted then! do you recollect when you were embracing the lamp post and crying

> 'Here's the post 'she loved so much'
> Where I was often planted'

Did you squint then? I (you well know to your grief) could give you enough in this strain. Tell Geo Willm[17]: George and William Latimer. got his letter

I will expect a scrap from you shortly.

Yours &c. | Ginty

NB | Tell Fin Malone I will 'touch him on the raw' next week See Cuddys vocabulary of witticisms.

P.S. | As you have all the 'copy rights' of my productions[18] you may as well have this among the rest. Tho' I cant exactly call it such for it graced the last page of Mary's[19] album a fortnight ago you take care and dont turn Albemarle St. Murray[20] on me one of these days!

'Sweet it is to gaze upon'
This Album's bright and lovely pages
To turn them over one by one
And view the muses in their rages
Here one will sigh – there one will cry
My Mary dear how can I leave thee
Bereft of thee
Where shall I flee
Oh woe is me – nought can relieve me!

Another there will sigh and moan
And say his heart is nearly broken
Then leaves when he sets off to roam
A sweet 'forget-me-not' – a token –

The token of his lasting love
Which ne'er shall die which nought shall relieve
But still remain
A heavenly flame
To light his soul to lovely Mary!

And gazing thus those pages oer
A thought has struck me quite perplexing
Oh! all ye skilled in woman's core
Unravel this? – for me tis vexing!
Pray tell me how – those lovesuits all
So sadly here within reflected
Of love the elite
Long loud and sweet
Where all – aye every one – rejected?

Ye do not know! Ye cannot tell!
No more can I – so let's try guessing!
The blockheads knew not how to seal
Their loves with kisses and caresses
But sighed and loved and madly roved
In search of pens and 'midnight taper'
What's in the name?
Without the game?
Where's all this love? Why on the paper

This paper love I do not like
It's like the paper easily blasted
Like it tis easily kindled bright
And just as easily quenched and wasted!
It should be written somewhere else
Where it would be more secret – safer
Deep buried in
The breast within
The soul the pen – the heart the paper

Then lovely Mary tell me this
Suppose one thought of such a writing
Could you deny his soul the bliss
So heavenly pure so sweet – delighting
Say would'st thou give – suppose he found
The Pen the words, the seal, the taper
Could you refuse
To such a muse
The one thing wanting then – the Paper!

I have written another piece in it in Don Juan style,[21] in answer to a most bitter piece called 'man's love' you will probably have that in my next
RI MS JT 1/11/3586
LT Transcript Only

1. *the 'continuation'*: see letter 0174, in which the enclosed poem ended 'To be continued (?)'.
2. *ween*: think or suppose (*OED*).
3. *Jack*: John Tidmarsh.
4. *Friday and Saturday*: 18 and 19 November 1842.
5. *the Fair of Ingleton*: The Yorkshire village of Ingleton held an annual fair, originally for the trading of leather and oatmeal, starting every 17 November (E. Hargrove, *The Yorkshire Gazetteer* (Knaresborough: Hargrove & Sons, 1806), n.p.).
6. *Baker*: see letter 0132, n. 5.
7. *nearer Kirkby than Hornby*: Kirkby Lonsdale is a town in Westmoreland (now Cumbria), Hornby a village 7 miles further south in Lancashire. Both are situated close to the River Lune.
8. *S–r*: Sapper.
9. *score*: a notch or groove made in a piece of timber or metal to allow another piece to be neatly fitted into it (*OED*).
10. *that monkey ... rubbed off*: Ginty may be referring to the brass Royal Engineer Grenade which, since 1825, was worn by Royal Sappers and Miners on the tail of their full dress coatee (Royal Engineers Museum http://www.re-museum.co.uk/research/corps_history/symbols_and_songs/), and may also have been worn by civil assistants working for them. The distinctive insignia of a nine-flame grenade might be rubbed off using a score in a piece of metal (see n. 9).
11. *Jim*: Phillip Evans.
12. *Bessie*: see letter 0115, n. 17.
13. *Claris*: not identified.
14. *Ellen*: from Kinsale.
15. *Miss Ronayne*: possibly the daughter of Dominick P. Ronayne; see letter 0099, n. 7.
16. *spalpeen*: a low or mean fellow; a scamp, a rascal (*OED*).
17. *Geo Willm*
18. *'copy rights' of my productions*: Ginty is using an earlier meaning of the word, indicating the original manuscripts from which printed copies could be taken (*OED*).
19. *Mary's*: Mary Edwards.
20. *Albemarle St. Murray*: the publisher John Murray (1778–1843), whose home and office were at 50 Albemarle Street in London's Mayfair. Ginty may be alluding to Murray's notorious decision, following the death of Lord Byron in 1824, to burn the manuscript of the poet's personal memoirs, which he had sent to the publisher several years earlier.
21. *Don Juan style*: in the style of Byron's *Don Juan* (1819–24), a long satirical and sometimes scandalous poem in seventeen cantos.

From John Chadwick 30 November 1842 0177

Kinsale | November 30th 1842.

Dearest Tyndall

I fully expect that by this time you have completely expunged me from your memory as a faithless and forsworn friend and devoted the manes[1] of our old cronyship to the shades of Erebus[2] to dwell with devils and damned spirits through the countless ages of eternity as Bagby[3] used to say. I received your poetical morceau[4] with much satisfaction. It quite breathed the spirit of a Tyndall – half ludicrous half sentimental for such was the composition of your mental parts. The spirit of poetry being either extinct or never having had a place in my soul, you must only expect a mere prosaical answer to yours, though according to musical rules sweet sounds should be responded to by others as sweet so as to form a concord. But n'importe;[5] in plain doggrell,

> My excuse must be,
> That 'the sisters three'[6]
> Are no friends to me,
> So my answer to thee,
> Is prose sans[7] poetry.

That's what you may call knocking a spark out of lead and as you have thrown the pearls of your imagination before swine, it is surely time for me to send my responsive grunt – And, to begin in the middle, I must according to law call up the witnesses for the defence. You begin your last letter with an animated exordium about my forgetfulness and want of friendship. Well I looked into my soul and the retrospect of the past told me that I was sinless, so with all the confidence of outraged virtue I repel the accusation with due respect to you. I sent a parcel of letters to England, one for you, another for George Latimer, and another for Richard Hunter to whom I directed them according to the way he told me, namely, Prescott[8] near Liverpool, enclosing the others in his with directions to forward them to you. Now I suppose that he must not have received them. I sent them a few days after I received your first letter.[9] They were a glorious windfall of nonsense to whoever got them their united weight might have either sank through the post bag (though indeed I did not hear of any post bag explosion at that period) or the sublimity and airiness of the thoughts and ideas contained in them may have soared through the top and disappeared in a smoke. At all events, God forgive the penny post,[10] which I hope will not serve me another trick with this present production. The head and front of my offending hath this extent: No more than a month has slipped over my poor head since the receipt of your last and without answer But my dear friend, for such I will still call you, I will not write to say farewell, for though surrounded by semi savages and unkindred spirits my soul still cherishes the memory of times I spent with you. God knows the only thing I long for here is the possession of a friend such as you were, who could think as I thought and feel as I felt and know and enter into feelings which are quite above the comprehension of those around me. But I must be content with the blessings I possess without seeking for those unattainable. I hope to be able to pay you a visit some time next summer if you remain so long a sapper, if not I may meet you on our own soil. Your description of the mess is very spirited and clever, quite a hallucination of past times. But I am surprised that the daughters of merry England do not please you better if they are better if they are not so exact and guarded as the females on this side, they are at least more – free and bauld,[11] meeting you half way so irresistably as to charm you out of all contenance and to a young gent like you, little more than adolescent, and not much burdened with precise morals – in short not quite a Joseph[12] – such she divinities ought to be very tasty. But I leave you to your morality and in the main I must agree with you that (using a liquorish simile)[13] the iced wine of Ireland is more palatable than the hot lascivious vintage of England. Before you draw a ticket in the matrimonial lottery, for the sake of auld Ireland shift your orb over here and take one of the native plants out of the wild sweet briery fence you mention, and even if you get a few thorns in your carcase in the pursuit it will repay you for your trouble. I will say no more but leave the matter to your own good sense. The reason I did not sooner reply to your last – Illness, long illness. About a week after I received yours I got a very heavy cold and rhumatism in the head which kept me in tortures for some weeks. I need hardly say that I was not in much humour for writing during the time, but at last thank goodness by the help of cooking myself well I have got rid of it and Jack's the lad again. And now to give you the first fruits of my recovery and to let you know all the wonderful events that have happened in this wonderful part of the globe since I last saw you. In the first place it has been raining here for the last three weeks without intermission and we're not full yet. Business is rather dull here and monotony is as much the rider of the day as ever. My partner[14] is quite well and sends her love to you – that is, the little she has to spare. We had a visit from Mr Harris from America, the person to whom John Fudge[15] went. John is conducting his business and makes himself very useful, likes the place very much. Harris gives a flaming fine account of America he says he would engage me 500 a year profit on a capital of 1000 which would be a pretty fair return. John Markham[16] came back from Sydney and brought a fine account of the fortunes he had passed. His journal[17] was very amusing, he has returned there for the purpose of settling, taking his wife[18] and children. George Markham[19] and sons are going there in spring[20] and one of the sisters. Robert Markham[21] who was in the dragoons came home and died after a long illness. We have a Mr Scott[22] here now as preacher he is a respectable one for he has taken Mr Chamber's[23] house on Dean's

Quay, keeps his horse and car and has a very nice woman to wife, is an excellent preacher. The Donaclifts[24] are all well, Sophy as usual thatching[25] the ladies. Her brother has been here this some time. The Batemans[26] and Goods[27] are all well and your ci-devant landlady's family[28] the same – every house in Kinsale stands on its own bottom as usual. We had a great tea meeting here lately, Mr Scott paid the expenses, and we got our Belly full gratis, which was a great matter those hard times. There has been the usual number of individuals coming on the earth, going under ditto, and <u>conjointing</u> in marriage. The scarlatina[29] has been very prevalent here lately. My little responsibility[30] is quite well, she has been on a visit to Clonakilty[31] this some time, she will remain there until after Christmas. She is growing the size of a whale. Myself and consort will pass the Christmas at Clonakilty and the Abbots[32] will come here on little wee Christmas.[33] William Fudger who is my wife's brother will be here he is in the 65 Reg.,[34] in the band, which reg., is lying somewhere near you.[35] I intend if I can to go to Dublin on the end of April and from thence to Liverpool and if possible I will pay you a visit I know you will excuse me for asking you to read this for Latimer and I am sure he will not think me so much in fault as I may have appeared; if I can I will write him a few lines but I hate repetitions which must occur if I write at length to you and him. Poor George and me are as divided in sentiment as the North is from the South, so that there cannot be much community of sentiment between us, which must be always the case between the children of this world and those of the next.[36] He was a great crony of mine until he was snatched as a brand from the burning,[37] and then there was a gulf sunk between us over which I unfortunately could not pass but was obliged to remain in the wilderness of sin and will bad as I am look forward to escaping thence even though I should wander for forty years.[38] God help us, we know the right and the world the flesh and devil holds us in the wrong by every fibre of the heart. I have no associate here now and if I do walk it is by myself. I received your last letter on a Saturday and I did not like to profane it by opening it during the hurry of business, so I kept it inviolate until next day, took a walk to the coast at the mouth of Oyster haven[39] and there, opposite the Sovereign's Islands with the booming sea roaring and its surf dashing beneath me I read it and verily the Water Spirits[40] shouted admirable! – And I there registered a vow that whatever change or fortune might betide me I would hold fast to faith and revive our old association. So my dear friend do not think me capable of breaking a vow made amid nature's thunders. At the time I would rather have had you with me than any gratification that the world could afford me. Dear fellow, if you think my friendship worthy of a further trial you will find that in future I will not be a remiss correspondent. So I will expect a letter from you shortly and send me a more detailed account of old chums. So as the paper is shook I must remain your friend in heart if not in <u>deed</u> and wishing you the compliments of the season and a good appetite for the English roast beef and pudding[41] and

remain yours very very sincerely | J. Chadwick.

RI MS JT 1/11/3499–500
LT Transcript Only

1. *manes*: the spirit or shade of a dead person (*OED*).
2. *Erebus*: in Greek mythology a place of darkness in the underworld where the dead were received immediately after dying.
3. *Bagby*: probably Francis Begbie (b. *c.* 1808), a Private in the 13th Company of the Royal Sappers and Miners who worked alongside Chadwick in the 4th Division, C District of the Irish Ordnance Survey (NAI OS/1/19).
4. *your poetical morceau*: A morceau is a short literary or musical composition (*OED*), so this is likely the poem 'To Chadwick' composed on 1 November 1842 (RI MS JT 8/2/1/8–9).
5. *n'importe*: no matter (French).
6. *'the sisters three'*: the Muses; see letter 0174, n. 11.
7. sans: without (French).
8. *Prescott*: see letter 0150, n. 15.
9. *your first letter*: letter missing.
10. *penny post*: The Uniform Penny Post, which fixed a standard rate for the carriage and delivery of letters weighing less than half an ounce between any two places in Britain and Ireland regardless of distance, was established by an Act of Parliament in 1840.
11. *bauld*: Scottish and Irish variant of bold (*OED*).
12. *Joseph*: the husband of Mary, the mother of Jesus. The Roman Catholic doctrine of Mary's Perpetual Virginity entails that Joseph never had sexual relations with his wife.
13. *liquorish simile*: a simile relating to liquor or alcoholic drinks.
14. *My partner*: Chadwick's wife; see letter 0148, n. 12.
15. *Mr Harris from America ... John Fudge*: not identified.
16. *John Markham*: Thomas John Markham (b. *c.* 1817) who had left Liverpool in June 1841, travelling to Sydney as ship's surgeon on the *Livingstone*, and arriving in October. Because an elder deceased brother had also been called Thomas, in Kinsale he was known as John.
17. *His journal*: published as *Dr Markham's Antipodean Voyage: The Journal of Dr Thomas John Markham on his 1841/42 Voyage as Ship's Surgeon from Liverpool to Sydney & Return*, ed. by T. Winter, (Melbourne: Destination Universe, 2010).
18. *his wife*: Elizabeth (née Donaclift).
19. *George Markham*: probably John's younger brother George (1822–64).
20. *going there in spring*: rather than Sydney, the Markhams relocated to Liverpool; see letters 0199 and 0205.
21. *Robert Markham*: probably John and George's elder brother Robert (b. *c.* 1815); or possibly the Robert Markham who, from 1819, was Cornet by purchase in the 2nd Regiment of Dragoons, and was then promoted, in 1823, to Lieutenant in the 9th Regiment of Dragoons (*A List of the Officers of the Army and of the Corps of Royal Marines* (London: War Office, 1824), p. 136).
22. *Mr Scott*: probably Reverend Dr Robinson Scott (d. 1883), a Methodist preacher or-

dained in 1835, who was appointed to the Bandon circuit in 1839 before being made governor of the Old Connexial School in Dublin in 1845.

23. *Mr Chamber's*: probably James Chambers, principal coast officer and assistant collector, Customs House, Kinsale (*Slater's*).
24. *The Donaclifts*: probably Samuel Donaclift, a shipwright who worked on Kinsale Harbour, and his wife Sophia (née Leggatt).
25. *thatching*: not identified.
26. *The Batemans*: probably Michael and Mary Bateman, publican and shopkeeper, of the Glen, Kinsale (*Slater's*).
27. *Goods*: probably William Good, boot maker, of Main Street, Kinsale (*Slater's*).
28. *your ci-devant landlady's family*: Chadwick presumably means Mrs Cotter, Tyndall's erstwhile landlady at Pine Street in Cork, so this is possibly William Henry Cotter, the comptroller of the Customs House in Kinsale (*Slater's*). Ci-devant means former, especially an erstwhile noble following the French Revolution (*OED*).
29. *scarlatina*: scarlet fever (*OED*).
30. *My little responsibility*: Chadwick's baby daughter, now nine months old.
31. *Clonakilty*: a small town in County Cork.
32. *the Abbots*: not identified.
33. *little wee Christmas*: traditional Irish name for the Feast of the Epiphany on 6 January.
34. *65 Reg.*: the 65th Regiment of Foot, an infantry regiment of the British Army.
35. *which reg., is lying somewhere near you*: Since returning from Canada in 1841, the 65th Regiment was stationed at the Salford Barracks, where it was deployed in containing Chartist riots in Lancashire.
36. *the children ... the next*: adaptation of Luke 16:8.
37. *a brand from the burning*: Zechariah 3:2.
38. *wander for forty years*: like the Israelites wandering in the desert before entering the promised land (Numbers 33:38).
39. *Oyster haven*: a sea-inlet on the coast of County Cork, east of Kinsale. The Sovereign Islands sit just outside the inlet.
40. *Water Spirits*: possibly a reference to the beautiful female sirens of Greek mythology.
41. *the English roast beef and pudding*: see letter 0157, n. 5.

To John Tyndall, Snr 3 December 1842 0178

Preston, | Wednesday Dec. 3rd, 1842.

My dear Father,

Expecting every day to hear from you, I was induced to defer writing myself until now. I have made enquiries respecting Col Austin[1] of Fishwick Hall[2] and Mr Marshall[3] of Penwortham Hall.[4] I find that they are both acquainted with Captain Tucker and am inclined to think that a word from either would serve me much. It would at all events induce the Captain to give me a good testimonial in the event of my leaving him.[5] He, poor man, is greatly afflicted at present. He had a brother, a Captain in the Navy,[6] who was shipwrecked and lost on the coast of France about a month ago[7] I don't know whether the body has been found or not.[8] Having thus disposed of the more important portion of my ideas,

I'll relate a rich incident which took place here last week. I saw it in Saturday's Chronicle.[9] A poacher happened to go into a tavern which he was in the habit of frequenting, he had a basket on his arm and in it a hare covered up. Having called for a glass of ale he proceeded very quietly to sip it. The landlord came into the kitchen where the chap was seated, and – knowing that he had the hare along with him – thought he, 'It would be a great job if I could pin it.' He chanced to have a dead cat in the house and, unnoticed by the poacher, he slipped the hare out of the basket and put the cat in its place. When the fellow had taken the ale he took up the basket and proceeded home. 'Eh!' said he to his wife 'I've brought thee a hare – a fine puss' – she uncovered the basket and exclaimed 'a cat'! 'Yea and a vonny one too' said he, little suspecting the trick that had been played on him. The wife at length lugged it out by the tail. 'I'm done' said he, and again basketted the cat and proceeded once more to Preston. He went straight to the public house. The landlord seeing him coming kept out of the way. Well, the fellow entered the kitchen as before, and called for another glass of ale; while the waiter was away getting the beverage, the fellow went over to the kitchen fire and looking into a boiling pan he saw a smoking leg of mutton. To haul it out was the work of a minute, and in he popped the cat to supply its place. Shortly after the maid brought the master and mistress a bowl of broth each. The fellow remained until he heard the former complaining of the broth and the loving spouse recommending ketchup to improve the flavour, when he thought it high time to cut his stick. That day himself and his family fared sumptuously on the leg of mutton.

Love to mother and Emma | Your affectionate son | John.

RI MS JT 1/10/3283
LT Transcript Only

---

1. *Col Austin*: The *Preston Chronicle* reported that 'Colonel Austin' was one of the 'justices of the peace for this county [i.e. Lancashire]' ('Extensive Moutre Case', 20 April 1839, p. [2]), and later listed him as a member of the Board of Guardians administering the Poor Law in the city ('Board of Guardians', 15 August 1846, p. 6).
2. *Fishwick Hall*: Fishwick is a town on the River Ribble, a mile south-west of Preston. Fishwick Hall was originally the seat of T. R. Shaw, but was presumably now owned by Colonel Austin (S. R. Clarke, *The New Lancashire Gazetteer* (London: Henry Teesdale, 1830), p. 52).
3. *Mr Marshall*: William Marshall (1796/7–1863), director of the Lancashire and Yorkshire Railway Company.
4. *Penwortham Hall*: Penwortham is a town on the River Ribble, 2 miles south-west of Preston. Penwortham Hall, originally called Penwortham Lodge, was built in a neo-classical style in 1801 by John Horrocks, owner of the largest cotton firm in Preston, whose son Peter sold it to William Marshall in 1829.
5. *in the event of my leaving him*: to go to a job on the Tithe Survey in London; see letter 0173.
6. *a brother, a Captain in the Navy*: Captain William Tucker RN (1795–1842), who joined

the Navy in 1810 and, by the time of his death, had risen to commander of HMS *Iris* and senior officer of Her Majesty's ships and vessels on the western coast of Africa ('Memoir of the Late Captain Wm. Tucker, R.N.', in *The Naval and Military Sketch Book* (London: Hugh Cunningham, 1845), pp. 404–7).

7. *shipwrecked ... about a month ago*: On 12 November 1842 the *Reliance*, an East Indiaman carrying a cargo of tea bound for Gravesend, ran aground off the coast of Boulogne with the loss of more than a hundred lives. Captain Tucker had joined the *Reliance* at St Helena as a passenger to return to Britain because of illness ('Loss of the Reliance', *The Times*, 17 November 1842, p. 5).
8. *I don't know ... found or not*: On 23 November 1842 *The Times* reported that 'only 12 bodies of the 109 persons lost' had 'been picked up', and noted: 'Numbers of the crew sank immediately, and were entirely lost sight of amidst the breakers; Captain Tucker was among this number' ('Loss of the Reliance', p. 3).
9. *Saturday's Chronicle*: 'A Cat-astrophe', *Preston Chronicle*, 3 December 1842, p. [2].

From John B. Edgeworth 15 December 1842 0179

Cove[1] | Dec. 15th 1842

My dear Tyndall

I dare say you have been surprised at my not writing to you for so long a time but the truth is that I have been rather busy lately reading for and putting in my examinations in college add to that, that I have as great a hatred to writing as ever a good orange man had to the Pope[2] I dont suppose I wrote half a dozen letters since I heard from you last – I have now something to ask you to do for me the first little spare time you have it is to draw for me a block of houses with yards walls and a street and a shed all coloured as the Cork City streets and also written on a slip of paper the colours you use in washing them in and if you would send me a river coloured, with mud banks at the sides the way it was drawn in the Cork plans I hope I am not giving you too much trouble – I have left Cork and am now staying at Cove with my Father – Circumstances have occurred since I last wrote which prevent me joining at present the English survey of which I am very sorry as I expected much improvement and pleasure from it. Major Waters has left Cork I suppose you heard that he has been ordered to Barbadoes;[3] I believe I never thanked you for your long account of the Preston riots[4] but I now give you my best thanks they afforded me a great deal of pleasure Will you send me just one line in acknowledgement of this and tell me if you have time to do the little or rather great job that I want, remember me to all that remember me at Preston

and believe me | very sincerely and faithfully | yours | J. B. Edgeworth.

RI MS JT 1/11/3560
LT Transcript Only

1. *Cove*: see letter 0068, n. 3.

2. *a hatred ... to the Pope*: Members of the Protestant Orange Order, founded in 1796 in County Armagh, were notorious for their animosity towards 'Popery'.
3. *Major Waters ... ordered to Barbadoes*: After Major Marcus Antonius Waters left the Ordnance Survey in Cork in November 1842 he served with the Royal Engineers in the West Indies from 1843 until 1849.
4. *Preston riots*: see letter 0164, nn. 2 and 3. The letter to Edgeworth with Tyndall's 'long account' of the riots is missing. Much later, Tyndall recalled of the riots: 'processions ... filled the streets – crowds of shiftless and hungry men who had been discharged from the silent mills. In their helplessness and misery they had turned out, so that their condition might be seen of all. Well, in Lune Street, down which we could look from our office, the tumult one day became unmanageable. Heated by its own interaction and attrition, the crowd blazed out into open riot, and attacked the bakers' shops. Soldiers had been summoned to meet this contingency. Acting under orders, they fired upon the people, and the riot was quelled at the cost of blood' ('On Unveiling the Statue of Thomas Carlyle', *New Fragments* (London: Longmans, Green, 1892), pp. 392–7, on pp. 392–3).

From John Lilly 24 December 1842 0180

2 Belview Terrace | Pendleton | Manchester | 24th Dec. 1842.

Dear John

It was with feelings of the most unfeigned pleasure I had the honour of receiving your letter[1] quite unexpected. But even that pleasure had its attendant pain, for in the said letter I see recorded in plain and legible characters another instance of that boy Willy's[2] dereliction from the path of virtue. What will become of him? Let us, having the past in view, dive as it were into the dark vista of futurity and let imagination fill up the picture and lo what a spectacle is there pourtrayed in indelible and imperishable characters! My mind recoils from the thought and as Virgil[3] has it (si bene recordor)[4] 'Vox haesit in faucibus: comæ staut in capiti'[5] But avaunt[6] such thought being only calculated to imbue the mind with gloomy forebodings quite foreign to our subject – however I beg you will keep a strict watch over that boy. I had two letters from Bloomfield – he appears well pleased with his change. He was when I last heard from him drawing one of the large plans of York. I have reason to think he is still the same Bob. I also had a letter from our much respected host at Sarsfield Court,[7] he was then quite well and was going on a visit to the Queens Co.[8] From the sketch you have given me and from other quarters it appears that you have not a very choice selection in your office.[9] I can without any reserve say quite the same. I really envy you, for in Preston you have some two or three respectable ones, but lo such a collection of boors as we have gotten[10] no one ever before saw together: You may form such an opinion of them as you can when Mr McKerlei[11] designated them as like a parcel of Irish pigs and sent two or three of them home to wash their hands. Of this satis est.[12] you may fancy what my feelings are when, ever and anon, I recall to mind the pleasant nights spent in Cork when the lively temperament of Tyndall set us all in a roar aided by that other gay soul Bloomfield. And never till memory ceases to do her duty shall the night at Sarsfield's Court be effaced from my recollection, when John T. with stentorian voice[13] such as threatened a total annihilation and devastation of that part of our corporeal frame called the lungs would maintain that 'you'r wrong Sir' after having crushed with the force of argument that emporium of knowledge Mr. Moorough.[14] You ask my opinion of the English ladies and in answer I beg to say that your description of them perfectly coincides with what I am in the habit of hearing, but I cannot give my opinion for you must be aware that I am no admirer of the fair sex but I shall bow with the utmost deference, as Willy was wont to express himself, to your opinion for I know no other more competent to give one. You who were the centre of attraction of some 5 or 6 of the belles of Cork, contrary to the laws of your gravitation; but such a phenomena is easily accounted for when we recollect that bright halo which emanated from and surrounded you. I would advise you to be very careful lest those same ladies (English) whom you describe as so warm, would not eventually be taking improper liberties: I also recommend not to let them engage you so much as to cause you to turn from that high, exalted, but still arduous, pursuit after knowledge which has hitherto characterised your life and the fruits of which are daily evinced in the respect and admiration which you elicit from all who have the happiness and honour of your acquaintance. I hope and expect that our letters shall be neither few nor far between but that a mutually profitable series of letters shall be the result. You will please excuse this attempt at a letter but we shall in time improve. Have you heard how the Cotter family[15] are? Please remember me to Willy and ask him if he would be kind enough to write when not immediately tripping it on the light fantastic toe in the mazes of a quadrille or Masourka[16] or else.

I am yours sincerely | J. Lilly.

Direct to me | 2 Belview Terrace | as on first page.

RI MS JT 1/11/3748
LT Transcript Only

1. *your letter*: letter missing.
2. *that boy Willy's*: probably William Hunter, who was regularly referred to as Willy and had a taste for mischief and womanizing; see, for instance, letter 0150.
3. *Virgil*: Publius Vergilius Maro (70–19BC), an ancient Roman poet best known for *The Aeneid*, a Latin epic of the origins of Rome.
4. *(si bene recordor)*: if I remember well (Latin).
5. *'Vox haesit ... in capiti'*: Lilly slightly misremembers the line 'arrectaeque horrore comae et vox faucibus haesit' (his hair stood on end with horror, and his voice stuck in his throat) from *The Aeneid*, IV.280, which describes Aeneas's fearful response to a warning from the god Mercury.
6. *avaunt*: be gone (*OED*).

7. *Sarsfield Court*: a large house near Glanmire in County Cork. Charles Putland, Jnr was recorded as the occupier between 1847 and 1864, holding the house from his father Charles Putland, Snr (*Griffith's Valuation*). Either father or son may be the 'host' Lilly recalls.
8. *Queens Co*: Queen's County, now County Laois.
9. *your office*: see letter 0171, n. 2.
10. *as we have gotten*: in the Divisional Office of the 4th Division of the English Ordnance Survey in Manchester.
11. *Mr McKerlei*: John Graham McKerlie (1814–1900), who joined the Royal Engineers in 1833 and was stationed in Mauritius for several years. On returning to Britain, he was employed on the English Survey, where he served in the 4th Division, and commanded the Divisional Office in Manchester. After reaching the rank of Colonel, he was appointed a Commissioner of the Board of Public Works in Ireland in 1855.
12. *satis est.*: is enough (Latin).
13. *stentorian voice*: loud and powerful, like that of a stentor; see letter 0169, n. 15.
14. *Mr. Moorough*: not identified.
15. *the Cotter family*: the family of Mrs Cotter, Tyndall's erstwhile landlady at Pine Street in Cork.
16. *a quadrille or Masourka*: types of dance, both having a lively tempo.

From William Ginty [27] December 1842 0181

Kirkby Lonsdale | Westmoreland | Tuesday evening.

My dear Tyndall,

A few moments before I started to Sedbergh[1] on last Saturday I received yours.[2] I'm glad you liked the 'Album's pages'[3] so much – as nearly all the poems in the album were original (are truly original) I felt myself in duty bound to be an original too – but in good truth I have no ambition to shine amongst the 'Poetical Luminaries' of Westmoreland.[4] I am sure the same album would amuse you more than any you ever seen – for at least one ½ hour. You'd 'start, for Soul is wanting there'![5]

I spent a most miserable Xmas[6] in Sedbergh. I never want to visit that city of the wilds again. I was ready to curse myself on Sunday evening for not accepting 'HOWA JOE'S'[7] invitation to dinner. What do you think they gave me for dinner at Tidmarsh's and Latimer's – <u>A piece of salmon</u> a pretty dish for a winter's day and a Xmas day (it was a 'spent fish'[8] of 15lbs weight that they got for 3d per lb.)[9] But 'oh, be de holy, be de holy!' the lodgings exceeded my most sanguine expectations of rustic simplicity and unsophisticated hospitality. I got the one third- of a bed of chopped straws of about 1 inch long which ever and anon dealt most unmerciful wounds on my outward man thro' the thread-bare sackcloth which formed a most suitable reconnoitring ground for a few millions of the <u>bloody</u> 'black horse'.[10] After spending an hour or so in this position, I very deliberately got up and made my clothes supply the place of a suit of armour.

I smoked, read and strode about the room until the farthing rush-light shed its last halo within the precints of the old saucepan (which in the true spirit of invention was made to supply the place of a candlestick) and then with a heavy heart I betook myself once more into the scene of strife and bloodshed. But sleep was out of the question – the rain dashed furiously against the shattered window casement, the wind moaned horribly thro' the antique chimney funnel, add to this the piteous lamentations of Tid[11] and Bill[12] as they held their post mortem examinations over a few of the discomfited enemy, and you will allow that I was in no enviable situation. Monday morning saw me tramping the road to Kirkby thro' rain and storm – alas too fully persuaded that I had been 'too far north'

To lie on straws– to gaze around one's cell
To slowly trace the ragged bed-clothes scene
Where things that own not man's dominion dwell
And mortal leg hath ne'er escaped their spleen
To feel them climb upon thy ribs unseen
A wild wild flock that never needs a fold
Alone with these to yell the piteous 'keen'[13]
This is not solitude; tis but to hold
Converse with Sedbergh's charms and feel its stores unrolled.

As Jim[14] is 'pouring aloft a flood of melody' o'er the beauties of Child Harold[15] you had better recommend this new edition to his soaring attention such a critique as him will admit that adversity hath its charms as well as solitude.

So Mr Taylor is enjoying all the sweets of Christian ennui in L.Pool.[16] And you wanted me to insult the benign young proteus[17] in theology by writing to him. Well! well! If I required evidence to prove that you were in Co.[18] Carlow when I embarked at Cork[19] this would be most irrevocable decisive, but He was there, and well I recollect how I then admired his 'meekness of spirit'[20] and 'Christian humility' as he paced the quarter deck and poop of the 'Ocean'[21] during our passage to Passage[22] "The scene and hour &c.' Enough! 'Well I will dream that we may meet again.'[23]

I wouldn't doubt Messrs Hunter and Latimer for an exposé of my affairs in Kirkby[24] – but I 'reckon nowt[25] on it'

The rose will cease to blow
The Eagle turn a dove
The streams will cease to flow
Ere I will cease to love.[26]

My soul's bright satellite sheds her luster at present in Kendal[27] I feel indebted to you for your kind offer with respect to your Kinsale correspondence. Remember me to Chadwick that's all! Tell the little beggar that I said that he should read

'Young's night thoughts'[28] set to the music of his nasal organ. – Heavens! What a beautifully sublime and harmonious contrast! – Snores verses Sublimity.

I had a letter from Payne a few days ago in which he says that you sent him a very fine description of your mess in poetry, 'but' says he 'George Latimer's nose was the most prominent feature in it'

I wish this Houri[29] of mine would come home. I feel most uncomfortably lonesome. Kirkby is no longer a lovely spot at least until she returns.

I love the brilliant courtly scene
I love the groves delightful green
The fountain and the bright cascade
The rose-wreathed bower and gentle shade
But palace fountain grove or grot
Can never charm where she is not.

Jack[30] is not in Hornby![31] He wrote to you on Sunday last[32] and of course gave you all the particulars of his situation. They can buy very excellent cheese for 2½ and 3d per lb.!!! Do you ever write to 'Xina'[33] You and her would do well together now that you are so deep read in the French, let me see Mr and Mrs T.[34] begs to announce to the nobility and gentry &c. Oh it would do remarkably well!!! Ahem! Mr Pedagogue ye would surely eclipse the luminarys of modern times in the Belle Lettres[35] and the fine arts. Faith Tyndall I'm serious!!!

While I was in Sedburgh the lads there received a communication apprizing them of their removal to L.Pool in a few days. Its curious I have not yet got such, it looks as if they intended to leave me out (but that I dont care for) and yet it is very improbable, as they are surveying and I measuring they of course require earlier notice – that may account for it, but if it is really a fact that I am to be left out – so let it be with all my heart

Oh! there is sweetness in the mountain air
And life that bloated ease can never hope to share[36]

I am beginning to grow a little sleepy and the post is nigh closing and so I will bid you good night. Remember me to the mess

Ever yours | Ginty.

RI MS JT 1/11/3587
LT Transcript Only

---

1. *Sedbergh*: a town in the West Riding of Yorkshire (now in Cumbria), about 10 miles from Kirkby Lonsdale.
2. *yours*: letter missing.
3. *the 'Album's pages'*: possibly related to 'Mary's album' with 'bright and lovely pages' mentioned in letter 0176. Gift albums of anthologized poetry, to which purchasers could add their own, were a fashionable publishing format in the early nineteenth century.

4. *the 'Poetical Luminaries' of Westmoreland*: Ginty is probably alluding to the Poet Laureate William Wordsworth (1770–1850), who lived in Rydal, Westmoreland, and, from 1813, served as Distributor of Stamps for the county.
5. *'start, for Soul is wanting there'*: Lord Byron, *The Giaour: A Fragment of a Turkish Tale* (1813), 93.
6. *Xmas*: Christmas.
7. *'HOWA JOE'S'*: possibly Joseph Collins.
8. *'spent fish'*: having recently spawned, and thus in a poor condition (*OED*).
9. *lb*: pounds; a unit of weight equal to 16 ounces.
10. *the* bloody *'black horse'*: presumably black horse flies (*Tabanus atratus*).
11. *Tid*: John Tidmarsh.
12. *Bill*: William Latimer.
13. 'keen': an Irish funeral song accompanied with wailing in lamentation for the dead (*OED*).
14. *Jim*: Phillip Evans.
15. *Child Harold*: Lord Byron's *Childe Harold's Pilgrimage* (1812–18).
16. *L.Pool*: Liverpool.
17. *proteus*: a sea or river god in Greek mythology.
18. *Co.*: County.
19. *when I embarked at Cork*: on 20 May 1842; see letter 0144.
20. *'meekness of spirit'*: adaptation of 1 Corinthians 4:21, 'spirit of meekness'. Meekness is a Christian virtue, like 'Christian humility' that follows.
21. *'Ocean'*: see letter 0143, n. 32.
22. Passage: see letter 0162, n. 2.
23. *'Well I will ...* meet *again'*: Byron, *Childe Harold's Pilgrimage*, II.ix.77.
24. *Kirkby*: Kirkby Lonsdale.
25. *nowt*: nothing (Yorkshire dialect).
26. *'The rose will ... cease to love'*: a ballad by the American composer Robert Guylott, first published in the 1830s.
27. *Kendal*: a market town in Westmoreland, about 14 miles from Kirkby Lonsdale.
28. *'Young's night thoughts'*: *The Complaint: or, Night-Thoughts on Life, Death, and Immortality* (1742–5) by Edward Young (1681–1765).
29. *Houri*: nymph of the Muslim Paradise, applied allusively to a voluptuously beautiful woman (*OED*).
30. *Jack*: John Tidmarsh.
31. *Hornby*: see letter 0176, n. 7.
32. *He wrote to you on Sunday last*: letter missing.
33. *'Xina'*: Christina Tidmarsh.
34. *Mr and Mrs T.*: Mr and Mrs Tidmarsh.
35. *Belle Lettres*: elegant or polite literature (*OED*).
36. *'Oh! there is ... hope to share'*: Byron, *Childe Harold's Pilgrimage*, I.xxx.349–50.

1843

From Thomas Charles Higginson28 January 1843 0182

January 28th. 1843.

My dear Tyndall

When I left England I thought I had a friend in Mr Tyndall whom neither time nor space could change. I am afraid I have been deceived. I hope I have not but it looks very like it not to have received any letters during the time I have been in this country.[1] I must acknowledge that when I was leaving England I wrote you a letter stating that when I arrived here I would write to you via Leighlin Bridge,[2] but I then considered it better to delay doing so until I heard from you as there are so many Tyndalls there it might not have reached you. I am afraid nothing you can say in extenuation of your crime can erase the bad impression which is fixed in my mind. Mind you, although I have lost – (unconscious of the cause) the space in your affection which I flatter myself I once occupied, still the memory of the friendship which did exist between us can never be obliterated from my memory and, although years may pass without the pleasure of my seeing you, although it may not be mutual, still I think (although it might give pain to write me a letter now and again) you should suffer that much for old friendship's sake. I did not think Tyndall capable of such a thought as 'Oh he is now an officer what does he care about corresponding with me' Such things ought to be above the thoughts of John Tyndall and especially of me. I would have written this by the last Mail but was attacked with fever – I do not know what news to send that would interest you – The war for which we were sent out here[3] is all settled and here we will remain for three years. On our arrival here we lost 28 men by cholera I also got it but, thank God, soon recovered. To a European this place[4] looked very curious. The officers bungalows are built in a line only one story high, with gardens of about an acre each round them. The native city at a short distance cannot be seen the houses are so small, being just like mud cabins in Ireland, and so closely planted with trees that it has the appearance of a wood. The climate of the place is delightful for half the year, the nights are so cool that we are very glad to put on as much clothing as possible and in the day time except out in the heat of the sun I wear cloth trowsers. What do you think, I met Wm. Brunnan[5] that was discharged from the Survey in Middleton[6] at the time of the reduction; here he is enlisted in the company's service.[7] I just met him by chance being at a court martial he conducted a prisoner to it – George Holmes in his last letter told me you were in Skipton[8] so I will direct this there. I dare say you won't be sorry to hear that I am pretty well off here. I have £18 per month. I am studying very hard the Hindoostanae language.[9] I am now only 5 months at it and can speak it fluently. I will pass an examination in May, and then I will have £10 a month more as Interpreter of our Regt.[10] Remember me to Jim, Tid,[11] Ginty, Cuddy and all the old friends I had on the Survey

Your ever affectionate but neglected | friend T.C. Higginson

P.S. I am sorry I cannot pay the postage here

RI MS JT 1/11/3733
LT Transcript Only

---

1. *this country*: India.
2. *a letter stating ... write to you via Leighlin Bridge*: letter missing.
3. *The war for which we were sent out here*: The First Anglo-Afghan War fought between the British East India Company and Afghanistan from the spring of 1839 to the summer of 1842. The heavy British casualties suffered during the retreat from Kabul in January 1842 led to reinforcements being rushed to India, although by the time they arrived a shift in British foreign policy meant that the war had been abandoned.
4. *this place*: the military cantonment to the east of Poona (now Pune), a city in the Bombay Presidency of the East India Company; see letter 0190.
5. *Wm. Brunnan*: William Brennan, a civil assistant in the 5th Division, C District of the Irish Ordnance Survey. He joined the Survey in March or April 1839, and was discharged in January 1841 (NAI OS/1/16–18).
6. *Middleton*: see letter 0032, n. 9.
7. *enlisted in the company's service*: in the Honourable Company Army, the private army of the East India Company.
8. *Skipton*: see letter 0096, n. 3. George Holmes was probably in error, as there is no record of Tyndall spending any time in the town.
9. *the Hindoostanae language*: the main language of northern India, incorporating Hindi and Urdu.
10. *our Regt.*: the 78th Regiment of Foot, Royal Artillery, which Higginson joined on 10 April 1842 (W. H. Askwith, *List of Officers of the Royal Regiment of Artillery From the Year 1716 to the Year 1899*, 4th edn (London: Royal Artillery Institution, 1900), p. 172).
11. *Jim, Tid*: Phillip Evans, John Tidmarsh.

From William Ginty 30 January 1843 0183

110 Pitt St. Liverpool | Monday Night

My dear Tyndall

Now! by the merry peal of 'Bob Major'[1] that thunders its loud echos from yonder tower of St. Nicholas,[2] this is too bad. By all the raven plumes, glossy ringlets, rosy cheeks, merry hearts, and laughing eyes that issue in babylonish confusion from the massive portals of its hallowed aisles, I like the quondam Cupid[3] have been 'treated foully'. God Forgive you Tyndall! what could have tempted you to such a cruel and heartless act – as to expose an unfortunate devil in such a public manner.

'On leaving Westmoreland by W. Ginty'[4] On my soul, your motives for this caper is so entirely incomprehensible and mysterious to me at present that I hardly know what to say or think. When I saw my name in the paper this morning, like Robinson Crusoe 'I stood as one thunderstruck'.[5] but faith we four have come to this unanimous conclusion that you are 'rather insane'. Though I must say your damnest cunning letter[6] to Jack[7] did not savour much of insanity – Heavens how you have praised it! My composition. (I've been desired to ask you this question: Could you get no one to praise you but yourself?) However I have a good deal to write and I must cut this subject short. EXPLAIN! EXPLAIN!! EXPLAIN!!! You want the paper back – as you got but two and one of them you sent where it will not be returned from. not to Kirkby, I hope? However Sinnet's there. I need not say that I will contradict it very shortly – not through the paper tho! but faith I will contradict it – forgery – forgery! defamation! action! Law! Law! Law!

Apropos – it is undoubtedly the cleverest piece I have seen of yours. There are some of the finest expressions of ideas in it I have ever seen. What a damned fool you were to disgrace the piece by crowning it with my name! A thought has struck me, by heavens, to the very core. As I live, I have sat and read the 'Preston Chronicle' in the Wheat Sheaf Inn, KY Lonsdale[8] – It is taken in there with some others, and She[9] is there. This may be fine sport for you, but in good truth my disgust for the person drives any ambition which I might have to be thought the Author of such a very Gallant piece into air 'thin air!' Well! Well!

But enough. I suppose you expect something about the Temperance blow out:[10] –

That theme, dear Sir, I must decline
Yet faith I'm ever since admiring
Its happy traits, tis thine! tis thine!
Oh powers! were I but you or Byron,[11]
I'd paint each worthy's glowing hue
I'd wreak my thoughts in such expression

And give to each and give to you
The father of an Irish 'dressin'[12]
–being such an utter stranger to the Dramatis personæ[13] it would be a task for which, I confess, I am alas too conscious that I would be most infernally inadequate!

A trifling circumstance took place in the office last Friday night and has been on the tapis[14] up to this evening, which sheds a most discreditable halo over the Ordnance Confederacy at Liverpool. AN ATTEMPT at ROBBERY in the ORDNANCE SURVEY OFFICE. On Friday evening last, after the minions of the Survey had retreated to their homes, the desk of Corpl. Lyndall[15] (only a letter) who has the payment of the department, was very curiously forced open with the poker. But very fortunate for him he had removed some 20 pounds to another chest or press in the forenoon and so the villain was disappointed. Of course it was reported to Hamley on his arrival next evening and he kicked a queer shine about it – had the opinion of a policeman on the spot and held a regular Court Martial on a Mr Finnican[16] who, poor fellow, was so unfortunate as to go to the office that evening for a newspaper he forgot in his drawer up stairs. It was proved he was not there 5 minutes – However after examinations, cross examinations, and cross cross examinations, nothing would do our gallant commander but to have two constables sent for to day to escort the poor young fellow before the Major[17] (luckily no constables could attend to make a public exposure).[18] However the magistrates laughed at the grounds of suspicion, and said there was more grounds for suspecting the messenger and his wife and their lodgers and visitors!!! I can say nothing as to the probable result – but this I can say. I wish to heavens I was out of this. Finnican is a general favourite.

Being in a roving humour on last Wednesday evening, I resolved to mingle for a few minutes among the Beaus[19] and fashion of Liverpool and so adjourned to Lord Street. I was not there long when I took a most hearty fit of laughing on meeting Mr D. Dempsey[20] promenading thro' the midst of a sabaoth[21] of angels, with a cigar stuck in his face while the iron-shackled foot acted the part of a pair of cymbals on the devoted pavement which seemed conscious of the load of consequential dust that it was doomed to sustain.

Tell George, Bill[22] got two letters, one from him and one from home. Direct to the office as the Mails are delivered during office hours. Excuse this Jack – I'll send you a better one next 'time'.

(How the devil is Mr Marquis's if not his name it should be his title.) Give my respects to all the Milesians[23]

Yours &c | Ginty | A Patlander[24]

Dick[25] and I are fairly incorporated among the draftsmen. We are plotting a plan each, also Taylor – all are in one room. Kennedy[26] seems to be a very nice gentlemanly fellow. They have another office taken, this one is not large enough and so those that are to be employed at the town are to be all in one house.

Write soon.

Evans owes me a letter. Kennedy is a Protestant. I heard he was a papist. There are 11 protestants in our room – 3 of which are Orangemen[27] – without one black sheep – that same is a trifling consolation. Send me a copy of that last piece of yours in the Sentinel.[28]

Mr. Tidmarsh was here yesterday, he expressed his sorrow that Preston was so far out of his way as to prevent him from going to see you.

RI MS JT 1/11/3591
LT Transcript Only

---

1. *'Bob Major'*: a traditional composition for church bells first rung in 1725.
2. *yonder tower of St. Nicholas*: the Church of Our Lady and St Nicholas, the Anglican Parish Church of Liverpool, whose tower, built in 1815, was the tallest building in the city at this time.
3. *quondam Cupid*: presumably Phillip Evans; quondam indicates the former holder of an office or position (*OED*).
4. *'On Leaving Westmoreland by W. Ginty'*: 'On Leaving Westmorland. By W. Ginty', *Preston Chronicle*, 28 January 1843, p. [4].
5. *like Robinson Crusoe 'I stood as one thunderstruck'*: the moment in chapter eleven of Daniel Defoe's novel *The Life and Adventures of Robinson Crusoe* (1719) when, as the lone castaway records, 'going towards my boat, I was exceedingly surprised with the print of a man's naked foot on the shore, which was very plain to be seen on the sand. I stood like one thunderstruck, or as if I had seen an apparition'.
6. *your damnest cunning letter*: letter missing.
7. *Jack*: John Tidmarsh.
8. *KY Lonsdale*: Kirkby Lonsdale.
9. She: Mary Edwards. The poem that Tyndall published under Ginty's name is addressed to 'My Mary', one of the 'Windermere Beauties'.
10. *Temperance blow out*: not identified.
11. *Byron*: George Gordon, Lord Byron (1788–1824).
12. 'dressin': drubbing, beating; chastisement, castigation, by blows or words (*OED*).
13. *Dramatis personæ*: the characters of a drama or play; the actors in a drama (*OED*).
14. *on the* tapis: on the table-cloth, under discussion or consideration (*OED*).
15. *Corpl. Lyndall*: Corporal Samuel Lyndall (b. 1802/3) of the 16th Company of the Royal Sappers and Miners. He had enlisted in January 1827, and worked as a Private in the 3rd Division, B District of the Irish Ordnance Survey, before being promoted to Lance Corporal in August 1839. Lyndall was transferred to England in November 1841, and was there promoted to 2nd Corporal in January 1842 (NAI OS/2/16–18). He was the paymaster of the 5th Division of the English Survey in Liverpool, before joining the 1st Division in Preston (NA OS 3/412).
16. *Mr Finnican*: not identified.
17. *Major*: possibly the Master-General of the Ordnance, George Murray, although his military rank was actually Lieutenant-General, having been promoted from Major-General in 1814.
18. *(luckily no constables could attend to make a public exposure)*: certainly the robbery was

not reported in the *Liverpool Mercury*.

19. *Beaus*: men who give particular, or excessive, attention to dress, mien and social etiquette (*OED*).
20. *Mr D. Dempsey*: Denis Dempsey, a civil assistant in the 2nd Division, C District. He seems to have joined the Irish Survey in November 1837, and was transferred to England in May 1842 (NAI OS/1/14–19).
21. *sabaoth*: Hebrew word for armies or hosts (*OED*).
22. *George, Bill*: George and William Latimer.
23. *Milesians*: the last invaders of Ireland in Irish mythology, an Iberian race of Gaelic speaking Celts descended from the warrior king Míl Espáine.
24. *Patlander*: Irish person (*OED*).
25. *Dick*: Richard Hunter.
26. *Kennedy*: probably Michael Kennedy, a civil assistant in the 1st Division, C District, who seems to have joined the Irish Survey in August 1839 and been transferred to England between July and November 1841 (NAI OS/1/16–18).
27. *Orangemen*: members of the Protestant Orange Order founded in 1796 in County Armagh.
28. *that last piece of yours in the Sentinel*: possibly the 'bit of a rhyme' Tyndall sent to the editor of the *Carlow Sentinel* sometime in January 1843, but which seems not to have been published nor even acknowledged; see letter 0190.

From James Sinnett 2 February 1843 0184

Kirby Lonsdale, Feb. 2nd, 1843.

Dear John,

I received your letter[1] and George's[2] this morning which occasioned me no small surprise. I wondered what the Postmaster's daughter was looking at so – she kept my letters in her hand and looked at me so hard – to my utter astonishment two were directed to Sergt. S.,[3] one was C's[4] the other was your amendment on George's motion. Here's a pretty affair, said I, some Preston gentlemen are quizzing me a bit. I looked at the L'pool (H.M.S.)[5] there I figured as I have done for many a long year. However, I hurried out and managed to get open one of them, and believe me that I felt more at the kind manner and the friendly promptitude with which it was communicated than the news itself. I can also say that I thank you all and feel proud of the warm friendship of such men as you three: but what do you suppose I was about last night? No less than clearing poor Ginty of the foul stigma of inserting some 'Verses on leaving Westmoreland'.[6] (I wish I could see John Tyndall's visage while he reads this. Surely, as he would say himself, it will be 'incarnadined'[7] up to the ears.) But pardon this digression; the v___n[8] who penned those verses, anxious that 'Mary' should have the full benefit and fearful no doubt that the Preston Chronicle would not penetrate so far north as Kirby actually sent a copy here directed to Miss E.,[9] and lest the said Miss E. should not notice the effusion of his muse, scored round poor Ginty's name with his pen, by which her eye was at once attracted – and then there was a scene. I received a letter same post from poor maligned G.,[10] in which, after a good deal of small news in G's own peculiar style, I came to the pith of the epistle – and that was a request from poor G. that I would declare the said verse[11] to be a forgery. Thus the bane[12] and antidote came bobbing against one another in the same post bag. Unfortunately, the bane had sufficient time to operate before the antidote could be applied ... and although G. may be cleared of actually writing and inserting the verses, yet how is he to get out of the scrape of letting the wicked wag at Preston[13] know Miss E's name, both Christian and surname? All the hope we have is the faint one that no other copy of the fatal P.C.[14] will arrive in Kirby. If there comes one more the whole affair will be as common as the handle of the town pump, for the acquaintance between 'M' and W.G. is so well known that the affair will be out at once, and then picture to yourself the result. Now, poor G. has named the person whom he believes to be the perpetrator of this sad business, and sorry I am to say that I was forced to agree with him: and I came to this conclusion from the recollection of the following circumstances: When looking over various tyros[15] in the offices at Cork and Kinsale, I would occasionally observe a nice bit of wood, again a good bit of bog, a nicely shaded lake, or a good capital letter. I used at first innocently to say – Well, that is nicely done. But the reply would invariably be, 'Oh, Sir, it was John Tyndall done that'. This happened so often that at length I gave over remarking anything nice, being assured that I might as well at once give John the credit. Now, as poor G. asked my opinion as to the perpetrator of this nefarious business, I thought the best thing I could do in reply was to relate the above circumstances and express my opinion that if the Editor of the Preston Chronicle[16] was asked who it was wrote those pretty verses, he would at once say 'Oh! Sir, it was John Tyndall done that'. Now, John, I may be wrong but much I fear I am right. I would bid you make answer, but I see not how the mischief can be remedied; but you have occasioned no small stir in Kirby Lonsdale, I can assure you. I was applied to for G's address, and as I could not refuse so reasonable a request I suppose poor G. is about to reap the benefit of your poetical labours. But, dear John, are you in earnest when you say 'you are a reprobate' and 'were I a Christian'? I hope you will never be the former, and soon be the latter. <u>It is something to be a Christian, but John, dear, you are an honest man and that is what a reprobate never was yet</u> ... Well, I have written a good deal to-day – a long letter to Baskin, one to G., one to C., and yourself last but not least: I am now thinking of Evans, who was to crush a tumbler to-night – at least, he promised to do so, he told me one day in Cork before he left.

May the Lord bless you all, says your | faithful friend, | J. Sinnett.

RI MS JT 1/11/3863

LT Transcript Only

1. *your letter*: letter missing.
2. *George*: George Latimer.
3. *Sergt. S.*: Sinnett means himself; having previously been promoted to the rank of Corporal in March 1839, it is likely that he had only recently been made a Sergeant.
4. C: probably John Chadwick, who worked closely with Sinnett in the 4th Division, C District of the Irish Ordnance Survey.
5. *L'pool (H.M.S.)*: the postmark on the stamp, meaning Liverpool On Her Majesty's Service.
6. *the foul stigma of inserting some 'Verses on leaving Westmoreland'*: see letter 0183.
7. *'incarnadined'*: reddened (*OED*).
8. *v___n*: probably villain.
9. *Miss E.*: Mary Edwards.
10. *G.*: William Ginty.
11. *the said verse*: 'On Leaving Westmorland. By W. Ginty', *Preston Chronicle*, 28 January 1843, p. [4].
12. *bane*: poison (*OED*).
13. *the wicked wag at Preston*: Tyndall.
14. *P.C.*: the *Preston Chronicle*.
15. *tyros*: beginners or learners in anything; those who are learning or who have mastered the rudiments only of any branch of knowledge; novices (*OED*).
16. *Editor of the Preston Chronicle*: Lawrence Dobson.

From William Ginty 4 February 1843 0185

110 Pitt Street | Saturday evening Feb. 5th 1843.[1]

My dear Jack

This pencil production of yours[2] is ultra-astonishing. What has driven such a notion into that head of yours. Why man, 'I never blamed you, never!' Surely you could not infer such from my last letter[3] – but I can guess Sinnett has 'come Paddy over you'.[4] I also sent a letter to him, in answer to which he said Miss E.[5] had got the paper and was half crazy and that he thought all the water in the Lune[6] would not clear me, and advised me to write to Miss E. Since that I have sent him a congratulatory letter with one enclosed for her –

I took his advice you see. Not to rectify her opinions nor to appease her anger but, in good truth, I did not wish her to imagine that I thought so much about her as to lavish such encomiums[7] on her beauty. Enclosed I send you a P.S., of Sinnett's,[8] you had better read it before you go further, as it will throw a light on what here follows. (Read it)

Here follows the answer which I sent to this as well as I can recollect it – it will shew you that I have said nothing to Sinnett that could lead him to think I was offended with you I have no vengeance to wreak on Tyndall. No! No!

Vengeance for what? Is it for praising my late inamorata?[9] Why man I have done so myself until my vocabulary of love plaudits were drained to the very bottom. (But in good truth I can't see what there is in it that discomposes that quondam[10] charmer of mine so very much. I only wish some fair one would devote so much of her time in praise of my charms – but people's tastes differ, and I'm only sorry Jack Tyndall has wasted so much of his 'sweetness on the desert air'.[11] Yes boy'o!!! – by heavens I'm sorry that that genuine worth which marked it has not received that admiration which is its just reward.

I wrote to her telling her that I was not the author – that I was sorry any friend of mine had annoyed her, but that I could confidently and positively say that it was not done with that intention – That it was merely in jest – jests of such kind being a common occurrence amongst us 'wild Irish men', that I could see nothing in it to offend, and that I thought there were many young ladies in these kingdoms who would likely go half-distracted with joy if such an exquisitely beautiful and gallant piece was sacrificed at the shrine of their beauty and concluded hoping that she would attribute it to no other motive than that of an Irishman holding forth in the exuberance of his nationality!!! Now it only remains for me to say on this subject that I regret you have for an instant imagined that I was displeased at-all with 'the transaction'. No boy O! if I had you here I would give you a hearty shake of an Irish fist for managing the affair so completely. But read this which I extract from the 'Mercury' of yesterday[12] (we get the paper regularly)

'same day (Monday last) at Kirkby Lonsdale Wm. H. Owen Esq.[13] Solicitor of Gorey[14] Ireland. To Mary only daughter of the late Capt. Edwards[15] of Liverpool'.

Wonders will never cease! I hope it is true[16] – yet fear she is now codded[17] in downright earnest – it must mean her – her father was a Captain I have heard, and I recollect her asking me if I knew where Gorey was! Sinnett's silence on the matter leads me to imagine it a cod, as he wrote on Tuesday evening, and I'm very certain it would be no easy matter to get 'spliced' in Kirkby without every little dog in the town knowing it, and again I wonder any fellow would be such a cod as to pay for the insertion of the announcement – if this is a jest I protest the jester is a cruel one.

Read again (Wonders, it seems, are only beginning)

'To Correspondents' 'Ordnance Survey – Would Veritas[18] favour us with a call? The subject requires great discretion'.

By the powers Tyndall – the moon of survey agitation is about to burst through the clouds which have long shrouded her — God speed thee 'Veritas'! thy name and motto is a good one and you need never make it blush to make the mean R.E. monopolists[19] tremble for their public security. Hurrah! Hurrah!

3 cheers for Veritas – glory to him. May the sun of prosperity cast her brightest rays even on the ould stump that scrapes the first line of it!!!

Next Friday's paper will probably contain it.[20] You shall have it But i' faith, father John, who knows but you and Veritas are in the singular No.[21] – if so – More power boy O! As you are not here to 'favour them with a call', if you commission me I'll go, aye even if they had a 'Board of Survey' assembled to mark my presence. and if this wont do, blessed if I wont give discharge the second in the character of VENGEANCE. my motto: 'I speak as unto wise men, judge ye what I say'.[22] Holy poker how I'd wollop em. I'd make old Tom knock t'other hand off[23] thumping the table with rage. I'd make deaf Harry[24] hear the postman's bugle in Liverpool as he sat conning over it in Preston. I'd make W.G.H.,[25] dance a quadrille to the air of 'devil take hindmost'[26] as he ran with the paper in his fist from here to the Mercury office. Aye I'd make the whole confraternity run as so many devils would run from Father Tom McGuire with a bucket of Holy Water on his head. I'd - - - - do the devil and all!!!

Liverpool is teeming with illustrious country men of ours at present. O'Connell, McNeill, Caughy, Fielding, Old Nangle (from Achill) and Nolan!!![27] The writing of this alone prevents me from hearing the 2nd last this evening gratis in the Music Hall. Dan said a few nights ago, at a great 'free trade banquet',[28] that no greater insult could be offered him than to class him and McNeill as countrymen – he says he has made Liverpool the hot bed of Bigotry and that if he thought there was a more bigotted place in the world he would go visit it as a 'National curiosity'!!! McNeill is nearly worshiped by the Protestants – you should go at least 1½ hour before service time to get a seat in his church.[29]

I asked Hamley for a recommendation[30] yesterday – but to no purpose – faith I astonished him with talk on my past grievances and future prospects on the O.S.[31]

I came here very much prejudiced against Mr Kennedy[32] but indeed I like him very well now. Dick[33] and I who worked his map agree in saying – he is a very decent gay fellow and an Irishman.

No more of your pencillings of penance I want scribblings of 'Nationality'

Cant write more – its near post time and I must away.

Yours ever sincerely | Ginty.

Best respects to the Mess. The Capt.,[34] was here to day I'll be apt to pounce on a summat[35] in the Mercury.

Adieu.

RI MS JT 1/11/3592
LT Transcript Only

---

1. *Saturday evening Feb. 5th 1843*: Ginty seems to have written the wrong date, as the Saturday was 4 February.

2. *pencil production of yours*: letter missing.
3. my *last letter*: letter 0183.
4. *Sinnett has 'come Paddy over you'*: see letter 0184.
5. *Miss E.*: Mary Edwards.
6. *the Lune*: the River Lune which runs through Kirkby Lonsdale.
7. *encomiums*: formal or high-flown expressions of praise (*OED*).
8. *Enclosed I send you a P.S., of Sinnett's*: enclosure missing. P.S. is an abbreviation of post scriptum, Latin for written after.
9. *inamorata*: female lover or sweetheart (*OED*).
10. *quondam*: former holder of an office or position (*OED*).
11. *'sweetness on the desert air'*: T. Gray, 'Elegy in a Country Churchyard' (1750), 56.
12. *the 'Mercury' of yesterday*: 'Marriages, Births, and Deaths', *Liverpool Mercury*, 3 February 1843, p. 35.
13. *Wm. H. Owen Esq.*: William Henry Owen, son of George Owen.
14. *Gorey*: a market town in County Wexford.
15. *Capt. Edwards*: Captain Robert Edwards.
16. *I hope it is true*: According to the Kirkby Lonsdale parish records, the marriage did indeed take place, although it is dated 7 June 1843 rather than in February. A year later it resulted in the birth of a daughter, Mary Anne.
17. *codded*: imposed upon, deceived (*OED*).
18. Veritas: the goddess of truth in Roman mythology.
19. *the mean R.E. monopolists*: the Royal Engineers who proposed to take on the work of the Tithe Survey, thus depriving private surveyors of the ability to compete for contracts for the work; see letters 0188, n. 10 and 0195, nn. 15 and 16.
20. *Next Friday's paper will probably contain it*: no such letter appeared in the *Liverpool Mercury* for Friday 10 February 1843, nor in any subsequent issues.
21. *in the singular No.*: Ginty means that Tyndall might be Veritas.
22. *'I speak as unto wise men, judge ye what I say'*: 1 Corinthians 10:15.
23. *I'd make old Tom knock t'other hand off*: Colonel Thomas Colby, who had lost his left hand when a pistol exploded in it in 1803.
24. *deaf Harry*: Captain Henry Tucker, who 'throughout his successful career ...had been extremely deaf' (J. H. Andrews, *A Paper Landscape: The Ordnance Survey in Nineteenth-Century Ireland* (Oxford: Clarendon Press, 1975), p. 246).
25. *W.G.H.*: Lieutenant William George Hamley.
26. *'devil take hindmost'*: a proverbial expression first recorded in print in Francis Beaumont and John Fletcher's play *Philaster, or Love Lies a-Bleeding* (1611).
27. *McNeill, Caughy, Fielding, Old Nangle (from Achill) and Nolan*: Hugh Boyd M'Neile (1795–1879), an evangelical preacher and demagogic opponent of Roman Catholicism; James Caughey (1810–91), a Protestant revivalist preacher recently returned from Canada; Fielding Ould (1801–69), evangelical minister of Christ Church in Liverpool; Edward Nangle (1800–83), an evangelical clergyman in the Church of Ireland who established a Protestant colony in Achill, an island off the west coast of Ireland; Thomas Nolan, another evangelical Liverpool clergyman who was part of the Protestant 'Irish clique' led by M'Neile (F. Neal, *Sectarian Violence: The Liverpool Experience, 1819–1914* (Manchester: Manchester University Press, 1988), p. 48).
28. *a great 'free trade banquet'*: The banquet, organized by the Liverpool branch of the Anti-Monopoly Association, was held at the Amphitheatre in Queen's Square on Tuesday 31 January 1843 and was attended by *c.* 2,000 delegates. O'Connell's remarks about

M'Neile were not reported in newspaper accounts of his speech at the banquet.

29. *his Church*: St Jude's Church on Hardwick Street, Liverpool, where M'Neile was the perpetual curate from 1834 to 1848.
30. *a recommendation*: for a pay rise.
31. *O.S.*: Ordnance Survey.
32. *Mr Kennedy*: see letter 0183, n. 26.
33. *Dick*: Richard Hunter.
34. *The Capt.*: presumably Captain Henry Tucker.
35. *a summat*: a something.

From William Ginty 7 February 1843 0186

Hurrah! Hurrah let others send
And sing and sigh their valentines
My name mongst such shall never blend
A nobler lay is mine.

Tis stuff – trash – nonsense – folly – madness
Curst 'paper love' I dearly hate you
I'd sooner sit in silent sadness
Than be consigned to idly prate you

Ha! Ha! Ha! around me sits
Three love-sick wights[1] – each at a ditty
One puzzles his belaboured wits
To sing the praise of one Miss Hetty[2]

Poor wretch e'en now that deep drawn sigh
That frown bespeaks thy soul's deep ardour
You little dream that bye and bye
Your theme will float in Kinsale harbour

See 'paddy'[3] too! oh powers just view
That all-absorpt-in-love-care phiz[4]
'The course of smooth love ne'er ran true'[5]
You surely can't apply to his

Poor devil! You should bear in mind
That your dear labours – loves ta
A few short days will se consigned
Secreted in some stinking privy

The <*word(s) missing*> a scholar in the art
With his fond task he seems delighted
He looks around – then with a start
Throws down his pen and crys afrighted

'Call ye this love — unhappy pair
That spreads such gloom upon your faces

Then mine's not lost – so to the rear
I'll quick return some 30 paces'

A such like fate they ever meet
Theirs is the doom 'beyant all tellin'
But zounds! I fain your ears would greet!
Pray have you sent a one to Ellen[6]

If so let me predict its fate
And deem me not a scoffing churl
When I beg leave to simply state
Fer 'raven plumes' twill help to curl

But no – that's wrong – they're natures own
Some other fate must meet your letter
Perhaps twould shroud some 'yalla soap'[7]
In truth I hope it wont meet better

And now adieu Sir valentine
May every lovers luck attend you
If sighs should burst your heart in time
All I've to say is 'devil hand ye'

I s'pose you thought you'd got a lob[8]
When native 'Carlow' met your eyes
Jist look how nate[9] I did de job
And be in future 'penny wise'[10]

RI MS JT 1/11/3593
LT Transcript Only

1. *wights*: creatures (*OED*).
2. *Miss Hetty*: not identified.
3. *'paddy'*: not identified.
4. *phiz*: face (*OED*).
5. *'The course of* smooth *love ne'er ran true'*: W. Shakespeare, *A Midsummer Night's Dream*, I.i.134.
6. *Ellen*: seemingly 'the lady of the raven plume' from Kinsale mentioned in letters 0151 and 0155.
7. 'yalla soap': a common soap made of tallow, rosin and soda (*OED*), which give it a distinctive yellow colouring.
8. *lob*: a nugget of gold; a 'lump' of money (*OED*).
9. *nate*: variant spelling of neat.
10. 'penny wise': careful with regard to small expenditures (*OED*).

To John Tyndall, Snr [14] February 1843 0187

Preston, | Tuesday night

My dear Father

I am just after returning from hearing a lecture by the Curate of the parish church of Preston in the lecture room of the Mechanic's Institution[1] of which I have been a member these three months. We have had a very interesting course of lectures during the winter – one every week.[2] There are much greater facilities here for the acquirement of information than in Ireland,[3] and as far as my time and circumstances permit I avail myself of all.

Yours[4] together with the 'Weekly Message'[5] were received yesterday. You were right in your conjectures as to whence it came. Poor Wright's hand is well known to me. I was surprised and grieved to hear of his father's death. You never mentioned it to me before. I cannot say how Joe Payne is. I'm not inclined however to go the whole way with you in believing the report of his sickness to be a scheme. You put me in mind to refund your two pence which you paid on the receipt of the American paper.[6] Did you not read it? If so was not that, in part at least, compensation and the remaining part of the debt, if any remains, I intend to liquidate by permitting you to read the Illustrated London News.[7] I send you inside this paper two penny publications,[8] they afford you a specimen of a class of papers with which this part of the country teems. They have a most astonishing effect on the public mind. Some Statesman has remarked that if he were allowed to write the songs of the people he cared not who made laws for them.[9] From a conviction similar to this has sprung a locust crowd of such publications as these I now send you. You will easily recognise the hooked noses of Peel and Wellington[10] among the group of heads, the 'gent' at the tail is Harry Brougham.[11] I suppose you have read about the division that has taken place between him and the 'League'.[12] If so you will immediately comprehend his position. I run the risk of losing my paper in sending these. I don't think they are allowed to go post free.[13] If they be found out I dare say postage will be charged for them. If so don't pay it. When you read the paper just enclose it as follows:–

Mr Haughton[14]
Main Street
Cincinnati
United States.
for Mr Wright

You will have to pay 2d for this, but I solemnly assure you that I'll refund it to you perhaps in my next letter – if I had stamps to that amount by me at present I'd send them – You ask me whether you may expect to see me in May, you say that a trip to Ireland would not be much more expensive than the one I took from Cork. Why man! my exchequer was nearly annihilated by the shock it received from that trip. Now I'll make a bargain with you – how many pairs of boots and shoes will you give me if I go? – There's a chance for you. However joking apart its very probable that I will take a trip over to you in the course of the summer. I must make some calculations on the matter, its very likely the result of them will be favourable to my going.

Your affectionate son | John.

RI MS JT 1/10/3285–6
LT Transcript Only

---

1. *a lecture by the Curate ... the Mechanic's Institution*: This was a lecture on 'Mental Philosophy' given on Tuesday 14 February 1843 by the Reverend J. P. Simpson ('Institution', *Preston Chronicle*, 18 February 1843, p. [3]). The parish church of Preston was St John the Evangelist, in Church Street; the Mechanics' Institution, also known as the Institution for the Diffusion of Useful Knowledge, was founded in 1828 in a small two-story building on Cannon Street, which contained a library, museum and lecture room. It moved into much grander, neo-classical premises on Avenham Walk in 1850.
2. *We have had a very interesting ... one every week*: Tyndall later recalled: 'In 1842, and thereabouts, it was my privilege to be a member of the Preston Mechanics' Institution – to attend its lectures and make use of its library. A learned and accomplished clergyman, named, if I remember aright, John Clay, chaplain of the House of Correction, lectured from time to time on mechanics. A fine earnest old man, named, I think, Moses Holden, lectured on astronomy, while other lecturers took up the subjects of general physics, chemistry, botany, and physiology' ('Address Delivered at the Birkbeck Institution on October 22, 1884', *New Fragments* (London: Longmans, Green, 1892), pp. 224–47, on p. 226).
3. *much greater facilities here for the acquirement of information than in Ireland*: The educational reformer J. W. Hudson observed that 'In Ireland, Mechanics' Institutions have never prospered for any lengthened period, owing to a variety of causes, not the least of which has been a strong political and sectarian animosity that has paralised committees and alienated members', while Irish 'reading-rooms ... *from the want of books and magazines* ... became mere gossiping rooms and were given up' (*The History of Adult Education* (London: Longmans, Green, 1851), pp. 213–14).
4. *Yours*: letter missing.
5. *'Weekly Message'*: probably the *Dollar Weekly Message*, a newspaper published in Cincinnati, Ohio between 1841 and 1843 (Library of Congress http://chroniclingamerica.loc.gov/lccn/sn85042260/). William Wright, who sent the newspaper, had emigrated to America in March 1841 and settled in Cincinnati.
6. *two pence which you paid on the receipt of the American paper*: While the Post Office (Duties) Act of 1840 decreed that printed British newspapers could be sent for free within the United Kingdom, foreign printed newspapers brought into the country by packet boats or private ships were liable to a charge of 2*d.* (3 & 4 vict c 96. XLII).
7. *the Illustrated London News*: a weekly newspaper, founded in 1842 and costing 6*d.*, that used wood engravings to, for the first time, provide graphic and realistic illustrations of current news events.
8. *penny publications*: From the early 1830s, and especially following the success of the Society for the Diffusion of Useful Knowledge's *Penny Magazine*, the word 'Penny' was

used regularly in the titles of newspapers and periodicals intended for a mass reading audience, which, collectively, were known as penny papers or the penny press.

9. *Some Statesman ... write the songs of the people he cared not who made laws for them*: The Scottish politician Andrew Fletcher (1653–1716) observed that 'a very wise man ... believed if a man were permitted to make all the Ballads he need not care who should make the Laws of a Nation' (*An Account of a Conversation Concerning a Right Regulation of Governments* (Edinburgh: n.p., 1704), p. 10).
10. *Wellington*: see letter 0146, n. 11. Although celebrated for his military prowess, and continuing as Commander-in-Chief of the British Army, the so-called 'Iron Duke', who was Prime Minister from 1828 to 1830 and again in 1834, provoked radical ire as a reactionary opponent of political reform.
11. *Harry Brougham*: Henry Peter Brougham, 1st Baron Brougham and Vaux (1778–1868), a Scottish lawyer and Whig politician who served as Lord Chancellor between 1830 and 1834.
12. *the division that has taken place between him and the 'League'*: Although Brougham was in favour of free trade and the repeal of the Corn Laws, his relations with the Anti-Corn Law League had become increasingly strained. Finally, in a debate on the Corn Laws in the House of Lords on 9 February 1843, Brougham stated 'I cannot ... suffer the mention of the Anti-Corn-law League, and its proceedings ... without expressing my entire and hearty concurrence with the disapprobation expressed at many of the proceedings of that body ... I desire to express my most pointed disapproval of the means taken by some of the persons connected with that association' (*Hansard*, HL Deb 9 February 1843, vol. 66, c. 292).
13. *I don't think they are allowed to go post free*: see n. 6.
14. *Mr Haughton*: Richard Haughton, William Wright's employer; see letter 0146.

From William Ginty [after 14 February] 1843 0188

O.S.O.[1] 33 Windsor Street[2]

Dear Tyndall

If I was writing to any other man but you on the subject which here follows, my letter would commence with the strictest conjurations to secrecy. I suppose you can guess all now.

The flame of opposition and revenge has been kindling in my bosom this last week – and I think I may properly say that at present it deserves the name of conflagration. I have taken it into my head since 'Veritas' sounded a false alarm[3] – that I will at least 'raise the wind'. To come to the point, I think the subject of the Tithe Survey as carried on by R.E.,[4] influence is a home-thrust, a death-wound – We all know that Tucker or the Govt.[5] have been at a considerable loss by his undertaking to execute it at 4d per acre.[6] This in my opinion will take well, when we consider how much Tucker's proposal annoyed the Civil Engrs.[7] and Surveyors of this Co.[8] And in the 2nd I dont think it will interfere with what you may intend to lay before the public. Of course if it is even an atom of your subject matter I renounce it altogether, willingly – proud of leaving it to a man of more discernment and ability.

I waited on and obtained an interview with the Editor of the Mercury[9] this evening. I laid the whole circumstance before him, told him all about the jeopardy of survey affairs. – reminded him of the death-blow that was made at the professional rights of the Lancashire Engrs. & Surveyors,[10] and then and there asked his advice and opinion on the matter – He seemed to relish the idea much – advised me (altho' unnecessary) to assert, or rather state nothing but facts. So now you see how the matter stands, and will favour me with your opinion, and also make a kind of a rough calculation of what the Tithe Plans cost the Capt. or the Govt. I have made such and find that it exceeds the 4d. an acre considerably.[11] Are you quite certain that it was 4d. an acre he proposed to do (and done) them for? I heard the principal Draftsman here say last week that the Exn.[12] & Computation alone cost more than 5d in this district. I asked the Editor about 'Veritas', he says that a person signing himself such[13] sent him a letter all about the young mens pay – but in his opinion it would do them more harm than good, it was so outrageous in its finding fault with officers &c.&c. And so he said he would not insert it until he favoured him with a call and made him completely write it over again – it is in all probability a good job that it never met the public eyes. I suppose it would have as much effect as a cannon loaded with feathers. The Editor promised me that he never would give up my name unless forced to it by law. I told him of course that dismissal was inevitable if ever a whisper of it came to the ears of the Engs[14] officers. And now, Father John, what say you? Let me have your opinion, as soon as possible, and dont forget if you have e'er a good hint be sure and let me have it. Sinnett is under orders for Hull to Perambulate. I suppose you have heard of Nelson's desertion.[15] He is missing since the 9th Inst.,[16] and the sappers here are ordered to keep a close eye to the out-bound American vessels. Tell Geo. Latimer I had a letter from my brother and that Harrison[17] and him are still enjoying their synecure.[18] As I write this in the office you will excuse its carelessness – it is now breakfast time.

Yours &c. | Ginty.

Thursday evening

Since I wrote the above I had a letter from Sinnett, he knows nothing yet of the route he is about to get. Jack[19] got a letter from you. So you got the valentine. Did my forgery deceive you for any time? If you had to pay only 1d., I have at all events deceived the postmaster. Turner told Hamley yesterday that Latimer and I were the two best Surveyors in his party – a lie of course – and one I'm afraid that will hunt us out to the field again. So Cupid[20] got a valentine with: How I would like to be along side of him when it met his view. We sent one to Joe Collins. Dick[21] forwarded the direction – and enclosed was a drawing of an old soldier cut out of the 'Gallery of Comicalities'[22] with the sentence 'A Private Still'[23] under it – but he was match enough for us – he crosses out his own name and address and without breaking the seal addressed it back to me with 3d on it.

I of course was not ass enough to release it. all I wonder at is how he could guess I had a hand in it.

Yours | Ginty.

RI MS JT 1/11/3595
LT Transcript Only

---

1. *O.S.O.*: Ordnance Survey Office.
2. *Windsor Street*: in Toxteth Park, Liverpool.
3. *'Veritas' sounded a false alarm*: see letter 0185.
4. *R.E.*: Royal Engineer.
5. Govt.: Government.
6. *his undertaking to execute it at 4d per acre*: to produce maps for the Tithe Survey, which, at the required scale of 1 inch to 3 chains, the equivalent of 26⅔ inches to 1 mile, would be likely to cost much more than *4d.* per acre; see n. 11.
7. *Civil Engrs.*: Civil Engineers.
8. *this Co.*: the county of Lancashire. On the annoyance of Lancashire's private surveyors, see letter 0195, n. 16.
9. *Editor of the Mercury*: John Smith.
10. *the death-blow that was made ...Lancashire Engrs. & Surveyors*: The proposal that the work done by private surveyors for the Tithe Survey be taken over by the Royal Engineers on the English Ordnance Survey.
11. *it exceeds the 4d. an acre considerably*: Colonel Thomas Colby's proposed cost had actually been 1*s.* 6*d.* per acre; see letter 0195, n. 15. It is not known whether Captain Henry Tucker was proposing to undercut this, or whether Tyndall's information was erroneous, although the latter seems most likely.
12. *Exn.*: Examination.
13. *a person signing himself such*: Tyndall later rebuffed the claims of 'some evidently unauthorized person, who styles himself "Veritas"' when the latter wrote to the *Liverpool Mercury* criticizing Tyndall's analysis of the ills of the Survey ('Ordnance Survey. Letter III', *Liverpool Mercury*, 27 October 1843, p. 359).
14. *Engs*: Royal Engineers.
15. *Nelson's desertion*: James Nelson, a Private in the 13th Company of the Royal Sappers and Miners, seems to have already had disciplinary problems while working on the Irish Ordnance Survey, as his pay was reduced from 9*d.* to 4*d.* in September 1839. He was transferred to the English Survey in April 1842 (NAI OS/1/16–19).
16. *inst.*: abbreviation for 'instante mense', Latin for this month.
17. *Harrison*: not identified.
18. *synecure*: a position that requires hardly any work but which provides a salary (*OED*).
19. *Jack*: John Tidmarsh.
20. *Cupid*: Phillip Evans.
21. *Dick*: Richard Hunter.
22. *'Gallery of Comicalities'*: a comic newspaper founded in 1832 and costing 3*d.* which reproduced caricatures already published in *Bell's Life in London*.
23. 'A Private Still': Despite having enlisted in 1822, Joseph Collins remained a Private in the Royal Sappers and Miners. His lack of promotion was perhaps attributable to disciplinary action taken over his absenting from duties in the 1830s (NAI OS/2/12).

From Thomas M. Foy 27 February 1843 0189

Manchester | 27th. February 1843.

My dear John

Independent of my long silence I think it is high time to acknowledge the receipt of your kind letter[1] which I received last week. When I look back to the date of our parting at Cork[2] and reflect on the length of time that has elapsed without either seeing (or writing to) each other, for my part I feel the most poignant sensation of shame and regret. I will not aggravate the impropriety of my omission by amusing you with useless and childish excuses – but confess that an unaccountable negligence and foolish habit of procrastination have made me so inattentive. I entreat of you not to consider me as deficient in friendship for you – for I feel convinced that our long acquaintance and perfect knowledge of each other will be sufficient grounds to exhonorate me from that charge – I am happy to hear that you are so very comfortably situated – but your low rate of pay I am sure preys upon your mind in this country – I wish you were here as there is a fine field for employment for a young man of your abilities – and I have no doubt it would be in my power from my extensive respectable connexion in Manchester to be able to procure you a good situation.

I seem to hear you say – what are you doing for yourself? 'Wait a wee' as the Scotchman says. I have read and studied more since I came to Manchester than all the rest of my days put together, and believe me I find the effects of it – Since I received your letter I have been out every evening – perhaps you are not aware that I have an uncle and a host of cousins in this town. They are all manufacturing chemists[3] – My uncle carries on one concern in one part, his eldest son who is married in another part and within ½ a mile of the office.[4] And two others of my cousins in another, this opens a great connexion of acquaintances for me. In fact they are all so kind and so fond of me that they are fighting to get me to spend the evening with them. I have many things to communicate to you, but am resolved to devote this letter to apology alone – I will write to you again in about a week. Remember me kindly to all my friends in Preston – and

believe me | my dear John | your affectionate friend | T. M. Foy

RI MS JT 1/11/3576
LT Transcript Only

---

1. *your kind letter*: letter missing.
2. *the date of our parting at Cork*: 10 June 1842; see letter 0147.
3. *I have an uncle and a host of cousins in this town. They are all manufacturing chemists*: The adjective manufacturing distinguished those chemists who, unlike pharmaceutical chemists, were concerned with the industrial applications of chemistry, as well as the manufacture of chemical products. In the 1830s there were some thirty-three manufac-

turing chemist firms in Manchester, and the Chemical Society's records reveal that the 'largest regional concentration of manufacturing members was in the north-west', where 'most of the major companies employed at least one trained chemist' (R. Bud and G. K. Roberts, *Science Versus Practice: Chemistry in Victorian Britain* (Manchester: Manchester University Press, 1984), p. 104). There are no records of any manufacturing chemists in Manchester at this time called Foy, so Foy's uncle and cousins were presumably on his mother's side of the family, with a different surname.

4. *the office*: presumably the Divisional Office of the English Ordnance Survey's 4th Division, located in Manchester.

To John Tyndall, Snr [late] February 1843 0190

My dear Father

It is now a long time since I last wrote to you[1] and I have been expecting in vain an answer this week back, but as I see no sign of anything from you I now lift the pen to put you in mind of your debt.

Has the Sentinel ceased to exist?[2] If not the conduct of some person connected with it is rather unaccountable. Three weeks ago I sent a bit of a rhyme to it and in the same letter I enclosed a sixpence requesting that the paper might be sent to me on the following Saturday. My letter[3] was directed to the Editor[4] – and I have never heard any thing of it since. Now I could get a Journal from any part of the united kingdom by writing for it. I have written to London for papers, and my letters were always promptly attended to. I cannot account for the observance of a different line of conduct in Carlow – they should at least have returned my money – the moment I finish this I shall write to the Editor about it. I shall wait for a few days for an answer from him, in the mean time you need say nothing about the affair, but if I do not get an answer, you must write to my uncle Bill[5] and desire him to obtain an explanation.

Send me in your next letter a nice sprig of a Shamrock[6] –I want to send it to America to Wm Wright, you cant imagine the interest which a sprig

Of the chosen leaf
Of Bard and Chief
Old Erin's native Shamrock[7]

creates in the breast of one of her sons who is far far away from the home of his childhood. I have always sent one to my friend Higginson who is stationed at present in Poona in the East Indies.[8] I saw his name in the army list[9] yesterday.

Dont forget to send me Mr Ryan's address

Your affectionate son | John.

Did you send the London News[10] to Wm Wright? if so my account with you stands thus: –

Little John to Big John	L	S	D[11]
To postage paid on Wm Wright's letter	0	0	8
To Do on paper sent to Wright-	0	0	2
	0	0	10
Cash--	0	1	0
Remaining due to little John--	0	0	2

I'll draw on you for the balance at some future period
I shall expect a reply to this very shortly | J.T.

RI MS JT 1/10/3287
LT Transcript Only

1. *since I last wrote to you*: letter 0187.
2. *Has the Sentinel ceased to exist?*: The difficulty in obtaining the *Carlow Sentinel* might be attributable to the ill health of its proprietor Henry Malcolmson (see letter 0108, n. 3). Malcolmson, who had begun the paper in 1831, finally decided to stand down on 4 March 1842, with the editor Thomas Harris Carroll then also becoming the proprietor ('To the Subscribers of the "Carlow Sentinel"', *CS*, 5 March 1842, p. [3]).
3. *My letter*: letter missing.
4. *the Editor*: Thomas Harris Carroll.
5. *my uncle Bill*: William Tyndall, who was acquainted with Malcolmson; see letter 0108.
6. *sprig of a Shamrock*: Derived from the Gaelic seamróg, the shamrock is a three-leaved sprig of clover that became a symbol of Ireland after Saint Patrick took it as a metaphor for the Christian trinity.
7. *'Of the chosen leaf ... | ... Old Erin's native Shamrock'*: T. Moore, 'Oh, the Shamrock' (1821), 15–17. Erin is a romantic name for Ireland.
8. *Higginson who is stationed at present in Poona in the East Indies*: see letter 0182, n. 4.
9. *army list*: the official list of serving officers in the British Army published monthly, quarterly and annually since 1740.
10. *the London News*: the *Illustrated London News*; see letter 0187, n. 7.
11. *LSD*: pounds (livres), shillings and pence (denarii).

To John Tyndall, Snr 3 March 1843 0191

Preston, Thursday | March 3rd, 1843.

My dear Father

Here I sit enveloped by inkstands, paint cups, and scribbled paper while before me the glorious sun climbs the eastern sky shedding many a bright beam on the page destined for your perusal. You seemed not to like the Illustrated London News,[1] very well I'll send it to you no more. You have exhorted me to patience in your last.[2] In some cases this might be needful, but it is not so at present. Repeated collision with the world has deprived my character of those sanguine traits, those high hopes which usually swell in the breast of youth. There is no fear of my being in the least impatient. I say this tho' I am acquainted with the fact of the Colonel's having paired off with some other member.[3] I saw his name in the Mail.[4] I also feel gratefully assured that Mr Bredin will not neglect my interests. I wrote to him some time ago thanking him for his kindness. Do you know did he receive my letters?

I am sorry to hear that poor Mrs Payne is not improved in circumstances. Joe[5] owed me a trifle when he left Cork which he has honourably refunded since I came here. I suppose Emma[6] is in Cork also. The blast which swept over that family was indeed a withering one – all the result of imprudence. There's no use however in talking of that now. I had a letter from poor Foy[7] yesterday. He is in Manchester at present. He has a great number of respectable friends there, and contrives to spend his time very agreeably among them. I hope as the weather has become somewhat milder that your next letter will be something longer than your last. How is John Murray? Has Mrs Treacy's[8] family left Leighlin yet? Have you and John Mooney come on terms yet?[9] I write this in a hurry so you therefore excuse whatever blunders you may meet. Debby[10] tells me that John McAssy[11] of Carrickfergus[12] has written a long letter to the Bishop of Down[13] on Apostolic Succession.[14] It appeared in the Christian Examiner,[15] could you procure the paper for me? If not, find out where the Christian Examiner is published and send me word. My love to my mother and Emma.

Your affectionate son | John Tyndall.

RI MS JT 1/10/3288
LT Transcript Only

---

1. *Illustrated London News*: see letter 0187, n. 7.
2. *your last*: letter missing.
3. *the Colonel's having paired off with some other member*: In the parliamentary practice of 'pairing' whereby MPs from opposite parties agree to both be absent from a specific division in the House of Commons, Colonel Henry Bruen paired off with William Sydney Clements, 3rd Earl of Leitrim (1806–78), MP for County Leitrim, and did not vote in the division on the Corn Laws on Friday 17 February 1843 ('Imperial Parliament', *Freeman's Journal*, 20 February 1843, pp. [3–4]).
4. *the* Mail: possibly the *Waterford Mail*, which Tyndall had previously sent to his father; see letter 0117, n. 13. The only other Irish Protestant newspaper with this title was the *Dublin Evening Mail*.
5. *Joe*: Joseph Payne.
6. *Emma*: presumably Emma Payne, although not further identified.
7. *a letter from poor Foy*: letter 0189.
8. *Mrs Treacy*: presumably the wife of John Treacy, whose insolvency and subsequent departure to Australia might be a reason her family was leaving Leighlin Bridge; see letters 0007, 0047 and 0050.
9. *Have you and John Mooney come on terms yet?*: They had fallen out when Mooney apparently took an active part in the Roman Catholic boycott of Tyndall, Snr's shoemaking business following the Carlow election in the summer of 1841; see letters 0084 and 0111.

10. *Debby*: Deborah McAssey.
11. *John McAssy*: Reverend John McAssey, a Methodist ordained in 1838 who was minister at the Carrickfergus Congregational Church.
12. *Carrickfergus*: a town in County Antrim.
13. *Bishop of Down*: Richard Mant (1776–1848), Bishop of Down, Connor and Dromore.
14. *Apostolic Succession*: the succession which links the Christian Church to the original twelve Apostles of Christ; for Roman Catholics this is through the direct lineage of Bishops from Saint Peter, while for Anglicans it is through a continuity of commitment, beliefs and mission. Methodists similarly emphasize the continuity of Christian experience.
15. *Christian Examiner*: the *Christian Examiner and Church of Ireland Magazine*, a monthly periodical founded in Dublin in 1825 to promote evangelical causes in the Anglican Church of Ireland. It was notable for its strident anti-Catholicism.

From John B. Edgeworth 6 March 1843 0192

Cove | March 6th. 1843

My dear Tyndall

Many thanks for the very beautiful specimen you sent me[1] It is exactly the sort of thing I wanted – I do not in the least wonder at your not being able to send it to me before Indeed I should think from your account of your time that you have not one spare hour of daylight in the week I am at present staying at Cove[2] so that I did not receive your packet[3] as soon as I should had you directed it to Cove, as it lay in Cork till I went up there – I have very little time to give up to doing the plans as my examinations in college are so near that I must devote most of my time to reading for them however I have it all nearly done except the writing – There was a large ship of war in Cove harbour which I went to see with a party – the ladies of our party were hoisted up on deck in an arm chair – She was the first man of war I ever saw and certainly was one of the most wonderful sights. She came to carry out the first battalion of the 45 Regiment[4] there were 420 rank and file add to that 650 men belonging to the vessel herself besides the officers of both the Regiment and the vessel and think of how much provision she must carry to feed such a multitude for a voyage of 9 weeks or perhaps three months.[5] Her name was the 'Thunderer'[6] and she was bound for the Cape of Good Hope.[7] Her midship men[8] were the most gentlemenlike and nicest young lads I ever saw nothing could equal their kindness and attention while showing the vessel.

There also put in here a Frigate the 'Calliope'[9] 26 guns from China she put in here for water of which she had run short but I did not go to see her she sailed leaving two of her officers behind who were obliged to follow her or rather to get to their destination before her which was Portsmouth in the Bristol Steamer[10] and take the railroad from that, they expected to be there before her Of course you have heard of Major Waters (or as he was called in the office 'the Major' very emphatically) being sent to Barbadoes.[11] I was very sorry at his being sent abroad but he was very fortunate in getting so favourable a station as Barbadoes which is a very good climate. However he expects soon to be promoted and then he will return home again. The first time you are writing to Mr Eivers will you tell him that I would have answered his kind letter but that I am quite ignorant of his direction as he told me that he was about to move in a very short time after he wrote to me

I shall now bid you for the present good bye | and believe me | very sincerely and truly yours | J. B. Edgeworth

J. Tyndall Esq.

RI MS JT 1/11/3561
LT Transcript Only

---

1. *the very beautiful specimen you sent me*: presumably the drawing of 'a block of houses with yards walls and a street and a shed all coloured as the Cork City streets' that Edgeworth requested in letter 0179.
2. *Cove*: see letter 0068, n. 3.
3. *your packet*: letter missing.
4. *the first battalion of the 45 Regiment*: the Headquarters Battalion of the 45th Regiment of Foot.
5. *a voyage of 9 weeks or perhaps three months*: The voyage to the Cape of Good Hope lasted until 3 May 1843, so was in fact less than eight weeks.
6. *'Thunderer'*: HMS *Thunderer* was a two-deck 84-gun second rate ship of the line, launched in 1831.
7. *Cape of Good Hope*: The 45th Regiment was sent to the Cape of Good Hope in southern Africa to assist in establishing the Queen's Authority in Natal, which George Napier, the Governor of Cape Colony, proclaimed a British Colony on 12 May 1843, nine days after the *Thunderer* arrived.
8. *midship men*: naval officers serving a three-year apprenticeship before being eligible for promotion to Lieutenant.
9. *'Calliope'*: HMS *Calliope* was a 28-gun sixth rate ship of the line, launched in 1837. Between 1841 and 1842 she served at Canton during the First Anglo-Chinese War.
10. *the Bristol Steamer*: passenger and freight steamship service between Cork and Bristol run by the Bristol General Steam Navigation Company.
11. *Major Waters ... being sent to Barbadoes*: see letter 0179, n. 3.

From William Wright 11 March 1843 0193

Cincinnati | March 11th 1843.

My dear Jack

I wish I was possessed of a small sprinkle of your genius that I might convey to you in a proper garb the pleasure I experienced on the receipt of your always but now doubly dear letter.[1] It was new life to me to know that the same emotions feelings and generous confidence still existed between us unimpaired by either time or distance and the same thoughts and actions which bound us in early years are still cherished and blooms as freshly in our bosoms as our native Shamrock[2] upon its native earth. I hope they will know no blight upon this side the grave and I trust we shall meet again in those blest abodes where no such thing as change is known. You seem to think that I have looked on the wrong side of the page of our boyhood and that I had passed over all the mishaps, troubles, and perplexities of that period unheeded. Now with all due deference to your opinions, I must say that it would be a great piece of folly in me to have done the contrary, for the dark side of the present is quite enough without ransacking the past for perplexities. I hope we may always experience the same pleasure in looking at the actions of our manhood that we do in beholding those of our earlier years. You also seem to imagine that the scenes and employments of our young days would not have the same attractions for us now that they had then. Granted, still Master Tyndall you must not think that I am either incapable or indifferent of enjoyments of that sort. Perhaps its because I am the father of a family that you would make the insinuation. But I say married men against old batchelors any day. You may regard that epithet lightly, but you had better look sharp; you have numbered nearly as many years as I have and have not much time to lose. Do not pass by this little bit of advice but take advantage of it and profit by it. One reason why I wish to get you in the traces is that already I am ahead of you in regard of family and if you do not make extraordinary haste I shall lessen the chances of your overtaking me. A wee bit about this gossoon[3] of mine. He has taken pretty much after your style tall and lovely as a lark and oftener in mischief than any other of his age I ever saw, in fact a real harum-scarum chap and if his mischief keeps pace with his years and stature I do not know what I shall do with him. So I find you have had your share of

> accidents by flood and field
> of hair breadthscapes[4]

I am glad you had so providential an escape from your visit to the wreck of the illfated Killarny.[5] The terrors of the land are nothing when compared with those of the mighty deep. You had difficulties to encounter by land also, I wish I was near you when you were on that shooting scrape and I would have slipped a corn dodger[6] in your fist. Oh its fine receipt for allaying those unpleasant feelings which one experiences when the stomach is empty, particularly when accompanied by a slice of raw pork. I was sorry to see that them ere gulls gull'd you.[7] But if you had borrowed my Kentucky rifle[8] it would have made them keep at a respectful distance – they are death on anything within 200 yards, those who are expert with them can take down a squirrel at that distance without making a hole in the skin that is by shooting them through the eyes. I have had some very pleasant days out shooting and fishing last fall. Pigeon shooting is very pleasant at that season of the year. They arrive in immense quantities to feed upon the Beech nuts and stop here some two or three weeks and sometimes stop as long as 6 or 8 weeks according as the weather holds, if it sets in cold they immediately take their departure for the South I have been told by persons in whose veracity I can rely that some seasons they arrive in such numbers that at night when they roost it is very dangerous to travel through in consequence of the falling of the limbs of trees borne down by their weight. I suppose you would like to hear a little about Cincinnati, its situation commerce &c. It is bounded on the South by the Ohio river and encompassed on North East and West by hills, so that it is situated somewhat like the pit of a theatre, the river occupying the place of the stage. The leading streets run nearly due North and South crossed by others which run East and West. The streets are numbered (as almost all the new cities in States are) from No 1. up in numerical order, so that it is easy for a stranger to make out the place he wants without much enquiry. Cincinnati being the largest city West of the mountains,[9] it is the great depot for supplying North parts of old Kentucky, Indiana Illinois territory of Iowa and South part of Ohio with manufacturies of all kinds, receiving and shipping the produce of these States which they receive in return for these manufactures. And as for pork why we beat the world all hollow. Imagine, if you can 300,000 hogs slaughtered in this city in 2 or 3 months as they are obliged to confine the year's killing to the cold season. Its astonishing what a number of hands are employed in this one business Coopers employed the whole year making Barrels and kegs, The number of light boats[10] which it gives employment to conveying it to N. Orleans[11] and returning laden with the productions of the sunny South, particularly groceries. It is no unusual sight to see ten Lt boats[12] puffing away at our wharfs preparing to start up or down the river laden to the guards with freight, passengers, and their luggage, as they can accommodate 300 passengers easily independent of freight. The building of the boats gives a great deal of employment to our mechanics, its greatly facilitated by the snags, which are plenty in the Mississippi and Ohio rivers. They are large trees which are carried away by the freshets[13] off the banks and placed in the channel. It frequently happens that where a boat runs on her downward trip, when returning in exactly the same place she strikes a snag and is sunk. Loss of life is small compared with what one would expect, as they generally have steam enough up to carry the boat to either side of the river after the accident occurs. I have seen in one of last week's papers the report of 4 boats snag'd each of which was worth 30,000 Dollars or £6000 independent of their cargo. In fact all the crack boats of West have been lost this winter. You would like I suppose to hear something respecting your humble servant I am engaged in the dry goods[14] grocery and 21 doz.[15] other trades such as the shoe trade Queens ware[16] and hard ware, in fact it would puzzle you to come into our stores and ask for an article we have not got. Business is carried on very differently here from what it is at home. There is a son-in-law and son of my uncles in company along with me. We trade under the cognomen of Wright and Company and I am happy to add are doing a very safe business, although I do not much relish a partnership concern, as one of the parties are generally imposed upon I expect as soon as our limitation expires to commence on my own a/c. My wife,[17] Sanders and his wife[18] attend the dry goods department; and my cousin and I the other – or grocery &c.

March 21st. the interval between the dates of this letter has been big with events; a sad catastrophy occurred by the explosion of a pork house which was on fire[19] by which a number of our most respectable men hurried to untimely graves.[20] As it is likely you will see the particulars in your newspaper[21] I have omitted to mention them. Another event which at present creates a great deal of curiosity and anxiety is the appearance in the Heavens of a bright beam Light extending from the Hudson to the belt of Ocean its breadth is in appearance about 1½ to 2 degrees. It makes its appearance about an hour after sunset and remains visible till ½ past 8 o'clock and gradually sets in the same places as the sun. Our learnd folks suppose it to be the tail of a comet and late discoveries in the East of its body seem to settle it in their opinion that it is one.[22] In the East there is a great deal of excitement on foot in consequence of the preaching of a Mr Miller[23] who asserts that the world terminates on this year,[24] and this appearance of a heavenly body is a great source of alarm at any time, it is doubly so now, to a great number of individuals who believe in the above gentleman's doctrine. It is computed that he has one million followers in the East and numbers of them are scattered throughout the republic[25] I myself am a little inclined that way I believe it is just as likely to occur this year as for 10 years hence as occur it must sometime (may God in his infinite mercy prepare us for it and we need not care then when it does come). Another prevention which I met with was we purchased a large lot of dry goods which were sunk in the Ohio river and hastening the sale of them occupied my attention almost day and night as we wanted to get them off hands before they were injured; the ducking did not injure the material at all but it spoiled the sale of them. The moment I got my hands loose I set to finish this epistle you must not imagine that these are the only lines I pen'd to you since I received yours, I assure you I have written some 4 or 5 letters and in consequence of their bad construction I consigned them to the flames. I now forward you this one although no better than its fellows, for I find if I was to wait until I would write one as it should be or as I would like it to be, it would be a long time before you would receive one from me. However if I have not conveyed my ideas to you as they should be, I trust you will attribute it more to the head than heart, for be assured you occupy the most exalted position there of any of my comrades in early life I would now make a request for you before I close and that is to loose no time in letting me have a letter from you. So farewell I hope we may meet again

I remain your attached friends | W. Wright.

P.S. Remember me to your Father and Mother I hope they are in good health. Also your uncles Caleb, Jack[26] and William's family You will much oblige me by letting my Father[27] and Mother know that you have heard from me and that we are all in good health, in fact enjoy as good health as at home, I never felt the least change or inconvenience since I came here except the heat in summer which is very great. This is an unusually late Spring here the ground is quite bound up with frost although at this season of the year it has been customary to have the market stocked with new vegetables I could not raise a bunch of clover for a Shamrock on Patrick's day[28] I felt the loss of it very much as I am proud of being a Paddy.

Once more good bye | W. Wright

I wish I could slip in a small piece of paper as I have got a great deal more to say. But it is impossible, as we are obliged to pay double postage for the smallest enclosure, they only allow a single sheet for one postage it is very high compared with what it is with you.[29]

Direct your next to | William Wright | Cheapside | Lower Market Street | Cincinnati | State of Ohio.

RI MS JT/1/5/1804–7
LT Transcript Only

1. *your always but now doubly dear letter*: letter missing.
2. *our native Shamrock*: see letter 0190, n. 6.
3. *gossoon*: boy (*OED*).
4. *accidents by flood and field | of hair breadthscapes*: W. Shakespeare, *Othello*, I.iii.134–5.
5. *your visit to the wreck of the illfated Killarny*: Tyndall had made a perilous trip with Phillip Evans and John Tidmarsh to view the half-submerged wreck of the steamship *Killarney* at Barry's Head, east of Kinsale, on 8 August 1841; see letters 0049, n. 1 and 0091, n. 11.
6. *corn dodger*: an American cake or bread made of the meal of Indian corn, baked very hard (*OED*).
7. *them ere gulls gull'd you*: to be gulled is to be duped, deceived or befooled (*OED*). In his handwritten narrative of the trip to Barry Head, Tyndall remarked: 'here the cormorant and the gull find a safe retreat the cliffs were totally covered with both as we passed by unacquainted with man they are heedless of his power seldom has a human foot dared the frowning steeps that form their safeguard' (RI MS JT/8/1/2, pp. 60–3, on p. 62).
8. *Kentucky rifle*: the nickname of the American Longrifle, a gun with a much longer barrel than those used in Europe.
9. *the mountains*: the Appalachian Mountains.
10. *light boats*: low open boats used for moving goods to and from ships, also known as lighters.
11. *N. Orleans*: New Orleans.

12. *Lt boats*: light boats; see n. 10.
13. *freshets*: small streams of fresh water (*OED*).
14. *dry goods*: American name for the class of merchandise comprising textile fabrics and related things; articles of drapery, mercery, and haberdashery (*OED*).
15. *doz.*: dozen.
16. *Queens ware*: earthenware ceramics (*OED*).
17. *My wife*: Sarah Maria Wright (née Neale).
18. *Sanders and his wife*: not identified.
19. *explosion of a pork house which was on fire*: The explosion and fire occurred on Sunday 25 February 1843 in the smokehouse of the pork-packing firm of Pugh and Alvord on Walnut Street in Cincinnati.
20. *by which a number of our most respectable men hurried to untimely graves*: The *New World* reported that the 'appearance of the explosion was awfully terrific, and its consequences fatal to several most estimable citizens ... Of the number that were positively ascertained to be killed on Sunday were Joseph Borsall, Caleb Taylor, of the late firm of Woodnut and Co., H. S. Edmunds, pork packer, J. S. Chamberlain, pattern maker, John Ohe, a German laborer, and two or three other German laborers' ('Great Fire in Cincinnati', 11 March 1843, p. 310).
21. *it is likely you will see the particulars in your newspaper*: The explosion does not seem to have been reported in any British or Irish newspapers.
22. *Our learnd folks suppose it to be the tail of a comet ... settle it in their opinion that it is one*: The Great Comet of 1843 (C/1843 D1), which was visible between 5 February and 19 April, developed a tail *c.* 108,000,000 miles long three weeks after its perihelion passage. The American astronomers who contributed most to the determination of the comet's identity were Joseph Stillman Hubbard (1823–63), of Yale College, and Benjamin Peirce (1809–80), Perkins Professor of Mathematics and Astronomy at Harvard University.
23. *Mr Miller*: William Miller (1782–1849), a farmer and Baptist layman in Low Hampton, New York, who used Biblical prophecy to preach that the Second Coming of Christ was at hand.
24. *the world terminates on this year*: In 1822 Miller had proclaimed 'I believe that the second coming of Jesus Christ is near, even at the door, even within twenty-one years,–on or before 1843' (S. Bliss, *Memoirs of William Miller* (Boston: Joshua V. Himes, 1853), p. 79).
25. *he has one million followers in the East and numbers of them are scattered throughout the republic*: The cult of Millerism grew rapidly in the early 1840s, led by Joshua Vaughan Himes in Boston, whose Adventist newspaper *Signs of the Times* helped spread the movement across the country.
26. *Jack*: not identified, although Wright might possibly mean Tyndall's father's younger brother Jacob Tyndall (1804–79).
27. *my Father*: Wright was unaware that his father had died recently; see letter 0187.
28. *Patrick's day*: 17 March.
29. *it is very high compared with what it is with you*: Unlike the Uniform Penny Post in Britain and Ireland (see letter 0177, n. 10), American postage rates, known as 'zone rates', differed according to the distance between the post office where a letter was sent and its destination.

From William Ginty 14 March 1843 0194

Tuesday evening

My dear Tyndall

I received the transcript of your 'Sober Seriousness'[1] from Tidmarsh this afternoon and may I be struck serious if it has not sobered me to the very lowest ebb of pump-water sublimity What in the name of common sense does this harangue mean. 'Tyndall's compliments' (how infernal polite!); 'Singular circumstance' (very singular 'pon my soul!) 'Serious cause' (serious effect too!); 'quite ignorant' (strange coincidence!); 'Discontinue correspondence' (symptoms of lunacy!); 'State the fact' (monomaniac I suppose!); 'Sober seriousness' (raging madness!); 'Speedy explanation' (A hopeless case I fear!). Now let me to counsel – My Lords and Gentlemen of the Jury. I the accused Ginty – prisoner at the bar of your most honourable court of conscience am indicted under the unmanly and un-Milesian[2] charge of 'SILENCE'. Gentlemen this is a crime for which my abhorence is as ungovernable as my innocence is unquestionable. If, gentlemen, I was guilty, and could show cause or provocation for the hideous crime, you would not longer detain me – but gentlemen it is the extreme reverse. I am not guilty – neither have I had cause – nor have I been silent. So fully confident am I of your unanimous acquittal – that I shall only detain you while I call on one evidence. Postman Mailbags Esquire:– I am letter stamper at Liverpool; I recollect the evening of the 28th ultimo,[3] I remember stamping the letter alluded to by the prisoner. I could not suspect him of the accusation – I know the prisoner's handwriting well. The Court:– Gentlemen of the jury, you have heard this conclusive evidence together with what the prisoner has urged in his defence. You can have no hesitation in acquitting the prisoner. A Juror – My Lord we are of opinion that the prosecutor is no longer a safe member of society – as no doubts can remain of his insanity – we would recommend your Lordship to commit him to the Ass--ylam. The Court. This is beyond our Jurisdiction, as we have not any indictment – Gentlemen I await your verdict. The Jury. NOT GUILTY GINTY.

> 'The court awards it and the law doth give it'.[4]

We were honoured by a visit from whats left of the Colonel[5] to day – but you had him yesterday – So let that pass. We had a precious fire blaze here last week at Fawcett's and Preston's foundry[6] where Chadwick is bound. On Tuesday morning last about 2 o'clock the glare of the flames aroused us altho' at least 3 minutes walk from here and 3 streets between us and it. I could hardly persuade my half asleep senses that the windows were not of red-stained glass, so brilliant were the heavens illuminated, every star was eclipsed and the yellow-silvery moon looked as if she got a cup of ermine thrown in her face– from the foundry it spread like fire to some adjacent cotton stores, on the whole it was all that I wished it 'great glorious, and free'.[7] Just imagine what it was, when Willy[8] was first to move for an adjournment to the scene of blazes – exclaiming with a half yawn – half snore 'oh if Tyndall was here'!!! We remained witnessing the operations of some 20 or 30 engines until the bell tolled the ominous five.[9] There was a great deal of damage done[10] and even this moment there are some engines playing on it.

Hamley took it into his head last week to place me under the necessity of risking the receival of 2s and 2d per diem[11] at some short future period. The fourth time may have the charm. Devil admire and recommend them to his prime-minister[12] – if ever I turn prayer vendor[13] I'll be 'at their service'.

What am I to do with this eighteen pence of yours? I thought you would have commisioned Willy as a stock-agent.[14] Beware Tyndall – there is great talk of a reduction![15]

Jack[16] and Bill Latimer are a distinct mess now, five together was too much, particularly when them 5 were mad devils, and more particularly when them 5 devils were 5 Irish mad devils – they are now in St. Georges Terrace![17] We are about leaving here, it is nearly 10 minutes walk from 33 Windsor St.[18] Is it a fact that you and Willy and Jim[19] caught George[20] one evening about to prostrate himself before the shrine of Miss Mullhall's[21] beauty!

Of course you know and must be well aware what my opinions are regarding the valentine and wild effusion which you sent me last week.[22] Your curses were delightful. One feels pleasure in being cursed so. I would have written on Sunday evening but I went to church and wrote the following while the themes of the preacher were yet tingling in my ears. You shall see what circumstances can do even with me. I would have written last night but I went with some others to hear a favourite piece in the 'Theatre Royal'.[23] Mrs Fitzwilliam and Mr Blackstone have just left the 'Liver'.[24] 'Oh she's a charming woman and he is a most wonderful man'. But now for my Sunday night's work!–

Should curses thus your pen control
  And thus consign immortal soul
Where flaming billows madly roll
  For that dull valentine!

Hear me ye powers of earth and air
While I vociferate a prayer
That such may ever be his share
  Who wildly wished it mine!

May fiendish temper's fearful ire
Which jealous souls alone inspire
His haughty bosom never fire
Nor curses scape his tongue!

Calmed be that passion – dire and dread
Which wreaked its bitters on my head
Its horror hushed – its vengeance dead
Its very echos gone!

Gone like a ripple on a wave
Lost like a hope within the grave
May soothing Lethes[25] fondly lave[26]
Athwart his burning breast!

And when his glass of life hath run
– His griefs forgot – his sorrows done –
– The Goal of goals well surely won
A calm celestial rest!
÷ ÷ ÷ ÷ ÷ ÷ ÷ ÷ ÷ ÷ ÷ ÷ ÷ ÷ ÷ ÷
Yes beauteous Ellen[27] – twas from you
That all this fiery vengeance grew
As grows the nettle in the dew
That's dripping from the rose!

From you the frown – the angry curse
Unconscious sprung for him to nurse
And shower on me with bitter force
Scum of his heart's hot throes!

From you – but no! that thought must rest
The name alone disturbs this breast
Tis graven here! – by grief impressed
By time but firmer fixed!

By sorrow nurtured – hope caressed –
In dreams alone an angel-guest
Fond dreams! to life the only zest
A cup of sweets unmixed!

On – on may roll fast flitting years
While every hour my spirit sears
And memory's tablet stains with tears
For thee and thee alone!
But even tho' Time his hoary pall

Shrouds o'er thy beauties one and all
That spirit still will on thee call
And hail thee for its own!

When love's forgot – when hope is gone
Thy name shall guild his every song
Thy praises glitter on his tongue
Thy image in his breast!

And when the cold sepulchral stone
Rears its dread o'er beauties flown
The soul that wished thee once its own
Will wish thee purer rest!

The clock struck 10 just as I came to the centre of this page, so that delays this a post. I suppose by the time this reaches you, your ire will be raging in all the horrors of Ethnic splendour. There was one little circumstance which attracted my attention very particularly about that valentine – How came my name under it and at the right hand side? as if I wrote it – and to make my suspicions more substantial – you then drew the pen thro' the 'W. Ginty'. You know what I mean! I say Tyndall – you dont say the Goddess Muatta[28] controled my pen to night at all events

Yours very sincerely | Ginty.

Either read or give the enclosed – to the 'wee bit of a thing'[29] (Oxford St., to wit)[30] We shall see the effect. It is now nearly three months since I wrote to him and he never answered it.

Lilly came in to see me this evening in company with Taylor. Lilly has resigned![31] He remains here for a few days with his uncle on his way to the land of his fore-fathers.

O'Neill's journal went with mine[32] – The only two out of 140 civilians and 20 or 30 sappers --!!! --!!! So that in 70[33] we may expect another!

RI MS JT 1/11/3596–7
LT Transcript Only

1. *the transcript of your 'Sober Seriousness'*: letter missing.
2. *un-Milesian*: un-Irish; see letter 0183, n. 23.
3. ultimo: of last month (*OED*), so 28 February.
4. *'The court awards it and the law doth give it'*: W. Shakespeare, *The Merchant of Venice*, IV.i.300.
5. *the Colonel*: Colonel Thomas Colby.
6. *fire blaze here last week at Fawcett's and Preston's foundry*: On Tuesday 7 March 1843 a huge fire partially destroyed the Phoenix Foundry of the Fawcett and Preston Engineering company, which stood between Suffolk and York Streets in Liverpool.
7. *'great glorious, and free'*: T. Moore, 'Remember Thee!' (1821), 5.
8. *Willy*: William Hunter.
9. *the bell tolled the ominous five*: The *Liverpool Mercury* reported that the 'flames raged with great fury till about six o'clock, when a plentiful supply of water having been obtained, the efforts of the Fire-police were of some avail' ('Partial Destruction of Messrs. Fawcett and Preston's Foundry, by Fire', 10 March 1843, p. 82).
10. *a great deal of damage done*: The *Liverpool Mercury* reported that 'the ruins had the appearance of a desolated town, so extensive was the establishment ... Happily, no lives were lost at this fire, nor have we heard of any person being seriously injured. The disaster, however, cannot fail to be most calamitous in its effects' (ibid.).
11. *per diem*: per day (Latin).
12. *his prime-minister*: presumably Robert Peel, then serving his second term as Prime Min-

ister.

13. *prayer vendor*: Ginty presumably has in mind the Roman Catholic clergy whose practice of selling indulgences, a temporal substitute for prayer in the remission of sins, helped provoke the Reformation.
14. *a stock-agent*: a dealer in livestock (*OED*).
15. *there is great talk of a reduction*: In a District Order issued on 9 May 1843 Captain Henry Tucker stated: 'Warning was given some time since that in consequence of the smallness of the grant of money for carrying on the survey it would be necessary to make a reduction of the force employed on the Detail Survey' (RI MS JT 8/1/4b).
16. *Jack*: John Tidmarsh.
17. *St. Georges Terrace*: on Comus Street in central Liverpool.
18. *33 Windsor St*: the location of the Ordnance Survey Office in Toxteth Park, Liverpool, south-east of St George's Terrace.
19. *Jim*: Phillip Evans.
20. *George*: George Latimer.
21. *Miss Mullhall*: not identified.
22. *the valentine and wild effusion which you sent me last week*: letter missing.
23. *a favourite piece in the 'Theatre Royal'*: Opened in 1772, the Theatre Royal on Williamson Square was the first theatre to be built in Liverpool. On the evening of Monday 13 February 1843, it hosted 'MRS. CLARKE'S BENEFIT', featuring 'this favourite actress and highly talented lady' in 'a performance ... of a most attractive character' with 'one of the fullest and most fashionable houses of the season to reward her exertions' ('MRS. CLARKE'S BENEFIT', *Liverpool Mercury*, 10 March 1843, p. 84).
24. *Mrs Fitzwilliam and Mr Blackstone have just left the 'Liver'*: Fanny Fitzwilliam (1801–54) and John Baldwin Blackstone (1802–79), two of the most popular actors of the period. Ginty is likely referring to their appearance in Blackstone's own opera *The Pet of the Petticoates* (1841) at the Liver Theatre on Church Street. The Liver opened in 1829, when the Pantheon Theatre, the second theatre built in Liverpool, was renamed. The owners of the Theatre Royal and the Liver were in dispute over the latter's unlicensed status, a controversy which, later in 1843, led to the passing of the Regulation of Theatres Act.
25. *Lethes*: see letter 0132, n. 1.
26. *lave*: wash or bathe (*OED*).
27. *beauteous Ellen*: from Kinsale.
28. *Goddess Muatta*: There is no goddess Muatta in any world religion, so this is perhaps a mistaken reference to Mutata, an alternative name for Ino, one the four daughters of the Greek god Comus.
29. *the* 'wee bit of a *thing*': not identified, but LT's handwritten insertion suggests this was 'proby. [Phillip] Evans who was small'.
30. *(Oxford St., to wit)*: Tyndall had lodged at 18 Oxford Street in Preston; see letter 0164.
31. *Lilly has resigned!*: John Lilly had become disillusioned with the Survey's low pay, as well as the 'dull and stupid monotony of plotting and drawing', although he had no alternative job to go into when he returned to Ireland and instead occupied himself with gardening; see letter 0206.
32. *O'Neill's journal went with mine*: presumably the 'Journal of Progress and Weekly Report', which, after they were submitted, were used to compile the 'Monthly Report'; see letter 0038, n. 3.
33. *in 70*: possibly in 70 days' time, as Ginty had previously complained of having to wait for

the return of 'Progress Journals | Of which no tidings yet are come'; see letter 0143. In fact, Ginty's journal was returned after a 'six weeks "leave of absence"' (i.e. 42 days); see letter 0200.

From Archibald McLachlan22 March–1 April 1843 0195

25 Byron Street[1] | begun on 22nd March, ended on 1. April[2]

Col. C. and Cap. T.[3] were at York – it appears that the O.S.[4] does not get on so well as could be wished. Col C. wishes to curtail or abridge the mechanical details, in surveying – calculating – registry &c. Nothing is to be done but what is indispensible – yet here, in Leeds, 5 feet[5] – every little insignificant walk and flower bed – almost every tree has to be fixed like a Trig point,[6] mere ornament takes up double the time of buildings – everything is done as if delay and expense were the only objects in view – just as if we were young ladies following a favourite amusement. Mere trifling with work. The Colonel made a speech to the gentn.[7] of the office – to the following effect –'I am sorry to inform you that I shall be under the necessity of reducing the establishment, for want of funds – yet I hope to have it in my power to employ in a co-lateral branch, all those who may be discharged'. This co-lateral branch is supposed to be the City of London[8] – Yet it is certain that all who have been discharged, and all who shall be discharged – are discharged, not for want of funds, but for some fault, or default – Now these, when re-employed, will on this account be re-employed at a reduced rate of pay!!! At York, some of the 'Board men'[9] are employed on the Boundary survey[10] 14/. to 20/. 'per diem'[11] – With them the orders are so stringent, so annoying &c. that one man named McLoughlin[12] who is styled Cap. McLoughlin on account of his being in the Royal Military Academy[13] as 'teacher of military drawing', he has resigned, and now must, from necessity, resign! At Leeds, we have last week got an order, to parade at the Royal Exchange, chains, sketches &c., this is the very centre of the town, and we are every morning the subjects of remark, our work lies South of our lodgings – and we have to travel more than a mile North to parade, then from parade we have to travel 2 miles South again to commence work! This was told to Cap: B.[14] He said. I can't help it, I have got orders for that purpose (of parading) So you may easily tell how this will end. It is reported that Col. C. some time ago, did propose – or say he could, make maps for the Tithe at 0/6 more than the 6 inch cost – i.e. – for 1/6 per acre.[15] – Query. – Does not Capt. T. suspect some of the old hands to be at the bottom of, or assisting in some way, the late resistance made at Preston, on the Tithe affair!![16] And does he not wish to get rid of the older men so that, should any fair offer be made of Tithe maps, none of these busy bodies will be in existence to answer any questions of private surveyors, or give any information as to the nature of the O.S.? Did you see the Parliamentary news of 28th?[17] Has

Sir Robert[18] proposed the Survey of London, and by whom is it to be done – It has struck me that if the Weekly Despatch[19] would insert some letters to Sir Robert during the progress of the Bill,[20] they would be of use; again I suppose that an address in a small pamphlet, sent to Land Surveyors of Lancashire and York,[21] would be a more effectual method of calling the attention of the parties interested. A small pamphlet, sent to Landsurveyors, Engineers, Agents, Members of Parliament, boys of the survey – present and past – This I think would be a good plan, as Newspapers are so easily lost. – If published in one paper only, others, perhaps, would not copy, or only on a limited scale. Which would be best – Newspapers, or pamphlets? In a pamphlet one could go more into the subject without daring the patience of anybody. In a Newspaper, several letters would be necessary, and perhaps tire the Editor – Besides, in a pamphlet one could say what an Editor would be afraid of publishing, for fear of a prosecution. – In a pamphlet, it would commence at 1825, follow the Irish Survey from beginning to end, then commence the English Survey and go into the politics of the affair, so that professional men would have a view of the whole subject – Will you let me know your thoughts on this – Can you find out the true cost of an officer of Engineers, that is, his allowance on the Irish, and English survey, there are so many accounts of this that I am uncertain of the truth. Does he receive pay, and half pay – and what is it? – allowance for Ho:[22] Horse, servant, travelling, wine &c. &c. Also the true cost of one Sapper. Cap B. valued him at 3/.: Is the Military pay included in the Grant for the survey, or is it possible to be included in the general estimate for R.S. and M.?[23] Calculate the cost of the Irish survey from 1825 to the end of 1830 – keeping distinct Cap: R.E[24] Lieut R.E. – R. S. and M., C.A.,[25] cost of chains, theodolites and scales &c., paper and books. In 1829 Col. C. said in Committee of the House of Commons that the cost up to that time (end of autumn) was £77, 000, this must be wrong. – Our Field work is still carried on as if it was to be laid down on a large scale.

Yours very truly | McLachlan

No.1 –

Preface to The Ordnance Survey of Ireland.

Section 1.

This pamphlet is designed merely as a preface to the History of the Ordnance Survey of Ireland; to enquire into, or dispose of, much preliminary matter which could not be embodied in the general History. It is also printed and distributed to a limited circle, with the view of calling the attention of those who are now, or have been at any former period employed on the O.S. And they are hereby respectfully requested to refresh their memories on the several points touched on in this preface, to put their thoughts on paper in the form of notes, giving as near as possible, names, places, dates &c. – and to refer to official documents, official letters, or their own personal correspondence – At no distant day, their notes will be called for (in a manner hereafter to be communicated) and persons giving important information shall be entitled to a copy of the History, gratis. –

Two questions may very reasonably be asked – 1st Why was the History not published sooner, at least the merits of the case ought to have been discussed during the progress of the Survey, when the incompetence of the parties were first discovered, when the R.E. officers were on the spot, could answer all questions and account to the public for all apparent failings, and when the evils detailed could be remedied. The 2nd question – Why is this History published now, all interest in the affair having long since ceased – why publish now, when it may be supposed no good result will follow!

To the first question and to all such, under every variety of form it may be replied, that during the progress of the Survey from 1825 to 1831, it was proposed to discuss publicly the merits of the case; but to this it was objected (and some members of Parliament took this view of the question)– That the sums of money voted were so great, and the incompetence of the parties so manifest, their professional practical knowledge so utterly contemptible and worthless, that should a public exposure be made, the Board of Ordnance would be obliged to abandon the work, it was also said that a Survey of some sort was necessary and although the O.S. would not answer the purpose for which it was designed, yet still it would be better than any previous Survey, and with all its faults, would be found of great National advantage. An answer to the second question will best be found in a careful perusal of all this preface; but it may be well to observe here, that you cannot judge of results until certain experiments have had a fair trial – The whole case has assumed a new character. Arguments applicable to 1827 cannot fully apply to 1843. From 1825 to 1831 the experiment was made altogether at the public expense. The experiments of 1843 are of a more baneful kind, affecting the very professional existence of a large body of private practitioners. In 1843 an attempt is made to recover a professional character forfeited between the years 1825 and 1831. Let it not be forgotten that something approaching to Collusion is suspected. The difficulties encountered by land holders and Land Surveyors in Tithe maps coupled with the recent extraordinary proceedings, subjects the character of the R.E. to much injurious suspicion, and certainly their present line of conduct requires some explanation – The O.S. both of Ireland and England has hitherto escaped public animadversion, this has emboldened the Ordnance officers to proceed much further than the nature of the office will warrant; it remains to be proved whether they will persist in their present course –

It is said that their Superintendent[26] solicits examination and enquiry: If he feels interested in having himself and those under him placed in their proper position in scientific society, he has now an opportunity of doing so, before the publication of the History, as sufficient time will be allowed for any counter history to appear. But if at the end of 12 months no notice is taken of the Preface, then it may fairly be concluded that the facts are indisputable, and admitted to be so, by competent judges. And we shall proceed with the History of the most extraordinary proceedings (in a scientific point of view) ever presented to public notice this or any other public!

This first section is concluded with a repetition viz all those who are in possession of important information are requested to put it on paper, and when the proper time arrives it shall be called for

B.

Criticise this, like an Editor – in a few words – not the grammar, but the points. The grammar is beneath notice. Cut and slash right and left, all in 10 lines. In a few days

Yours very Truly | B. | Did you get my last posted Friday?

Retrospective View of the Ordnance Survey of Ireland. And an examination into the present state of the Ordnance Survey of England

---

Address to the Master General of the Ordnance.[27] 1843.

The retrospective view of the Ordnance S. of I.[28] and examination of the present state of the O.S. of E.[29] is respectfully laid before the Master General of the Ordnance, who, from his high character and public station, cannot be influenced by that class of causes, which may actuate others. He has high and important public duties to perform, and works of public interest committed to his care.

He is addressed with confidence, and respect, because it is known, that when any statement is fully and fairly laid before him, touching the public business over which he presides such statements have always received a full and impartial consideration.

With respect to the O.S. of I. the Master G. will remember that from 1825 to 1830 regular official returns were made of the progress and expences of the work, and he will recollect how grievously he was disappointed in 1830 in the Actual state of the work,[30] notwithstanding all the official returns.

The Master General is by his very position necessarily far removed from the scene of actual operations, and can know from official returns only, the state of the work, and the object of this pamphlet, is to communicate information which cannot be obtained from official returns.

The Master General is not aware that one third of the money expended on the O.S. of E. is lost by mismanagement – nor is he aware that money which has been voted by Parliament for the purposes of the Ordnance Survey of the British Isles, has to a certain extent, been appropriated to an undertaking never contemplated by Parliament, nor the B. of Ordnance: And that this misdirection of public money is still persevered in – this is a line of conduct which appears to require explanation.

The Master General is well aware that the vigorous prosecution of a public work, or a work in which many are employed, depends materially on the opinions, or feelings, which those employed entertain of their situation, circumstances affecting their minds have a corresponding effect on the progress of the work, causing either energy or apathy.

Distress and discontent exists among the men employed on the O.S. of E. and to such an extent as materially to retard the progress of the work, this is now become too evident, by the tardy, sleepy slothful manner with which operation drags its slow length along – it is evident by the manner in which all speak of themselves and their superiors, retrospective views are taken, calculations made, and inferences drawn from what are considered to be indisputable facts – and the result is increased dis-satisfaction – It is believed by the men employed on the Ordnance S. that the M.G.[31] takes an interest in the advancement of the work, and by consequence, an interest in the welfare of all those upon whom the advancement almost entirely depends – this consideration tends to increase the bitterness of disappointment in many of those engaged in the work, they have left their home, are strangers in the place of their birth, they have spent the health and vigour of youth, endured all the fatigues of a National Survey in the heat of summer and cold of winter – are now unfitted for any other employment, are very useful in their present one, and have no wish to leave it – And it is a melancholy fact, that after 10 or 17 years arduous exertions many of the best hands have not received one half the promised remuneration – and the disappointment is still further increased when it is reflected on, that sufficient funds are placed at the disposal of the Board of Ordnance – <u>and these funds are mismanaged</u>. As enough has been said to convince the Master General that this pamphlet is the result of pressing grievances – and as a lengthened detail in this place would be inconsistent with the limits of it, the Master General is here earnestly and respectfully requested to peruse the contents and to cause an examination, to see 'whether those things are so'.[32] And it is earnestly hoped that he may have it in his power to remove the cause of discontent

B.[33]

The reply of X[34] was received to day – Wednesday – It is always good to take 2 views of a question. The 1st was intended to start the proposition – how shall the thing commence – by an appeal to the Public or the Parliament?

X. will compare the tone of the 2 views given by B. – B thinks that neither the one nor the other of these views is good, but something of a middle course should be taken – D[35] is very cool and cautious on these subjects and can point out all objectionable expressions – and the points – most likely to draw attention and examination from certain parties.

B has given his views of the opening chapter, and shall say nothing more on that head – of course X and D will consult and prepare the 1st section, as 'tis not prudent to let much matter accumulate.

The other chapters will appear in 'rapid succession'.

B.

To me at Mr. Shoutesmith Plumber & Glazier.[36] Bishopshill. Senior.[37] York.

The last from X has been received – the enclosure of Y[38] will soon be returned.

Y is a new comet!! I hope he may show a long fiery tail high in the zenith – with nucleus, nebulae, and nebulosity!! A comet as Milton describes. frightening and perplexing monarchs with a fear of change!!![39]

The introduction of Y will cause a change in the arrangement. B is of opinion that his two chapters on the prospectus should be thrown on one side until we see what Y has to say. The preface to any book tho' placed at the beginning is always last written; and perhaps this would be judicious in the present case – let these two chapters lay by till the conclusion and perhaps something good on the point may come from Y.

Should X not put this question to Y for his future consideration? viz. Whether the thing should be addressed to the Master General; To the Irish Members of the House of Commons; or, to the Civil Engineers and Land Surveyors of Great Britain and Ireland? B will think on this subject.

It is quite clear that the Building must be erected by X. B is a good drudge, and perhaps Y may be so too. X is the architect, and assisted by the cool head D, cannot fail to do justice both to himself and the subject. B and Y should act simultaneously, and to ensure this – X should demand from both, a full and complete examination into the several points – allowing, 3 days to each point to the consideration of each point – each point to take up one or more letters – each letter to have a broad margin like the present sheet, for the convenience of X for references – The following are some of the leading points. X will follow up the items in future letters. The first point is nearly completed by B. But Y should as soon as possible reply to the first point also – NB do not allow matter to accumulate.

1st State the cause or causes of the O.S. of I. – Give something on the Parliamentary preliminaries – name the parties examined before the Committee of the House of Commons – State the object to be attained by the Survey – (the Parliamentary objects) The estimated time and expenses. Who was entrusted with the execution of it. Were the parties entrusted qualified by previous practice, Reason for entrusting the details of the work to the parties employed, from whom did they receive their orders?

2nd Valuation: State the reasons of it – or causes – To whom was that part of the work entrusted – reasons for entrusting them. State their previous professional career – from whom did they receive their orders?

3rd Describe the nature of the valuation,[40] its extent, its objects, state the reasons why good maps are necessary, from whom did the practical details proceed, from Mr G.[41] or the Board of Ordnance: or from some R.E. by the Boards order?

4th Describe how maps are used on the O.S. of I. by the valuators. Describe a good map fit for the above purposes.

5th Enquire into the Parliamentary or Board of Ordinance's meaning attached to the words 'Townland Survey'.[42] Do they mean the outlines or boundaries merely, or otherwise.

6th Enquire into – whether Colonel C. and Mr G. had a correspondence or mutual understanding on their relative duties as Land Surveyor and Land Valuator and the nature of the maps required – or – whether the O.S.I. was begun without any previous understanding and consequently each following (separately) the orders he had received from their employers. Whether Colonel C, knew or did not know the nature of the future valuation – whether he knew or did not know the orders Mr G. had received.

7th Enquire into – a book called "The Colonel's Blue Book".[43] Who is the real author, Col. C. or the Board of Ordnance. Examine its contents – when was it written –what was the object in view by writing it – give an opinion on its contents.

And so going over every item, down to the present hour. A closely written letter may embrace perhaps 3 points – if so, each point should not be jumbled, but kept distinct by a broad black line – all for the sake of ready reference.

With respect to contents of the last from X: it is good, keep it in reserve: B intends to follow the order of events – just as they have occured, not an item, not a single pin point shall be overlooked, or passed over in silence.

Perhaps it would not be well to insert anything in newspapers at present, it would awaken suspicion of the authors – every action, every expression would be watched, the post office would be narrowly watched too, and any stray letter opened as in the case of Captain Fenwick!![44] The explosion will take place before any serious mischief is done on a large scale – Let the electric fluid burst like a clap of thunder!! – let them sleep in fancied security and when the proper time arrives pitch it into them before they can have time to recover their self-possession. Take it easy. The thing must be ready in about 2 months – and then letters can be put in the papers of Cork, Dublin, Belfast, Liverpool London &c.!! Push Y, tell him to make an exertion for one month and all is well – from this out B will work like a 'good un'

It is proposed for future determination, that is, let it be made a matter of general reflection and to be determined, when the other points have been got over: First

T.O.[45]

1st Whether it would be prudent to address a short and pithy letter of one page or so, to the Master General.

2nd Whether a short one should not be addressed to Civil Engineers and Land Surveyors.

3rd Whether one should not be addressed to the Irish Members of Parliament.

4th One to Newspaper Editors

5th One to the gentlemen now employed on the Ordnance Survey called Civil Assistants.

6th One to the Royal Engineers themselves – giving a heavy push on the ribs!!

Each to be short and emphatical, of not more than ½ or a whole page

---

B is puzzled to dispose of, properly, all the materials on hand – X will see the necessity of following the order of events and the subdivision of items – tho' they appear a formidable array, yet they can in the end be reduced by X to a very small compass. A good margin is indispensible – and also the different heads or points should be separated by a broad clear black stroke – X will soon find the benefit of this.

Keep up your heart – and make Y exert himself, for there is something in him. B.

I will continue the list of items myself and you can put them to Y.

The reason why old Officers are sent off and young ones come on – are – that in Ireland where a great number of them were employed, a good deal of jealousy was shown by other R.E. who never were on the O.S. – but got an extra journey abroad while all the favoured one remained at home on the O.S. receiving all the nice pickings &c. Some grumbling was the consequence about the year 1836. Again a good many young ones received all their college course except surveying, to learn that part of it they are sent to the O.S. Never mind when we come to the conduct of the Royal Engineers towards <*words missing*> of a raking then I'll eat my hat!!! On this subject I have as much as would sink a ship!!!!!!!!

Friday

Dear X –

The last came to day – It is known! that the Council of Leeds debated over the Motion, whether a map of the Town with a series of levels, should be requested of the Ordnance Survey. The debate ended in a flare up! – A publisher and engraver was at that time making a copy and engraving of the town of Leeds on 3 chains,[46] (taken from the tithe map) The Council gave the Engraver £300, to drop the undertaking, and destroy the contract, which was done. The Council have lately had another debate, the consequence, that a map of the town on the scale of 5 feet to one mile should be requested from the O.S. The town to pay the expense – which will be about £300.[47] The request was made through Colonel C. and succeeded. Captain T. was at York a few days ago, inspecting the progress of the maps of Leeds. 21 sheets to be done, and no draftsmen!! It was settled that 6 should be sent to York, some to be drafted from Cap. T. and some from Cap. D.[48] It will be nearly a 6 months job – So, whoever goes to York may reckon on 5 months stay. In a few days X shall have another letter on the proper subject, in the mean time he is directed to consider the following points – Is not the Royal Military College of Engineers at Woolwich, supported at the Public expense? – that is to say – the students at the College do not pay the Professors, or support themselves, as to board, lodging, expense of experiments. Who pays the Professors, and the expenses of the establishment? Is it included in any Grant, or, included in the general yearly estimates? How many R.E. officers are there in existence, paid by the Board of Ordnance. Recollecting all our public works, projected, or paid by the Government, such as, Harbours, Docks, Roads, – Bridges – Have any of the great works, showing both Theoretical and great practical knowledge, Have any of these been done, or designed, by the R.E. since the year 1815? – Has not the O.S. of Ireland been applauded in the public prints – and have not the officers and men been eulogised – Has not all the work of York and Lancaster, surveyed and drawn for the scale of 2 inches, been thrown aside as useless? – Has not all the Hill Sketching done in Ireland been thrown aside as useless, and a new system introduced, called contouring?[49] The 6 inch has superseded the 2 inch – also contouring superseded the Hill Sketching.

What is the reason why the county and city of Dublin 6 inch scale, is not published yet.[50] X is requested only to think over these things in his own mind, to refresh his memory.

Those who go to York, or some of them, may be at first sent to examine on the ground. The Survey of London has not been proposed yet, it was delayed by other business, look to the newspapers – Very soon, another letter may be expected

Yours very truly | B. | care of Mrs Buckle[51] | 4 Bishop Hill senior

RI MS JT 1/11/3763–72

1. *Byron Street*: in Leeds, on the north side of the River Aire.
2. *begun on 22nd March, ended on 1. April*: This seems to be a sequence of letters and enclosures that were sent to Tyndall, presumably in a single package, at the beginning of April.

3. *Col. C. and Cap. T.*: Colonel Thomas Colby and Captain Henry Tucker.
4. *O.S.*: Ordnance Survey.
5. *here, in Leeds, 5 feet*: In late 1842 Colby oversaw a switch to a scale of 5 feet to 1 mile, as opposed to the usual 6 inch scale, for producing detailed plans of populous districts that might assist local authorities in making sanitary improvements. By April 1843, though, the Board of Ordnance was questioning under whose authority Colby had made this change, which would inevitably cost more, and requesting that he seek subventions from the local authorities involved. The Ordnance Survey map of Leeds, at the large scale, was published in 1850.
6. *Trig point*: a trigonometrical point, a fixed surveying station.
7. *gentn.*: gentlemen.
8. *supposed to be the City of London*: see letter 0154, n. 3.
9. *'Board men'*: presumably personnel of the Board of Ordnance.
10. *Boundary survey*: the part of the Ordnance Survey charged with detailing county, city, town, parish and other administrative boundaries; the appointment of boundary surveyors was the responsibility of justices at local quarter sessions.
11. *'per diem'*: per day (Latin).
12. *McLoughlin*: not identified.
13. *the Royal Military Academy*: founded in Woolwich in 1741 for the training of commissioned officers of the Royal Artillery and Royal Engineers.
14. *Cap: B*: probably Captain John Bailey of the Royal Engineers, who was stationed in Yorkshire.
15. *Col. C. some time ago ... for 1/6 per acre*: On 13 June 1842 Colby proposed to make maps at the Tithe Commissioners' preferred scale of 26⅔ inches to 1 mile for an additional 2*d.* per acre above the cost for a scale of 6 inches to 1 mile (which was 1*s.* 4*d.* per acre), adding £68,000 to the cost of the Survey of the six northern counties. The proposal, which would have transferred the English Ordnance Survey from the 6 inch to the 26⅔ inch scale and produced maps usable by both the Tithe and Ordnance Surveys, was rejected by the Treasury in January 1843, who insisted on a return to the 6 inch scale.
16. *late resistance made at Preston, on the Tithe affair!!*: Private surveyors resisted the proposal that their work for the Tithe Survey be taken over by the English Ordnance Survey, and in the autumn of 1842 a 'successful bid by the Divisional Officer at Preston to execute a tithe survey led to friction with local private surveyors, and a ruling from the Ordnance that Colby and his men were not to tout for custom' (R. Oliver, 'The Ordnance Survey in Great Britain 1835–1870', PhD thesis, University of Sussex, 1985, p. 119).
17. *the Parliamentary news of 28th?*: nothing relating to the Survey of London was discussed in either of the Houses of Parliament on 28 March 1843.
18. *Sir Robert*: Robert Peel.
19. *Weekly Despatch*: the *Weekly Dispatch*, a London-based Sunday newspaper of independent radical politics that, in the early 1840s, had a circulation of *c.* 60,000.
20. *progress of the Bill*: presumably the Health of Towns Bill, which was first presented to Parliament in March 1843 and under the provisions of which the Survey of London was begun, although not until the Bill was finally passed as the Public Health Act (11 & 12 Vict. c.63) in 1848.
21. *Land Surveyors of Lancashire and York*: private surveyors of those counties; see n. 16.
22. *Ho*: House.
23. *R.S. & M.*: Royal Sappers and Miners.
24. *R.E*: Royal Engineers.

25. *C.A.*: Civil Assistants.
26. *their Superintendent*: Colby, who had been made Superintendent of the Ordnance Survey in 1820.
27. *Master General of the Ordnance*: George Murray.
28. *S. of I.*: Survey of Ireland.
29. *O.S. of E.*: Ordnance Survey of England.
30. *how grievously he was disappointed in 1830 in the* Actual *state of the work*: Murray was not actually Master General in 1830, and instead it was probably William Carr Beresford (1768–1856), the Master General between 1828 and 1830, who had expressed disappointment.
31. *M.G.*: Master General.
32. *'whether those things are so'*: adaptation of Acts 17:11.
33. *B*: McLachlan.
34. *X*: Tyndall.
35. *D*: not identified, but stationed in Preston according to LT, and therefore presumably a civil assistant in the 1st Division (LT, 'Biography', vol. 1, p. 76).
36. *Mr. Shoutesmith Plumber & Glazier*: not identified.
37. *Bishopshill. Senior*: St Mary Bishophill the Elder, a parish in the city of York.
38. *Y*: William Ginty; see letters 0201 and 0226.
39. *A comet as Milton describes ... perplexing monarchs with a fear of change!!!*: It is in fact an 'eclipse' that 'with fear of change | Perplexes monarchs' in John Milton's (1608–74) *Paradise Lost* (1667), I.597–99.
40. *the valuation*: the Irish land valuation exercise, begun in 1830, that used maps provided by the Ordnance Survey to determine both the value and the occupancy of all lands and buildings in order to assess liability to pay the poor rate; it was published as *The General Valuation of Rateable Property in Ireland* – or *Griffith's Valuation* – between 1847 and 1864.
41. *Mr G.*: Richard Griffith (1784–1878), the Irish Commissioner of Valuation.
42. *'Townland Survey'*: Townlands are ancient Irish land units; in 1825 Griffith was charged with establishing the boundaries of townlands, following which the Ordnance Survey both surveyed these boundaries and measured the acreage of each townland so that accurate rateable values could be ascertained.
43. *a book called 'The Colonel's Blue Book'*: The *Instructions for the Interior Survey of Ireland*, which was issued in installments between May and October 1825. These were subsequently lithographed, under seventy-four headings, to form a single document that became known as 'The Colonel's Blue Book'.
44. *any stray letter opened as in the case of Captain Fenwick!!*: see letter 0226.
45. *T.O.*: presumably Turn Over.
46. *on 3 chains*: at a scale of 1 inch to 3 chains, the equivalent of 26⅔ inches to one mile.
47. *The Council have lately had another debate ... will be about £300*: see n. 5. In May 1843 the 'Leeds town council ... enquired about the cost of a tracing of the Ordnance 1:1056 [5 feet scale] plan of the city, for use in laying out sanitary schemes. The council was deterred from pursuing the idea on account of the high cost' (Oliver, 'Ordnance Survey in Great Britain', p. 132).
48. *Cap. D.*: Captain Robert Kearsley Dawson.
49. *a new system introduced, called contouring*: The introduction of contour lines to show altitude on maps more accurately than impressionistic sketching and shading began in 1832–33, with the new system becoming the norm following the Irish Railway Commis-

sion of 1836–8, which used Ordnance Survey maps to determine potential railway lines.

50. *the county and city of Dublin 6 inch scale, is not published yet*: The 6-inch Ordnance Survey map of Dublin would be published in February 1844.

51. *Mrs Buckle*: not identified.

To John Tyndall, Snr [March / April] 1843 0196

Preston, | Monday night

My dear Father

Yours[1] came to hand this morning and I hasten to reply to it, not indeed because of your having asked me any questions which demand prompt replies but simply that I have a little spare time. Here I sit alone in my comfortable room, the last embers of a dying fire smoulder to my left. Evans and Latimer my other messmate are wandering where I know not, and surrounded by the circumstance of unbroken solitude my thoughts naturally revert to poor Leighlin – with all its faults I love it still – but I must check myself. I was just about entering upon a strain so pathetic that had I continued it I would most infallibly be guilty of having made my father weep like a child.

Yesterday was a most delightful day, I strolled more than once along the green banks of the clear Ribble,[2] viewing the scenery of each side. Everything looked blooming – the birds carolled on the boughs, the boughs were covered with fresh green leaves or scented buds, 'the early promise of the spring'. Yet notwithstanding the heat of the sunbeams which glanced upon us during the day we had a most tremendous shower of hail in the evening, accompanied by some vivid flashes of lightening and booming thunder – You ask me in earnest when I think I will go home. I have just paid my subscription to the Preston Institution[3] for the ensuing quarter, and when it has expired why then I'll be talking to you somewhat serious about the matter. I am sorry I have no plants to mould[4] here I'd like to exercise my muscles at such work, perhaps I'd eat some of those which Kit[5] has moulded. You imagine the small papers I sent you[6] to be poison – Millions think otherwise. I am glad to hear that my mother and Emma were in Ballybromwell.[7] They ought to go out often, it would serve my mother very much to take a trip out to the country every month at least. So Emma got the rhyme that I wrote to Debby.[8] When I go home I'll shew you some more specimens of rhyming. Ginty and I have kept up a constant fire since I came to England. Nearly all our letters are in rhyme. You'll laugh at some of them. You ask how is the public mind in England. In this portion of it, the public tongue calls aloud for a repeal of the corn laws.[9] The English are a stern people, and will not suffer what they conceive to be their rights kept from them. They have plenty of firebrands among them to keep their minds in a blaze. I sent Emma a couple more specimens of their publications. Tho' I did not mention Mrs Hennessy[10] in my letter her bankruptcy was surprising news to me, has Pat Kelly[11] come to much loss by her?

RI MS JT 1/10/3289
LT Transcript Only

1. *Yours*: letter missing.
2. *Ribble*: The River Ribble runs for 70 miles through Lancashire and North Yorkshire.
3. *Preston Institution*: see letter 0187, n. 1.
4. *mould*: to cover (the root or stem of a plant) with earth (*OED*).
5. *Kit*: not identified.
6. *the small papers I sent you*: see letter 0187, n. 8.
7. *Ballybromwell*: see letter 0121, n. 4. Tyndall's mother was presumably visiting members of her family, the McAsseys.
8. *Debby*: Deborah McAssey.
9. *In this portion of it, the public tongue calls aloud for a repeal of the corn laws*: The manufacturing districts of Lancashire were at the fore of the agitation against the laws, maintained from 1815 to 1846, imposing protectionist import duties on foreign grain, and keeping the price of bread artificially high. Much of the funding for the Anti-Corn Law League, which was led by the Manchester businessman Richard Cobden, came from middle-class donors in Lancashire who favoured free trade and were frustrated at having to pay high wages to enable their employees to pay for bread. The workers themselves were resentful that the Corn Laws were in effect a subsidy to wealthy farmers, and some Chartists (see letter 0164, n. 2) supported the Anti-Corn Law League, although relations between the two remained uneasy.
10. *Mrs Hennessey*: not identified, but possibly related to Michael Hennessy.
11. *Pat Kelly*: not identified, but possibly one of the Kellys mentioned in letters 0001 and 0008.

From William Ginty 1 April 1843 0197

33 Windsor Street[1]

My dear Jack

Your exquisite production[2] greeted my eyes in due time. Oh you bigot! – however it is first rate. The 'Requiem of common sense' has called forth my most extravagant admiration. You done well to finish it; <u>look out for it shortly in the P.C.</u>[3] I would send you 3 or 4 every week on an average, only for the same reason – I cant finish. This day I tried two – no go tho', one was a capital subject and I may finish that. 'Tis sweet to be watching her eye'– Expressive is it not? eh?

I think it very probable that we may see Sinnett here ere long. Scott[4] is on leave and Lowrie[5] is for Canada, along with 3 others from this Division.[6] I hope Sinnett may come at all events.

I send you this week's Mercury with this, <u>vide</u>[7] 'Ordnance Survey' in the column to correspondents.[8] What think you of this? Here now the ice is broken and let us see what a fine splash you'll make. Some fellow seems to be inquisitive on this subject. Just set the poor fellow's mind at rest – it demands an answer.

I will write in the course of the week. I made an application last week for a Clerkship of £70 a year. I got no answer yet. I think it is a failure. No word of the journals! Yet[9] – When you squint over this paper please to put a label on it for 'auld' Ginty[10] Carrickonshannon Co Leitrim.[11] Willy[12] is beginning to run foul of a black eyed Susan[13] that shakes the feathers for us. Jack's father[14] gave him a very nice little watch when he was here last and talks of removing him from his admirable occupation. Did you read that affair for Evans that I remodelled for his especial use.[15]

You must excuse this, damn me if I have either pen or ink within the room and as I soon must trot to 75 for cash[16]

Good bye | Yours | Ginty

RI MS JT 1/11/3599
LT Transcript Only

---

1. *33 Windsor Street*: the location of the Ordnance Survey Divisional Office in Toxteth Park, Liverpool.
2. *Your exquisite production*: letter missing.
3. P.C.: the *Preston Chronicle*, although Tyndall's 'Requiem of common sense' did not appear in the newspaper; see letter 0199, n. 1.
4. *Scott*: Corporal William Scott (b. 1811) of the 16th Company of the Royal Sappers and Miners, who had worked in the 5th Division, B District of the Irish Ordnance Survey before being transferred to England in July 1842 where he superintended work in the Divisional Office of the 5th Division of the English Survey in Liverpool (NAI OS/1/17–19).
5. *Lowrie*: Corporal William Lowrie (b. 1816) of the 16th Company of the Royal Sappers and Miners, who had worked in the 1st Division, B District of the Irish Survey before being transferred to England in February 1843 and working in the 5th Division of the English Survey in Liverpool (NAI OS/1/16–20).
6. *this Division*: the 5th Division of the English Ordnance Survey.
7. vide: see (Latin), used to direct the reader's attention to a specific point in a text.
8. *'Ordnance Survey' in the column to correspondents*: 'To Correspondents. ORDNANCE SURVEY', *Liverpool Mercury*, 31 March 1843, p. 106, which stated: 'The gist of *A Constant Purchaser's Note* seems to be the question why this expensive national survey could not include the necessary data, such as the precise contents of fields, &c., to serve in arrangements of tithe commutation. We are given to understand that the *scale* of the great survey is not sufficiently large to answer this purpose completely, and that, therefore, *special sketches* are made, for such purposes, for parties paying for them'.
9. *No word of the journals! Yet*: presumably the 'Journal of Progress and Weekly Report' (see letter 0038, n. 3) whose late arrival Ginty had also bemoaned in letter 0143.
10. *'auld' Ginty*: Ginty's father William.
11. *Carrickonshannon Co Leitrim*: the town of Carrick-on-Shannon in County Leitrim, in the north of Ireland.

12. *Willy*: William Hunter.
13. *black eyed Susan*: not identified.
14. *Jack's father*: the father of John Tidmarsh.
15. *that affair for Evans that I remodelled for his especial use*: not identified, but possibly a poem addressed to a woman.
16. *trot to 75 for cash*: not identified.

From Archibald McLachlan 10 April 1843 0198

25 Byron Street, Leeds | 10th. April 1843.

Dear Mr Tyndall,

I have delayed thus far in hopes to have something to say of the Act: for the S. of E.[1] passed in July 1841, but it has not arrived yet. Now Mr T. you have youth and energy on your side, as for me both are gone – worn out. Repeated disappointments and failures have super-induced a cautiousness amounting to apathy, I have often said, suppose I do it and no practical good results, I shall only blame myself, and perhaps others will do so too, for having prejudiced the cause by a weak abortive effort.

It must be done – and, old caution says, in such a manner as not only to cut, to wound pride, but also to intimidate indirectly, That they may fear more what has not been said, than what has been said – to appear to keep certain disclosures in reserve – that fear of this disclosure may cause a change of action – if their present course is still persevered in, then 'Lay on Macduff'.[2] They themselves are to blame for the public damnation of their public character – In this view of the case, the College at Woolwich[3] should not be touched nor the S of Ireland[4] mentioned. Only beginning with the Survey of England and ending with the Tithe Maps.[5]

But if you think this course would not be a 'good un' then you must begin with the Survey of Ireland at 1824 – and a power of writing you will have! I fear much whether the N. paper[6] would not tire of the yarn – yet, if it could be done, the public would have a better view of our machinery, tho' perhaps it would not in the end have such a good effect on the Royal Engineers. – All must be done so that the writer or writers cannot be even conjectured. No phrase or technical term should escape or no expression used that could identify the writer with a certain locality. The pronoun I should not be used – You shall speak as a King or Editor!! and say, we assert this – we are informed &c. &c. - - - Perhaps it would be better not to use I or we – but write like a stately historian – 'It has been affirmed' 'certain documents containing the following'. Be sure that nothing escapes you that may lead to a conjecture that some of the C.A.[7] have been concerned or R.S. and M[8] – that no clue – no analogy can be made – so they cannot tell what the D- -[9] to make of it!!

Let me know whether you wish to begin with the S. of I. or merely with the S. of E. – You shall have a 'full, true and particular account' in a short time – You must re-arrange it, add good language and facts – and the 'Lord speed you'. Did you note the words C.T.[10] said to the gent:[11] how did he recommend him to proceed, did C.T. recommend any form of words to be used? Was it to the Board, or to Colonel C.[12] that the application[13] was to be made. Note: What has become of Aldcliffe Township[14] on 3 chain scale[15] – has any faults been found with Standish[16] areas, have the maps passed the Commissioner?[17] No Field books were sent with them to the Tithe office. – Cap: Bailey[18] up to a month ago (I got this from a Field man) surveyed all field work for 3 chain scale – this is annuled, they are now to survey with a view to 6 inch[19] only except villages.

Are Meresmen[20] sworn? The Town of Windsor is to be done on 5 feet scale (special survey)[21] two of the sappers left us on Saturday for that place. I spoke of a small pamphlet in my letter as being better calculated to induce ultimate success than Newspapers. I spoke to my printer, he says a pamphlet of this size &c. Type cost for printing £2.10.0 per sheet – i.e. 28 pages of a common Methodistical penny tract[22] – A pamphlet of 40 pages printing, paper &c. 200 copies would cost here in Leeds something more than £4.0.0.

Think on this.

The matter – or facts concerning the O.S.[23] which I have by me, were, so far as they go, arranged for Colonel C: and not for the public – So that considerable alteration must be made. – You will peradventure find it necessary to strike out many of the statements and substitute your own – I can pretend to give you only the raw material, you must apply your steam power on this – Do you know any steady old friend on the O.S. in some other division who could give you a private history &c. If a great deal is said of Lancashire and nothing of York it will look suspicious – I remember seeing some 3 chain scales sent from Preston to some other office, but could never find out whether they were ever used for Tithe maps &c. You must write before long

Yours truly | A. McLachlan

RI MS JT 1/11/3773–4

---

1. *Act: for the S. of E.*: the Ordnance Survey Act (1841 c. 30 (Regnal. 4 & 5 Vict)); see letter 0150, n. 6.
2. 'Lay on Macduff': 'And damned be him that first cries "Hold, enough!"'; W. Shakespeare, *Macbeth*, V.x.33–4.
3. *the College at Woolwich*: the Royal Military Academy; see letter 0195, n. 13.
4. *S of Ireland*: the Survey of Ireland.
5. *the Tithe Maps*: see letter 0195, n. 15.
6. *N. paper*: newspaper.
7. *C.A.*: Civil Assistants.

8. *R.S. & M.*: Royal Sappers and Miners.
9. *D- -*: Devil.
10. *C.T.*: presumably Captain Henry Tucker.
11. *gent*: gentleman.
12. *Colonel C.*: Colonel Thomas Colby.
13. *the application*: probably for the surveying of a parish or township at a scale of 1 inch to 3 chains, the equivalent of 26⅔ inches to 1 mile; the 'nearest the Ordnance came to a comprehensive 1:2376 [3 chain] scale of northern England, at any rate in the 1840s, was in surveying 23 parishes or townships in Lancashire and Yorkshire at that scale, in response to specific local requests' (R. Oliver, 'The Ordnance Survey in Great Britain 1835–1870', PhD thesis, University of Sussex, 1985, p. 119).
14. *Aldcliffe Township*: a village in Lancashire, south-west of Lancaster.
15. *3 chain scale*: see n. 13.
16. *Standish*: a village in Lancashire, north of Wigan.
17. *the Commissioner*: the Tithe Commissioners; see letter 0173, n. 7. McLachlan is probably referring to Robert Kearsley Dawson, who was Assistant Commissioner.
18. *Cap: Bailey*: Captain John Bailey; see letter 0195, n. 14.
19. *they are now to survey with a view to 6 inch*: a scale of 6 inches to 1 mile; see letter 0195, n. 15.
20. *Meresmen*: persons appointed locally to ascertain the exact boundaries of a parish (*OED*).
21. *The Town of Windsor is to be done on 5 feet scale (special survey)*: Windsor, in Berkshire, was the first location in the south of England to be mapped by the Ordnance Survey. The survey, at the large scale of 5 feet to 1 mile (see letter 0195, n. 5), was conducted at the request of the Commissioners of Woods and Forests, with, crucially, the support of Prince Albert, who had originally expressed the desire for such a survey. It began in April 1843 and concluded in March 1844.
22. *a common Methodistical penny tract*: as produced by organizations such as the Religious Tract Society, founded in 1799, and numerous local Methodist Tract Societies.
23. *O.S.*: Ordnance Survey.

From William Ginty 18 April 1843 0199

Tuesday evening

Dear Tyndall,

I could not for the life of me scribble you anything sooner and even this moment I dont know how I will make this what might be called a note.

I see I took a rise out of you about the Preston Chronicle.[1] I did not send and indeed your letter requesting me not to do so[2] would have been too late did not the same thought strike me, and even I couldn't be 'cock <u>shure</u>' that the editor of the P.C.[3] was not a P.P.[4] And so I thought better to wait for a more favourable opportunity. I have <u>one</u> piece even should you never send another that will do right well! ahem! Andy Ingram[5] passed this way on yesterday on his way to Ireland, he was discharged in consequence of the reduction.[6] John Davy is dead. I'm afraid it will break poor old Davy's heart for he was exceedingly fond of his children.[7] Nelson has returned and delivered himself up.[8] It is said here that Capt. Tucker is to have charge of the London and Windsor job[9] and that a Lt. Stanley[10] is to replace him. Can you vouch for this if truth I guess you will visit the great metropolis. I think Griffin is broke.[11] I seen a letter last week from Manchester including him as Pt. Griffin amongst those who were for Windsor.

Just power thy vengeance never sleeps.

My journal has not yet made its appearance. Parker and Gibbons[12] have more pay than me at present![13] I got a letter from Jim[14] for a wonder, he spun a yarn about old times, Bessie[15] and the fatal bridge! poor devil that affair still haunts his thoughts, for my part I have scarcely a recollection of those things, it puzzled me to bring to mind who Bessie and Sally[16] and Maryanne[17] were when I was reading it! –

I believe I put the saddle on the right ass about the Westmoreland affair[18] in Dr. Parker's letter.[19]

Geo. Markham[20] and his brother and the whole family are residing here now. Chadwick will be over next month! I suppose he will take a trip to Preston. I say Tyndall would you like to see Ellen[21] again – I intend going over in the first 'areil carriage'[22] that soars that way. 3 to 2 of the editors of the English papers are of opinion that it will succeed. The Mercury says 'a pig may fly but it is a very unlikely bird',[23] – however I'm blessed if I would not like to see and kiss Ellen. Out of all my enamourers I value her only at present, I cant say how long it may be so. Still I can think of her and of her charms

A thousand pictures in my memory live
I would not part with for a monarch's throne[24]

You can tell Jim to look out for an answer on or about the 1st of August 1843.

Remember me to Mr Marquis and Geo.[25] A verbal message from Westmoreland last week was – a request for me to write to Miss E.[26] dont you wonder at a married woman to ask for the like – as in duty bound I'd see her very well damned first! We have the plans of Liverpool in preparation for old Tom Colby he has been promised and expects them by the 1st of June – he wants them to lay before the Government as specimens, but I am certain the 1st of August will not find them finished!

I am far enough – I cant go on 'I'm almost sorry that I ere began'[27] what a thing it is to be obliged to write about nothing Excuse this damned insipid rigmarole of bladdering[28]

Yours sincerely | Ginty.

I would make you a present of ½ a dozen good cigars if I had any means of sending them, and will if you only send a messenger or a hint of a means

Yours | Ginty | Smuggled of course.

RI MS JT 1/11/3600
LT Transcript Only

---

1. *I took a rise out of you about the Preston Chronicle*: Ginty is probably referring to Tyndall's 'Requiem of common sense', of which he told him 'look out for it shortly in the P.C.'; see letter 0197.
2. *your letter requesting me not to do so*: letter missing.
3. *editor of the P.C.*: Lawrence Dobson, editor of the *Preston Chronicle*.
4. *P.P.*: not identified, although usually the abbreviation means Parish Priest, so Ginty may have been concerned that the *Preston Chronicle*'s editor was a Roman Catholic. Dobson was not, and in the following month the paper published a poem probably by Tyndall, under another of his pseudonyms Wat Ripton, which mocked 'the great Dan O'Connell, with blarney galore' ('An Hibernian's Song to ____ ____', *Preston Chronicle*, 20 May 1843, p. [4]).
5. *Andy Ingram*: Andrew Ingram, a civil assistant who worked in the 5th Division, C District of the Irish Ordnance Survey. He joined the Survey in May 1839, and was transferred to England in January 1842 (NAI OS/1/16–19).
6. *the reduction*: see letter 0194, n. 15.
7. *it will break poor old Davy's heart ...fond of his children*: Ginty's prediction may already have been fulfilled, as James Davy, a Corporal in the Royal Sappers and Miners who had previously suffered ill health, was discharged from the army on 11 April 1843 after being found 'worn out and permanently unfit for military service' (NA WO 97/1149).
8. *Nelson has returned and delivered himself up*: see letter 0188, n. 15.
9. *the London and Windsor job*: presumably the superintendence of the putative Survey of London (see letter 0154, n. 3), and the actual Survey of Windsor (see letter 0198, n. 21).
10. *Lt. Stanley*: Lieutenant William Stanley of the Royal Engineers, who was in charge of the 1st Division of the English Ordnance Survey, and commanded the Divisional Office in Preston.
11. *I think Griffin is broke*: David Griffin had been acting Lance Corporal in the 13th Company of the Royal Sappers and Miners while on the Irish Survey, but was now seemingly demoted to his original rank of Private.
12. *Gibbons*: James Gibbons, a civil assistant who had worked in the 5th Division, C District of the Irish Survey. He had joined the Survey in November 1838, and was transferred to England in January 1842 (NAI OS/1/15–19).
13. *more pay than me at present!*: John Parker was earning 2*s.* per day at the time of his last pay rise on the Irish Survey in July 1840, while Gibbons' pay had risen to 1*s.* 6*d.* in January 1841.
14. *Jim*: Phillip Evans.
15. *Bessie*: see letter 0115, n. 17.
16. *Sally*: possibly 'Hunter's Sally' mentioned in letter 0115.
17. *Maryanne*: possibly Maryanne Kearney mentioned in letter 0097.
18. *the Westmoreland affair*: possibly a reference to the contretemps caused by Tyndall's poem 'On Leaving Westmorland'; see letters 0183 and 0184.
19. *Dr. Parker's letter*: not identified.
20. *Geo. Markham*: see letter 0177, n. 19.
21. *Ellen*: from Kinsale.
22. *'areil carriage'*: The aerial steam carriage was patented in 1842 by William Samuel Hen-

son and John Stringfellow. In 1843 the Aerial Transit Company funded an extensive publicity campaign in the press, but this did not generate sufficient investment in the firm to actually construct the proposed flying machine.

23. *The Mercury says 'a pig may fly but it is a very unlikely bird'*: 'The Aerial Steam Carriage', *Liverpool Mercury*, 7 April 1843, p. 119.
24. *A thousand pictures ... | ... monarch's throne*: W. H. Leatham, *A Traveller's Thoughts: or, Lines Suggested by a Tour of the Continent* (1837), xvii.1–2.
25. *Geo.*: George Latimer.
26. *Miss E.*: Mary Evans; on her marriage, see letter 0185, n. 16.
27. *'I'm almost sorry that I ere began'*: Lord Byron, *Don Juan* (1819–24), I.cxv.920.
28. *bladdering*: blathering; talking foolishly or nonsensically (*OED*).

From William Ginty 23 April 1843 0200

110 Pitt Street | Liverpool | Sunday evening | 23rd Apl 1843.

Dear Tyndall

Rather curious is it not that I did not write to you ere this. How stands your Barometer of 'sober seriousness'?[1] You would have one from me 3 days ago but for Larry Eivers, who has just arrived from the great metropolis. He is walking about here these last 2 or 3 days at his own request, (Jack[2] says he will write soon) – to improve his health previous to his removal for the examination of a Township on the Tithe scale[3] about 10 or 12 miles from here. Larry's soliloquys on London are rich – in fact I never knew him before – his passionate description of the churchyard rendered immortal by Gray's Elegy[4] is indeed highly amusing, the more so as this of all poetry and poems was Larry's greatest favourite! He says it is a lovely spot for reflection and that there is a cenotaph erected on the spot where Gray was supposed to have composed the Happy piece! He went out to Windsor[5] to Eton College;[6] on the whole he seems to have been very merry while away and is well pleased with his trip. Tell George[7] that Larry seen McDermott[8] and says he is in very promising circumstances.

And now may I ask what fiend possessed you when you talked of sending pencils in a letter – only just imagine what a pretty enclosure!!!

Bladdering[9] ahoy! my present position is an unenviable one – beside me sits a little black headed devil with a big arm and great 'splaw' feet,[10] blowing the fire and asking me fifty questions about Ireland, 'Oh dear how I should like to go there' she cries as she rivets her vacant stare upon my sour looking phiz.,[11] and tries to look tender at me. 'Indeed', rejoins I 'the passage is but a mere trifle' – 'ah' she cries 'what would strangers be doing there alone' 'Enjoying themselves of course' says I – 'The English generally go there to learn good breeding. I wish some more of them would go' and may I be canonized, if she has not taken the hint and buried her snub nose in the mazes of a white handkerchief which might be very safely used for a table cloth – so now I am free, she has just left the room and 'I am all alone' By the way you can just book in your remembrances that Sunday evening the 23rd of April 1843 is a 2nd edition of the Deluge at Liverpool.[12]

Pangs pierce my heart your rigmarole
Electrified my very soul
Could not your fancy find a song
That claimed more ardour from your tongue
Than this of Ellen[13] – is that name
For ever doomed to be your theme
Could not your muse inspire a lay
More genial to your heart's decay
Than that which to your soul must bring
Affection's keen and poignant sting
The pang of hopes so quickly crushed
The woe of lays for ever hushed
The dirge of extacies which rung
Their brightest octaves from your tongue
The 'Passing-bell' of loves young dream
To rack your bosom as a flame
The requiem of a bliss once cherished
Which now is gone – for ever perished
Called you her 'ours' – this cannot be
In adoration we agree
But no – not further – well you know
She never loved nor looked on you
As one with whom she would entrust
The purest heart that ere from dust
Was generated to control
The feelings of immortal soul
Call her no more then 'ours' for I
Alone found favour in that eye (!!!)
Methinks I see it even now
Flash forth its loveliness – her brow
Meantime in frowning beauty lowers
A contradiction to that 'ours'
Alas sad thought – twelve moons at least
Their aerial path have slowly traced
Since last I gazed upon that face
Yet here its portrait finds a place
Aye in my heart – a deep regret
Shall teach me never to forget
Flow sorrow flow! dark dreamer I
Must ever hail that form as nigh
Flow sorrow flow – a life of sighs
Must oft pourtray those loving eyes
So soft their glances – bright their hue
Outrivalling heaven's cerulean[14] blue
Yes 'clouds of care' may darkly roll

Their dismal phantoms round my soul
And waves of sorrow dash their yeast
Of rippling horrors o'er my breast
My buoy of life – of hope – must be
My lovely Ellen – thoughts of thee!

My journal after six weeks 'leave of absence' has safely returned[15] with the consolatory moiety of 2 pence sterling to commence from the 1st of May A.D. 1843.[16] Curses light upon their miserable souls. Is it not fit to try the philosophy of a Bolingbroke![17]

There are some of this office party[18] going to Hull shortly. Exclusively plotters.[19] Turner is going and I think Bill Latimer will go too – he seems not to dislike the chance of the change!

I intend having another trial at getting to Preston, others have got like applications granted and I don't see why I shouldn't – I will however try, when these London and other arrangements are made, for in the sober seriousness of good truth I am most heartily sick of Liverpool or rather the Survey office at Liverpool. I can find no congenial spirit among them – 'No heart that beats reply to mine'.[20] With very few exceptions they are a set of drones – boors and nincompoops. 'We'll talk of this anon'[21]

RI MS JT 1/11/3601
LT Transcript Only

1. *'sober seriousness'*: see letter 0194, n. 1.
2. *Jack*: John Tidmarsh.
3. *the Tithe scale*: the scale of 1 inch to 3 chains, the equivalent of 26⅔ inches to 1 mile, requested by the Tithe Commissioners.
4. *the churchyard rendered immortal by Gray's Elegy*: 'Elegy Written in a Country Churchyard' (1751) was reputedly composed by Thomas Gray (1716–71) in the churchyard of St Giles' Church in Stoke Poges, a village in Buckinghamshire.
5. *Windsor*: see letter 0198, n. 21.
6. *Eton College*: the public school founded in 1440.
7. *George*: George Latimer.
8. *McDermott*: possibly either the J. McDermott or M. McDermott who had been civil assistants in the B District of the Irish Ordnance Survey.
9. *bladdering*: blathering; talking foolishly or nonsensically (*OED*).
10. *'splaw' feet*: splay feet.
11. *phiz.*: face (*OED*).
12. *a 2nd edition of the Deluge at Liverpool*: On Friday 28 April 1843 the *Liverpool Mercury* reported that 'On Sunday there were a succession of thunder storms from the southward, accompanied in some instances by a heavy shower of hailstones' ('Thunder Storms', p. 139).
13. *Ellen*: from Kinsale.
14. *cerulean*: pure deep blue, azure (*OED*).

15. *My journal after six weeks ...safely returned*: see letter 0194, nn. 32 and 33.
16. 2 pence sterling ... *1st of May A.D. 1843*: Ginty was earning 2*s*. a day after rejoining the Irish Ordnance Survey on 7 October 1840. It is not known whether he was awarded any further rises before 1 May 1843.
17. *the philosophy of a Bolingbroke!*: Henry St John, 1st Viscount Bolingbroke (1678–1751) enjoined that, although it was not possible to prove it rationally, it could be assumed that God is omnipotent and omniscient and always does what is best.
18. *this office party*: from the Divisional Office of the English Survey's 5th Division.
19. *Exclusively plotters*: see letter 0163, n. 10.
20. *'No heart that beats reply to mine'*: Lord Byron, *The Giaour: A Fragment of a Turkish Tale* (1813), 1289.
21. *'We'll talk of this anon'*: adaptation of 'We'll talk of that anon' in W. Shakespeare et al., *Sir Thomas More*, IV.iii.84.

From William Ginty 27 April 1843 0201

33 Windsor Street[1] | Thursday forenoon

My dear Jack

I received your note[2] a short time ago since. I must say you have for once been rather mysterious. I do not perfectly understand your intentions. Do you intend to write or rather publish in a pamphlet, a kind of correspondence as if it had taken place merely in the exchange of friendly sentiments between a few of the Survey men – to be published as if never written to be published. and to be written as if intended to be published? Do you understand me? I hope I understand you –

Now as regards an estimate of the Tithe Plans expense[3] I fear I can furnish nothing on that head without running a risk of a complete exposure – These fellows from whom only I could collect a few facts, cannot keep their own thoughts in secret, much less ours. I can at all events state the general opinion in very strong terms and without a shadow of doubt. As there are a dozen going from here Saturday evening I expect something to talk about on that head. I wish you had written at more length on the matter. I feel exceedingly anxious to taste the marrow of it. What you mean by 'private address'[4] or the necessity of it I cannot fathom. And why I am to address you as 'X'[5] I cannot see the drift of either. I suppose it was near post time when you set about it and so had not time to be more explicit. I had a letter in the early part of the week from Marquis – requesting a favour in the poetical line. I may want your never sleeping tongue for this. It is on the death of a young man in Ceylon,[6] a friend and relative. I cannot bring myself to attempt it. Elegiac pieces are too deep in sentiment for a paltry rhymer – and if my present opinions don't change I'll refer him to you. At all events, I'm sure I can't do it.

We had a grand match of foot-ball here this week, between an equal number of the surveyors and office men. The office men were the best – but not one game was won during the whole evening. I hurted my arm severely by a fall, but it is mending rapidly, we are to have another shortly.

I hope to hear from you shortly – take a pencil and a sheet of paper and scribble away for a while – you need not lose time in writing with ink.

Yours | 'Y'[7] | 110 Pitt Street[8]

If you suspect anything like a suspicion – direct to my lodgings. Some rascal might purloin it here. I really think his life would pay the forfeit – However direct to 110 Pitt St.

McKerlie[9] was here yesterday. His large plans he will have drawn merely with red and black as the small scale.[10] Hamley says the coloured plans are like waistcoat patterns.

Original:

'We plotters' says a surveyman 'pay little attention to divine writ!' 'How's that?' says a comrade 'Because we' add 'house to house and field to field'.[11]

RI MS JT 1/11/3601–2
LT Transcript Only

1. *33 Windsor Street*: the location of the Ordnance Survey Divisional Office in Toxteth Park, Liverpool
2. *your note*: letter missing.
3. *the Tithe Plans expense*: see letter 0188, n. 6.
4. *'private address'*: presumably the address of Tyndall's lodgings rather than that of the Ordnance Survey Divisional Office in Preston.
5. 'X': the pseudonym Tyndall used to avoid detection.
6. *Ceylon*: an island in the Indian Ocean (now called Sri Lanka) that had become a Crown Colony of the British Empire in 1815. The deceased young man has not been identified.
7. *'Y'*: the pseudonym Ginty used to avoid detection.
8. *110 Pitt Street*: Having begun the letter at the Ordnance Survey Divisional Office, Ginty presumably completed it back at his lodgings.
9. *McKerlie*: see letter 0180, n. 11.
10. *His large plans ... small scale*: The large plans are presumably those at the scale of 1 inch to 3 chains, the equivalent of 26⅔ inches to 1 mile, while the small scale is that of 6 inches to 1 mile, at which level of detail a limited range of colours such as red and black would be sufficient.
11. *little attention to divine writ! ... add 'house to house and field to field'*: an allusion to Isaiah 5:8, 'Woe unto them that join house to house, that lay field to field, till there be no place, that they may be placed alone in the midst of the earth!'.

From John B. Edgeworth 4 May 1843 0202

Dublin | May 4th 1843

My dear Tyndall

I am quite ashamed of not answering your very kind note and enclosure[1] long before this and the only excuse I have to offer is that my examinations in college were to come off a few days after that on which I received yours and I was so very busy that I had very few moments to spare. However now they are over and I hasten to thank you very sincerely for the specimen which you sent me[2] it is exactly what I wanted and will serve for a standard whenever I am doing a map.

I am now staying for a week in Dublin and then I will go to the County Longford where my direction if you think it worth your while writing to me will be

Kilshrewly
Edgeworthtown[3]
Ireland

and be assured I shall always be very glad to hear from you of your happiness and to serve you whenever I may be able. We have had nothing but wet weather all winter in Cork with the exception of a very few fine days but here I believe they have had very good weather – When Mr Taylor[4] was in Cork I went to the Methodist shop[5] with him where he expected I should be instantly converted but I returned from it more than ever attached to the true church[6] in which opinion I still continue. I hope Mr Eivers was quite well when you heard from him and also all those with whom I was acquainted when in Cork Remember me to such as remember me and believe me to be

your very sincere | friend | J.B. Edgeworth

RI MS JT 1/11/3562
LT Transcript Only

---

1. *your very kind note and enclosure*: letter missing.
2. *the specimen which you sent me*: see letter 0192, n. 1.
3. *Kilshrewly | Edgeworthtown*: an estate in the midlands of Ireland around 65 miles north-east of Dublin, owned by the powerful Anglo-Irish Edgeworth family, after whom the townland of Edgeworthtown was named. At this time, the estate was administered by the novelist Maria Edgeworth and her brother Francis. John B. Edgeworth was presumably a member of this family.
4. *Mr Taylor*: see letter 0168, n. 4.
5. *the Methodist shop*: The Methodist community in Ireland had split into two factions in 1818, so this was either the Church of the Wesleyan Methodists on St Patrick's Street, Cork, which had opened in 1805; or the Primitive Wesleyan Methodist Chapel in French Church Street, which was built in 1712 as a Huguenot church. As the Primitive Wesleyans saw themselves as loyal members of the Established Church and did not themselves

administer the sacraments or hold services during 'Church hours', Edgeworth's response suggests it was the St Patrick's Street Wesleyan Church he attended with Edward Taylor.
6. *the true church*: presumably the Anglican Church of Ireland.

From William Ginty [Early May] 1843 0203

110 Pitt Street[1] | Monday evening

My dear Tyndall

I received your letter with the order enclosed[2] on Saturday, it is useless to dwell on the sincerity of my thanks for your kind and prompt compliance. My feelings I am quite unable to express. To Friendship, however I owe this much, to you Tyndall and to your memory for ever shall my heart turn with the warmest throbs of affection and gratitude.

Much as I hate even the name of the Ordnance Survey – much as I loathe its cruel villainy and injustice, I leave it with regret – need I say why? Yet 'I will dream that we may meet again'[3] it must be so! – it shall be so

> Earth does not hold a secret glen
> So secret but we meet again[4]

Yes Tyndall we meet again, perhaps in after years, perhaps with hoary heads – to talk of bye gone days. Circumstances <u>cannot</u> be always unfavourable, and when removed from the perfidy of ordnance influence things will take another and a happier turn enough of this! It makes me gloomy. I have not yet received the letter from my brother, stating the final arrangements of the matter, a post or two will however place it within my grasp,[5] this has been the cause of my not writing yesterday. I met with an accident of a very serious nature on Saturday morning and I must say I am rather fearful of the consequences. Getting from behind one of the drawing tables with a plunge! my leg came in contact with an iron hook or spike a little under the knee where it entered leaving a very ugly wound of at least ¾ inch deep I found it exceedingly difficult to extract it and in a few moments afterwards I swooned away – fainted sir as dead as a hatched! not the loss of blood for it bled none. I wished it had, it is on this all my fears are based, lest the rust or oxide of the iron might cause mortification.[6] I have it poulticed,[7] and at present it looks very inflamed. I say Tyndall wouldn't I cut a heavy swell with a stick foot!!! If it had bled the rust would have oozed out along with it!

(I quote) 'You'll say I'm mad' – Not I by heaven. Truth is not madness and for this and this only might hearts of 'living fire and flame' envy the riches of illustrious fools! but let that pass, such things must be. I envy not the lordlings pride <u>while</u> glorying o'er the monotony of his stupid and insipid existence – born to drag a load of wealth into the dreaded pomp-destroying tomb! They taste not of the happiness of natures own free and unfettered!

You said nothing on the grand project[8] – let me hear how you are getting on in your next. What about the fund? I would like to establish it among the few genuine hearts that are here, ere I leave for my father-land. I will return you the monody[9] when I copy it. I consider it first-rate. Making all just and reasonable allowance for the difficulty in praising one – not a friend – nor yet a hero of a history or a tale

Yours | Ginty.

RI MS JT 1/11/3605
LT Transcript Only

1. *110 Pitt Street*: in central Liverpool, close to the docks.
2. *your letter with the order enclosed*: letter missing.
3. *'I will dream that we may meet again'*: Lord Byron, *Childe Harold's Pilgrimage* (1812–18), II.ix.77.
4. *'Earth does not hold a secret glen | So secret but we meet again'*: slight misquotation of W. Scott, *The Lady of the Lake* (1810), II.xxxv.823–24, which has lonesome glen in the first line.
5. *I have not yet received ... within my grasp*: Ginty's confidence was in fact misplaced, and the job offered by his brother would prove unsatisfactory; see letter 0222.
6. *mortification*: death of part of the body, especially of an extremity; gangrene (*OED*).
7. *poulticed*: the application to the skin of a moist mass of a substance with a soft, pasty consistency, usually by means of a bandage or dressing, in order to promote healing, reduce swelling or relieve pain (*OED*).
8. *the grand project*: presumably the project to issue a protest against conditions and pay on the Ordnance Survey; see letter 0201.
9. *monody*: a lyric ode sung by a single voice (*OED*).

From Thomas Harris Carroll 10 May 1843 0204

Carlow | May 10th. 1843.

Dear Sir,

In reply to your letter of the 7th[1] inst[2] I beg to inform you that the lines to which you refer[3] were omitted as I believed them not altogether suitable and the received was of so trifling a nature I did not think it necessary to notice it until something occurred that might be interesting to you in which case it was my intention to forward you a paper.

I shall be happy to hear from you frequently and if you could turn your attention to graphic sketches of public men, which I know you are capable of giving, in the humorous style, I will be most happy to forward you a paper regularly and to acknowledge your services.[4] There is Steel,[5] O Connell and many others who figured in Carlow you might take up such public characters and apply your talents to a delineation of their political morals Let me know whether you could try your hand in this line and you will oblige

Yours truly | Thomas H. Carroll.

RI MS JT 1/11/3491
LT Transcript Only

---

1. *your letter of the 7th*: letter missing.
2. *inst*: abbreviation for 'instante mense', Latin for this month.
3. *the lines to which you refer*: possibly the poem that Tyndall subsequently published, under the pseudonym Wat Ripton, in the *Preston Chronicle* as 'An Hibernian's Song to ____ ____' (20 May 1843, p. [4]).
4. *graphic sketches of public men ...acknowledge your services*: No such sketches of public men appeared in the *Carlow Sentinel* during the following weeks and months.
5. *Steel*: see letter 0066, n. 4.

From John Chadwick 10 May 1843 0205

May 10th. 1843. | Auld Kinsale

My very dear Tyndall

The receipt of your last[1] spoke to my tortured conscience like the stinging of bees and I hasten to make some poor amends for indeed what you justly call my unkind silence. But my dear fellow you judge me at least a degree too harshly. God knows I esteem and reverence the memory of that portion of the span of my existence which I whiled away in your society – and it is a sorrow to me to think that there were not more of the human family cast in the same mental mould that you were. Indeed in my course through life I may truly say that I never met a masculine individual that I could regard with such esteem and friendship as I did you, and such being the fact I allow that there is the less excuse for treating your friendship with neglect – curse the word, I never did treat it with the like, for in my inmost soul you have and ever will hold the first place among friends. Never dare to accuse me again of disregard, for by so doing you naturally insult yourself. forgive me my trespasses as you would wish to be forgiven.[2] My friend, I am sinking into a gulf of apathy in this place. I am like an animal tied by a rope to a stake which can only feed in a ring circumscribed by the length of its tether and my whole soul and thoughts are fixed within that circle and the little world of its own which it bounds still looking inward to the support of family, the maintaining of business, and so forth, yet I would now fain become as eccentric as a comet[3] and fly off at a tangent in writing to you. My cacoethes scribendi[4] is like a galled[5] post horse[6] which would fain lie on his road for ever, but when roused by a few skelps[7] in the gear, buckled to and whipped into a sweat, draws along as kind as any other jade. There is a complimentary comparison for you. You see I have got so far out of nothing, but I trust that my nothings will be only enough to smooth down the bristles of your anger, and that the heart of the fayre and courteous childe Walter[8] will again expand towards poor me. Past times seem to me like the memories of buried years and indeed the great cause why the cold chain of silence has hung o'er me so long is that I feel such a change in the tenor of my life and such a separation from old friends and old pursuits that I feel like as if I had been transplanted to the dog star[9] or set up shop in the moon; which of course would prevent me from keeping up a correspondence with the world I had left. But now a truce to nonsense, and let me tell you the events of the time past which have been but few. As to myself, I have been ill and well since in my turn, and am getting on in the same track, accompanied by my wife.[10] My little one is a fine stout child; she is able to walk and say some words and is of course a wonderful child. All the good folks here are well and hearty, all hugging fast to the world still and keeping above ground. Sophy Donaclift will be married in a few weeks to the Captain of a vessel which came here to be repaired.[11] He is a very nice gent and a German. I bought a boat from him a very fine one, 22 feet long wide in proportion built of oak and copper fastened, and all I want is to have you or some one like you here to keep me alive. I have not a male companion in the town now, all the Markhams[12] are now living in Liverpool, having sold their property here[13] and flitted. I was amused with the account you gave me in a former letter[14] of the memoir of the Survey snott's,[15] and was tempted to say with the preacher 'Lord, What is man that Thou are mindful of him'[16] for I am sure that it must be the intervention of their good angel that kept some of them from the gallows. Do not suppose I allude to any of your small family, I only mean the upper crust or off scourings of the Survey – I thought to have gone over among you all some time this summer, but fear that neither time nor the way and means will permit, and as the ariel machine[17] has not come to the perfection of travelling and we cannot go to the North Pole for a penny, I fear I must put it off for some time longer. I had no right to expect any news in your last but I trust in your next letter that you will let me know the aspect of affairs relating to the Survey and whether you intend to continue on it, and how you get on with the woman and how many children were fathered on you since, and whether you have reared whiskers as yet. And had you any diversion during the winter? It would be the delight of my heart to see you and visit you, but for God's sake do not judge of my friendship by my letter writing. it would indeed be very scant if you were to do so. I believe it was Brutus that forgave Cassius because he inherited the failing which gave the offence from his parent.[18] There is a noble precedent for you, for my sin is also a family failing, handed down through a long line of illustrious ancestry who were all bloody bad letter writers. I have at this time to write letters to six different relatives and think the best way would be not to write to any of them for fear one would be jealous of the other. I always feel a sober, quiet, delight in thinking of you, but it requires one of your fiery blow up letters to spur me to the perpetration of a billet. So now you know my weak point I think I am spinning a good yarn this turn. Give the enclosed to Latimer, remember me to all who may care for me and joined by the holy stone and all the other interesting localities in love, I am truly yours from Alpha to Omega[19] Amen

J. C. Chadwick

All friends desire to be remembered and send their best wishes au revoir

I am just after getting a lend of a book called Pencellings by the Way, by N. Willis Esq.[20] which is quite a God-send and as I am eager to devour it this must be my last postscript | farewell | write soon | good Bye.

and mention every thing remarkable. God bless you

We have Seymour's company[21] from Cork here at present and the height of fun – let me know how far Preston is from Liverpool and the best way of getting there, is it any way connected by railway?

RI MS JT 1/11/3501–2
LT Transcript Only

---

1. *your last*: letter missing.
2. *forgive me my trespasses as you would wish to be forgiven*: adaptation of the Lord's Prayer, 'And forgive us our trespasses | As we forgive them that trespass against us' (*Anglican Book of Common Prayer*, 1662).
3. *as eccentric as a comet*: the degree to which the elliptical orbit of a comet deviates from a perfect circle.
4. *cacoethes scribendi*: compulsive need to write (Latin), derived from a line in the *Satires* of Juvenal: 'Tenet insanabile multos scribendi cacoethes' (the incurable desire for writing affects many).
5. *galled*: irritated, vexed, unquiet, distressed (*OED*).
6. *post horse*: a horse kept at a post-house or inn for the use of post-riders, or for hire by travellers (*OED*).
7. *skelps*: strikes or blows (*OED*).
8. *fayre and courteous childe Walter*: likely referring to Tyndall's poetic pseudonym Walter Snooks (see letter 0109), with overtones of Lord Byron's *Childe Harold's Pilgrimage* (1812–18), but possibly also an allusion to the medieval troubadour Walther von der Vogelweide (*c.* 1170–*c.* 1230), who had recently been brought back into prominence in ballads recovered by the German poet and antiquarian Ludwig Uhland.
9. *the dog star*: Sirius, the brightest star in the night sky.
10. *my wife*: see letter 0148, n. 12.
11. *Sophy Donaclift will be married in a few weeks ... came here to be repaired*: The marriage may not have taken place, as a year later, in a letter from 9 July 1844, Chadwick told Tyndall, 'Miss Donaclift is not yet married but will be in time to Barry the tutor at Mr Perdue's' (RI MS JT 1/11/3508).
12. *the Markhams*: see letter 0177, nn. 16–21.

13. *their property here*: The Markham's family home had been Rose Abbey, Kinsale.
14. *a former letter*: letter missing.
15. *snott's*: applied to persons as a term of contempt or opprobrium (*OED*).
16. *'Lord, What is man that Thou are mindful of him'*: Psalms 8:4.
17. *ariel machine*: see letter 0199, n. 22.
18. *Brutus that forgave Cassius ... offence from his parent*: Marcus Junius Brutus (85–42BC) and Gaius Cassius Longinus (85–42BC) were Roman senators and leaders of the conspiracy against Julius Caeser, who reputedly fell out during the military upheavals following Caeser's assassination in 44BC. In William Shakespeare's *Julius Caeser*, Cassius asks 'Have you not love enough to bear with me | When that rash humour which my mother gave me | Makes me forgetful', to which Brutus forgivingly declares 'from henceforth, | When you are over-earnest with your Brutus, | He'll think your mother chides, and leave you so' (IV.ii.172–7).
19. *Alpha to Omega*: the first and final letters of the Greek alphabet.
20. *Pencellings by the Way, by N. Willis Esq.*: *Pencillings by the Way*, 3 vols (London: John Macrone, 1835), a collection of letters to the *New York Mirror* about his experiences in Europe by the American writer Nathaniel Parker Willis (1806–67).
21. *Seymour's company*: not identified.

From John Lilly 15 May 1843 0206

15th May 1843.

Dear Tyndall

It is with feelings of shame and remorse that I commence this letter after so long a delay but what with travelling and other incidental circumstances I have been prevented not from thinking of you but from committing those thoughts to paper For believe me amid all the changes of life[1] memory ever and anon recurs with pleasing sensations to some scene to wit the night at Sarsfields Court,[2] where you, the mainspring of joy and beau ideal[3] of merriment, took some prominent part and scarcely a day has elapsed since the receipt of your last animating epistle[4] that I have not reproached myself with neglect and ingratitude towards you in not writing: On my journey here I stopped 10 days in Woodside[5] on the opposite side of Liverpool I was surprised to see that boy Willy[6] who from all appearance is not a bit improved as regards morality notwithstanding his proximity and daily intercourse with his able tutor[7] and if he does not alter he will become a very bad boy. I have indeed left the Survey and must say without much regret you may ask why not after being nearly four years connected with it:[8] my answer is contained in the following reasons 1st the emoluments arising from it are barely sufficient for to keep our corporeal tabernacle[9] in a fit state to fulfil the avocations of life without one cheering hope or bright anticipation of a sufficiency for premature age occasioned by the dull and stupid monotony of plotting[10] and drawing: 2nd I found that I could not make a proper use of the time called my own, owing to a want of perseverance and decision in the performance of the many resolutions made but to be broken: and in this you have the advantage for I am glad to hear that you still sedulously continue your studies which tend to the cultivation of the mind and will eventually lead to a higher and more exalted line of life than the Survey can boast of: and last not least, being separated from friends (among whom I hope I am not presuming too far in enumerating you) whose conversation &c tended to make the otherwise dull days pass smoothly by, I came to the resolution of quitting a place which had no charms for me: I have as yet laid out no line of conduct for myself, yet I am still at home but expect to get a situation of which you shall hear more anon. My chief amusement is gardening. I thought I would feel more lonesome than I do. I had a letter some time ago from Furnby in which he states Ben[11] having had two narrow escapes, one from the bursting of the gun and the other from a blow of a sledge:[12] he is getting on fast at the building: there is nothing else you would care to hear. I was going to put in a few scraps of French, but knowing you are such a good scholar I was afraid: if you should write again to me I will thereby know whether my previous neglect is forgiven

I am your sincere friend | J. Lilly | Portalington[13]

RI MS JT 1/11/3748
LT Transcript Only

---

1. *all the changes of life*: Lilly had resigned from the Ordnance Survey in Liverpool in March 1843, and, according to William Ginty, had then gone 'on his way to the land of his fore-fathers [i.e. Ireland]'; see letter 0194.
2. *Sarsfields Court*: see letter 0180, n. 7.
3. *beau ideal*: perfect type or model (French).
4. *your last animating epistle*: letter missing.
5. *Woodside*: a small settlement across the River Mersey from Liverpool Pier Head. It was probably here that Lilly stayed 'a few days with his uncle', as Ginty reported in letter 0194.
6. *Willy*: William Hunter.
7. *his able tutor*: presumably Ginty.
8. *nearly four years connected with it*: Lilly had actually joined the Irish Survey in January 1839, but he might be thinking of his transference to the C District Examination Office in October of that year (NAI OS/1/16–18).
9. *tabernacle*: place of abode (*OED*).
10. *the dull and stupid monotony of plotting*: see letter 0163, n. 10.
11. *Furnby ... Ben*: not identified.
12. *sledge*: a large heavy hammer usually wielded with both hands (*OED*).
13. *Portalington*: Portarlington is a town on the border of Queen's County (now County Laois) and King's County (now County Offaly).

From William Ginty 16 May 1843 0207

Tuesday evening | May 16. 1843

My very Dear Tyndall

Your two letters[1] came into my clutches in proper time – The P.S.[2] to the first astonishes with a vengeance – Miss Edwards![3] in – Preston! – what could she want there? – What did she want with you? – if your meeting was accidental certain am I you would have said so! – but it was not 'an hours tete a tete' is far from the result of accident – but you say 'more of this anon' so say I – I feel particularly anxious to learn what formed a subject for the 'hour' – I suppose you acquitted yourself honourably of the charge respecting the 'lines by W Ginty'[4] – she is well aware it was you (and not me) that wrote them! – 'Honour bright' Tyndall – my conduct towards her has been the reward of her merit – but halt thee Ginty – !Whither so fast! I am in blessed ignorance of this proceeding

And musing thus the matter o'er
I needs must own it is perplexing
Oh all ye skilled in womans lore
this for me tis vexing
Pray do explain that 'tete a tete'
From Tyndall's pen so late ejected
An hours chit chat
And all to that
Ye cannot guess – but I suspect it

I recollect at one time in Kirkby one evening I was writing a letter to you – Miss E was present when I was directing it – She took a pencil out of her pocket and copied the address I laughed at her of course and asked her her motive to no purpose tho' – she laughed at me in return – so that this enabled her to trace out your reverence without inquiry.

– She seems to have some kind of foreknowledge! Call you her handsome – interesting &c. &c. hope you have not fallen in love with her – Oh, Jack! – No advantage taken during the tete a tete en solitaire?[5] Eh! Just explain this affair when you have time – to me at present it is utterly incomprehensible – she has a large languishing loving eye has she not? One that could 'sublimate the soaring soul above the sublunary system of solidities, or precipitate it into the dismal depths of Dis[6] — darkness and desperation'.[7] A letter from Sinnett lately informs me that she has come of age[8] and has left her uncle and aunt in Kirkby – 'She has' continues Sinnett 'taken the reigns into her own hands' 'where she will drive to is another question'. You are resolved to keep me busy, what between book-keeping and other projects I am a man of business now indeed. You shall have the enclosure by next post and a letter to the Society.[9]

Yours Ginty.

Letter from Skipton[10] today says Col Colby has informed Capt. D.[11] to his great surprise, that 200 men must be discharged[12] – What about this London job?[13] – Is it not a pity to let the opportunity which this contest affords slip? – one subject would do all – their skill at making estimates – Tithe Plans to wit.[14]

RI MS JT 1/11/3603
LT Transcript Only

1. *Your two letters*: letters missing.
2. *P.S.*: abbreviation of post scriptum, Latin for written after.
3. *Miss Edwards!*: Mary Edwards.
4. *'lines by W Ginty'*: see letter 0183, n. 4.
5. *tete a tete en solitaire*: solitary (i.e. on their own) head to head (French).
6. *Dis*: a fictional city in Dante's *The Divine Comedy* containing the lower circles of hell.
7. *'sublimate the soaring soul ... darkness and desperation'*: a line from a humorous letter by 'Palælogus Modestus' published in the Liverpool-based *Kaleidoscope; or, Literary and Scientific Mirror*, 1 n.s. (1821), p. 183, and subsequently reprinted in *Chambers's Edinburgh Journal*, 7 (1839), p. 287.
8. *come of age*: turned twenty one, the age of legal majority in nineteenth-century Britain.
9. *the enclosure by next post and a letter to the Society*: letter missing.
10. *Skipton*: see letter 0096, n. 3.
11. *Capt. D.*: Captain Robert Kearsley Dawson.
12. *200 men must be discharged*: presumably as part of the general reduction in civil assistants and labourers on the English Ordnance Survey; see letter 0194, n. 15.
13. *this London job?*: not identified, but presumably concerning the putative Survey of London; see letter 0154, n. 3.
14. *their skill at making estimates – Tithe Plans to wit*: see letters 0188, n. 6 and 0195, n. 15.

From Thomas Charles Higginson 25 May 1843 0208

Camp Kurrachee[1] | May 25th 1843.

My dearest Tyndall

You cannot imagine the delight with which I received yours[2] and more so when it appeared you had not received mine[3] at the time you began. It showed that you at last determined to write to me but I think when you did receive mine and found the reason I did not write to you and that it was not for lack of love or friendship you might not have been so careful of a sheet of paper but have torn it up and not used the cutting expressions which are therein written. I suppose you wanted to show me the state of your mind before you received my letter, you need not have done so, as I knew by my own what your feelings were. I knew the mistake which you laboured under, such the expressions sound so much harsher in my ear than I am convinced you intended them for – such as – no I will not mention any of them, I know if even I did, they are so badly cherished in your breast towards your poor old bedfellow Harold[4] you scarcely remember them.

I know Wallace[5] too well to think he would harbour a bad opinion of me. No more on the subject. I must finish this part with the promise of not again letting a month pass if possible without giving you an account of my adventures and protestations of eternal affection. Now I must begin and give you a short account of all that has befallen me since I started from England – my voyage out[6] is almost erased from my memory, suffice to say it was very favourable but very unpleasant as we had on board a young lady, daughter of Col. McNiel,[7] who is a disgrace to her sex, she annoyed all the officers in every possible manner. My uncle[8] coming round the Cape in a storm got his arm broken which did not get well for some months. On the 30th July we arrived in Bombay after a passage of three months and a half. I suffered very much for want of exercise being accustomed to so much on the Survey. On my arrival I immediately went ashore. I was greatly disappointed on seeing Bombay. I expected to see a beautiful city like Dublin, but instead a dirty place more like an Irish village than a city; The hotel a dirty miserable place. I, my uncle, and two other officers with two hundred men were stationed in a barrack outside the town for 10 days. The 86th Reg.[9] were there also, and of their men ten were dying every day with cholera; but none of our men died. Then we commenced a horrid march of one hundred miles through a miserable country and on the 18th of August we arrived in Poona,[10] which is rather a handsome place – that is, where the officers are quartered; but the native town is a miserable place, the huts are made up of any thing. The officers bungalows are all built in lines with handsome gardens before them but on arriving all were obliged to buy their own bungalows. My uncle bought one for himself and me during the winter months. It is a very pleasant place as the bigwigs came up from Bombay and then there were plenty of balls and parties; but the worst of them is that there is such a scarcity of females, none but officers wives. There were only about three or four unmarried ladies in all Poona. The first time I went I was so very much ashamed putting on the kilt[11] that I was almost afraid to dance but now I dont care much about it. I then took to study the Hindoostanee language[12] and in six months – on the 5th of this month – passed my examination. Out of 13 officers my name was third. The Col.[13] recommended me for an interpretership, which would have seven or eight pounds per month more; but they said I was too short a time in the service to obtain it. When I came down to Bombay for the examination I heard that our Regt.[14] was to be off to Scind[15] to fight the battles of her country, so after I passed I immediately started for Poona to get my things ready and arrived only two days before we began our march, which we did on the night of the 9th at 12 o'clock, in this country we always march at night as the sun in the day is so powerful it is impossible to be out of doors. We were always at the end of our march before the sun rose and then immediately pitched the tents. About half of the march was over when we received orders to take very little baggage and to leave all the women behind, so that after that we brought very little clothes. One of the men died by a sun stroke and we were obliged to bury him with his 'Martial Cloak around him'[16] as no coffins were to be obtained. We then embarked in small boats to come down a river into the harbour where we immediately embarked on board a large steamer which was to take us here. I got on shore for an hour, eat a good breakfast, had a pleasant chat for a short time with a couple of handsome girls who I suppose have come out on speck,[17] then bought a tent and went on board again when we began our voyage, and in 5 days arrived here. All slept on deck at night as the heat in the cabins was not bearable. When the vessel anchored off, this place is about 9 miles from the landing place, we disembarked into small boats but, to get on shore, for a quarter mile were obliged to walk up to our knees in mud, just as bad a place as was at sweet Youghal.[18] This place was taken three years since[19] and some of the young fellows who were at the taking of it have exchanged into this Regt. It is a most miserable place, just one plain of sand. The men have some kind of barracks but are all obliged to lie on the ground. I and two other officers are living in a kind of barn, the walls made of bamboo matting and the roof of some kind of reeds. I often pitied the people of Ireland on account of living in little mud cabins, I wish I had one of them here it would keep out the sand and dust. To day we were obliged to cut <*word(s) illeg.*> of the house and go into my tent which was near. A whirlwind of dust came and filled the house with sand and when we returned what a picture everything was – covered an inch thick with it. The hot winds begin at twelve o'clock in the day. Just imagine yourself before a furnace and the blaze almost touching your face, this wind is not very far short of it. The thermometer in the middle of the day is from 100° to 110°,[20] lying on the bed a drop of sweat is continually dropping from our chins. We are here expecting every day to receive the order to proceed to Hyderabad,[21] about 140 miles from this, where the Army is. When we go there I am afraid I shall not be able to write to you, as this country is in such as state, a detachment is always obliged to be sent with the dispatches, and monsoon sets in immediately and lasts for four months which stops all communication between Bombay and this. But every opportunity that offers I will write to you. This will not leave Bombay for England for another month. Remember me affectionately to Jim,[22] Tidmarsh, Ginty and all enquiring friends. I must close now, as I am afraid I shall be late for the post. Excuse all blunders I have not time to read it over – there is not much sentiment in it but you will see what Bony[23] has been doing in this country. I hope we shall be removed immediately as this is such a horrid place. If you write to old Mr. Manaway[24] remember me to him. He has never written to me and I know not where he is. George Holmes has not written to me for many months, although I have not forgotten him. Read this if you can I almost defy you

I remain now and for ever | Your affectionate friend | Harold of Old.

RI MS JT 1/11/3734–6
LT Transcript Only

1. *Camp Kurrachee*: the military cantonment at the seaport of Kurrachee (modern Karachi) in the south-west of the Sindh province.
2. *yours*: letter missing.
3. *mine*: possibly letter 0182.
4. *Harold*: Higginson's nickname.
5. *Wallace*: presumably a nickname for Tyndall.
6. *my voyage out*: The Royal Artillery's 78th Regiment of Foot arrived at Gravesend on 19 April 1842 and embarked six ships usually used for the transportation of convicts (*Mary*; *Morley*; *Columbus*; *Lady Lyndoch*; *Bodicea*; and *Bussorah*). They sailed direct to Bombay via the Cape of Good Hope ('The Army', *Morning Post*, 21 April 1842, p. 6). Higginson seems to have sailed on the *Lady Lyndoch*, which docked in Bombay on 29 August 1842 ('The Overland Mail. Shipping Intelligence', *Morning Post*, 8 November 1842, p. 2).
7. *Col. McNiel*: Lieutenant-Colonel Roderick McNeil (d. 1863) of the 78th Regiment of Foot.
8. *My uncle*: In a letter of 23 October 1843 Tyndall told his father of Higginson: 'His uncle is Captain in the same regiment, and thro' his interest young Higginson obtained a commission' (RI MS JT 1/10/3299–300).
9. *The 86th Reg.*: the 86th Regiment of Foot of the Royal Artillery.
10. *Poona*: see letter 0182, n. 4.
11. *kilt*: The 78th Regiment of Foot was originally a Scottish regiment, known as the Highlanders, and ceremonial regimental uniforms included kilts in the tartan of the MacKenzie Clan.
12. *the Hindoostanee language*: see letter 0182, n. 9.
13. *The Col.*: probably Lieutenant-Colonel Neil Douglas (1779–1853), the Commander of the 78th Regiment of Foot.
14. *our Regt.*: the 78th Regiment of Foot of the Royal Artillery.
15. *Scind*: (modern Sindh) an Islamic province of north-western India (now in Pakistan) in the valley of the Indus river.
16. *'Martial Cloak around him'*: C. Wolfe, 'The Burial of Sir John Moore after Corunna' (1814), 12.
17. *on speck*: on the chance of obtaining some advantage, gaining some profit (*OED*).
18. *walk up to our knees in mud, just as bad a place as was at sweet Youghal*: On 18 August 1840 Tyndall had told his father of 'wading thro' the mud with my shoes in my hands' on the shore in Youghal; see letter 0008.
19. *This place was taken three years since*: In February 1839 the seaport of Kurrachee was captured by HMS *Wellesley* and HMS *Algerine*, forcing the local rulers to agree to a new treaty with the East India Company.
20. *100° to 110°*: temperature in degrees Fahrenheit; approximately 38–43 degrees Centigrade/Celsius.
21. *Hyderabad*: a city on the River Indus. On 24 March 1843 troops from the British Army and the Honourable Company Army, commanded by General Charles Napier, defeated the forces of Sher Mohammad at the Battle of Hyderabad, enabling the annexation of the Sindh province.
22. *Jim*: Phillip Evans.

23. *Bony*: presumably another of Higginson's nicknames.
24. *old Mr. Manaway*: not identified.

From Joseph Payne 25 May 1843 0209

Skipton | May 25th 1843.

My dear Tyndall

'Promises and piecrusts &c.'[1] however such was not the case with me, but on turning the circumstance of the suggestion (I was about to make) over in my mind and of course finding the little effect (if any) it was likely to produce – emanating as it does from my poor numb-soul, I was just waiting until you would forget that ever hinted at a forthcoming mares nest[2] – but at you – Perhaps it is too presumptuous in me to speak on a subject which your more learned friends (as a favoured child of the muses)[3] must have hinted to you before now, and far more which your own mind must have suggested. Now Master John 'you know very well what I mean'; and, without infringing upon your more instructive hours, if the time you devote to scraps was given to the composition of a Poem such as you could write, then I would say with every one who has read your 'Reminiscences of Leighlin'[4] that things were going on as they should. But I need not tell you that the best scraps of the best authors serve for very little more than to grace a scrap book and though beautiful as they undoubtedly are you nor your friends would never read them with the one tenth of the zest that they would a Poem by John Tyndall – Do try it, you know the success that must attend it in every way. You know the patronage you can calculate upon and you know – what's far more – your own abilities. I have often this time past been thinking that such a thing was in contemplation, nay sometimes that it was in progress, but then when I consider Tyndall is learning French, I imagine that you say you are robbing yourself by even sparing a bit for the poor Sentinal.[5] And bye the bye, as I think of it, could you never by any chance just slip me the Sentinal Jim (of course) and you know I would be happy to see any thing that ever passed through Ln. B.[6] supposing it was only the blind borheen,[7] or Mrs Nowlans cat, aye or the red'un[8] either. We had your Captain[9] here today. He came to bring me 3 pence per diem[10] of an increase[11] (Luther's curse[12] on him and Tom Colby, it does not commence until the first of June) and to transact some other business, of more importance which I did not care to ask him about. It is conjectured that it is merely the dismissal of about 200 unfortunates;[13] but conjectures and reports of the Survey are as much to be depended upon as that which we know already to be a thundering jumped up lie. But you know a chance shot may kill the odd'un, and sometimes they may tell truth. However I hope such is not the case. Now Mr Baskin is almost afraid to write to you, he has neglected it so long – so you may believe me when I tell you that I did not quit laughing until I went to my bed on the day I received your last[14] when I thought of how Tidmarsh had the tears running down his cheeks at the idea of Ginty's second 'Farewell'[15] and at his furnishing the material and poor Ginty at seeing his 'name in print' unauthorised[16] and under such bloody storming colours his oaths newly coined for the special purpose of vowing his never before thought of modes of revenge on some person unknown, but Master John Topine[17] you were strongly suspected and you in your virgin quarters exculpated yourself in the eyes of the fair victim[18] not forgetting George[19] when the precious inspiration came softly stealing o'er your mornings what – yawning that last expression is quite poetical and original I believe in the application – Tyndall you may call me Ass, Fool, gull, and brand me with all other epithets by which sapient[20] philosophers are designated for the first article of this here epistle. I will be satisfied with all if it only tends in any way to set you at it write to me as soon as ever you conveniently can and tell me may I soon expect to see a 'Corsair' a 'Sara' a 'Bride of Abydoss' or a 'Lalla Rookh'[21] from you.

Thine ever | Payne

RI MS JT 1/11/3843–4
LT Transcript Only

---

1. *'Promises and piecrusts &c.'*: are made to be broken; the common phrase derives from Jonathan Swift's dialogue *Polite Conversation, &c.* (1738).
2. *mares nest*: an illusory discovery, especially one that is much vaunted and betrays foolish credulity (*OED*).
3. *muses*: see letter 0174, n. 11.
4. *'Reminiscences of Leighlin'*: this is possibly the poem Tyndall would submit to the *Carlow Sentinel*, albeit unsuccessfully, in letter 0217.
5. *Sentinal*: the *Carlow Sentinel*.
6. *Ln. B.*: Leighlin Bridge.
7. borheen: a small road (Gaelic).
8. *Mrs Nowlans cat ... the* red'un: not identified.
9. *your Captain*: Captain Henry Tucker.
10. *per diem*: per day (Latin).
11. *an increase*: Payne's last pay rise on the Irish Ordnance Survey before being transferred to England was in September 1840, when his daily rate had gone up by 3*d*., bringing his pay to 2*s*. (NAI OS/1/17). It is not known whether he received any further pay increases before the one mentioned in this letter.
12. *Luther's curse*: presumably the imprecation against the papacy made by the German reformist monk Martin Luther (1483–1546), 'I will curse and scold the scoundrels until I go to my grave, and never shall they hear a civil word from me. I will toll them to their graves with thunder and lightning. For I am unable to pray without at the same time cursing' (*The Catholic Encyclopedia*, 15 vols (New York: Appleton, 1907–12), vol. 9, p. 447).
13. *the dismissal of about 200 unfortunates*: see letter 0207, n. 12.
14. *your last*: letter missing.
15. *Ginty's second 'Farewell'*: Ginty announced that he was leaving the survey in letter 0203;

it was his 'second "Farewell"' because he had previously left the Irish Survey in the spring of 1839, when he went to England to work on the Tithe Survey, only to rejoin it in October 1840; see letter 0014.

16. *his 'name in print' unauthorised*: see letter 0183.
17. *Master John Topine*: Tyndall.
18. *the fair victim*: Mary Edwards.
19. *George*: George Latimer.
20. *sapient*: wise (*OED*).
21. *a 'Corsair' a 'Sara' a 'Bride of Abydoss' or a 'Lalla Rookh'*: *The Corsair* (1814) and *The Bride of Abydos* (1813) are poems by Lord Byron; *Lalla Rookh: An Oriental Romance* (1817) is a poem by Thomas Moore; 'Sara' is possibly 'The Eolian Harp' (1796) by Samuel Taylor Coleridge, which is addressed to 'My pensive Sara'.

To John Tyndall, Snr 27 May 1843 0210

Preston, Saturday

My dear Father

Your last[1] was received in proper time, it affords me much pleasure to find that you and my mother see things in their true light. I shall lose no time in profiting by the permission you have granted. With respect to going home, that's quite out of the question, it should indeed be my last shift, and I am not put to the shift at all as yet.

With this I send you a number of the Illustrated News[2] and a Supplement to the same. Its a beautiful production, as it is their anniversary number.[3] Will you present it to Mrs Steuart with my respects?

How are the Carlow Repealers[4] getting on? Is there any violence expected? From all I hear things have assumed a serious appearance in Ireland.[5] The present Government have a hard card to play.

I'm pressed for time at present and therefore must conclude

Your affectionate son | John.

RI MS JT 1/10/3311
LT Transcript Only

---

1. *Your last*: letter missing.
2. *the Illustrated News*: the *Illustrated London News*; see letter 0187, n. 7.
3. *their anniversary number*: this was the number for Saturday 27 May 1843, which included a supplement entitled 'A May Garland to Celebrate the Anniversary of the *Illustrated London News*'.
4. *the Carlow Repealers*: supporters of the Repeal Association which campaigned for the repeal of the Act of Union between Ireland and Britain.
5. *things have assumed a serious appearance in Ireland*: Daniel O'Connell declared that 1843 would be the 'repeal year', and from the spring he travelled across Ireland addressing mass meetings of Repealers. The Conservative government responded by passing the Irish

Arms Act, which restricted the manufacture and importation of arms and ammunition, and dismissing magistrates who attended Repeal meetings.

From Archibald McLachlan 1 June 1843 0211

Wednesday June 1st 1843

Dr.[1] X.[2]

The letter of Y. is enclosed[3] ----------- By this time you have received the conclusion of No. 1.[4] When I state a fact, you may rely on the assertion; you are to take the facts of quantity and quality of work done as strictly correct, or, rather, I always make the sum total less than the truth, and thus err on the right side – for instance I state the quantity done, I know from those at the Phenix Park[5] that I am nearly correct, also, from old acquaintances of long standing and who have made, in the olden times, similar calculations, I am bourn out that there were fully 14 counties finished, I remember having given to Mr Hume, a full true and particular account, forwarding it to the House of Commons getting his letter in reply &c. I, and another knave concocted the Report (in 1836), and also a map colored to show the quantity done, and we then made it 15 counties. We had also a good large pamphlet printed in Dublin 1834, the quantity done there was 15 counties, expenses £200.000. But the thief of a printer cowed, and would not give us a single copy, 'not for £50' – he was afraid of a prosecution. As to the error in the No.[6] of men employed, whatever error is + on one side will be– on another – that there were at one time 32 R.E.[7] officers on the O.S.[8] but how long they remained is uncertain, we are sure that 25 Lieutenants remained on it up to the time of the revision,[9] also 5 Captains – The number of laborers is understated, the real number is more near 1000 than 600. The conclusion to which I come is infalliable, but my mode of coming at it may not be strictly correct. I am not positive that 240 R.S.M.[10] were on it. But on the other hand, I have not included a number of Royal Artillery who were for some time employed. If I remember correctly Colonel C.[11] stated before the Committee 1829[12] that 10 Co:[13] were finished and expenses £77,000. I am positive as to the £77,000, but not sure of the 10 counties, Now at the lowest calculation from 1825 to 1829 or 3 years[14] the expenses would be £120,000, you can put confidence in what I state, because I am, bourn out, by others who know as much as I do. I would advise that you take just so much from my story as suits you, the rest throw away. And study brevity, for the lengthy part is yet to come. I am hereafter to take certain positions, and to challenge the parties to deny the facts if they can and this will be most serious stickler of all. I could refer to some letters, but then that would compromise the parties concerned (Bickett[15] for instance) to whom alone the secret was made known – You will of course consult D.[16] a very cautious cool gent: I forgot to tell you, I cannot recollect the time because I never saw the N.

paper,[17] but think it was in 1840 that T.C.[18] spoke of the pay of C.A.[19] in the House of Commons.[20] Also the Act for the S. of the 6 Northern counties of England,[21] ceases in January 1846 – the Act contemplates that the S. will be finished at the end of 1846 – Several questions you have asked in your letters shall be answered when we come to that point. – But I wish to keep the survey – the failure – revision – and conducting of the work separate from the pay of the men &c. Let the pay and work &c. get a chapter to itself – Look out for Babbages work on the O.S.[22] in some Mechanics Institution. another letter in 2 or 3 days.

B.[23]

RI MS JT 1/11/3781–2

---

1. *Dr.*: Dear.
2. *X*: the pseudonym Tyndall used to avoid detection.
3. *The letter of Y. is enclosed*: enclosure missing. Y is William Ginty; see letters 0201 and 0226.
4. *No. 1*: the 'Preface to The Ordnance Survey of Ireland' enclosed in letter 0195.
5. *the Phenix Park*: Mountjoy House in Phoenix Park, Dublin, the head office of the Irish Ordnance Survey.
6. *No.*: number.
7. *R.E.*: Royal Engineers.
8. *O.S.*: Ordnance Survey.
9. *the revision*: the revision and correction of inaccurate topographical details in maps made between 1825 and 1830 that began in the autumn of 1830.
10. *R.S. & M.*: Royal Sappers and Miners.
11. *Colonel C.*: Colonel Thomas Colby.
12. *the Committee 1829*: the Ordnance Committee on the Irish Survey, made up of Alexander Bryce, E. W. Fanshawe, J. T. Jones and Charles William Pasley, which reported in March 1829. The Committee was generally favourable towards Colby's conduct of the Survey, with Pasley noting in his diary 'our report went to give the responsibility on Colby again, giving him his own way as far as we could' (quoted in J. H. Andrews, *A Paper Landscape: The Ordnance Survey in Nineteenth-Century Ireland* (Oxford: Clarendon Press, 1975), p. 74).
13. *Co.*: Counties.
14. *from 1825 to 1829 or 3 years*: presumably a miscalculation on McLachlan's part.
15. *Bickett*: Joseph Bickett, a civil assistant who joined the Irish Survey in December 1826 and worked in the District Office in A District, becoming, by December 1839, the highest paid civil assistant in the entire District, at a daily rate of 4*s.* 6*d.* (NAI OS/1/8–16).
16. *D.*: see letter 0195, n. 35.
17. *N. paper*: newspaper.
18. *T.C.*: Thomas Colby.
19. *C.A.*: Civil Assistants.
20. *in the House of Commons*: There is no record of Colby making a statement on this subject in Parliament in either *Hansard* or *The Times*.
21. *Act for the S. of the 6 Northern counties of England*: the Ordnance Survey Act (1841 c. 30 (Regnal. 4 & 5 Vict)); see letter 0150, n. 6. The Act was renewed in 1846 and at five year

intervals thereafter, before being made permanent in 1922.

22. *Babbages work on the O.S.*: Charles Babbage (1791–1871), 'Remarks on the Statistics Contained in the Ordnance Survey of the Parish of Templemore', in *Report of the Fifth Meeting of the British Association for the Advancement of Science* (London: John Murray, 1836), pp. 118–19.
23. *B*: the pseudonym McLachlan used to avoid detection.

From Robert Martin [24] June 1843 0212

Midsummer[1]

My dear friend

Be pleased to accept my grateful thanks for the bottle of 'medicine' you were so kind as to send to me this morning I value it highly for its intrinsic worth, but I value more highly the friendly spirit which such an act exhibits. I esteem friendship in all men, but more particularly do I love to have the friendship of the good and discerning, for I then think they must see something better in me than I can discern in myself and this increases a desire in me to become what I seem to them to be I shall attend to your directions in the use of the medicine: when Mrs M.[2] becomes quamish[3] or low in spirit I shall administer a little of Mr T's[4] bottle and then give her a good shaking to rouse up her spirits I am taking to Ireland a very lean and delicate wife, but I hope to bring her back fat and healthy. She is more unwilling to go as she fears, in consequence of her extreme debility, that she will not have any pleasure on the journey, but I hope that the change of air, scenery &c. will renovate her constitution and cheer her spirits.

I am very sorry for your present debilitated state,[5] and if wishes, prayers, or benedictions would be of any service, you should have mine with all my heart. I often enquire at my own mind why it is that the worthy man is generally called to pass through more bodily suffering and privations, than the man who does not seem to have a particle more of intellect than the ass, or ox that he drives; and the only answer that satisfies me is that the just Governor[6] would not be just in giving to the man who possess extensive mental capacity, and intellectual enjoyment the same share of bodily strength and health as he who is deficient in mind, and I feel confident that the intellectual man, although he may suffer weakness and other privations, would not exchange places with the man who is only capable of goading a donkey. 'He whom the Lord loves he chastens'[7] – and 'God takes the good, too good on earth to stay, and leaves the bad too bad to take away'.[8] Although I hope you are good enough to be taken away, I hope you will be left to experience more of the enjoyments of this world than you have ever known. You have not experienced the sweets of matrimony. If you could get the genuine article it is even sweeter than the cogniac,[9] and he who made the woman ordered the man to marry her,[10] so, if you should go to the other world without obeying orders, you might get your ears pulled and pinched, and so you ought; yet I would not like to hear you made squeak.

Yours affectionately | R. Martin

RI MS JT 1/11/3757
LT Transcript Only

1. *Midsummer*: In Ireland the summer solstice was celebrated on St John's Eve and Day, 23 and 24 June, although Martin might also mean the longest day, 21 June.
2. *Mrs M.*: Mrs Martin, presumably Robert's wife.
3. *quamish*: variant spelling of qualmish; affected, or liable to be affected, with a qualm or qualms (*OED*).
4. *Mr T*: Mr Tyndall.
5. *your present debilitated state*: Soon after, in letter 0219, Tyndall told his father that his 'health [is] somewhat delicate – resulting from too much confinement during the winter and spring'.
6. *the just Governor*: an expression used by John Wesley and other Methodist theologians, who differentiated between God the Creator and 'God the Governor, the just Governor' who acts justly in meting out rewards and punishments to humankind (J. Wesley, 'Thoughts upon God's Sovereignty', in *The Works of the Rev. John Wesley, A.M.*, 4th edn, 14 vols (London: John Mason, 1841), vol. 10, pp. 347–9, on p. 349).
7. *'He whom the Lord loves he chastens'*: Hebrews 12:6.
8. *'God takes the good ... too bad to take away'*: a popular eighteenth-century epitaph of unknown origins (B. Richings, *A General Volume of Epitaphs, Original and Selected* (London: J. W. Parker, 1840), p. lxiv).
9. *cogniac*: variant spelling of cognac.
10. *he who made the woman ordered the man to marry her*: probably an allusion to Matthew 19:4–6, 'And he answered and said unto them, Have ye not read, that he which made them at the beginning made them male and female, And said, For this cause shall a man leave father and mother, and shall cleave to his wife: and they twain shall be one flesh? Wherefore they are no more twain, but one flesh. What therefore God hath joined together, let not man put asunder'.

To John Tyndall, Snr — 29 June 1843 — 0213

Preston, | June 29th, 1843.

My dear Father

I was quite prepared for the sentiments stated in your last[1] relative to my long letter,[2] I knew you would not relish it, and my mother poor woman, why I had more hope of radicalizing[3] even you than her. I shall make just one remark in connexion with our two last letters and that is that the God of protestantism never intended that it[4] should be established by unjust means.

I am not now employed in the office,[5] I have been engaged during the last week examining the town of Preston – that is – testing the Surveyors measurements. I was out in the country yesterday about seven miles from this. At about two o'clock I found myself getting hungry and by great good fortune I found two pence reposing in my pocket. I decided on getting a gill of milk. I went into a farmer's house and called for my drop. The milk cost me a penny and I devoted the balance on hand to the purchase of something less than a square yard of a comical kind of oaten bread called 'snap and rattle'.[6] I made a hearty meal on my two penny worth. It enabled me to work hard until five o'clock and then walk a distance of eight miles home.

With respect to my going to Ireland I am so circumstanced at present that I'm quite unable to form an idea as to when I can go there.

You might have heard me speak of young Tidmarsh very often, he will be going to the Cape of Good Hope shortly. I'm sure he ought to rejoice to find himself independent of the Ordnance Survey. He has some very influential friends out at the Cape and I doubt not that his arrival there will be crowned with all the success he anticipates. He is now about 21 years of age[7] – his mother doats on him – of course she feels the separation deeply, but she says she'd feel quite content if she saw me going out along with him.[8] Poor woman, she'd like to see us companions for ever.

Do you think a letter would reach William Tyndall if I directed it to Philidelphia? If he be an influential person I think it would.

Were you well acquainted with Robinson Tyndall?

Do you know what Debby[9] is doing in the Co.[10] Galway?

My love to my mother and Emma | Your affectionate son | John.

Mr William Tyndall Market Street, Philidelphia

RI MS JT 1/10/3290
LT Transcript Only

---

1. *your last*: letter missing.
2. *my long letter*: this does not seem to be letter 0196, and thus this 'long' letter is probably missing.
3. *radicalizing*: Tyndall later recalled that at this time he 'met some extracts' from the writings of Thomas Carlyle 'in the Preston newspapers. I chanced, indeed, to be an eye-witness of the misery which at that time so profoundly moved Carlyle ... With my memory of the Preston riots still vivid, I procured 'Past and Present', and read it perseveringly ... I found in it, moreover, in political matters, a morality so righteous, a radicalism so high, reasonable, and humane, as to make it clear to me that without truckling to the ape and tiger of the mob, a man might hold the views of a radical' ('Personal Recollections of Thomas Carlyle', *New Fragments*, pp. 347–91, on p. 348–9).
4. *it*: presumably the Kingdom of God on earth.
5. *the office*: the Divisional Office of the English Ordnance Survey's 1st Division.
6. *'snap and rattle'*: a dish from north-east England of toasted oatcakes crumbled in milk.
7. *He is now about 21 years of age*: According to an obituary, John Tidmarsh was born on 17 January 1824, so was in fact only nineteen at this time ('Death of Mr. John Tidmarsh', *Register* (Adelaide), 12 November 1906, p. 6).

8. *she says she'd feel quite content … along with him*: a desire reiterated in letter 0215.
9. *Debby*: Deborah McAssey.
10. *Co.*: County.

To John Tidmarsh 1 July 1843 0214

(Pour mon cher[1] Jack)

Dear Jack, ere the pennant[2] above thee is streaming,
A meteor tracing its path thro the sky –
While Remembrance o'er pleasures departed is dreaming,
I raise my blunt steel-nib to wish thee Good Bye!
And shall I appeal to the empty illusion,
That floats o'er Parnassus,[3] and raise with the throng,
The incense of prayer for the fancied infusion
Of light from the muse to embellish my song?
Away with the mock'ry – the language of feeling
Is fairest disrobed of the tinsel of art;
Its musical tones a deep pathos revealing
Bear on them the fair, sunny stamp of the heart!
Oh! peaceful and soft my dear Jack be thy pillow,
Encircled by thoughts of the fair native shore;
As riding secure o'er the foam crested billow,
Thou dreamest of Friendship and Tyndall once more!
How I think on the nights we have nestled together!
When the voice of 'discussion' waxed warm and shrill –
They are vanished and flown like a wind-borne feather,
Yet deeply does Memory cherish them still!
And oh! when he roams by the brink of the ocean
Which laves[4] that far shore with its wavelets of blue –
When his heart is alive to each tender emotion,
I feel that my Tidmarsh will think of them too!
And with them remember the green, sunny mountains
Of Erin,[5] which lift their proud summits on high;
Forget not her vales, nor her flowers, nor her fountains
Where the bright smile of Boyhood first gleamed in thine eye!
There are bosoms which love thee, whose full tide of sorrow
Is fanned as it flows by a thousand fond fears;
There are bright eyes and lovely which scarcely can borrow
From Hope a glad sunbeam to dry up their tears!
Yet, onward my Boy! may the balm-wafting pinion[6]
Of unalloyed happiness wave o'er thy breast;
When far, far away from the Sappers dominion
Your thoughts wander back to 'your own darling West'!
May the God of your fathers protect you in danger
While your course o'er the dark-rolling surges you steer
Till you press the bright strand of the Ishmaelite[7] stranger,

And the warm tones of welcome strike sweet on thine ear!
And now ere the pennant above thee is streaming,
A meteor tracing its path thro' the sky –
While Remembrance o'er pleasures departed is dreaming
Last chime of my song, Jack – God Bless you – Good Bye!

John Tyndall
Preston Lancashire
July 1st 1843.

It's hardly worth sending to you Jack, had I time I'd make it better – take it however as it is – it will perchance call me to your mind when a thousand surges roar between us[8] – every wish it contains is a bubble from the deepest fountain of my heart.

J.T.

RI MS JT 1/11/3873
LT Transcript Only

---

1. *Pour mon cher*: for my dear (French).
2. *pennant*: a long tapering or swallow-tailed flag (*OED*).
3. *Parnassus*: a mountain in central Greece that, in Greek mythology, was the home of the three Muses.
4. *laves*: washes, bathes (*OED*).
5. *Erin*: romantic name for Ireland.
6. *pinion*: wing (*OED*).
7. *Ishmaelite*: a descendant of Ishmael, one of Abraham's two sons. The Arabs claim to be descended from Ishmael and he is a patriarch of Islam (*OED*).
8. *when a thousand surges roar between us*: Tyndall believed Tidmarsh to be soon sailing for the Cape of Good Hope; see letters 0213 and 0215.

From Mary Tidmarsh 6 July 1843 0215

My dear Mr Tyndall

In addition to your usual kindness I have received your welcome favour[1] of the 2nd inst.[2] the perusal of which afforded me much comfort. When we have the approval of those whose opinion we esteem there is a secret comfort in sanctioning the act we are about to perform.[3] So it is with me and my dear boy,[4] the separation is to us the most painful we ever felt, but, when we consider that it will be I trust for his future advantage, it would be doing him very great injustice not to consent to his going out, particularly as he goes under very agreeable circumstances which I suppose he has put you in possession of. The persons he is going to are very extensive general merchants – two partners there[5] and two in London – they are also relatives of Mr T.[6] We entertain a hope that John may yet be sent to London on their business, if so we shall have an opportunity of seeing him.

I expect we shall have to part with James[7] also next summer he was written for several times by a nephew of Mr T's in Calcutta Captain Lyster[8] who trades to China. When he wrote last he was expecting a freight[9] to London; if he succeed in getting it he will come to Ireland and take him in his own vessel. Under these circumstances we could not for a moment hesitate, but again we dread the climate. However we have a year to think on this matter – many a change will take place in that time.

So you intend that the breezes of the next Spring shall waft you to another clime, you did not say where. How happy would I feel if it were to the Cape. What an addition it would be to my dear boy's comfort and happiness (and, without flattery) whose society I would prize for him. I do entertain a comfortable hope that we may be all there yet and say from my heart the Lord grant it. Christina and May[10] present their best regards to you, also (Mrs Cox) Mother. The last letter Christina wrote you came back under dead letter office cover.[11] We could not account for it, unless that she directed it wrong. With regard to the song she does not recollect it,[12] says it is of no consequence – it would be rather severe to horsewhip you for such a trifle. Many thanks for all the trouble you take in sending us so many papers. If you go to Carlow before you leave Ireland dont neglect calling if you go through Cork I need not say how happy we would be to see you. Dear Mr Tyndall will you write often to my dear boy giving him advice which I know he values and you will oblige

your sincere friend | Mary Tidmarsh. | Cork July 6th 1843 30 Merchant's Quay

RI MS JT 1/11/3868
LT Transcript Only

---

1. *your welcome favour*: letter missing.
2. *inst.*: abbreviation for 'instante mense', Latin for this month.
3. *the act we are about to perform*: sending her son John to the Cape of Good Hope in southern Africa.
4. *my dear boy*: John Tidmarsh.
5. *there*: the Cape of Good Hope.
6. *Mr T*: Mr Tidmarsh.
7. *James*: presumably James Tidmarsh, John's brother.
8. *Captain Lyster*: possibly the same Captain Lyster who, in the spring of 1854, was murdered onboard the merchant ship *Clarissa* by Sikh convicts he was transporting from Calcutta to Malacca ('Another "General Wood" Tragedy', *Straits Times*, 20 June 1854, p. 4).
9. *freight*: the paid service of transporting goods (*OED*).
10. *Christina and May*: Christina and May Tidmarsh.

11. *under dead letter office cover*: The dead letter office was a division of the Post Office dealing with undelivered letters. The office had the authority to open such letters, and, if the sender's address was included, return them.
12. *With regard to the song she does not recollect it*: not identified.

From William Wright 11 July 1843 0216

Cincinnati | July 11th 1843.

My dear Jack

It gave me great pleasure to receive a newspaper from you if it had been a letter it would have met a much heartier welcome as it is I am thankful for small favours and large ones in proportion. It is now a long time since I had the pleasure of hearing from you and I am very anxious to have a line from you I would like to learn from you how all our old acquaintances are getting along. I do not expect that you will be able to give me any account of a number of them. But as you are in a constant correspondence with your Father you must be in possession of a great many incidents that would interest me. So that Jack if you would condescend to take up your pen for five minutes and give me all the particulars you know of any of them you would place me under very great obligations to you I would also like to learn a little about yourself how many sprees,[1] rows, and love scrapes you have been engaged in – Oh I forgot myself for I should remember that you are too sober a man for anything of the kind with the exception of the latter, and as that is part of the nature of an Irishman I may reasonably expect that you have had your share of them. It would give me great pleasure to hear you signify an intention of emigrating to this country nay to this city for it possesses so many opportunities for a young man to acquire the requisite information to enable him to earn an honest independence. I do not mean to confine my remarks to mercantile pursuits. But the Law, Medicine &c. are within the reach of almost every young man here are several schools of medicine, law &c. whose charges and hours of attendance are so regulated that 2/3 of the classes are composed of young men who attend to other occupations in the interim and earn sufficient to defray the expenses of their board and studies. I will state a case which came within my own knowledge. The subject of it is a young Irishman of the name of Lindsay.[2] He resided for some time in Canada and while there became acquainted and married a young woman also from Ireland. They married in opposition of their friends and as it usually happens they were cast upon the world friendless pennyless and inexperienced but I saw him a few days after his arrival. At the time he was relating some of his misfortunes to a teacher of book-keeping with whom I was at the time learning, by accident I overheard the conversation it consisted of his having travelled very far and a great portion of the way on foot in search of employment and being unsuccessful in his applications, and it ended in his asking the teacher if he could give him some employment I was sorry when I heard the teacher say it was not in his power. A few days after his case became known to an Irishman of the name of Andrews,[3] and he interested himself about him and after a short time procured for him employment in a lawyer's office. After some time he commenced studying and will in the course of a few months receive his diploma to practice. That is only one case of the hundreds which happen in Cincinnati. I will give you another, A few years ago a very young lad was doing business in dry goods[4] stores his father had died some years previous and left him unprovided for. Some of his father's acquaintances procured for him the situation with dry good stores. Whilst there some law books fell into his hands and it struck him that he might obtain a knowledge of the law and he commenced studying and in a short time he went through a regular course of studies and is now a good lawyer in good practice. So that on the whole there are a great many advantages which we possess over the old country. But it is our children which possess them all – education is so abundantly supplied by the State to all children who are not able to pay for it.[5] It you should ever think of coming out make straight for Cincinnati and in it you shall find as hearty and as welcome a reception as you would wish. In fact I have been for some time looking forward to the day that I would be able to grasp your hand in the streets of this goodly city. I do not see any hindrance to your at once starting except that of leaving your honoured parents. Since I last wrote to you I have been informed of my poor Father's death.[6] It was a sore trial to me to be so far away from him in his last hours for I have ever been his favourite. His last days have been days of suffering and trial. May his all forgiving Saviour pardon his offences and give him that peace and happiness which the world and all its gifts could not bestow. I will now give you some particulars about myself I have been blessed with another lovely child a little girl – she is now a month old and is as large and plump as a child of three months. I haved called her Kate after my revered Mother. May she resemble her in her many virtues is my most ardent wish and prayer. I am getting along in business and at present am in treaty for an old and established Stand which my aunt is about giving up. If I succeed it will be a very fortunate move for me. It is a dry goods stores which she occupied, and made a vast deal of money in, within the last ten years. Since we commenced business here I and my wife have acquired a knowledge of the dry goods trade so that I can now measure a yard of calico[7] as well as weigh a pound of tea or a cwt.[8] of iron as I have done since I came here. It is not often that a person in the old country learns 2 or 3 different business in 2 years. But it often happens here that a man who is in the grocery line now will in two years hence be in the hard ware line or erecting a steam saw mill on the Arkansas river or a distillery on the Rio Grande in Mexico. But what will surprise and probably please you most about myself is that within the present month I have had an operation performed upon my left eye for squinting and it has succeeded admirably. I am now nearly free of any deformity and the surgeon has guaranteed that within 3 months it will be impossible to discover which has been defected. I will endeavour to give you a sketch of the operation;[9] you may recollect that my left eye turned inwards. The patient is placed upon his back and a large spectacles with the glasses out is fixed and made tight upon the face then small silver wire is placed under the upper eye lid and then made fast to the spectacles the same is done to the lower lid in order to keep eye open, the muscle is then taken up on a small instrument with a hook on the point. This muscle is then detached from the eye it almost immediately reverts to its proper position. The cause of squinting is occasioned by one of the muscles being too short or contracted when it is detached the eye naturally goes into its proper place after 14 days the muscle again attaches itself to the eye but further back and a uniform action takes place. What is very singular, there is very little pain and no loss of time occasioned by the operation. After it was performed it felt like as if there was a quantity of sand thrown into it, that was the only inconvenience I felt and that only for one evening. I am obliged to close on account of the post office closing.[10] I have not received a letter from you for the last 6 months you must on receipt of this immediately sit down and make me amends for it. I sent you and am sending with this some newspapers, please to let my mother know you had heard from me I am not writing to them by this packet[11] and I would like them to know I am well. Give my best respects and love to your Father and his family Sarah[12] joins with me in kind regards and good wishes for your welfare she expects one day to have the pleasure of grasping the hand of her husband's earliest and best friend in which wish I heartily join and remain

ever yours | William Wright

I have lately heard of some of your productions in poetry appearing in the public press I think you might at least have sent me a copy for inspection particularly as it related to some of our loved haunts[13] please do not put me to the trouble of asking for it again. I must not forget to return you my thanks for your newspapers I have received some of them were from my brother-in-law and I admire them very much adieu | W. Wright

You are now two letters in my debt and I don't give long credit – so pay up Jack.

RI MS JT/1/5/1808–10

LT Transcript Only

---

1. sprees: lively or boisterous frolics; occasions or spells of somewhat disorderly or noisy enjoyment (*OED*).
2. *Lindsay*: not identified.
3. *an Irishman of the name of Andrews*: not identified.

4. *dry goods*: American name for the class of merchandise comprising textile fabrics and related things; articles of drapery, mercery, and haberdashery (*OED*).
5. *education is so abundantly supplied ...not able to pay for it*: From 1838 the Ohio state government levied a direct tax that paid for all children to attend the common schools that had been established in 1825 on the model of the New York City public school system.
6. *my poor Father's death*: see letter 0193, n. 27.
7. *calico*: general name for cotton cloth of all kinds imported from the East, from the name of a city on the coast of Malabar that, in the sixteenth century, was the chief port, next to Goa, of intercourse between India and Europe (*OED*).
8. *cwt.*: abbreviation for centum weight, an American unit of weight equal to 100 pounds.
9. *the operation*: this seems to be an ocular myotomy, a surgical procedure for the correction of strabismus that was first performed successfully in 1839 by the German surgeon Johann Friedrich Dieffenbach, after which 'news of this procedure spread like a wildfire throughout the world with hundreds of cases being reported from London to Boston within months after Dieffenbach's first case' (G. K. von Noorden, 'History of Strabismology', in *Transactions of the 29th European Strabismological Association Meeting*, ed. by J.-T. De Faber (London: Taylor and Francis, 2005), pp. xxiii–xxviii, on p. xxiv).
10. *I am obliged to close on account of the post office closing*: Wright presumably intended to take the letter to the post office to send it.
11. *by this packet*: Wright's practice was seemingly to send bundles of letters, and newspapers, that were then divided when they reached Britain and Ireland.
12. *Sarah*: Sarah Maria Wright (née Neale), Wright's wife.
13. *some of your productions in poetry ... our loved haunts*: Wright may be referring to Tyndall's earlier poem 'Carlow', *CS*, 30 October 1841, p. [3], or possibly he had been told of the new poem Tyndall was preparing to submit in letter 0217.

To the Editor of the *Carlow Sentinel* 12 July 1843 0217

11 Duttons St Preston No 6[1]

Sir

Perhaps you'd be kind enough to allow the following a place in the Sentinel[2]
John Tyndall
The Ed[3] of the Sentinel

The day is gone, no golden beam
Now smiles upon the fair hill side,
And cheerless flows the darkling stream
By weeping flowers which fringe its side.
The day is gone, and darkness flings
Her mantle over crag and dell
Its caroll now no warbler sings,
Amid the brakes of Berkenfell:[4]
But higher notes convulsive rise,
The booming of the angry sea,
The thrilling war song of the skies
Now dwells in wildering melody.

There's something in the lofty feeling,
That swathes my soul with burning glow,
Too grand, too glorious for revealing,
Too high to grovel here below:
When fancy spreads her daring plume,
And curbless[5] wings amid the gloom,
Her welkin[6] sweeping flight to rise,
And mingle with her native skies!
Past hours into existence start,
And scenes long lived round my heart:
yes, burning thro' the gloom afar
Is seen the golden glowing star
of memory, whose brilliant rays
sheds glory over bygone days.
The days when boyhoods reckless joy
Rung forth unmingled with alloy;
When generosity and truth
Shone frontlets[7] on the brow of youth;
When with a swimmers dauntless pride,
I skimmed the barrows[8] placid tide.
Oh! these give memory a zest,
And sheds a halo round my breast.
The 'decent church'[9] still topples[10] there,
Where first my childhood lisped a prayer;
While round are strewn the graven[11] stones,
Chill records of the mouldering bones
Which lie beneath, as sadly waves
The rank grass o'er the lonely graves.
Ah! how I loved when smiling even
Blushed beauteous in the western heaven;
As rose the merle's[12] farewell note,
To clamber up the lofty mote,[13]
And from the fairy legioned mound,
To view the lovely scene around.
The ripple of the brooklet near[14]
Struck sweetly soothing on mine ear;
And rich in beauties varied dies,
The groves of Burgage[15] blessed my eyes.
Beyond, amid the stately trees,
Where softly crept the evening breeze,
Which as it sighed o'er lawn and bower,
Kissed perfume from each drooping flower,
'The Lodges'[16] appeared – beneath me rolled
The gentle barrow tinged with gold
How calmly would the sunbeams smile
Upon the castle's hory pile:[17]
Not always thus[18] – the sabres flash
Glanced lightning there – the commons crash

Rung thunder o'er the startled flood,
And stained its chrystal waves with blood!
But peaceful is his war seared brow
And silent are his thunders now:
The pall of centuries is spread
In gloom oer many crested head,
Where proudly waved the nodding plume,
Amid the battles deepest gloom:
They're gone – and now the ivy clings,
And many a songster safely sings,
Where erst the clarion blasted far
The thrilling notes of blood and war!
Oh! how would thoughts like these unbind
The trammels from my embryon[19] mind:
Aroused as by a glowing beam,
The young chrysalis ceased to dream
Assumed the wing and stretched its flight
Thro' scenes by fancy rendered bright.
'Twas then I ventured first to fling
My hand across the trembling string;
Tho' wild and broken was its tone,
I loved it, for it was my own,
Oft would my straggling bosom long
To wreak its thoughts on nobler song.
Hail! happy hours e'en now ye shine,
The brightest gems in memorys chime
Hail! blissful scenes, between us swell
The crested wave and towering fell
And now perchance the surging gale
Sweeps oer you wild with frenzied wail.
Hark how it sobs – its murderous breath
Now haply strews the shore with death.
The embers of a wasted fire,
Now quickly one by one expire,
A waning lamp in
Flings oer my page its dying blaze;
Admonished by the midnight chime
My wary pen I now resign
My couch to seek while wild and high,
The tempest sings my lullaby
(finis)

RI MS JT/8/2/1/20–1

---

1. *Editor of the* Carlow Sentinel [from title]: Thomas Harris Carroll.
2. *allow the following a place in the Sentinel*: The poem was not published in the *Carlow Sentinel.* On 15 July 1843 the newspaper featured Eliza Cook's 'Song of the Haymakers'

('Selected Poetry', p. [4]), on 22 July there was no poem, and on 29 July it carried Letitia Elizabeth Landon's 'We Might Have Been' ('Selected Poetry', p. [4]).

3. *Ed*: Editor.
4. *Berkenfell*: see letter 0174, n. 23.
5. *curbless*: without curb or restraint (*OED*).
6. *welkin*: sky (*OED*).
7. *frontlets*: ornaments or bands worn on the forehead (*OED*).
8. *barrow*: the River Barrow, which runs through Leighlin Bridge.
9. *'decent church'*: possibly either Wells Church in Leighlin Bridge, where Tyndall's uncle Caleb was a church warden, or St Laserian's Cathedral in Old Leighlin, where Tyndall's father would be buried in the family plot in 1847. A 'decent church' is presumably a Protestant one, and may relate to directives on ecclesiastical furniture in the English Church Canons such as 'A decent Communion-table in every Church' (Canon LXXXII).
10. *topples*: to lean over unsteadily, as if on the point of falling; to overhang threateningly (*OED*); both Wells Church and St Laserian's Cathedral are mediaeval structures.
11. *graven*: sculptured, hewn (OED).
12. *merle*: blackbird (*OED*).
13. *the lofty mote*: see letter 0146, n. 17.
14. *The brooklet near*: presumably Maudlin stream; see letter 0146, n. 18.
15. *Burgage*: a parish in County Carlow.
16. *'The Lodges'*: possibly Burgage House, the seat of the Vigors family.
17. *the castle's hory pile*: the remains of Carlow Castle, a Norman structure built between 1207 and 1213.
18. *Not always thus*: Carlow Castle was besieged in 1495 amidst the repercussions of a plot to seize the English crown, and again during the Confederate Wars of 1641–2, when more than 400 Anglo-Irish Protestants were trapped by Catholic Confederate forces and starved for three months. It was also besieged by Oliver Cromwell's New Model Army in 1650.
19. *embryon*: immature, unformed, undeveloped (*OED*).

To William Ginty 12 July 1843 0218

Preston

My Dear Ginty

In my last but one[1] I promised to revert to yours preceeding[2] at what was then a future time that time has become the present and I now lift my pen to redeem my pledge. the following was suggested by the convictions that appear, to exist between the lovers of your song and the idea of Ellen[3] – accordingly as the clothing of her idea if strewn less or more as on your page in the same proportion does your song lose or gain in power – when she fills it you're overpowering – of this enough

The aerial phantazies of youth
Robed in their own bright loveliness –
The visions clothed with seeming truth
Now melted into nothingness! –
Aye, all are vanished – yet not so

Behind the evanescent throng
There lingers still a holy glow –
A beam which 'gilds thy every song'!
Whence comes it? – does the scented gale
From distant Kirkby bear the prize?
Or does the heav'nly brightness dwell
In 'lovely Mary's'[4] diamond eyes? –
Ah! no – it shines upon the breast
Of every billow wild and high
Which rears aloft its foamy crest,
Rebellious to the darkened sky –
It smiles, where ripples gently lave
Each barque that spreads her snowy sail
On every rock – on every wave
Between 'New Babel'[5] and Kinsale!
Yes – there its nucleus dwells, to bless
Thy morning thought – thy midnight sigh
There clusters too 'the raven tress'
There radiates the lustrous eye –
The clouds of care may gloom and lour
In darkling masses round thy breast,
There is a ray of magic power –
As glorious sunbeam from the West!
Which calls to life thy buried love –
That 'cup of sweets' without alloy
And like a spirit from above
Gives vigour to the pulse of joy! –
*Oh! there are ideas which dart –
Like meteors thro' the midnight air
A gleam of glory thro' the heart
Where waved the banner of despair!
Visions of bliss untasted roll
Before the visionary ken[6] –
Destroy the canker of the soul
And bid the mourner smile again.
I know I touch a speaking string –
A string which quivers in thy core
And sounds responsive while I sing
Of days which shall return no more –
of whom? oh! God of poesy
My log should smoke and blaze and flame
And consecrations from on high
should sparkle round her sacred name!
of whom? – of Ellen – oh! I find
My swelling bosom's deep devotion

* 'oh! there are looks & tones which dart
An instant sunshine thro the heart' | Byron[7]

Unutterable – while my mind
Is crushed by mountains of emotion!
Not so with you – you grasp the lyre
And shake from it the dust of slumber
From her you catch the heavenly fire
And unholy wake the burning number!
I love to con[8] the glowing line
Where Ginty's restless spirit revels
And flings with fervours quite sublime
The fury of a dozen devils! –
I love to ponder o'er the lay
Where Ginty's muse does gentler duty
To wing his spirit quite away
Before the shrine of 'Mary's' beauty
But oh! a deeper – holier spell
Like music thro' the moonbeams streaming
Is riven thro' the fervid swell
And from the leader one is gleaming
When thoughts of lovely Ellen raise
The murmurs of this melting stream
And all 'the light of other days'
Around thy stanza smiles again!
Farewell thou bright ideal ray –
Which lent to life its happiest hue –
Thou gleam of heaven's own halcyon day[9]
Angelic hope – adieu! – adieu! –
oh give to him thy sunny smile –
I claim it not – that thought is flown
Upon my harp I lean the while
Its last sad twinkle dies – tis gone!!

RI MS JT/8/2/1/24–5

1. *my last but one*: letter missing.
2. *yours* preceeding: possibly letter 0200 which includes a poem addressed to 'My lovely Ellen'.
3. *Ellen*: from Kinsale, also known as 'the lady of the raven plume'.
4. *'lovely Mary'*: Mary Edwards.
5. 'New Babel': presumably Liverpool; see letter 0161, n. 7.
6. *ken*: range of sight or vision (*OED*).
7. *'oh! there are looks ... | ... sunshine thro the heart' | Byron*: the lines are not in fact by Lord Byron, but are instead from Thomas Moore's *Lalla Rookh: An Oriental Romance* (1817), IV.i.685–6.
8. *con*: pore over, peruse (*OED*).
9. *halcyon day*: a period of calm weather, anciently believed to occur about the winter solstice when the halcyon bird, a species of kingfisher, was breeding (*OED*).

To John Tyndall, Snr 24 July 1843 0219

July 24th, 1843.

My dear Father

Am I never to get a letter from you? it is now four or five weeks since I heard from you – this I deem rather extraordinary – I have had some tossing about since last you heard from me I have been employed for the last five weeks at field duties. I found my health somewhat delicate – resulting from too much confinement during the winter and spring – and so applied to get out. My request was granted and after having been a week employed at the town of Preston – and thinking the country air would be more healthful – I started for a village called Goosnargh,[1] situated about seven miles from this. Here I spent a fortnight. I was not alone, I had a companion – a Dublin Gent[2] along with me, and being congenial spirits we contrived to spend our time most happily. We made ourselves quite at home, and were treated with the greatest kindness. We made an impression on our host's family[3] which will not be quickly effaced. Our sprightly Irish manners were strongly contrasted with the plain matter of fact English farmers who dwelt in the locality. There's not an old woman within four miles of Goosnargh who does not know me. Many a gill[4] of warm fresh milk have I quaffed as I returned from my work in the evening, this, combined with snap and rattle[5] – a peculiar kind of oat bread – and the fresh country breezes has completely recruited me. I have returned to Preston with a constitution renovated and nerves restrung. I continue still at outside work and intend to do so as long as I can. Its preferable to breaking the chest over a drawing table from six in the morning to five in the evening. I hope you wont neglect writing to me on the receipt of this. I am very anxious to hear from you. Give my love to my mother and Emma

and believe me | Your affectionate son | John.

RI MS JT 1/10/3291
LT Transcript Only

---

1. *I started for a village called Goosnargh*: Tyndall made a detailed record of this sojourn in his Journal (vol. 13a, pp. iii–xi).
2. *a Dublin Gent*: William Marquis.
3. *our host's family*: John Barton, innkeeper at The General Elliot pub on Church Lane, Goosnargh, his wife and daughters, the youngest of whom, Elizabeth, Tyndall found particularly beguiling.
4. *gill*: a measure for liquids, containing one fourth of a standard pint (*OED*).
5. *snap and rattle*: see letter 0213, n. 6.

To John Tyndall, Snr 28 July 1843 0220

July 28th, 1843.

My dear Father

Ten minutes ago I received your letter[1] and the returned testimonials.[2] I am not cast down, I am not even disappointed, but a conviction is deepened which has occupied my mind these many months, and that is – that I have no business here. A voice tells me that with reference to human agency I must depend on my own exertions and on them alone. Think not that I slight the kindness of those who have borne testimony to my character. No, I feel as grateful to them as if their exertions had obtained me all I desired; and whereever my lot may fall I shall always carry with me a deep sense of my obligation to them. You advise me to write to the Dean,[3] if I did so it would be to express my heartfelt gratitude for what he has done for me, and not to ask him to do more. Any plan that I could adopt with reference to Somerset House[4] would I'm sure prove futile. I must wait the tide of events, encouraged by the hope that sometime or other I'll meet a favouring current on which to launch my barque – think not that I am cast down.

Why should the subject of America lie dormant? You should put me in possession of all you know concerning it. I shall do nothing without your concurrence. I shall write to Mr Conwill, surely he will tell me all he knows about the matter. I am very glad to know the street Wm[5] Tyndall lives in. In the present state of public opinion in America I have no doubt but Mr Ryan is attached to the repeal movement,[6] and this might make him careless of forwarding the interest of one whom he knows to be at least the son of an antirepealer. I dont expect that he will answer my letter. What do you think then of my writing to Wm Tyndall? I will consult you in everything. I shall propose nothing irrational, so that I feel assured you will assent to all I propose. I'm no way impatient, but I'm convinced that we should look the subject of going to America in the face, and by no means let it lie dormant. Shall I write to Wm Tyndall?

And now with respect to Emma. Its a project which often has struck me since my arrival in England, but never as a practical one. I am sure I'd feel most happy in her company. But circumstanced as I am I'm sure Emma would feel miserable here. How she would spend her time I know not. My circle of acquaintances in Preston is very limited, and entirely confined to units of the masculine gender. Again I say that I would be delighted to see her; and in sober seriousness I'll begin on the spot to repair a shattered exchequer in order to afford me the means of going over and dancing the dullness off her! My love to my mother

Your affectionate son | John

With respect to writing to Wm Tyndall send me a single line in reply immediately on the receipt of this – Dont forget. Dont put 'preapaid' outside stamped letters; when you do write the word, leave out the first 'a' and write 'prepaid'[7] – take a lesson! ha! ha!

John.

RI MS JT 1/10/3292
LT Transcript Only

1. *your letter*: letter missing.
2. *returned testimonials*: see letters 0134 and 0173.
3. *the Dean*: Dean Richard Barnard.
4. *Somerset House*: the map office of the Tithe Commissioners based at Somerset House, on the Strand in central London; see letter 0173, n. 7.
5. *Wm*: William.
6. *the present state of public opinion in America ...the repeal movement*: Many second-generation Irish Americans, as well as newer immigrants, organized American auxiliaries of Daniel O'Connell's Repeal Association, and the 'cause captured the imagination of a large element of the swiftly swelling ranks of the Catholic Irish' (B. Jenkins, *Irish Nationalism and the British State: From Repeal to Revolutionary Nationalism* (Montreal: McGill-Queen's University Press, 2006), p. 215). In Philadelphia, where Ryan lived, there was a Repeal Society, although it had split into two factions in the early 1840s over O'Connell's condemnation of slavery.
7. *write 'prepaid'*: Although prepaid ready-stamped envelopes and adhesive stamps had been introduced in May 1840, Tyndall, Snr evidently still preferred to add his own confirmation, even if his misspelling of 'prepaid' suggests he was struggling to adjust to the new terminology of the Post Office reforms.

From Archibald McLachlan[Late July] 1843 0221

25 Byron Street, Leeds | Sunday

Dr.[1] X[2]

The conclusion of No 1[3] is ready but not knowing the state of things with you I shall not send it till I hear from you. It is said that 32 men have been discharged in Preston. Here, 2 Surveyors discharged on the spot, 1 of 4 years the other 12 years standing. 3 reduced from being surveyors, to chainmen. 1 Surveyor reduced from 5/. to 3/6 – 1 do from 4/6 to 3/6 – 1 surveyor of 9 years, to be discharged on the end of the month. All chainmen reduced from 1/11 down to 1/6.

At Huddersfield 13, have been either reduced or discharged, perhaps partly both, the news of that is uncertain. None of the draftsmen have been touched, perhaps they will not, till Leeds is finished. Captain Tucker is reported to have said, that the detail survey was eating up all the funds, and leaving nothing for contouring[4] or engraving, that the grant for the S.[5] this year is less than last year by £29,000, that a great many will be discharged,[6] and among the very first shall be those who left the Ordnance Survey and went to the Tithe[7] – So, if this will be carried out, your humble servant must tramp in 2 months.[8]

None have been reduced or discharged in the office at York, up to the present time. The letter of Y[9] shall be returned – Will you give the enclosed note to Mr Higgins[10] should he not be discharged, if he has gone, then throw it in the fire.

Tis strange, none have been discharged solely on account of the general reduction, but some fault either past or present is brought to bear on their case![11]

Write soon | Yours truly | B.[12]

RI MS JT 1/11/3780

1. *Dr.*: Dear.
2. *X*: the pseudonym Tyndall used to avoid detection.
3. *No 1*: the 'Preface to The Ordnance Survey of Ireland' enclosed in letter 0195.
4. *contouring*: see letter 0195, n. 49.
5. *S.*: Survey.
6. *a great many will be discharged*: see letter 0194, n. 15.
7. *went to the Tithe*: The higher pay received by civil assistants employed on the Tithe Survey by private contractors had prompted many surveyors trained on the Irish Ordnance Survey to resign and take jobs on the Tithe Survey during the so-called 'tithe survey mania' of the late 1830s (T. W. J. Connolly, *The History of the Corps of Royal Sappers and Miners*, 2 vols (London: Longman, Brown, 1855), vol. 1, p. 321).
8. *if this will be carried out, your humble servant must tramp in 2 months*: McLachlan had resigned his position as a civil assistant in the A District of the Irish Survey in April 1838, before subsequently rejoining in February 1841 (NAI OS/1/15–18). It seems probable that during this period he was employed on the Tithe Survey in England.
9. *The letter of Y*: Y is William Ginty; see letters 0201 and 0226.
10. *Mr Higgins*: probably Martin Higgins, a civil assistant who had worked in the 1st Division, C District of the Irish Survey. He seems to have been transferred to the English Survey in autumn 1841 (NAI OS/1/17–18).
11. *none have been discharged ...bear on their case!*: In the District Order issued on 9 May 1843 Captain Henry Tucker declared 'I have received orders to make a large reduction by discharging the least efficient hands'. He demanded information of work rates from Division Officers, and urged: 'officers are requested to be very particular in stating the efficiency and character of each man' (RI MS JT 8/1/4b).
12. *B*: the pseudonym McLachlan used to avoid detection.

From William Ginty 1 August 1843 0222

33 Windsor Street, Toxteth Park | Liverpool, Tuesday morning | Aug. 1 1843

My dear Tyndall

Now for a long letter! I deposited my precious charge[1] last night in the post office for the Editor of the M.G. Chronicle[2] he has it this day. I sent a note marked 'confidential' along with it, giving my name and address! and requesting him to insert it, after briefly reminding him of his editorial secrecy &c. I beg him to state in his 'notice to correspondents' whether he could insert it or not. I told him, that if he would favour me by publishing it I would immediately transmit cash for 6 copies! and, if not, to return the letter at his earliest convenience! I said I dated it from Dublin merely to confuse suspicion! On the whole I think the affair is very well managed and likely to kick up a shindy.[3]

Before I received your last letter[4] I had supplied a short bar[5] about 'Liverpool' chiming in well after the fellows that were 'not fit to take any part in the undertakings'[6] And I did not want to cross it out and copy that 3 hour job! I altered the word 'hesitancy' – and why? not because it was improper! but in a thoughtless moment I asked Lowrie[7] (a corpl.) to look the dictionary for the meaning of it! you will agree with me that this piece of a 'slip' might give rise to a suspicion, so I think I done wisely to alter it!

Jack Tidmarsh arrived here on Sunday evening. He left his father after him in Manchester, where he spent four hours and never went to see any old cronies, the lazy blackguard. The mother is against him going to the Cape,[8] and some of the father's commercial friends strongly advise him to get him into a mercantile establishment here for one or two years before he would send him out. The father was to be here last night on his way to Ireland he will arrange that matter so that I think Jack won't go just yet. There is no doubt but he bids farewell to the Survey!

I received the final decision of my migratory project[9] a few days ago – a letter from my father desiring me not to accept it, says Hall[10] is a very bad paymaster, he owes my brother upwards of £12! That work would not do for me and so I am glad, very glad, I did not take my brother's advice 'and go over at once'. I hope for better! With regard to the cash which you so kindly lent me, I will return it in a few weeks with the thanks of a grateful Irish man. I have at present 30 shillings of it hoarded up in a fund to call upon any moment. I would not have used a rap[11] of it had I not been so sure of removing, so that I was just beginning to prepare for the trip and consequently spent some of it.

I hear there is to be a reduction of 25 from each division.[12] I infer from this that London is knocked in the head.[13] Your letter will be a 'coup de grace'[14]

Sinnett has sole charge of this establishment. He is very happy for the change. The more so as he has had an opportunity of gloating his eyes upon two exhibitions since he came here. I wish you were here to see the 'Polytechnic Exhibition at the Collegiate Institution'.[15] You can form no idea – not the most distant – of its superlative magnificence and splendour. Illustrations (practical ones) of all the trades and arts – Weaving, glass-blowing, printing, binding, calico-printing, combmaking, pinemaking, and about 5 and 30 different kinds of steam engines, all in full play! Gas cooking apparatus, patent stoves, egg-hatching apparatus, decorative gilding, nautical models (in hundreds), ship launch, miniature steamers in full action, locomotive on railway, Paisley shawl room, potters wheel, model of Hobart-town,[16] fountains and plants, portraits of great men, type composing machine, electrical galvanic and magnetic apparatus, silk-weaving, fringe-loom, autograph room, containing autographs of all the great men of the last two or three centuries, antiquities innumerable, curiosities without number, seal engraving, ivory carving, lithography, and copper plate printing, philosophical apparatus, centrifugal railway,[17] Chinese room, the 'happy family' – a great number of some of the most voracious and timid beasts and birds all together in one immense cage, in perfect amity, mercantile samples, and specimens, basket-making, the invisible girl! likeness cutting. Natural history (a complete museum), Gothic or Baronial Hall! Articles of taste and virtue. Sculpture gallery (some of Chantrey's);[18] West picture gallery !!! containing 420 oil paintings the best in existence! Some of the artists are: – Sir Joshua Reynolds! Maclise! Hogarth! (a portrait of himself painted by himself)! Patten! foreign viz Teniers! Claude Lorraine; Rubens! Vandyke! Salvator Rosa! Correggio! and Hans Holbein![19] &c. &c. &.c &.c. Sculpture gallery: – Rock &c Engravings 92 of the best ever printed (many by Hogarth!) Anatomical models and preparations! Such is man!!! Architecural drawings and models of all the principal buildings in the kingdom! East picture gallery – 110 oil paintings from the great fellows – Water color drawings – 114 of the most beautiful things ever any one seen! And more engravings! And all this for one sixpence – one kids age. There are about 30 or 40 men constantly working away at the different trades and professions. Add to this for another sixpence Dubufe's painting of Adam and Eve. 'The Temptation' and 'The Expulsion'[20] — valued, and going to be sold, for £20,000, As for these I am not ass enough to say a word about them! Oh holy poker!!!! Sinnett could not drag his heels away until the gas was put out at 10 o'clock, altho' he was there before 6!! and what think you of Latimer and Tidmarsh that never went to see the Polytechnic – I think they ought to be soused in the Mersey. It is now closed!!!

This piece of poetry[21] is admirable, I will copy your things sometime and send them to you.

My best respects to Geo.[22] Latimer, Jim Evans,[23] Marquis and Parker. Is Geo., getting fat?

I am happy to say that I am getting nearly as lank[24] as yourself!

I shall write when I see <*word(s) missing*> about 'The Letter'[25] I will send you 4 <*word(s) missing*> papers.

Yours very faithfully

RI MS JT 1/11/3606
LT Transcript Only

---

1. *my precious charge*: presumably an early version of letter 0228; see letter 0224.
2. *Editor of the M.G. Chronicle*: presumably the *Morning Chronicle*, a London-based Whig newspaper whose editor was John Black (1783–1855). Black had succeeded James Perry in 1817, and would himself be replaced as editor by Andrew Doyle in 1844.
3. *shindy*: row or commotion (*OED*).
4. *your last letter*: letter missing.

5. *bar*: Ginty possibly uses the term in the sense of a legal plea or objection (*OED*).
6. *'not fit to take any part in the undertakings'*: this statement appears in the penultimate paragraph of letter 0128.
7. *Lowrie*: see letter 0197, n. 5.
8. *The mother is against him going to the Cape*: Mrs Tidmarsh had previously given her reluctant assent; see letter 0215.
9. *my migratory project*: Ginty announced he was leaving the Ordnance Survey in letter 0203, but did not give details of his prospective project beyond noting that he was awaiting a 'letter from my brother, stating the final arrangements of the matter'.
10. *Hall*: not identified.
11. *rap*: a counterfeit coin, worth virtually nothing, but used as currency in Ireland during the eighteenth century at the value of an English halfpenny or farthing, owing to the scarcity of genuine copper coinage (*OED*).
12. *a reduction of 25 from each division*: In a District Order issued on 7 June 1843 Captain Henry Tucker requested that 'Division Officers ... warn the Civil Assistants & Labourers that a further reduction will be necessary and ... send to the District Office the names of 25 persons who are the least efficient and useful whom they recommend to be discharged' (RI MS JT 8/1/4b).
13. *London is knocked in the head*: the proposed Survey of London; see letter 0154, n. 3.
14. 'coup de grace': stroke of grace (French), a blow by which one condemned or mortally wounded is 'put out of his misery' or dispatched quickly (*OED*).
15. *'Polytechnic Exhibition at the Collegiate Institution'*: The Collegiate Institution was a day school for boys on Shaw Street in Liverpool. Beginning in June 1843, the school hosted the Polytechnic Exhibition, of which the *Liverpool Mercury* reported, 'This exhibition ... is the most extensive which has been presented in this town to public notice. The articles occupy forty rooms, some of which are very large' ('Polytechnic Exhibition at the Collegiate Institution', 7 July 1843, p. 226).
16. *Hobart-town*: the main conurbation on the island of Tasmania.
17. *centrifugal railway*: The 'Centrifugal Iron Railway' patented by Hutchinson and Higgins in which a single carriage rolls down a track on an inclined plane, traverses a central loop, and then ascends another inclined plane on the opposite side ('Centrifugal Railway', *Mechanics' Magazine*, 36 (1842), p. 360).
18. *Chantrey*: Francis Leggatt Chantrey (1781–1841), an English sculptor.
19. *Sir Joshua Reynolds! Maclise! ... Correggio! and Hans Holbein!*: Joshua Reynolds (1723–92), an English painter and first president of Royal Academy; Daniel Maclise (1806–70), an Irish painter; William Hogarth (1697–1764), an English painter and satirist; George Patten (1801–65), an English painter; David Teniers the Younger (1610–90), a Flemish painter; Claude Lorraine (*c.* 1600–82), a French painter; Peter Paul Rubens (1577–1640), a Flemish painter; Anthony van Dyck (1599–1641), a Flemish painter; Salvator Rosa (1615–73), an Italian painter; Antonio Allegri da Correggio (1489–1534), an Italian painter; Hans Holbein the younger (*c.* 1497–1543), a German painter.
20. *Dubufe's painting of Adam and Eve. 'The Temptation' and 'The Expulsion'*: two large oil paintings by the French painter Claude Marie Paul Dubufe (1790–1864) based on scenes from the Old Testament and John Milton's *Paradise Lost* (1667). The paintings were subsequently exhibited in Preston at the Bull Inn Assembly Room, where Tyndall saw them on 20 November 1843, recording in his Journal: 'went to see Dubufe's paintings of Adam and Eve in the temptation and expulsion – the former breathes the very essence of beauty almost divine – the latter puts forth all the attributes of grandeur and

awful sublimity' (vol. 13a, p. 4).
21. *This piece of poetry*: presumably that included in letter 0218.
22. *Geo.*: George.
23. *Jim Evans*: Phillip Evans.
24. *lank*: lean (*OED*).
25. 'The Letter': presumably letter 0228.

From Archibald McLachlan 6 August 1843 0223

Sunday

I will not be in time for this night's post – After reading the last,[1] I conclude it is of no use spending time with Ed:[2] this one shows the white feather already. Prosecution! – Tho' he does not know it, Colonel C.[3] or any other officer would rather be excused – rather than prosecute, the cure would be worse than the disease – Whatever is said – true or false, a prosecution is the very last thing they would think of. The Master General[4] might say, Colonel C. are those things so? and he would reply they are false my Lord – Then Colonel C. you should prosecute – No, no says Tom, he would make a defense – and – and – it, it is not him I fear 'tis his witnesses!!! No I will take my own way – I will intimidate and Bully the Editor, threaten an action, and demand the authors name – if he complies, then I drop the prosecution!!!!. Never think of a prosecution – no fear of that. But you have to fear that the Ed. to save himself may be induced to give up the names. From his letter I would conclude this to be very likely – I have tried Ed. and printer in the olden times, and found them both afraid of the same danger.[5] Tho' we see and know all about it, they don't know, and can hardly believe things to be so bad as represented – we are not public men – not weighty enough for an Editor's confidence!! If you can strike out all hard words and speak not affirmatively but negatively – putting your facts in the form of questions, peradventure the Ed would not baulk at the leap – If you can get it in conveniently, well, if not, let him go to the --- I don't much like his white feather. –

Burning Maps '/' It would be well, not to say 'by the fiat of the B. of or.'[6] but, 'an order was given that they should be burned, and they were burned according to order'. I don't know who gave the order – the order ran nearly thus:– 'The maps are to be burned in the presence of an Officer'. – It did not order them – it did not say, you will burn the maps, – nor did it say when, or where, or by what officer!! very cautious!! I remember it was in Colonel C's handwriting.

Tithe maps;[7] it would do as well to substitute. 'the expenses were much greater than the estimated cost' – That would be a term tame enough God knows.

Money 1st The money was drawn from the Bank, in the usual way. 6 inch money,[8] all the men entered on a sheet called Pay list. 'Ord: S. of Eng.' – Now were all the men entered on this P.L.[9] employed on the O.S.E.?[10] No; 6 inch men and 3 chain men[11] jumbled together, was it not so? ask D.[12] They were all paid from the money drawn from the Bank 'aforesaid' – Here we have a double falsehood viz: money drawn from the Bank as usual 2.8.6 inch money[13] drawn, but not for 6 inch purposes[14] – men entered on the P.L. of the 'O.S.E' and those men were not employed on the O.S.E. during the period expressed against their names; the O.S.E. is down by or under one Act – the Tithe under quite another,[15] but in this P.L. the 2 are jumbled together. There never was any money voted for Tithe purposes – but the men were paid every Saturday; then out of what fund? – the 6 inch, for there is no other, therefore – Q.E.D.!![16]

2nd Every P.L. certifies that the men were employed during the period expressed against their names – Now in such a week between the day of --- and the day of --- ABC is set down for 7 days, was he employed the 7 days? No – for the men were received extra time after they had long discontinued working extra hours – you will recollect this – and you will recollect too the Tithe S.[17] is not the O.S.E. – but the men were returned and paid as if on the O.S.E. – if they did not receive the O.S.E. money why were they put on the O.S.E. paylists?

3rd There is a great difference between the estimate and actual expense, has he,[18] or will he have to pay this difference out of his own fob?[19] Why should he? he is not liable. If the thing is a speculation of his own, then he should pay the piper. Has he acted according to orders received, or has he not? If he has acted without orders then No 1 and 2 – or 1st and 2nd – come slap in his face. If he has acted up to his orders – then you are justified in saying that money voted for one purpose has been applied to another – If one side be taken, then you trip him on the other.

I don't like the Editor's hesitation – he says – 'Beside as it stands, it would probably invoke severe legal proceedings. Are you prepared for such a result'. Now what does he mean by saying – 'are you prepared' &c. Does he mean, that if legal proceedings were threatened – he would call on his correspondent – or give up the name of the writer? If he is not proof against threats, it would be better to have nout[20] to do with him – Modify the thing and give one trial more – and only one. It is much better to club our halfpence – say half a guinea from five men – £2 10 – this would print a good long pamphlet of 12 pages, where we could speak out – What do you think could we collect? £2 10 in less than a week after the thing is done – or can we collect now £1. This would print one of 5 pages – and would be a strong beginning – and continue them periodically – Pray try it now – nothing of it is to be known except by the Elect. My cash is ready for the first letter – here you have 5/. and 5 every month, this can be done, will you try – I hate newspapers – having been so often disappointed. Printers too are shy – but there we could not be so much disappointed. Let us blaze right off ourselves – begin and continue it – Let Mr S.[21] take the benefit and use the facts. Come, let us begin, at once, begin where you will – at Ireland or England. If it were possible to consult Mr S. it would be prudent – but recollect, 'Who would be free, themselves must strike the Blow'.[22]

S. would be invaluable, should he take it up – but if the same difficulty be encountered in communicating with him as with an Editor I'm afraid it would be endless work – would it not be worthy of serious meditation to give S a trial – and if found solid pour the contents of the S. into him immediately – If not let us do our own work ourselves.

I know it would shock the parties to find the matter exposed in a series of letters or pamphlets, printed and circulated privately, – sent to all C.E.,[23] Professional men, &c. &c.

I think it would be expeditious, should you – (being in a central position) determine how – and what is to be done – calling on us only as adjuncts or offsets – or secondary parties – that all should be arranged planned and matured by you – calling on others as inferior agents. This I think would ensure expedition and strength – We unsettle you by our opinions – don't follow any parties. Hear and weigh all, but as a leader, follow your own judgement – You will hear soon again.

B.[24]

RI MS JT 1/11/3777–9

1. *the last*: letter missing.
2. *Ed.*: presumably either Joseph Wrightson (1796/7–1856), editor of the *Weekly Dispatch* from 1838 until his death (see letter 0195, n. 19), or John Black, editor of the *Morning Chronicle* (see letter 0222, n. 2).
3. *Colonel C.*: Colonel Thomas Colby.
4. *The Master General*: George Murray.
5. *I have tried Ed. and printer ... both afraid of the same danger*: see letter 0211.
6. *B. of or.*: Board of Ordnance.
7. *Tithe maps*: see letter 0195, n. 15.
8. *6 inch money*: money to produce maps at a scale of 6 inches to 1 mile.
9. *P.L.*: Pay List.
10. *'Ord: S. of Eng.' ... O.S.E.*: Ordnance Survey of England.
11. *3 chain men*: surveyors producing maps at a scale of 1 inch to 3 chains, the equivalent of 26⅔ inches to 1 mile; see letter 0195, n. 15.
12. *D*: see letter 0195, n. 35; D is described as 'a very cautious cool gent' in letter 0211.
13. *2.8.6 inch money*: not identified.
14. not for 6 inch purposes: The Tithe Commissioners requested that their maps be produced at the much larger scale of 26⅔ inches to 1 mile.
15. *the O.S.E. is down by or under one Act – the Tithe under quite another*: the Ordnance Survey Act (1841 c. 30 (Regnal. 4 & 5 Vict)) and the Tithe Commutation Act (1836 c 71 (Regnal. 6 & 7 Will 4)).
16. *Q.E.D.!!*: abbreviation of quod erat demonstrandum, Latin for which was demonstrated.
17. *Tithe S.*: Tithe Survey.

18. *he*: presumably Colonel Thomas Colby.
19. *fob*: a small pocket in the waistband of breeches used for carrying a watch, money, or other valuables (*OED*).
20. *nout*: nothing (Yorkshire dialect).
21. *Mr S.*: possibly John Short; see letter 0227, n. 9.
22. *'Who would be free, themselves must strike the Blow'*: Lord Byron, *Childe Harold's Pilgrimage* (1812–18), II.lxxvi.721.
23. *C.E.*: Civil Engineers.
24. *B*: the pseudonym McLachlan used to avoid detection.

From William Ginty 8 August 1843 0224

August 8th 1843

Dear Jack

This Editor chap[1] has been codding[2] us – not one word has yet appeared – and I think never will. The blackguard might have returned it.[3]

Well now the Mercury;[4] either rearrange the present letter[5] or tell me if I am to send it to him[6] in its original state. It will be necessary for me to copy it, as I will have to send a private note reminding him of his promise and explaining why he did not hear from me sooner; it is also probable he may require some questions answered, and as I am always at hand it will be better to have all in my writing. You need not (if you do alter it) be particular in the writing, so lose no time. Let me have it by Saturday or Sunday evening at the farthest.

Carey was here to day seeing the wife off to Ireland. The old rascal looked better than he ought. I did not know until last Sunday that the daily papers are not published on Sunday[7] – What a farce. Religious scruples of course, and yet they must print Monday's paper on that day, and leave Saturday's news a day late, whereas nothing remarkable ever occurs on Sunday – unless they would report sermons. (God forbid!)

How is Evans with regard to politics? I'm now within a trick of being a Repealer[8] – My shilling will follow of course. I'll keep that close tho'. Some of the Irish Tory papers say 'Repeal they will have if the Govt. don't coerce'[9] (oh!) Suppose John Tyndall and Jim Evans[10] chanced to be born Roman Catholics – would they be repealers?

'Civil and Religious liberty'
'More pigs and less parsons'
'Ireland for the Irish'[11]
Protestantism! Puseyism![12] Popery! Methodism! Presbyterianism! everything! O'Connell and Repeal! Hurrah!

W. Ginty

RI MS JT 1/11/3607
LT Transcript Only

1. *This Editor chap*: presumably John Black, editor of the *Morning Chronicle*; see letter 0222, n. 2.
2. *codding*: hoaxing (*OED*).
3. *returned it*: presumably the letter Ginty described as 'my precious charge' in letter 0222.
4. *the Mercury*: the *Liverpool Mercury*.
5. *the present letter*: presumably a version of letter 0228.
6. *him*: John Smith, editor of the *Liverpool Mercury*.
7. *the daily papers are not published on* <u>Sunday</u>: There were, of course, Sunday newspapers, beginning with the *Observer* in 1791, although they were separate from daily newspapers, with even the *Sunday Times*, begun in 1822, having no relation to *The Times*.
8. *a Repealer*: a supporter of the Repeal Association which campaigned for the repeal of the Act of Union between Ireland and Britain.
9. *'Repeal they will have if the Govt. don't coerce'*: The Prime Minister Robert Peel refused to enact coercive legislation in Ireland, fearing that it would only increase support for Daniel O'Connell. This stance of deliberate indifference toward the Repeal Association and O'Connell's mass meetings enraged reactionary Tories and Unionists within the Conservative Party.
10. *Jim Evans*: Phillip Evans.
11. *'Civil and Religious* <u>liberty</u>*'* ... | ...*'Ireland for the Irish'*: slogans used by Irish nationalists.
12. *Puseyism!*: the High Church doctrines of the Oxford theologian Edward Bouverie Pusey (1800–82).

To John Tyndall, Snr [August] 1843 0225

Thursday

My dear Father

According to my promise. Since I wrote my last[1] I have been in search of lodgings. I have travelled the most of Preston and I shall now tell you how things stand. In taking lodgings heretofore I had generally four persons to please – myself and three more – by collecting thus together we have always been able to procure lodgings pretty cheap; that is, by the time the amount was divided it seldom amounted to more than 3s a week for each. But now you see I am thrown out of my usual course, and lodgings will be extremely expensive I was in a certain place last night and the good woman asked me 12/- a week for two rooms, and declared that she had got that sum before now for them. Preston is a manufacturing town, the greater part of it is thickly studded with factories, I would not like to bring Emma into the vicinity of those[2] as the air is generally impregnated with the smoke thrown out from their enormous chimneys.[3] I shall now let you know one course I have thought of adopting. There is a man lodging with me, Evans, and Latimer at present, who joined us about two months ago on account of his wife going on a visit to Ireland. She is an exceedingly nice woman, I have been speaking to him about joining together and taking lodgings so that we all could sit together at the same table and thus Emma would have a companion during my absence. Mr Timms however could give me no decided answer until he had consulted Mrs T.[4] He writes to day on the matter. Now I want your answer. Mrs T. will be coming over in a fortnight. In the event of her agreeing with what I have proposed, would you think it too long to wait? or shall I take lodgings on the spot for Emma and myself?[5] Could she not go out to Ballybromwell[6] in the interim? Send me an explicit answer to this. I'm so bewildered that I scarcely know what I'm about

Your affectionate son | John Tyndall.

RI MS JT 1/10/3293
LT Transcript Only

---

1. *my last*: probably letter 0220.
2. *I would not like to bring Emma into the vicinity of those*: The suggestion that Tyndall's sister Emma might visit him in Preston was first discussed in letter 0220.
3. *the air is generally impregnated ... enormous chimneys*: Emma seems to have been less perturbed by the industrial landscape than her brother, telling her father, on 16 August 1843, that 'Preston is a nice clean town ... the suburbs ... are beautiful' (RI MS JT 1/10/3294), and later observing of Tyndall 'He is very much attached to the country, he hates the hum and bustle of the town, and would fain spend his life amid green fields and waving woods' (22 August 1843 RI MS JT 1/10/3295).
4. *Mr Timms ... Mrs T*: Emma was possibly referring to Mrs Timms when she told her father, on 22 August 1843, that 'John was fortunate enough to procure very nice lodgings for me. My hostess has some friends in town ... I am acquainted with none but persons of respectability ... After tea last night I was handed a glass of primrose wine made by my kind hostess. The English excel in making such things'.
5. *would you think ... for Emma and myself?*: Emma's own letters to her father indicate that her 'safe arrival in Preston' occurred before 16 August, suggesting that, at least initially, Tyndall took the option of taking lodgings just for themselves.
6. *Ballybromwell*: see letter 0121, n. 4.

From William Ginty [August] 1843 0226

Very dear X[1]

I am just after setting my signet[2] on the suggested letter.[3] It is a 'feeler'[4] with a vengeance! I highly approve of the scheme of this matter – it is a noble undertaking. 'Be ye strong therefore and let not your hands be weak for your work shall be rewarded'.[5] Yes rewarded. Reform of some kind must follow. I will now proceed to lay a few hints before you. 'Weigh and consider' and excuse the impertinence of a warm friend and enthusiastic admirer. Fancy me along-side on the sofa making the following propositions with all the gravity of a Fergus O'Connor.[6]

1st That the spirit of it be mild, emphatical, and decisive.

2nd That it embraces the perfections of the survey as well as the imperfections, giving an epitome of its general use and practical purposes so as to render some information to the public on this hitherto mysterious business – thereby creating an interest in their minds which will enlist their sympathies in our behalf.

3rd That it be printed in close type on light paper so as to remove any pecuniary inconvenience as to postage.

4th That the expenses be borne by a few trustworthy enthusiasts

5th That 'X' bills 'Y'[7] with his share of the expenses when ascertained.

6th That a copy be sent to every D.O.,[8] on the Survey, a copy to the Master General,[9] a copy to the Premier[10] and a copy to the illustrious Daniel O'Connell!!! (this may seem flighty and groundless, but I do think and urge this. Harm it cannot do. Good it may do, and it will undoubtedly kick up a shindy)[11]

7th That some of the principal newspaper Editors – radical and conservative – in England and Ireland be sent copies, so as to ensure editorial pieces which is most important.

8th That it pays a tribute to the memory of Captain Fenwick[12] – his baseness in opening and reading private letters to civilians[13] from others who had obtained situations on the Tithe Survey,[14] and then in the excellency of his despotism – discharging them for attempting to release themselves from misery.[15]

9th The recommendation farce[16] – the 'Forms' the 2 months suspense and the inevitable 2 pence.

10 That the pay of the Ordinance survey men be contrasted with that of the Dock labourers in Liverpool – 3 shillings – An immense number have made a 'strike' lately for 3-6[17]

11 Captain Tucker's excellent and judicious proceedings with regard to a petition sent to him by a number of Surveyors in this division[18] for to increase their pay shortly after coming to England which was to send away the journals[19] of 3 or 4 of the most inefficient in the party (who did not sign) and to obtain for them from 6d to 8d a day each out of pure derision to the petitioners. The few that were recommended, fortunately for themselves, were not at hand when the petition was got up, to sign it.

12 Lt. Hamley's late ungentlemanly and inhuman conduct in refusing a character[20] to a civilian who asked him for one – he being about obtaining a situation – after he had discharged him for over-sleeping (*) three! mornings in the month – his only crime – for which he in months previously checked his pay to an amount equivalent to nearly the treble of the time he was absent – Left with a few pence to take him to his friends and home but for the benevolence of a few comrades! (A moral character!)

The base and atrocious inhumanity of disallowing a sapper sent from here to Windsor[21] to take his wife she being afflicted and perfectly blind and left to take care of a large family of helpless children, when other men could be sent in his stead (*)

The case of Ingram[22] having to obtain relief from the relieving officer at Skipton[23] to take his family to their native country (a labourer?) and a subscription from the office men also.

The farce of the medical attendance – none being given beyond a dose of salts, unless the patient goes into the military hospital and pays 10d a day while there, and no pay allowed him while sick and while well not as much as will keep him in good health! A pretty institution for men in a strange country! and men carrying on the greatest scientific undertaking that ever honoured a nation!

The office pay here stands thus:– about 24 on 2 shillings 12 from 2s2d to 2s and 4d, 12 from 2/6 to 2/8 and a few more at 3 shillings and 3/6 and one at 5 and another at 5/3 – about 60 altogether. About 1 out of 8 of the strongest in the party exceed 2s. There are about 120 civilians in the division (one the most useless in the lot has 11 or 12 shillings a day)

I must conclude now in a hurry. I would not be surprised if you called me an ass for some of the remarks but they came into my head they got into pen. I will be able to say something more to the purpose in the course of the week. No hurry I should say there is an affair pending here at present which I guess and hope will be as disgraceful as possible. I am not acquainted with any of the persons you allude to. Make a good job, I am certain you can if you like.

Yours &c. | Y[24]

(*) See 12th--- one whole chapter or section is to be devoted to the grievances of the R.S. & M.[25]

RI MS JT 1/11/3604
LT Transcript Only

1. *X*: the pseudonym Tyndall used to avoid detection.
2. *my signet*: a wax seal on a letter.
3. *the suggested letter*: possibly letter 0228, which Ginty had arranged to send to the editor of the *Liverpool Mercury* in letter 0224.
4. *a 'feeler'*: a proposal or hint put forth or thrown out in order to ascertain the opinions of others (*OED*).
5. *'Be ye strong therefore and let not your hands be weak for your work shall be rewarded'*: 2 Chronicles 15:7.
6. *Fergus O'Connor*: Fergus O'Connor (1794–1855), an Irish nationalist and Chartist leader.
7. *'X' bills 'Y'*: Tyndall bills Ginty.
8. *D.O.*: Divisional Office.
9. *Master General*: George Murray.
10. *the Premier*: Robert Peel.
11. *shindy*: row or commotion (*OED*).

12. *a* tribute *to the memory of Captain Fenwick*: Captain Robert Fenwick had recently died in the West Indies, where 'yellow fever broke out at Bermuda in August [1843] and continued with unabated virulence and fatality until the middle of September. In that brief period ... it carried off ... Captain Robert Fenwick, R.E.' (T. W. J. Connolly, *The History of the Corps of Royal Sappers and Miners*, 2 vols (London: Longman, Brown, 1855), vol. 1, p. 426).
13. *civilians*: civil assistants.
24. *his baseness in opening and reading private letters ... on the Tithe Survey*: see letter 0195, n. 44.
15. *release themselves from misery*: Civil assistants employed on the Tithe Survey by private contractors received higher pay than surveyors on the Ordnance Survey.
16. *The recommendation farce*: probably related to the non-payment of recommended pay rises; see letter 0164.
17. *a 'strike' lately for 3-6*: The word strike only began to assume the meaning of a cessation of work on the part of a body of workers in the 1810s (*OED*); this particular strike was not reported in the *Liverpool Mercury* or other local newspapers.
18. *this division*: the 5th Division of the English Ordnance Survey.
19. *journals*: presumably the 'Journal of Progress and Weekly Report'; see letter 0038, n. 3.
20. *a character*: a character reference.
21. *Windsor*: see letter 0198, n. 21.
22. *Ingram*: probably Andrew Ingram; see letter 0199, n. 5.
23. *Skipton*: see letter 0096, n. 3.
24. *Y*: the pseudonym Ginty used to avoid detection.
25. *R.S. & M.*: Royal Sappers and Miners.

From William Ginty 14 August 1843 0227

Monday afternoon Liverpool
August 14th 1843

Dear Tyndall

Your two letters[1] came to hand in due time. The Yorkshire incognito's council[2] thoughts are too late. The Editor[3] feasted his bewildered optics on the grand Coup de Grace[4] this morning before either he or you feasted on the best 'Mocha'.[5] Just for the sake of unanimous agreement I hope he will advise the proceeding – if not he may stand alone in his inglorious forbearance!

I posted it last night, accompanying it with a letter reminding him of his promise to take such a thing into consideration, showing him the benefits that must arise from its publication, and after bladdering[6] 'summat'[7] about the gratitude of our honorable (?) community I told him if he required any questions answered or alterations made before next Friday, either by interview or letter, I would feel happy in attending to and defraying the expenses of any communication with which I might be honored. I would have went and seen him if I could; but in the first place he is not there in my time for the first 2 or 3 evenings of the week, & it is nearly 2 miles from the office, so that I could not go at dinner time.

But confound this bladdering – could I do it any better or more complete? – NO!!! (d'ye hear that?)

I will write to the London cove[8] this evening!

Mr. Short[9] has written another letter to the M.G.,[10] reporting Hamley for uttering a falsehood, in that he said to O'Neill that he (Mr S) delivered up his name with a view to trick him into a confession (which was unnecessary), when Mr Short actually refused most emphatically to do so. He also sent him a copy of the vexata questio[11] – lest Hamley should not – and added that he requested Mr H to give him his answers to each – which he did not. 'Thermophylea'[12] expects to walk! So he may I think, we must however remunerate him for his invaluable services, if such should take place! As brother Johnathan says, 'I calculate they are at a tarnation hard fix'.[13] This day Hamley called up another investigation about the 'Time' to know if it was really the case that the hours in Preston and other places were so much better than ours – 7 or 8 gave up their authors for the fact (of course) he is to write to Tucker about it. Now what is all this for? Just to 'Conciliate' and to give no room for an improved edition of the vexata questio. Lord pity his ignorance, and the devil help his distress, and may both lend their aid to crush the despotism of vagabonds like his Royal Corps![14] I wish to the fact of their being corpses it is a matter of course to expect a damned strict investigation on Friday! I will of course deny point blank – To the winds with all fanaticism and conscientious scruples – it is a Christian maxim that says 'do evil that good may come'![15] What a very accommodating thing pure Christianity is! Oh! my elastic conscience!!! Yes! again I repeat the propriety of petitioning & forwarding it thro' an M.P.,[16] giving them due notice, to save appearances and to gag them! Let them prevent us at their peril. We can have all ready before they get notice and then wing it heavenwards (!) before they can cancel (the game!)? This however wont do just now! Let us raise the wind! Agitate! Agitate! Persevere! Persevere! And then – crash goes the haughty edifice from its rotten butments![17] Let us cannonade[18] it with letters! When O'Neill's sentence comes we will undoubtedly get an insight into what they call the 'proper channel!' and in that channel I'll go neck and ears ....

There will of course be a general order maybe we wont drive a coach and six through it! They must speak out! And surely we will find out the 'soft place in their heads' we will have the rascals from the pinnacles of their naughty heedlessness! And instead of 'war to the knife'[19] and 'blood! blood! blood!' our battle cry will be 'War to the steel nib' Ink! Ink! Ink! Sir J. B. Y. Buller Bart.[20] it was that handed in Short's letter to Sir Geo. Murray. This fellow is no joke! He it was that moved the 'want of confidence' measure that threw out the late ministry!!!![21] Among the interest that Short can command is that of the lone member for Montrose – the incorrigible democrat Joe Hume!! I never spoke to O'Neill about the letter. It can do no good at all events until it is published. I never seen Short!

As for the 'deputation' I 'by myself I' dont admire it! It is expensive! and again I think few would have anything to do with it (I dont include myself among the majority) and I think Petition would do as well with the additional advantage of forcing them to investigate! If spoken of at all in the Parliament! Such must take place!

How is it you write such short letters? cant you curse them awhile? anything – so you but write a long one

Yours | Ginty

If the Editor of the Mercury should be any way scrupulous about inserting it, I'll regulate him! and coerce! Coerce! Coerce!

You dont know the effect the presence of an enraged Irelander has upon one of these John Bull Gents![22]

Corpl. Scott[23] – Hamley's locum tenens[24] – 'K.T.E.' (Knight of the Tyrannical Espionage!) goes about like a roaring jack ass offering his 'calm and collected opinion' that 'the Colonel[25] will be down on this Division'[26] I have his measure he had better 'look out' for squalls! The fact is these fellows imagine they can intimidate us into forbearance!

'Heigh! Diddle diddle!'[27]

He is cunning rascal and with his villainous commandant they form a very admirable and perfect compound of tyranny and trickery!

For pity's sake steer clear of methodists, they are so devilish conscientious in answering questions! You can tell friend Geo.,[28] if he asks you about the letter, to guess, and you can laugh – anything at all to let George off safe!

Damn you and your 'N.B.'[29]

RI MS JT 1/11/3608
LT Transcript Only

---

1. *Your two letters*: letters missing.
2. *Yorkshire incognito's council*: presumably Archibald McLachlan, who was stationed in Leeds.
3. *The editor*: John Smith, editor of the *Liverpool Mercury*.
4. *Coup de Grace*: stroke of grace (French), a blow by which one condemned or mortally wounded is 'put out of his misery' or dispatched quickly (*OED*).
5. *'Mocha'*: a type of fine quality coffee (*OED*).
6. *bladdering*: blathering; talking foolishly or nonsensically (*OED*).
7. *'summat'*: something.
8. *the London cove*: presumably John Black, editor of the *Morning Chronicle*; see letter 0222, n. 2.
9. *Mr. Short*: possibly John Short, who had worked as a civil assistant on the Irish Ordnance Survey from July 1836 until September 1838 (NAI OS/1/13–15), and may have subse-

quently gone on to work on the English Survey.

10. *the M.G.*: the Master-General of the Ordnance, George Murray.
11. vexata questio: disputed question (Latin).
12. *'Thermophylea'*: a location of hot sulphur springs in ancient Greece, here used as a nickname, presumably for Short.
13. *As brother Johnathan says, 'I calculate they are at a tarnation hard fix'*: Brother Jonathan is a personification of New England, and tarnation is an American slang variant of darnation or damnation (*OED*).
14. *Royal Corps*: the Corps of Royal Engineers; it was granted the Royal prefix in 1787.
15. *a* Christian *maxim that says 'do evil that good may come'!*: Romans 3:8.
16. *M.P.*: Member of Parliament.
17. *butments*: supports of an arch (*OED*).
18. *cannonade*: batter or attack with cannon (*OED*).
19. *'war to the knife'*: war to the last extremity, fierce or relentless war (*OED*).
20. *Sir J. B. Y. Buller Bart.*: John-Buller Yarde-Buller (1799–1871), Conservative MP for South Devon.
21. *the 'want of confidence' measure that threw out the late ministry!!!!*: In January 1840 Yarde-Buller moved a motion of want of confidence in the Whig Ministry of Lord Melbourne. Although the motion was defeated, Melbourne's Ministry lasted for only another eight months.
22. *John Bull Gents!*: see letter 0143, n. 21.
23. *Corpl. Scott*: see letter 0197, n. 4.
24. *locum tenens*: one who holds office temporarily in place of the person to whom the office belongs, or who undertakes another's professional duties during his absence (*OED*).
25. *the Colonel*: Colonel Thomas Colby.
26. this Division: the 5th Division of the English Ordnance Survey.
27. *'Heigh! Diddle diddle!'*: from the nursery rhyme 'Hi Diddle Diddle, The Cat and the Fiddle, The Cow Jumped Over the Moon'.
28. *friend Geo.*: George Latimer.
29. *'N.B.'*: abbreviation of 'nota bene', Latin for note well.

To Robert Peel via the *Liverpool Mercury* 15 August 1843 0228

From the LIVERPOOL MERCURY – Sept. 1st. 1843.

ORDNANCE SURVEY: ROYAL ENGINEERS.

Knowing nothing of the question treated of in the following letter, but being of the opinion that it calls for consideration, and is written in a fair spirit, we give it insertion on public grounds, and shall be glad to receive a fair answer to it, especially if the discussion shall sub-serve the interest of the public. Of the importance of the subject we need not say a word.[1]

Letter No.1,

To the Right Honourable Sir Robert Peel, Bart.,[2] &c.

Sir – It appears by the Parliamentary reports in the newspapers, that you have moved for leave to bring in a bill relative to a survey and the construction of maps, on a large scale, of the city of London.[3] I trust it will not be deemed an impertinent intrusion on my part to make a few observations in connexion with this undertaking.

It has not yet been formally declared who the parties are that will be entrusted with the execution of this important work; conjecture, however, has awarded it to the Royal Engineers, acting nominally under the Board of Ordnance. I have watched the proceedings of these men for years with a lively interest; early associations have, in some measure, excited that interest, and kept it alive. Being intimately acquainted with their modes of operation, and enabled thereby to form correct conclusions relative to their value as practical men, I now come forward to deprecate the measure which would entrust to them the survey of the city of London, and to contend for the propriety of its being left open to competition.

You, doubtless, will be furnished with an estimate of the time and expenses necessary for the execution of this work. I can readily believe that if estimates were the data which would determine your choice of parties, the Royal Engineers would at once crush the pretensions of all competitors. But permit me to ask: Have the estimates heretofore made by the Royal Engineers relative to works of this nature proved correct? It is most notorious that they have not. You are aware that the survey of Ireland has been executed by these persons. That undertaking was commenced in 1825, the estimated time for its completion being seven years, and the estimated expense £300,000. Did subsequent experience prove the truth of these calculations? No, it proved their utter fallacy. You, perhaps, will be surprised to hear that, far from bearing out the previous estimates, the scientific operations of those seven years were almost, if not entirely, useless: the maps, and other documents, proved so grossly incorrect, – so utterly unfit for the purpose for which they were designed – that the greater part of them were ordered to be destroyed. Thus, Sir, and for these reasons, were the materials which bore the impress of the concentrated talent of five captains and twenty-five lieutenants of the Royal Engineers, for the period of seven years, reduced to their primitive elements. The survey of the island was not completed till 1842,[4] and it incurred an expense little short of three quarters of a million sterling.[5]

But a specious objection may here be started. It may be said that the above would be unfair premises whence to draw conclusions relative to the present claims of the Royal Engineers, inasmuch as they were, at the time alluded to, comparatively inexperienced. I shall not wait to discuss the merits of this plea, but I shall meet the objection on its own ground. In 1841, an Act was passed for the survey of the six northern counties of England,[6] the estimated time for its execution being six years; two years of this time have already elapsed, and there is not one county finished yet. The expense, up to the present time, is about £120,000, or £60,000 a year; the quantity surveyed at present amounts to about 2,000,000 acres; and before the mapping and calculation of this will be finished, three years will have expired. We are here furnished with data to find the expense of this survey, and the time of its completion. The contents of the six northern counties are:—

Northumberland 1,197,440 acres
Cumberland 974,720 acres
Westmoreland 487,680 acres
Lancashire 1,130,240 acres
Durham 702,080acres
York 3,835,040acres
Total - - - - - - - - - - - - - - 8,327,200 acres.

Then, by a plain statement in the rule of three,[7] if 2,000,000 acres require three years for its completion, 8,327,200 acres will require above twelve years. So that though the Act contemplates that this survey will be finished in 1846–7, it will not be completed till 1853, and will incur an expense of £700,000!

You may now judge, Sir, what effect the experience of the Royal Engineers has upon their estimates. And will you, Sir, permit the continuance of such an expensive system at a period of such financial embarrassment as the present?[8] Let an examination be made before the House of Commons into the past history of the Ordnance survey of Ireland, and the history of the Ordnance survey of England which is at present in a state of progress; let persons be examined who will give plain facts as to the progress of the work and the outlay of the public money; let some of the most intelligent of the Royal Sappers and Miners and assistants, be examined before the House, and afterwards you can judge of the propriety of entrusting to the Royal Engineers the survey of London. Captain Boldero's[9] reply to a question lately asked by Lord H. Vane,[10] proves that there exists on the part of the Government gross misconception relative to the present state of the survey of the six northern counties. The proceedings of the Royal Engineers have ever been mysterious. In 1835, an Honourable Member[11] moved for an examination into the state of the Ordnance survey of Ireland, but Sir H. Vivian[12] assured the Honourable Member that the work was getting on rapidly, and that an examination would only cause dispirit. When this assurance was given, the Ordnance surveyors were plodding their weary way over a district, the survey of which had been previously executed so incorrectly as to need a complete revision.[13]

Since their arrival in England the Royal Engineers have adopted a line of conduct which strikes at the very vitals of a respectable profession: they have entered into a most unfair competition with the civil surveyors of England. I refer to their contracting for the surveying of townships for tithe commutation,[14] their estimates of which work are so low as to wither in the breast of the civil surveyor every hope of successful competition.[15] But, Sir, in point of accuracy, they may be ranked with the other estimates spoken of – they are utterly false. I challenge contradiction when I assert that the actual expense of the tithe plans executed by the Royal Engineers is far beyond the estimated expense, so that the public service must suffer from such undertakings.

Now, Sir, contrast the case of these men with that of a body of civil surveyors similarly circumstanced. The latter have a professional character at stake; the former need not rely on theirs. The Royal Engineers may calculate on their salaries, maugre[16] all their blundering; the civil surveyors have no such prop. The Royal Engineers have a ready salve for each mistake – the public purse is their panacea; the civil surveyors have nothing but their talents to bear them out, and a single error such as those which are so common with the Royal Engineers would injure their professional character for ever.

These, then, Sir, are the grounds on which I contend for the survey of London being left open to competition. I am satisfied that the civil profession are able to do it no less correctly, no less quickly, and more cheaply than the Royal Engineers.[17] Let Engineers and surveyors put in tenders for certain sections of the metropolis (all to be executed on a uniform system); let all parties be bound to a certain time; let the Ordnance surveyors put in their tenders also, but let them stand on their own merits – let them have no funds to fall back on, as heretofore, in case of failure, let their own pecuniary loss be the result of any error in their estimates – and the issue will prove that the civil profession are able successfully to compete with them.

I would not have you infer from what I have stated that the Ordnance survey is altogether wanting in point of talent; I am free to admit that there are many intelligent, respectable men connected with it; but I do contend that the treatment which these men receive is sufficient to wither all their energies. It has been the boast of individuals in high places, that they can procure the best draftsmen in Europe at the rate of two and sixpence and three shillings a day. I willingly admit that, at the present moment, there are first-rate draftsmen on the Ordnance survey, who are condemned to toil for this miserable pittance; but does not the individual 'glory in his shame'[18] who would thus make a boast of his ability to crush merit; and I appeal to you, Sir Robert, what interest will a man of talent take in that situation, the pecuniary emoluments of which amount to no more than half a crown[19] or three shillings a day, and even that much lessened by travelling and other expenses incidental to the migratory life of the operative Ordnance surveyor? I could abundantly prove the fallacy of a system of economy which would treat deserving men thus; and I could show that the best, and ultimately the cheapest, mode of proceeding would be to make a situation on the Ordnance survey worth holding, to pay the man of merit properly, and thus excite in him an interest in the work entrusted to him for execution.

Another clog on the proceedings of the survey is the employment of persons at a still lower rate of pay than the above, several of whom are utterly unfit to take any part in the undertaking.[20] The great majority of those employed belong to this class, and their misery beggars description. If the limits of a letter would permit me, I could detail to you the sufferings of numbers of unfortunate Irishmen, who, after having been lured from their homes by an ignis fatuus[21] hope, have been dismissed without a fault from their employment, and thrown upon the world houseless and penniless, in a country whose sympathies often run counter to the name of Irishman. I could show you the quondam[22] Ordnance surveyor craving at the workhouse door for a morsel of bread, and more than once repulsed as an imposter. I could show you the faithful servant of the survey, toiling along a miry road on public business, on a fearful night in December, the rain and snow beating in his teeth, until, overpowered with toil, he sinks and dies, while his lifeless body, instead of receiving decent interment, is thrown like a cumbersome clog into the grave, at the expense of the parish in which he lost his life! I know there are persons whose interest it would be to damn the credibility of one who, in their parlance, will doubtless be dubbed an 'anonymous scribbler'. With reference to such, I have only to say: Let them contradict me if they can. Why, I would ask, is there not a single Englishman employed on the Ordnance survey? – Because an Englishman would spurn the paltry meed[23] which the rulers of the survey offer for his services. The Ordnance surveyors are, to a man, natives of Ireland. I have before stated that there are many men of talent to be found among them; but the most unfair advantage has been taken both of the peculiar circumstances of these men and those of their country.

It may be thought that I have wandered from the proper topic in the course of this letter – that I have forgotten the subject with which I commenced; but this, Sir, is not the case. The revealing, though partially, of the machinery of the Ordnance survey, and the treatment which those employed on it receive at the hands of their Lords and masters, will, I trust strengthen the credit of what I have before stated – that the civil profession are able to execute the survey of London no less quickly, and more cheaply than the Royal Engineers; and that, therefore, the work should be left open to competition.

Spectator. | Dublin, August 15th., 1843.[24]

RI MS JT 1/11/3907–10

1. *Knowing nothing of the question ... we need not say a word*: This introductory paragraph, appended to Tyndall's letter in the *Liverpool Mercury*, was presumably written by its editor John Smith.
2. *Bart.*: abbreviation of the title Baronet.
3. *a survey ...of the city of London*: see letter 0154, n. 3.
4. *The survey of the island was not completed till 1842*: In fact it would not be formally concluded until 1847; see letter 0168, n. 1.
5. *three quarters of a million sterling*: £750,000.
6. *an Act was passed for the survey of the six northern counties of England*: the Ordnance Survey Act (1841 c. 30 (Regnal. 4 & 5 Vict)); see letter 0150, n. 6.
7. *rule of three*: a method of finding a fourth quantity, given three known quantities, which

bears the same relation to the third as the second does to the first (*OED*).

8. *a period of such financial embarrassment as the present*: In March 1842 Peel had made a 'full and unreserved disclosure ... with respect to the financial difficulties of the country' in the House of Commons, acknowledging that from 1837 to 1843 the government was running an aggregate deficit 'in six years, of 10,072,000*l*.' (*Hansard*, HC Deb 11 March 1842, vol. 61, c. 430).
9. *Captain Boldero*: Captain Henry George Boldero (1794–1873) of the Royal Engineers, MP for Chippenham and Clerk of the Ordnance from 1841 to 1845.
10. *Lord H. Vane*: Harry George Powlett, 4th Duke of Cleveland (1803–91), styled Lord Harry Vane, MP for South Durham. *The Times* reported that in the House of Commons on Monday 26 June 1843, 'Lord H. VANE asked the hon. and gallant officer (Captain Boldero) when they might expect to have the maps of the Ordnance survey of the six northern counties? Captain BOLDERO (who was very imperfectly heard) was understood to say, that considering the sums made available for the purpose, it would be five years before they could be ready. It would be three years before that of Durham could be prepared' ('House of Commons, Monday, June 26', 27 June 1843, p. 4).
11. *an Honourable Member*: Joseph Hume, MP for the Montrose Burghs. In March 1835 Hume told the House of Commons that 'It appeared to him desirable that the present Board of Ordnance should be entirely reduced, and a very different arrangement made. If the change which he urged was adopted, the savings under the two heads of Ordnance expenditure might be reduced, if not several hundred thousand, at least several thousand pounds. He proposed that the artillery and engineers should form an integral part of the array, as they did in every other part of the world' (*Hansard*, HC Deb 27 March 1835, vol. 27, cc. 322–3), although Tyndall may actually be referring to Hume's intervention a year earlier, when he told the House of Commons, 'The people of England had nothing to do, and ought to have nothing to do, with surveying the baronies and fields of Ireland for the mere purposes of local taxation', concluding that 'the Ordnance was an unnecessary and an enormous establishment, and that it ought to be reduced' (*Hansard*, HC Deb 21 March 1834, vol. 22, cc. 545 and 555–6).
12. *Sir H. Vivian*: Richard Hussey Vivian, 1st Baron Vivian (1775–1842), MP for East Cornwall and Master-General of the Ordnance between 1835 and 1841. Vivian's apparent assurance to Hume is not recorded in *Hansard* or *The Times*, although he later delivered a notable 'cut' to Hume in the House of Commons, declaring in January 1838 that 'his conduct ... reminded him of the case of a certain Mr. Martin, who prophesied that in seven days, York minster would be burned down; and well might he prophesy–for he was detected in setting fire to the scared edifice at the appointed time' ('Canada', *Quarterly Review*, 61 (1838), pp. 249–72, on p. 264).
13. *a district ... need a complete revision*: not identified.
14. *tithe commutation*: see letter 0169, n. 18.
15. *estimates of which work are so low ... successful competition*: see letter 0195, nn. 15 and 16.
16. *maugre*: in spite of, notwithstanding (*OED*).
17. *the civil profession are able to do it ... than the Royal Engineers*: On 6 June 1843 the private 'surveyors of the metropolis', led by John Bailey Denton and John O. Browne, formed an 'association ... for the purpose of proposing to contract with Government for the execution of the work', asserting the 'injustice of employing military staff for the execution of the survey of London and suburbs, in preference to educated and responsible men of the civil profession' ('Survey of London', *Builder*, 1 (1843), p. 216).
18. *'glory in his shame'*: adaptation of Philippians 3:18–19, '(... *they are* the enemies of the

Cross of Christ | Whose end *is* destruction, whose God *is their* belly, and *whose* glory *is* in their shame ...)'.

19. *half a crown*: 2*s*. 6*d*.
20. *unfit to take any part in the undertaking*: William Ginty's comments in letter 0222 suggest that he may have composed some of the remainder of this paragraph.
21. ignis fatuus: foolish fire (Latin), specifically a phosphorescent light that hovers over swampy ground at night and gives an illusory impression of a mystical presence.
22. *quondam*: former holder of an office or position, especially one who has been deposed or ejected (*OED*).
23. *meed*: wages, hire; recompense, reward, deserts (*OED*).
24. *Dublin, August 15th., 1843.*: As well as falsely claiming to have written it in Dublin, Tyndall may also have strategically postdated the letter, as in letter 0227 Ginty suggests that he posted it before 14 August.

From Archibald McLachlan 26 August 1843 0229

Aug. 26th | Saturday – 11 o'clock

Your two last and paper[1] came to day at 11 o'clock – I won't be in time for this day's post – You say in one, 'don't post on Saturday it will be too late'. One letter is dated Friday, the other Saturday, the postmark on both bears 25th August – is this the fault of post – were they both posted same day?

We will recollect that measures, not men are attacked – no man's private character is aspersed – the thing is public property, and in default of official documents, we can only make general statements – in every public scrutiny, accuracy in all and every part is not expected if the great point is proved all is done.

From 1827 to 1832 or 1833 I was attached to the E. District, and the 1st Division, under Lt. Bordes[2] – about 1833, (being then in Belfast) the Captain of the District was dispensed with and the Lt. of the 1st Division being the oldest, he became Acting Captain – or took charge of the E. District. And he had still charge of his old 1st division – Our office then, was both a division, and district office, all plans &c. past through our office. From 1830 to 1833, the patching system[3] had got a good trial, and found wanting – Certain parishes of our di.[4] and other divisions in the E. district had by this time been re-surveyed.[5] What was to be done with the old or original plans of 1825, 6,7 &c. – ? An order came about this time (1833) in Carbon's[6] handwriting that "The plans are to be burned in presence of an officer – it did not state what plans nor order any plans – nor any particular officer – it did not say – You will burn such and such plans, nor when, nor where! Soon after this a Regular Bonfire was made in the back yard, certain plans &c. &c. were burned – (some were nabbed by the boys!) but what plans or rather what parishes, I don't recollect – I can only recollect the names of a few* (you can have the names if you wish.) I believe the maps burned at that time were only such as had been re-surveyed and completed up to the date of conflagration, and not retained for further consumption! When we got to Drogheda[7] in 1834 or 1835 certain other plans, books plots &c. &c. were roasted to death. I believe these composed the parishes that remained after the first fire in 1833, being part of Down, Louth, Meath[8] – both of our own and other divisions! I don't say every parish was burned by us, perhaps some parishes were burned in the division offices, but this I can't authenticate, although I have an indistinct recollection of being told by some of the boys, that they fought about some old plots &c. &c. and a scrimmage about cutting the demesnes[9] out of some plans, to be kept as specimens &c. &c.

*You can have the names if you wish.

We leave Drogheda and go to the City – Dublin – and here I don't recollect to see any bonfire, nor ever heard of one in Dublin – Ask D[10] did he ever hear or see one in Dublin. But I do recollect old Books – Plots knocked about in some considerable quantum, what became of them I could never make out – Some were sometimes used (the back of them) to try lines on – and score points to see how the work would plot. But none of the plots or books were ever used in the new plans.* I never saw any plans of the Co.[11] Dublin – the first survey was merely plotted (lines) The plans were never drawn – i,e, we did not draw any plans in our original survey of Dublin Co. or City – we only drew plots.[12] Whether in the revision any of the plots were found available or not, I dont recollect, but am pretty sure they all shared the same fate whatever it was (We surveyed the City and we revised it, that is – re-surveyed it!!)

*I recollect that in the North (Down) some parishes were found pretty good and were cobbled[13] up – I say 'some', for the great Body of the work was Bad!

Who surveyed the Drogheda side of Navan all round by the Boyne[14] – ? I don't know! What became of the original plans? God knows. – All I know is they were drawn and returned in the Estimates[15] – (or returns rather) as finished; and after they were so finished we i,e, myself and others in company suffered many a ducking,[16] and days of sore fatigue in re-surveying the same ground!!! What became of the original books and plans* of Swords[17] I don't know, I mean the plots drawn in 1830? I don't know!! – but this I do know, that in 1836 Captain Tucker revised the same country, and had some dozen of ragged lads – 1/. 1/. 1/6 – 1/7!! So wretched that they stole potatoes from potato pits[18] – brought before the magistrates for theft, the Corporal in charge went and said they belonged to the O.S.[19] – The matter hushed up. They were cautioned to behave for the future – And if he could make such men cobble up good plans from bad old plots, it is more than better men could do, – !!

* We surveyed Swords and round it, we did not revise it, the revision was done by Acid.[20] We surveyed the city and we revised it, that is – Re-surveyed it!! –

All this is of our own district – And I will say, that taking it as a whole, the broad assertion stands unrefuted and unanswerable, that the thing was a bottle of smoke.[21]

Our district embraces, Down, Louth part of Meath and Dublin.

If you ask me to prove all this – you say so and so – this is assertion, but prove it. – I answer, I can't prove it, my witnesses are almost all gone – some dead – But I appeal to the original documents – Produce them. Where are the originals? that's the point – produce them and I am floored!!

Other Districts

In our own I revised ¼ of a parish, area of the parish 17,000.[22] – something more was done by others; it was all no go – We were just thinking to take in hand another at 7,000 – when just at that time our Governor[23] went on a visit – or – came by his knowledge through letters – I don't know which, but it was just after a visit to some place in the West that a new order was given, and a specimen of some plans done in the N.[24] West – all by lines,[25] (think this was 1832) and it was said by our Governor that all was done there by lines – but where? None of us could make out the name of the county!! – However it was tried by us, and succeeded. And not long after this we heard that they all did the work by lines – This was only hearsay at the time. But it shows that the work done by lines superseded the old plans.

We worked at Down – the adjoining Co. was Antrim. Lt. Aldrich,[26] with the celebrated Hughey McCluney![27] and others, surveyed parishes – When the revision commenced those parishes (mid way between Belfast and Lisburn) came to be revised, they were revised – re-revised – so said our celebrated Hughey, and after that, they were re-surveyed – Query, what became of the originals? – When at a certain period 2 districts were knocked into one, we got part of this Lt. Aldrich's work of the Co. Armagh to revise – we revised and re-revised – and some we burned!! – that is to say, part of Co. Down was useless, Do[28] of Co. Antrim, do Armagh. This fell under our own noses so we could not but see it. It could be proved by Carbon's evidence before the House of Commons that the 14 counties were re-surveyed &c. &c. But you ask, how can I prove that the originals were burned? I answer I can't prove it – I can prove even now, that certain documents of Co Armagh were burned – documents of Co Antrim to a great degree useless – by our proximity we could discover Down, Antrim, Armagh equally erroneous, – i,e, 3 out of 5 districts involved in the same scrape[29] of error. And there is no reason to conclude the other 2 were much better – they were all in error nearly equally so. They all introduced the system of lines about the same time – for what purpose? – Why use lines? I have a very bad memory, but can recollect being told of battered old maps in other districts – and hearing Captain

Right[30] say in our office that 'twas all the fault of the draftsmen &c. And of one of the valuators saying 'the new plans of Derry were beautifully correct'. When I recollect this and that our own district was full of intelligent men, fully as well paid as others &c 3 out of 5 districts in the same scrape under our own observation – the documents of one district nearly all burned – by order – the natural inference is they all shared the same 'common' fate along with the Captains of each district!! They have a uniformity dependent, in burning as well as botching. But if you put the question again – How do I prove the maps were burned? I answer, produce the originals and I am floored!! But I will bet you sixpence that Botching and Burning in one district is equal to Burning and Botching in another. What was the use of having two sets of maps – an old set – and a new?

'The scientific labour of the first 7 years ended &c. &c' The statement of 7 years is not strictly true – 7 years is the mean time[31] nearly – but you can't state any time without being open to objection, 6 years is too small – 7 years is correct nearly – or from summer of 1825 to about the end of 1832 – Some Captains went off sooner, some later, 7 years is as near as one can come to it.

Revision[32]

This word was adopted instead of re-surveying, they thought at first that a few alterations &c. &c. would be quite enough. Re-surveying is a bad word, revision is much better – The old maps were not entirely superseded There was a parish in Down called Dromore, surveyed by the R.E.[33] themselves, this was entirely superseded – as you would say, but I say not so, for it was cut up for diagrams – like many other maps – given to the surveyors that they might lay out their lines quickly by it – ! The old maps formed excellent Boundary sketches for surveyors! Query – Did you ever know any man, civil or military, that revised a parish in 1834.5.6.7.8 &c, I think that would puzzle you – and from what I heard old hands say of Tyrone – Athlone Sligo, Derry &c. &c. the maps might be superseded at the earliest convenience!!!

I recollect distinctly. No. 1

1st To hear the men from other districts speak of the drinking matches – judging – Examining of work – and that they had glorious times &c. &c.

2nd An officer of the West, say that his best draftsmen had no more than 2/. – plenty of field men at 1/. – 1/6.

3rd A complaint of our district, that our progress was small and our Cap: brag, and Lts. that the progress was smaller, but the maps better &c. &c.

3rd[34] One of the principal valuers, say, that when they tried the valuation[35] in Derry they never saw such bad work – villages in the wrong townland[36] – areas erroneous – the maps perfectly useless.

4th The same valuer says, at the same time with the above, that the new maps of Derry were beautifully correct, he would not be ashamed to present some of them as private surveys of Estates.

5th A new system of surveying, by lines, about the year 1832 or summer of 1832 – and this system was adopted in some other districts before it came to us in 1832 – or we had it at second hand.

6th A Captain Right or Wright say in our office Belfast 1832 or 3 – that this mess was all owing to, the fault of the draftsmen – and at that very time the said Right was re-surveying out of a few!!

7th Parishes in Antrim – revised – re-revised – and finally re-surveyed.

8th Parish of Armagh Do Do some burned.

You will recollect that our District was considered the best as to accuracy –

And the above 8 points refer to other Districts; and then say, whether our maps are to be burned, and the others retained, that an order applicable to us is not so to other Districts – Recollect the order has been applied to 2 ie A:E.[37] or rather the Co. of Armagh and Down were knocked into one – D will recollect this change – and some of Armagh was burned at Drogheda

Whoever denies the accuracy of the assertion 'That the scientific labours of the first 7 years ended literally in a bottle of smoke' will envolve himself in great difficulty with respect to the returns of Carbon – and the order that the maps should be burned – the dispensing with Captains. – He will have a choice of several evils, and may and would succeed partially, but this could only be done by admitting some other points, fatal to the whole argument he advances – Depend on it he will have a tough job – if he replies by flat denial, then I demand the documents!!!

Further

We know – Antrim was worse than – Down

----------- Armagh as bad as Antrim

----------- Derry worse than Armagh.

Here you have 4 counties out of 5 – or 4 districts out of 5 – the best is condemned – what is to become of the worst? Your statement extends to the whole not to every part – you are right on the whole – every parish was signed and countersigned and returned as correct – no exception – if then they certified that to be right, which was afterwards proved wrong – they need not cry out on Spectator[38] for being wrong to a certain extent, on a subject of old recollection!! We must teach them to have a fellow feeling and besides nobody can find the real truth from their official documents; and every man, speaking on this subject must be liable to error – only trivial ones – But will they be bold enough to deny – ? I question it much, tho you may expect somebody else will, on their behalf!!

Poor House

The gent who spoke to me at Leeds, came out of a certain house, I don't know his name, but recollect the place well, & should it come to blows, can soon find his name, the gentleman said 'they came to us', so there were more than one of them present – A few days ago, a certain discharged man was with me for a short time (2 days) an old acquaintance (don't mention this to any one) he tried to get enlisted, took the shilling!![39] – would not pass the doctor!! he said 'the soldiers in ____ Leeds know all about the O.S. – they told me they had enlisted nearly a dozen of them' – What was the cause why he could not pass the doctors? It was certain injuries received on the O.S. – poverty of food & over exertion; keep this talk to yourself I would not on any account it would be mentioned – He is an old old crony, and fairly broke down – never saw a worse case*!!

* never mention this

B.[40]

RI MS JT 1/11/3783–7
RI MS JT 8/1/4a/18–20

1. *Your two last and paper*: letters missing.
2. *Lt. Bordes*: Lieutenant George Francis William Bordes (d. 1845) of the Royal Engineers, who at the end of 1832 was put in charge of the A and E Districts of the Irish Ordnance Survey, which were administered from Belfast, before passing on the command to Captain Henry Tucker in the summer of 1837. Bordes was promoted from Lieutenant to 2nd Captain in 1837 (NAI OS/1/14–18).
3. *patching system*: possibly a reference to the 'patching up' of older maps in the system of revision and correction; see letter 0211, n. 9.
4. *di.*: district.
5. *re-surveyed*: In many cases it proved impossible, or excessively expensive, to revise and correct the topographical details of maps produced between 1825 and 1830, meaning that the areas had to be surveyed again.
6. *Carbon*: This seems to be a nickname for Colonel Thomas Colby, possibly relating to his experiments with using pure carbon in printing Ordnance Survey maps in the early 1830s (see *The Correspondence of Michael Faraday*, ed. F. A. J. L. James, 6 vols (London: Institution of Electrical Engineers, 1991–2012), vol. 2, p. 90).
7. *Drogheda*: a port town in County Louth.
8. *Down, Louth, Meath*: counties in the north-east of Ireland.
9. *demesnes*: private estates (*OED*)
10. *D*: see letter 0195, n. 35.
11. *Co.*: County.
12. *we only drew plots*: The information collected during surveying was transferred from field books to paper, but was not then used to generate more detailed plans.
13. *cobbled*: mended clumsily (*OED*).
14. *Navan all round by the Boyne*: Navan is a town in County Meath, through which the River Boyne passes.
15. *the Estimates*: the estimates of costs of surveying returned to the Ordnance Committee

in Parliament.

16. *ducking*: immersion in water (*OED*).
17. *Swords*: a town around 8 miles north of Dublin.
18. *potato pits*: shallow pits, usually covered with a mound of straw and earth, in which potatoes are stored in winter (*OED*).
19. *O.S.*: Ordnance Survey.
20. *the revision was done by Acid*: presumably using acid to remove inaccurate topographical details on the revised maps; see letter 0211, n. 9. The head office of the Irish Survey at Mountjoy 'was sufficiently well-equipped and well-staffed to serve as a laboratory', and acid was used in the process of making aquatint maps (J. H. Andrews, *A Paper Landscape: The Ordnance Survey in Nineteenth-Century Ireland* (Oxford: Clarendon Press, 1975), p. 241).
21. *a bottle of smoke*: a deceit or falsehood.
22. *area of the parish 17,000*: acres.
23. *our Governor*: presumably Bordes; see n. 2.
24. *N.*: North.
25. by lines: presumably contour lines; see letter 0195, n. 49.
26. *Lt. Aldrich*: Lieutenant Edward Aldrich of the Royal Engineers, who during the late 1820s and early 1830s was in charge of a Division in the A and E Districts of the Irish Survey. He left the Survey in 1832 (NAI OS/1/18).
27. *the celebrated Hughey McCluney*: Hugh McCluney, a civil assistant who worked in the A District of the Irish Survey, having joined in November 1826. He seems to have left the Survey early in 1836 (NAI OS/1/11–13).
28. *Do*: Ditto.
29. *scrape*: embarrassing or awkward predicament brought about by imprudence and thoughtlessness (*OED*).
30. *Captain Right*: Captain Charles Wright (1793–1866) of the Royal Engineers, who joined the Irish Survey in November 1824 and was in command of the D District, before taking control of the A District early in 1832. By April of the same year he had left the Survey, serving in other capacities in Canada and back in Ireland, and being promoted from Captain to Major Brevet in 1838 (NAI OS/1/2–8).
31. *mean time*: presumably the arithmetic mean or average of the time taken up by the Survey.
32. *Revision*: see letter 0211, n. 9.
33. *R.E.*: Royal Engineers.
34. *3rd*: presumably an erroneous repetition on McLachlan's part.
35. *valuation*: see letter 0195, n. 40.
36. *townland*: see letter 0195, n. 42.
37. *A:E*: A and E Districts.
38. *Spectator*: the pseudonym used by Tyndall in letter 0228 and his four subsequent letters to the *Liverpool Mercury*.
39. *took the shilling!!*: Until 1879 new recruits to the British Army and Royal Navy received a payment of 1*s*., so taking the King's or Queen's shilling became a common expression for enlisting as either a soldier or sailor.
40. *B*: the pseudonym McLachlan used to avoid detection.

To John Tyndall, Snr 30 August 1843 0230

Preston | August 30th, 1843.

Dear Father

You tell me in your last letter[1] that I have omitted to mention the name of the 'head man' at Somerset House[2] in accordance with the Dean's[3] wishes. I hope the Dean does not imagine that this omission was caused by neglect or a want on my part of appreciating his very great kindness. The department of Somerset House to which my application referred is, as the Dean has stated in his testimonial, 'a map office under the superintendence of the Tithe Commissioners'.[4] The names of these Commissioners I know not; there are I think four of them.[5] I could obtain their names by writing to some of the London Journals. I have done so before in similar cases. I however have reason to believe that an application <u>at present</u> would not be attended with success. A Captain of the Royal Engineers a short time ago applied to Captain Dawson, Assistant to the Commissioners, in behalf of a young gentleman from Limerick who has been on the Ordnance Survey as long as I have. The answer of Captain Dawson was, that at present nothing could be done. I have this from the lips of the person on whose behalf the application was made; besides I would think it best to wait for some news from America and let that determine our future proceedings. I had a letter some time ago – the fifth of this month[6] – from Wm Wright. I quote from it the following: – 'It would give me great pleasure to hear you signify your intention of emigrating to this country, nay to this city, for it possesses so many opportunities for a young man to acquire the requisite information to enable him to earn an honest independence'. Again:– 'If you ever should think of coming out, make straight for Cincinnati, and in it you will find as hearty and as welcome a reception as you could wish for, in fact I have been for some time looking forward to the day when I will be able to grasp your hand in the street of this goodly city. I do not see any hindrance to you at once starting, except that of leaving your parents'. I could quote much more in this strain – Let this suffice. He tells me he has been blessed with another lovely child – a little girl, a month old, but as plump and large as a child of three. From the tone of William's letter I gather that the wildness of his youth is quite gone that he is a settled – nay a religious man. Tell his mother what I have told you. If I knew her address I would write to her on receiving his letters, he always desires me to do so. I have written to him asking him many questions, he no doubt will give me punctual answers. I received a paper from him to day <*rest of letter missing*>

RI MS JT 1/10/3296
LT Transcript Only

1. *your last letter*: letter missing.
2. *Somerset House*: on the Strand in central London.
3. *the Dean*: Dean Richard Bernard.
4. *Tithe Commissioners*: see letter 0173, n. 7.
5. *The names of these Commissioners … I think four of them*: There were in fact three Tithe Commissioners, William Blamire (1790–1862), the Chair of the Commissioners, Thomas Wentworth Buller (1792/3–1863) and Richard Jones (1790–1855). Tyndall was subsequently told the names of the three Commissioners by Archibald McLachlan, who warned him 'observe the 3 distinctions – Blamire Esq – he is for the land owners – Bullar R.N. is for the Admiralty – and Jones is for the Church – each party is to watch and preserve the interests of his caste' (18 October 1843 RI MS JT/1/11/3813–7).
6. *a letter some time ago – the fifth of this month*: letter 0216.

BIOGRAPHICAL REGISTER

This register contains the names and biographical details of people who are mentioned three or more times in the letters included in this volume. Biographical information about people mentioned once or twice is contained in the notes appended to the appropriate letter(s).

Conventions

In transcribing the letters the editors have generally retained the spellings of family names given in LT's transcripts. However, in a number of cases a variety of spellings were used by correspondents; for example, the family name of Thomas Bunbury MP was often spelled 'Bumbury'. Where there is a preferred spelling – for example, as used in official documents or biographical dictionaries – the alternative is placed in '( )' brackets. Thus, 'Bunbury (Bumbury), Thomas' is employed below. Where there is no preferred spelling, the alternatives are given in alphabetical order, separated by a '/': thus, 'Davey/Davy, James'. Where known the dates of birth and death are given in '( )' brackets.

Nicknames and pseudonyms were widely used by Tyndall and his friends. For example, Tyndall was known as Walter Snooks (or W. S. for short), John Roberts went under the name Sam Weller, Phillip Evans was called Jim and William Ginty signed himself 'Whiskerando' in letter 0082. Where such nicknames occur in the letters, the person they refer to is given in an accompanying note.

Ranks in the British army mentioned in letters, with abbreviations (in ascending order):

Private	Pt.
Lance Corporal	LCpl.
Corporal	Cpl. or Corp.
Sergeant	Sgt. or St.
Second/First Lieutenant	Lt.
Captain	Cap.
Major	Maj. or Mr.
Colonel	Col.
Major General	MG.
General	Gen.

Alexander, John (1764–1843), a wealthy and successful mill-owner who developed a thriving business milling wheat, oats and other grains at Milford, 3 miles north of Leighlin Bridge close to the River Barrow. He also purchased an estate at Milford and in 1799 constructed an impressive manor house. In 1840 the Milford estate occupied 1600 acres and accommodated ninety tenants. Widely appreciated as a good employer and an improving landlord, he was highly respected, both by fellow Protestants and by Catholics, and was considered by John Tyndall, Snr, as a suitable patron for his son (S. Kinsella, 'Landlords, Politicians, Entrepreneurs: The Alexanders of Milford, Co. Carlow in the Nineteenth Century', *Carloviana*, 58 (2009), pp. 70–83).

Barnard (Bernard), Richard Boyle (1787–1850), the second son of Francis, Earl of Bandon. After studying at Cambridge and being called to the Bar, he returned to the family seat and served as MP for Bandon Bridge, County Cork, from 1812 to 1815. He subsequently vacated his parliamentary seat and was ordained as a priest. In 1820 he became rector of Glankeen, County Cashel, and two years later he was appointed Dean of Leighlin, at the Cathedral Church of St Laserian in Old Leighlin. The many references in these letters to 'the Dean' indicate that Tyndall and his father considered Barnard to be a useful and reliable patron.

Baskin, Robert, joined the Irish Ordnance Survey on 9 November 1837, working as a civil assistant in the 4th Division, C District at Portarlington in Queen's County, and then, from 1840, in the Examination Office at the C District headquarters at Cork. His main duties were examining content registers and admeasuring. He was transferred to the English Survey on 5 August 1842, where he worked in the 1st Division and was stationed at Skipton.

Bates, David, a civil assistant in the 5th Division, C District of the Irish Ordnance Survey. His main duty on the Survey was plotting, although during 1837 he was with one of the 5th Division surveying parties, and later undertook calculating, registering, computing and examining work. He left the Survey in March or April 1839, probably to work on the Tithe Survey in England which paid much better wages than the daily rate of 1*s*. 8*d*. he was receiving in Ireland. He tried to rejoin the Irish Survey in January 1840, but was told there were no vacancies. He then asked about being enlisted in the army, but was told he would need to go to Woolwich to do so, after which he went to England with his brother George.

Bergan/Bergin, William, of Buncree, a Catholic tenant of Henry Bruen's, who, having voted for the Protestant and Conservative Bruen in the 1841 General Election, was attacked on 12 September 1841 by a pro-Repeal mob in Old Leighlin but was rescued by John Tyndall, Snr.

Bloomfield, Robert (Bob), joined the Irish Ordnance Survey on 12 September 1840, working as a civil assistant in the Examination Office at the C District headquarters at Cork. He was transferred to the 3rd Division, C District on 14 February 1842, with his principal duty being drawing. By August 1842 he was back at the Examination Office in Cork, where he worked alongside John B. Edgeworth, although he was transferred to the English Survey in the following month, where he was involved in the surveying of York.

Bredin, James (*c.* 1790–1872), rector of Myshall and Nurney and precentor of Old Leighlin.

Bruen, Henry (1789–1852), the second son of Henry Bruen (1741–95) who had served as an MP in the Irish Parliament and purchased properties in Counties Carlow, Wexford and Kildare, his Carlow properties alone covering 20,000 acres. His son, Henry, who resided on the family's Oak Park Estate on the northern outskirts of Carlow, was the Colonel of the Carlow Regiment of Militia. A wealthy Protestant, Tory landowner and man of influence, he represented Carlow County in Parliament from 1812 until 1831 and then stood again in 1835, but the result was contested. He returned to Parliament later that year after a contested by-election but was defeated in the General Election of 1837. The letters in this volume relate to the By-Election he won in December 1840, following the death of the standing Liberal MP, and his re-election at the hard-fought General Election in July 1841.

Bunbury (Bumbury), Thomas (1775–1846), born into a leading Protestant family with an estate at Lisnavagh, Rathvilly, County Carlow, while his mother was an heiress from Dublin. After attending Oxford University, Bunbury spent much time in Bath and London. However, in 1837 he joined with Henry Bruen in fighting the Carlow County parliamentary election, in which they were both defeated. At the General Election in July 1841 both he and Bruen gained seats in Parliament. Bunbury retained his seat until his death in 1846.

Butler, Mr, a school inspector who visited John Conwill's National School at Ballinabranna and assessed the pupils. He was replaced by Mr Shean for the February 1841 inspection. He may be the Edward Butler who is listed as Head Inspector of National Schools in *The Fifteenth Report of the Commissioners of National Education in Ireland: for the Year 1848* (Dublin: H.M.S.O., 1849). Also listed is a James R. Butler, who was inspector for District 18 (Droghedra, County Louth).

Carey, James (1807/8–1860) from Dungannon, County Tyrone, enlisted in the Corps of Royal Sappers and Miners at Belfast in December 1825 and rose through the ranks, becoming a Sergeant on 1 March 1839. A miner by trade he worked in the 4th Division, C District of the Ordnance Survey of Ireland under Lieutenant Robert Fenwick. With the departure of a number of personnel by June 1841, Carey was transferred to the 3rd Division and was in charge of the

Kinsale detachment during Tyndall's time there. Together with Joseph Collins and several other servicemen he was transferred to the English Survey on 20 May 1842, where he worked in the 4th Division. After an accident at the Ordnance Survey office in Manchester, in which he sustained a fractured skull, he was discharged in July 1849.

Carroll, Thomas Harris (d. 1861), gained his MA at the University of Dublin and then served as editor of the *Carlow Sentinel* for more than thirty years. Following the death of Henry Malcolmson in 1843, he also became the paper's proprietor. The *Cavan Observer* described him as 'a consistent Christian, a good man, a gentleman, a scholar, and an able journalist' (30 November 1861, n.p.).

Chadwick, John C., joined the Irish Ordnance Survey on 16 October 1838, working as a civil assistant in the C District, initially in the 4th Division under the command of Lieutenant Robert Fenwick, where he was stationed at Portarlington in Queen's County. His duties included plotting and drawing. He was transferred to the 3rd Division within C District in September 1840 when Fenwick took it over, but was back in the 4th Division by the beginning of 1841. He moved between the two Divisions thereafter. Chadwick resigned from the Survey on 26 February 1842, probably because of the birth of his daughter, and settled in the town of Kinsale in County Cork, where he worked as a private surveyor.

Colby, Thomas Frederick (1784–1852), the son of a Major in the Royal Marines, he trained at the Royal Military Academy, Woolwich, before joining the Royal Engineers in 1800. He was soon working for William Mudge's Ordnance Survey, becoming permanently attached in 1802. Despite an accident with a pistol, which led to the amputation of his left hand, Colby had extensive experience as a surveyor, including extending the meridian line from Dunnose, Isle of Wight, into Scotland. Following the death of Mudge in 1820, he lobbied for and obtained the post of Superintendent of the Survey, which he sought to improve. After Parliament's decision in 1824 to initiate the Survey of Ireland, Colby took charge of the project and decided against employing civil surveyors; instead he gave the task to the Royal Corps of Sappers and Miners. He began by carefully measuring an 8-mile base-line at Lough Foyle and by making improvements to the surveying instrumentation. Being a stickler for accuracy, progress was slow and expensive and the first county map – Londonderry – was only published in May 1833. In 1833 he joined Henry de la Beche in making a geological map of the west of England and by 1838 the Irish Survey was mainly overseen by Lieutenant Thomas Aiskew Larcom. Under pressure to complete the Survey of Britain and to trim its budget, Colby became increasingly involved in arguments over the scale of maps produced by the Survey. He left the Survey in April 1847, having been promoted to Major-General, and died six years later.

Calder (Colder), Alexander (b. 1808), born in Dunbar, near Edinburgh, and enlisted in the Corps of Royal Sappers and Miners on 16 July 1829. In 1839 he was promoted to 2nd Corporal, to 1st Corporal on 1 February 1843, and to Sergeant on 14 January 1846. A carpenter by trade, he worked in the 3rd Division, C District of the Irish Ordnance Survey under Lieutenant William Delves Broughton until he was transferred in January 1841 to the 5th Division, under the command of Lieutenant Paul Whittingham. Now based in Cork, he supervised the office party that included Tyndall. In September 1841 he was transferred to England. He was discharged on 12 April 1853 owing to chronic rheumatism, and because of his exemplary service was recommended for a Silver Medal and all the gratuities for which he was eligible.

Collins, Joseph (b. 1803), born in Whickham, County Durham, and enlisted in the Corps of Royal Sappers and Miners on 19 February 1822. He was either a blacksmith or a miner by trade, and worked in the 5th Division, C District of the Irish Ordnance Survey, under the command of Lieutenant George Wynne. Although he had earlier gained a reputation for absenteeism, he worked with Tyndall, James Davey and others in County Carlow and in early 1841 he superintended the field parties within the Division. He was transferred to the English Survey in May 1842 and was discharged on 8 July 1845, owing to chronic rheumatism. He later served with distinction in the Crimea, was severely wounded and invested with the Order of the Legion of Honour.

Commins, Bridget, possibly the Bridget Cummins who was baptised on 7 December 1823 at Bagenalstown and whose parents are listed as James Cummins and Anne Nolan (Irish Church Records).

Commins, Maryann, probably a sister of Bridget Commins.

Conwill, John (1802–80), educated in his father's school at Rathornan and taught for eleven years at the National School at Bagenalstown, where he employed the Lancasterian system of teaching in which advanced students were involved in assisting those below them, allowing children in large classes to receive individual tuition. Following the opening of a National School at Ballinabranna in 1833, he was employed as its schoolmaster, a position he held until September 1847, when he transferred to the National School at Leighlin Bridge. In 1840 the school at Ballinabranna attracted 236 pupils (both male and female), mostly Catholics, and Conwill's annual salary was £20. As his letters to Tyndall demonstrate, he possessed an impressive range of attainments, including poetry, mathematics and English grammar, a subject on which he sternly opposed the local school inspectors. Although Conwill was a Catholic, he and Tyndall developed a close personal relationship. Conwill was Tyndall's academic mentor and Conwill viewed Tyndall as his star pupil.

Conwill, Mrs Catherine (*c.* 1774–1862), mother of John Conwill.

Cotter(s), Mrs, possibly Mary Cotters (née Mulcahy), the baptism of whose son Laurence took place in St Peter & Paul's, Cork in June 1855 (Irish Church Records). *The County and City of Cork Post Office General Directory 1842–3* (Cork: F. Jackson, 1843) lists a Mrs Cotter as a 'lodging house keeper' at 6 Pine Street (p. 19).

Cuddy, Martin, a civil assistant who worked with Tyndall in the 5th Division, C District of the Irish Ordnance Survey. He joined on 30 April 1838 and was mainly involved in registering and examining work on the ground. In May 1841 he was transferred to the 2nd Division under Lieutenant Charles Edmund Wilkinson in Bandon, where he was later joined by Phillip Evans. On 8 April 1842 he was transferred to the English Survey.

Davey/Davy, James (*c.* 1800–73), joined the 13th Company of the Royal Sappers and Miners in 1822, was promoted 2nd Corporal on 1 September 1834 and 1st Corporal on 10 January 1838. A mason by trade, he worked in the 5th Division, C District of the Irish Ordnance Survey and headed the team that Tyndall joined at Leighlin Bridge in April 1839. In October 1840 he was transferred to the English Survey, first to York and subsequently to Woolwich. At Leighlin Bridge he suffered throat and neck complaints and was discharged from the army in April 1843 owing to illness, possibly exacerbated by the death of his son John. Greatly admired by Tyndall and his father, Davy was clearly strong-willed and had been described by Major Marcus Antonius Waters as 'steady and zealous' (NAI OS/1/14). Yet he had a somewhat chequered career; in 1834 he was caught franking a personal letter in the Survey Office, court marshalled and found guilty. Franking was the privilege of sending and receiving letters free of charge and was used by official bodies and MPs. Davy had abused the privilege by signing for a private letter as if it were an item of official Survey correspondence. Additionally, in 1837 he was suspended for almost two weeks for careless work and he was also investigated for speaking disrespectfully of Waters.

Davey/Davy, John (d. 1843), the son of Corporal James Davey, a civil assistant in the 5th Division, C District of the Irish Ordnance Survey, having joined in January 1840. His main duty was plotting. After his father went to England in October 1840, John was transferred to the 4th Division under the command of Lieutenant Robert Fenwick. He was then transferred to York on 24 June 1841, where he worked under the command of Captain John Bailey, his father having requested that his son be transferred to England to be closer to him earlier in June. He died, while still employed on the English Survey, in April 1843.

Davey/Davy, Mrs, the wife of James Davy and mother of John Davy.

Dawson, Robert Kearsley (1798–1861), educated at the Royal Military Academy, Woolwich, he joined the Royal Engineers in 1815, rising to the rank of Captain by 1837. He served as assistant to Colonel Thomas Colby on the trigonometrical branch of the Ordnance Survey of Great Britain in Scotland, before being transferred to the Map Office at the Tower of London in 1822. He rejoined Colby in Ireland in 1829, where he supervised the sketching of hills in the northern counties. In 1836 he was recalled to London as Assistant Commissioner to the Tithe Commutation Commission, based at Somerset House, in which role he determined that Tithe Maps should be made at a scale of 1 inch to 3 chains, the equivalent of 26⅔ inches to 1 mile. His subsequent proposal that this large scale should be adopted nationally by the Ordnance Survey was rejected by Parliament on the grounds of cost.

Dobson, Lawrence (1792/3–1865), served as editor of the *Preston Chronicle* from *c.* 1832, when he succeeded Isaac Wilcockson, until his death.

Donaclift, Sophy, daughter of Samuel Donaclift, a shipwright who worked on Kinsale Harbour, and his wife Sophia (née Leggatt). She may have married a German maritime captain in the summer 1843, or a tutor named Barry in the following year. She is possibly also the Sophie Donaclift who was born and lived in Kinsale until her marriage to Henry McIntosh, later Headmaster of the Methodist College, Belfast. As Sophie McIntosh, she published short stories describing the lives of the peasants and fishermen of Kinsale that were collected together as *The Last Forward and Other Stories* (1902) (T. O'Toole, ed., *Dictionary of Munster Women Writers 1800–2000* (Cork: Cork University Press, 2005), p. 132).

Edgeworth, John B. (b. 1806), a Private in the 13th Company of the Royal Sappers and Miners, having enlisted on 25 December 1826. He was an acting Lance Corporal on the Irish Ordnance Survey, where he worked in the 3rd Division, C District. A mason by trade, his duties on the Survey included surveying, laying down lines, plotting and levelling. He left the Survey in autumn 1841, although he had rejoined it by August 1842, and was working in the Examination Office at the C District headquarters in Cork. It seems likely that he was a member of the influential Anglo-Irish Edgeworth family from County Longford, and was related to the novelist Maria Edgeworth.

Edwards, Mary (b. *c.* 1822), daughter of the late Captain Robert Edwards from Liverpool, lived with her uncle and aunt in Kirkby Lonsdale, and, despite having a putative romance with William Ginty, married William Henry Owen, an Irish solicitor, in June 1843.

Eivers, Laurence/Lawrence (Larry), joined the Irish Ordnance Survey on 15 June 1829, working as a civil assistant in the C District. By the late 1830s he was working in the 4th Division under the command of Lieutenant Robert Fenwick, where he was receiving 5*s.* a day at the beginning of 1838. By late 1840 this had increased to 6*s.*, and the rapid rate at which his pay rose suggests that he was very important to the work of the Survey, particularly in drawing and examining ground work. He was transferred to the 3rd Division, C District when Fenwick took it over during the summer of 1840, but with the work of the 3rd Division winding down by the winter he was sent back to the 4th Division, moving between the two Divisions thereafter. No doubt kept on for his skills and experience, Eivers was one of just twelve men, including Tyndall, left in the 3rd Division by June 1842. Once its work was completed on 5 August 1842, he was transferred to the English Survey, but, by April 1843, he had moved to a better paid position on the Tithe Survey.

Ellen (the 'lady of the raven plume'), a love interest and romantic ideal of both Tyndall and William Ginty, whom they presumably met while stationed in Kinsale in County Cork in the summer of 1841. Although she remained a regular subject of Tyndall's and Ginty's letters and verse, Tyndall's ardour seems to have waned by July 1842. That Ellen and the 'lady of the raven plume' were the same person is established by Ginty's poem in letter 0186.

Evans, Phillip Deighton (Jim) (*c.* 1822–1906), the son of Richard Evans of Carlow, he joined the 5th Division, C District of the Irish Ordnance Survey on 28 July 1838 in Leighlin Bridge. He appears to have already known Tyndall and may also have been a pupil of John Conwill. His primary duty on the Survey was drawing and his salary rose from 1*s.* to 2*s.* 9*d.* in the summer of 1840. He and Tyndall were close friends both at work and in their leisure time and they lodged together. In the winter of 1841–2 Evans was moved to the 2nd Division and on 20 May 1842 he was transferred to the English Survey, together with William Ginty and others. He joined Tyndall in making the complaint of mismanagement against the Survey's commanders and both men were dismissed from the Survey in November 1843. At the time of his marriage in 1861, Evans was described as a 'Temporary Foreman of Works War Department' and he gave an address in Limerick (Irish Church Records). He died in Liverpool in 1906.

Faulkner, Hugh (*c.* 1773–1845), of Rathvinden Cottage, Leighlin Bridge, from a family that owned an estate at Castletown, County Carlow. An established member of the local Protestant gentry, he was magistrate and had served as High Sheriff of Carlow in 1831.

Fenwick, Robert (1801–43), a Royal Engineer who had enlisted early in 1825. He progressed to 1st Lieutenant in May 1825, 2nd Lieutenant in January 1826 and 2nd Captain on 14 September 1841. He became dangerously ill in 1833 and in January 1835 was assaulted by a Reverend William Maguire while sketching ruins at Castleboy, County Galway. Fenwick had considerable experience working on the Irish Ordnance Survey, including being in charge of the 3rd and 4th Divisions, C District, before taking command of the 5th Division at Kinsale after Lieutenant Paul Bernard Whittingham's departure in July 1841. Later that year he was based at Grassington, Yorkshire. After leaving the Ordnance Survey, he commanded the 4th Company of Royal Engineers in Bermuda, where he died of yellow fever in August 1843.

Ford, a pupil of John Conwill's at the National School at Ballinabranna.

Foy, Thomas M. (b. 1814/5), enlisted in the Corps of Royal Sappers and Miners on 25 February 1833 and commenced employment on the Irish Ordnance Survey in February 1834. A miner by trade, he worked in the 5th Division, C District under Lieutenant George Wynne and Corporal James Davy, and by August 1840 he was an acting Lance Corporal in charge of one of the surveying teams. He was later charged with being absent without leave, being married without permission and with embezzling £19 11*s*. 3*d*. of public money. For these misdemeanours he was reduced to the rank of Private and on 27 January 1841 he was sentenced to a fortnight in prison in Cork Barracks. His wife died later that year. According to Tyndall, Foy was 'very much changed since his imprisonment' (letter 0117) and, on 27 May 1842, he was promoted to Lance Corporal. A fortnight later he was transferred to the English Survey, where he worked in the 4th Division and was stationed at Manchester. His friendship with Tyndall later soured over an unpaid loan.

Ginty, William, Snr, an architect of Carrick-on-Shannon, County Leitrim, and the father of William Ginty. He was sometimes known as '"auld" Ginty' (letter 0197).

Ginty, William Gilbert (*c*. 1820–66), born at Carrick-on-Shannon, County Leitrim, and joined the 5th Division, C District of the Irish Ordnance Survey on 1 February 1836, working as a civil assistant under the command of Lieutenant George Wynne. His duties included plotting, registering, laying down lines, tracing for ground examinations, and writing and drawing. He left the Survey in the spring of 1839 to work on the English Tithe Survey where private contractors paid higher wages, but came back to Ireland in October 1840. A close friend of Tyndall's, they shared lodgings until he was transferred to England on 20 May 1842, where he joined the 5th Division in Liverpool, and worked in the field in Lancashire and Westmoreland, particularly in Kirby Lonsdale. As well as swapping poetry on their respective love interests, Ginty advised Tyndall on the contents of his pseudonymous letter to Robert Peel protesting the Royal Engineers' conduct of the English Ordnance Survey, and it was he who arranged to have it published in the *Liverpool Mercury*. Subsequently he became a prominent Manchester businessman, served as honorary secretary of the Manchester Athenaeum and was involved in various civil engineering projects, including the construction of the Liverpool-Bolton railway. In 1854 Ginty moved to Rio de Janeiro in Brazil, where he worked as an agent of the Baron de Maua, contributing to such projects as the construction of the imperial gasworks and a railway which ran from Rio to Tijuca Serra. For his contributions to civic projects in Rio, he was awarded a knighthood in the Order of the Rose of the Brazilian Empire.

Griffin, David (1802–81), born in Derry he enlisted in the 13th Company of the Royal Sappers and Miners on 28 December 1825. As an acting Lance Corporal, he worked in the 5th Division, C District of the Irish Ordnance Survey, under the command of Lieutenant George Wynne. A miner by trade, his duties included plotting and drawing, laying down work and completing plans. He was transferred to the English Survey on 20 May 1842, although he seems to have been demoted to his original rank of Private when he joined the Survey of Windsor in April 1843. He was subsequently promoted to 2nd Corporal on 10 April 1844, Corporal on 15 October 1845, and Sergeant on 10 January 1849. Griffin was discharged from the army in April 1850 as a consequence of an injury to his left leg, and the loss of his right eye and weakening of the left. The leg injury was the result of a fall while on the Irish Survey in 1829.

Griffith, Matthew (1802/3–65/6), a farmer from Newtown who was made High Constable of Idrone East in March 1842.

Hamley, William George (1815–93), born in Cornwall as the son of Vice-Admiral William Hamley, he enlisted in the Royal Enginneers at the age of eighteen, quickly rising to 2nd Lieutenant by August 1833, and 1st Lieutenant by September 1836. He worked on the English Ordnance Survey between 20 April 1840 and 15 December 1846, and was in command of the 5th Division, based at Liverpool, at the time that William Ginty was stationed there in 1842 and 1843. Before leaving the Survey, he was made 2nd Captain in May 1845. Hamley also served abroad in the West Indies (1837–40 and 1848–52), Gibraltar (1855–6) and Bermuda (1863–7), drawing on these experiences to write numerous adventure novels including *Captain Clutterbuck's Champagne* (1862). He had become Major General (Brevet honorary) when he retired with full pay of £600 in January 1873.

Hennessy, Michael (sometimes called 'Classic Hennessy' and 'Professor Hennessy'), the master of the National School at Graigue, a suburb of Carlow. In his letters to Tyndall, John Conwill contradicted Hennessy's views on grammar and mathematics.

Henry, William (Billy), a friend of the Tyndalls from Leighlin Bridge.

Heydon, Edward, the father of Francis and William Heydon. Roman Catholics, the Heydons lived close to the Tyndalls on Bridge Street, Leighlin Bridge. He encountered financial difficulties in March 1841 (letter 0051).

Heydon, Francis, one of the sons of Edward Heydon of Leighlin Bridge. It does not appear that he was employed by the Ordnance Survey of Ireland, although he may have been involved in survey work.

Heydon, William, one of Edward Heydon's sons from Leighlin Bridge and well known to the Tyndalls. He joined the 5th Division, C District of the Irish Ordnance Survey as a civil assistant and worked with Tyndall from about August 1839 until he was discharged from the Survey on 12 June 1841.

Higginson, Thomas Charles (*c.* 1821–98), a civil assistant, originally from Dublin, in the 5th Division, C District of the Irish Ordnance Survey, having joined on 8 May 1839. His duties were primarily surveying in the field party under Corporal James Mulligan, working in Counties Kilkenny, Tipperary and Cork. He was transferred to the English Survey in September 1841, where he worked in the 1st Division, stationed near Skipton in north Yorkshire. In the spring of the following year, Higginson left the Survey and, with the assistance of an uncle, enlisted in the Royal Artillery's 78th Regiment of Foot on 10 April 1842. He was sent to India, although arriving, in August 1842, too late to serve in the First Anglo-Afghan War. Instead he participated in the annexation of the Sindh Province, and was made Lieutenant in July 1845. By the early 1850s, Higginson had returned to Britain, being stationed at Aldershot and Shorncliffe near Folkestone, where he served, from April 1856, as Paymaster for the 16th Regiment of the Royal Artillery. He was promoted to Captain in 1861, and Major in 1866, before retiring, as Lieutenant-Colonel, in October 1881.

Holden, Mr (d. 1842), a resident of Leighlin Bridge.

Holmes, George, a civil assistant who joined the Irish Ordnance Survey on 26 June 1839 and worked in the 5th Division, C District, initially in Corporal James Mulligan's surveying party under the command of Lieutenant George Wynne. During the spring of 1841 he worked closely with Tyndall before being transferred to the English Survey in May 1842, where he joined the 5th Division and was stationed in Liverpool.

Holmes, J., organised the public debate at Youghal in April 1841 at which Tyndall defended the Roman Catholic religion (letters 0057, 0058 and 0059). He may have been the secretary of the society that held this debate.

Hume, Joseph (1777–1855), radical MP for the Montrose Burghs and inveterate campaigner against extravagance in public finances, 'whose criticisms of Ordnance spending were practically an annual Commons ritual' (R. Oliver, 'The

Ordnance Survey in Great Britain 1835–1870', PhD thesis, University of Sussex, 1985, p. 113).

Hunter, Richard (Dick), a civil assistant who worked in the 4th Division, C District of the Irish Ordnance Survey, under the command of Lieutenant Robert Fenwick, having joined the Survey on 7 May 1838. His duties included plotting and drawing, on which he worked closely with his brother William. He was transferred to the English Survey on 20 May 1842, where he was stationed in Liverpool as part of the 5th Division, working in a field party at Prescott in Lancashire.

Hunter, William (Willy), a civil assistant who worked in the 4th Division, C District of the Irish Ordnance Survey, under the command of Lieutenant Robert Fenwick, having joined the Survey on 22 April 1839. His duties included plotting, calculating and drawing. Although he worked closely with his brother Richard, the two were separated when Richard was transferred to the English Survey in May 1842. A month later William was engaged to help complete the surveying of Cork and Kinsale, working alongside Tyndall. He was transferred to the English Survey on 5 August 1842, where he retained his notoriety amongst his peers for mischief and womanizing.

Jackson, James (1790–1871), served in the Peninsula War against Napoleon, at the Battle of Waterloo and in both India and Arabia. At the time of these letters he was Lieutenant-Colonel in the 6th Dragoon Guards (the Carbineers), which was moved to Carlow in July 1841 to quell any disturbances during the hotly-contested election. He was later promoted to General and was knighted.

Latimer, George, joined the Irish Ordnance Survey on 6 January 1836, working as a civil assistant in the 4th Division, C District, under the command of Lieutenant Robert Fenwick, at Carrick-on-Shannon in County Leitrim. His duties included plotting, calculating and drawing, with the latter his primary responsibility. He remained on the Irish Survey until the completion of the work in C District, working with Tyndall at Cork and Kinsale. He was transferred to the English Survey on 5 August 1842, where he joined the 1st Division and was stationed, with Tyndall, in Preston.

Latimer, William, a civil assistant who worked in the 4th Division, C District of the Irish Ordnance Survey, under the command of Lieutenant Robert Fenwick, having joined the Survey in *c.* January 1840. He was moved to the 3rd Division, which was also under Fenwick, in August 1840, but soon returned to the 4th Division. His duties included calculating and plotting, on which he worked closely with his brother George. The two were separated, however, when William was transferred to the English Survey on 20 May 1842, where he joined the 5th Division and was stationed in Liverpool.

Lilly, John, a civil assistant working in the C District Examination Office at Cork, having joined the Irish Ordnance Survey on 28 January 1839. He was initially employed in the 4th Division under Lieutenant Robert Fenwick, but was transferred to the Examination Office on 13 October 1839, where his work mostly consisted of plotting. He was briefly transferred to the 4th Division, C District on 21 February 1842, although he was back in the District Examination Office by April. He was transferred to the English Survey in the summer of 1842, where he joined the 4th Division and was stationed in Manchester. Disillusioned with the Survey's low pay, Lilly resigned abruptly in March 1843, although he had no alternative job to go into when he returned to Ireland and instead occupied himself with gardening.

Linedale, Anthony, a civil assistant probably from Dublin who joined the Irish Ordnance Survey at Leighlin Bridge in either 1839 or 1840. He worked as a plotter, surveyor and draughtsman in the 5th Division, C District, initially under the command of Lieutenant George Wynne. In the spring of 1841 he worked closely with Tyndall, but appears to have suffered a significant illness and then absconded on 17 May 1841. He soon returned to County Carlow and married Ann M'Gee on 29 May. Tyndall considered Linedale to be an ungrateful and vindictive colleague.

Little, James, a coach agent and driver in Fermoy. He married Sarah James on 22 October 1840 at Dunleckny (Irish Church Records). As Deborah McAssey stayed with Mr and Mrs Little when visiting Fermoy and Mr Little was known to Tyndall and his father, it is possible the Little family were distant relatives.

M'Gee, Ann, sister of John M'Gee, who married Anthony Linedale at Dunleckny in County Carlow on 29 May 1841.

M'Gee (or McGee), John, one of the Carlow maltsters who in 1833 complained to Parliament about the high excise duty on malt, Carlow being 'one of the largest malting districts in Ireland' (*Parliamentary Papers* 31 (1833), pp. 98–9). A friend of the Tyndall family, he was later charged for non-payment of excise duty on grain and was declared bankrupt in July 1840. On 24 August 1841 he appeared before the Court of Bankruptcy in Dublin, in order to discharge a debt of £461 0*s*. 9*d*. (*Perry's Bankrupt Gazette*, 7 August 1841, p. 8).

M'Gee, Celia/Cecilia, probably the sister of John M'Gee. In *Slater's* (1846) she was described as the postmistress at Leighlin Bridge.

M'Gee, Helen (née Hayden), married John M'Gee in 1835.

Maguire, Father Tom (1792–1847), a Roman Catholic parish priest from Ennismagrath in County Leitrim who was celebrated for defending Catholicism by entering into bitter polemical disputes with Protestant evangelicals.

Maher, Father James (1794–1874), from a comfortably prosperous farming family connected to several leading clergy in the diocese of Kildare and Leighlin. After attending a Quaker school at Ballitore, County Kildare, he studied at Carlow College and later at Rome, where he was ordained in 1821. He then served as a curate in Carlow and a parish priest in Leighlin Bridge and subsequently in the united parishes of Goresbridge and Paulstown.Following a period as Administrator of the Carlow Cathedral, he served as Professor of Theology and Sacred Scripture at Carlow College (1837–41) before becoming parish priest of Carlow-Graigue, where he remained for most of the remainder of his life. Maher strenuously supported the cause of the downtrodden Irish Catholics and in the hard-fought election of 1841 he was a leading advocate of Repeal and a staunch opponent of Henry Bruen and the Protestant landowners. (D. McCartney, 'Fr. James Maher, P.P.', http://www.rootsweb.ancestry.com/~irlcar2/Fr_James_Maher.htm [accessed 24 March 2014]).

McMahon/M'Mahon/Mahon, Father John (d. 1850), from County Laois, trained at Saint Patrick's College, Carlow (1831–2) and at Saint Patrick's College, Maynooth (1832–6). Described as 'the curate of Leighlin' (*CS*, 24 July 1841, p. [2]), he served the parishes of Leighlin (1837–41), Hacketstown (1841–3) and Mountrath (1843–50) (J. McEvoy, *Carlow College 1793–1993* (Carlow: Carlovian Press, 1993)). Tyndall's father considered that Father McMahon was one of the ringleaders of the O'Connellite agitation that attacked Caleb Tyndall's house during the 1841 election campaign.

Malone, Denis (Finn), a civil assistant in the 1st Division, C District of the Irish Ordnance Survey, having joined in September or October 1838. His primary duty was writing boundary registers and boundary remarks. He was one of only three men remaining in the 1st Division in April 1842, and was transferred to the English Survey in the following month. He joined the 1st Division and was stationed in Preston, where he later worked with Tyndall. In September 1842 his brother took over the 'Select Classical and Commercial Academy' in the city, and was employed by Tyndall to give him one-to-one tuition in French.

Marquis, William (b. 1820/1–95), from Dublin, the son of the architect George Marquis and younger brother of the landscape painter James Richard Marquis, he won a premium in Architecture in 1835 at the Royal Dublin Society's School of Drawing. On 1 December 1837 he joined the Irish Ordnance Survey, working as a civil assistant in the A District headquarters under the command of Captain Henry Tucker. His primary duty was drawing, but he also undertook examining work at various times during 1839. In this he appears to have worked closely with Archibald McLachlan. He left Ireland in August 1841 when the work in the A District was all but complete, and was transferred to the English

Survey. He worked in the 1st Division and was stationed at Preston, where he later accompanied Tyndall on a restorative sojourn to the village of Goosnargh in June and July 1843. He is recorded in the 1881 Census as a widowed civil engineer, living in Kirkdale, Lancashire.

Martin, Robert (Bob), a civil assistant in the A District of the Irish Ordnance Survey, commanded by Captain Henry Tucker. Having joined the Survey in February 1827 he became one of its most senior assistants and an accomplished draughtsman, so that by July 1841 he commanded a salary of 4*s*. a day. In the spring of 1840 he was based in the Waterford region in the 4th Division, A District, under Lieutenant Thomas R. Lyster, and later that year the Division was transferred to Clonmel in County Tipperary. In August 1841 Martin left Ireland for the English Survey.

McAssey, Deborah (b. 1805), the daughter of John McAssey and Mary McAssey (née Smith/Smyth), was Tyndall's cousin on his mother's side of the family, although the precise relationship is not known. She lived at Ballybromell, a village in the parish of Fenagh, County Carlow, and by April 1847 had emigrated to Wisconsin with several other members of her family.

McAssey, Mary, wife of Patrick McAssey, she was wounded on 27 June 1841 by Caleb Tyndall when he shot into an allegedly hostile crowd that had approached his house. According to *Griffith's Valuations* a 'Mary Mackessy' leased a house and yard on Bridge Street, Leighlin Bridge, and also owned two other properties on the same street, one of which was occupied by 'Mary Mackessy's lodgers' (p. 180).

McCarthy, Carberry, the husband of Mrs McCarthy, Tyndall's landlady at Strand Road, County Cork. He is not listed in either *Slater's* or *The County and City of Cork Post Office Directory for 1842–3* (Cork: F. Jackson, 1843). A Carberry McCarthy was, however, described as 'a clerk in the establishment of Quarry and Wilson', patent saw-mills (*Cork Examiner*, 5 August 1842, p. 2).

McCarthy, Mrs, Tyndall's landlady at Strand Road, County Cork. Tyndall's claim that Mrs McCarthy was a Quaker (letter 0119) is not confirmed by the records of the Cork Quaker Meeting. She, like Tyndall's mother, may have been descended from Irish Quakers.

McLachlan, Archibald, a civil assistant from County Down who joined the Irish Ordnance Survey on 28 May 1827, and worked initially in the A District at Dublin under the command of Captain Henry Tucker. His duties included comparing plans and drawing. In 1834 he worked in the 1st Division, E District under Lieutenant George Francis William Bordes, where news that his daily rate of pay was to be increased to 4*s*. 3*d*. raised concerns among the Survey's administrators. Bordes was directed to send an account of the actual service of each civil assistant and military man in his Division to Colonel Thomas Colby, so that he could apportion an appropriate sum for wages. In 1836 McLachlan was employed in Dublin by the radical MP Joseph Hume to collect statistics on the Irish Survey that might be laid before Parliament. McLachlan resigned his position on the Ordnance Survey on 18 April 1838, probably to work on the English Tithe Survey where private contractors were paid higher wages, but rejoined the Irish Survey on 12 February 1841, returning to the A District. He was once again working under the command of Tucker, and was based at Waterford, where he appears to have worked closely with Robert Martin. He was transferred, along with Martin, to the English Survey in August 1841, where he was stationed in Leeds. Whilst there McLachlan provided Tyndall with details of the earlier mismanagement of the Survey, and the current pay and conditions of its civil assistants, that became the basis of his pseudonymous letters to Robert Peel in the *Liverpool Mercury*.

Mooney, John, a boyhood friend of Tyndall's from the Leighlin Bridge area with whom he discussed philosophical issues, together with Mooney's father. However, following the highly divisive July 1841 parliamentary election, Tyndall resented Mooney's interventions to prevent Roman Catholics from dealing with Tyndall's father.

Mulligan, James (b. 1804), from Roscommon, enlisted in the 4th Regiment of the Dragoon Guards in August 1824 and two years later transferred to the the 13th Company of the Royal Sappers and Miners. A blacksmith by trade, he worked in the 5th Division, C District of the Irish Ordnance Survey under Lieutenant George Wynne and was principally involved in surveying. Following Corporal James Davy's transfer to Woolwich in October 1840, Mulligan superintended the office party before himself being transferred to England in May 1841. Although he was suspended for a month in October 1836 for 'negligence in preparing the accounts and pay lists of their parties' (NAI OS/2/14) he was subsequently described by Major Marcus Antonius Waters as 'steady and of good promise' (NAI OS/1/14). He was promoted to 2nd Corporal on 1 March 1839, Corporal on 1 December 1839 and later Sergeant; but owing to illness he was discharged on a full pension in 1846.

Murray, George (1772–1846), born in Crieff, Perthshire, he entered the army in 1789, when he was commissioned ensign in the 71st Regiment of Foot, and had reached the rank of Captain by 1794. He excelled as a Quartermaster-General, and was appointed to the Duke of Wellington's army in the Peninsular War against Napoleon, although a brief posting in Canada meant that he did not fight in the Battle of Waterloo. In 1819 Murray was made Governor of the Royal Military College, Sandhurst, and was elected MP for Perthshire in 1824, also becoming Lieutenant-General of the Ordnance in the same year. He was made Master-General of the Ordnance during Robert Peel's first ministry between December 1834 and April 1835, and again in the second Peel ministry of September 1841 to June 1846. Perhaps the most respected soldier in Britain after Wellington, he died only a month after leaving the post of Master-General.

Murray, John, joined the Ordnance Survey of Ireland in December 1839 and served as a civil assistant in the 5th Division, C District. Primarily involved in registering and calculating, in 1841 he worked in the office party with Tyndall, Phillip Evans and Joseph Payne. In April and May 1842 he was correcting and completing registers, before being transferred to the English Survey on 20 May 1842.

Murray, Mr and Mrs, lived in or near Leighlin Bridge and were the parents of William Murray.

Murray, William, from Leighlin Bridge, worked as a civil assistant in the 5th Division, C District of the Irish Ordnance Survey between May and October 1840. He was then transferred to the English Survey, being part of the 'first draft that went to Yorkshire' (letter 0134).

Newton, Beauchamp Bartholomew (1798–1850), a son of Colonel Philip Newton of Dunleckny Manor, County Carlow, an established Protestant landowner. He was a Lieutenant in the 34th Foot Regiment, owned an estate at Rathwade, Bagenalstown and served as High Sherriff of Carlow in 1850.

Newton, Henry (1800–63), a son of Colonel Philip Newton of Dunleckny Manor, County Carlow, an established Protestant landowner. He lived at Mount Leinster, Borris, County Carlow and served as a Poor Law Guardian.

Newton, Isaac (1642–1727), a natural philosopher and mathematician of outstanding ability who rose to the Luscasian Professorship of Mathematics at Cambridge (1669), Master of the Mint (1699) and President of the Royal Society (1723). Using some of the many Newtonian textbooks John Conwill would have introduced Tyndall to various aspects of Newton's science; see, for example, the problems discussed in letters 0166 and 0167.

O'Connell, Daniel (1775–1847), from an established Roman Catholic family with lands and a mansion (Derrynane House) in County Kerry. O'Connell was partially educated in Flanders before studying law in both London and Dublin and commencing his practice as a barrister in Munster. Enthused by the political radicalism of the 1790s he became increasingly active in movements that advocated Irish nationalism and sought to remove the political, social and economic disabilities under which Roman Catholics laboured in Ireland. The Catholic Association, which he founded in 1823, pressed strongly for Catholic emancipation. Despite his landslide election victory in County Clare in 1828, he could not take his seat in the House of Commons because of the oath which excluded Catholics. Only after extensive political wrangling, and despite Robert Peel's determined opposition, he was able to take his seat after the Catholic Emancipation Act was passed in 1829. Except for two short periods he remained an MP until 1841. Another of O'Connell's major campaigns was for the repeal of the union between Ireland and Great Britain. The Repeal Association, founded in 1840, attracted large numbers to its public meetings and inspired extensive agitation by the Catholic population to achieve its goal. The July 1841 General Election proved a major test for the Repealers and, although O'Connell lost the contest at Dublin City, he was actively supporting his son, Daniel, Jnr, who contested Carlow County. Thus the hard-fought Carlow election, which is discussed in several letters in this volume, was a major focus for the Repealers and their Conservative opponents. Although the Repealers were not successful in the 1841 election, O'Connell was the most prominent Irish nationalist of the period and his supporters referred to him as 'The Liberator' and 'The Great Emancipator'.

O'Connell, Daniel, Jnr (1816–97), the youngest of Daniel O'Connell's four sons he became the proprietor of the Phoenix Brewery in Dublin. Despite his father's support he failed to be elected as one of the Repeal movement's candidates for Carlow County in 1841 but later served as MP for Dundalk (1846–7) and Tralee (1853–63).

O'Neill, John, a civil assistant in the 5th Division, C District of the Irish Ordnance Survey, having joined on 7 October 1839. He worked in the office party with Tyndall, where his primary duties were registering, writing and correcting. He was transferred to the English Survey on 5 August 1842, where he worked in the 5th Division and was stationed in Liverpool.

O'Neill, Terence, the proprietor of the Swan Hotel, Leighlin Bridge. He had been elected a member of the Loyal National Repeal Association of Ireland in September 1840 and hosted the County Carlow meeting of the Repeal Association on 19 September 1841, the event that prompted Tyndall to write his poem 'The Leighlin "Orators" – or, the Late Repeal Meeting' (*CS*, 16 October 1841, p. [3]).

Parker, John, a civil assistant who worked in the 5th Division, C District of the Irish Ordnance Survey, where his main duties were plotting and writing boundary registers. He joined the Survey on 5 November 1838 and was transferred to the English Survey on 8 April 1842.

Payne, George, probably a relative of Maria Payne (and thus of the Tyndalls) and was possibly the George Payne, son of Thomas and Mary Payne, who was born on 29 May 1820 (Irish Church Records).

Payne, Joseph (Joe), a civil assistant in the 5th Division, C District of the Irish Ordinance Survey. He joined the Survey on 1 April 1839, the same day as Tyndall, and worked closely with him. He was initially employed in plotting but subsequently in drawing and correcting. On 8 April 1842 he was transferred to the English Survey. He is possibly the Joseph Payne, son of William and Jane Payne, who was born on 5 November 1821 in Dunleckny, County Carlow (Irish Church Records).

Payne, Maria, a cousin of Tyndall's who lived at Bagenalstown. The family connections have not been positively determined, but she was probably born either on 17 November 1818, the daughter of Thomas and Mary Payne, or on 4 April 1820, the daughter of Edward and Mary Stiles (née Tyndall) (Irish Church Records).

Payne, Mrs, probably Maria Payne's mother or mother-in-law. She seems to have lived in Bagenalstown but travelled to Youghal and Cork, possibly to be close to her sons, Joseph and Thomas Payne, who worked for the Ordnance Survey of Ireland.

Payne, Thomas (Tom), a civil assistant in the 5th Division, C District of the Irish Ordnance Survey, where he worked alongside Tyndall. He joined the Survey in May 1839 and his principal duties were registering and calculating. He left the Survey later in 1839, only to return in May 1840. He was transferred to the 3rd Division in April 1841 and then to the English Survey on 8 April 1842.

Peel, Robert (1788–1850), born in Bury, the son of a printed calico manufacturer, he became an MP in 1809 and soon entered the cabinet, in May 1812, as Irish Secretary, in which role he strongly supported the Protestant ascendancy. Peel returned to England in 1818, serving as Home Secretary from 1822 to 1827, and again from 1828 to 1830, overseeing a transformation of policies for law and order. He also enacted the emancipation of Catholics in 1829, a measure forced upon him by unrest in Ireland led by Daniel O'Connell, but which left Protestants feeling betrayed by their former champion. Following the upheavals of electoral reform, Peel became Prime Minister in December 1834, although his ministry lasted only until April of the following year, with his Irish Tithe Bill having just been introduced to Parliament. After six years in opposition to successive Whig governments, Peel became Prime Minister again in September 1841, moving quickly to tackle the economic recession and the huge deficits run up by the Whigs. In the budget of 1842, Peel announced the reintroduction of income tax at 7*d.* in the pound. Ireland, however, was excluded because of concerns over the collection of such a tax, and instead the Irish duty on spirits was raised. In the same budget, he also reduced the sliding scale in the duty on imported corn, and in 1846, partly prompted by the Irish famine, finally repealed the Corn Laws altogether. In Ireland, Peel's appointment in 1841 of an ultra-protestant Lord Lieutenant gave new momentum to O'Connell's campaign to repeal the Act of Union, peaking with the so-called 'monster' meetings of 1843, which led to the passing of the Irish Arms Act restricting the manufacture and importation of arms and ammunition. Peel, however, refused to enact further coercive legislation in Ireland, fearing that it would only increase support for O'Connell, and instead he introduced conciliatory measures including the creation of three new Irish universities. With both the repeal of the Corn Laws and his distaste for coercive measures in Ireland enraging reactionary Tories and Unionists within the Conservative Party, Peel resigned in June 1846, dying four years later after falling from his horse.

Ponsonby, Frederick George Brabazon (1815–95), from an influential Whig family with an extensive estate and grand house at Piltown, County Kilkenny, and other lands in Ireland. His father, John William Ponsonby, 3rd Earl of Bessborough, had represented County Kilkenny in Parliament from 1826 to 1832 and served as Lord Lieutenant of County Kilkenny (1838–47). Frederick was educated at Harrow and Trinity College, Cambridge, and then trained for the law, being called to the Bar at Lincoln's Inn in June 1840. He was adopted as the Liberal candidate to replace Nicholas Aylward Vigors, one of the Liberal MPs representing Carlow County, who had died on 26 October 1840. Widely viewed as an outsider who had spent most of his time in England and who was not sympathetic to the cause of Repeal, he lost the seat to Henry Bruen, the Conservative candidate. Famous also as a cricketer, Ponsonby later became an enthusiastic advocate of agricultural improvement and after succeeding to the title of the Earl of Bessborough in 1880 he headed the commission (generally known as the Bessborough Commission) that examined the rural economy of Ireland.

Read/Reid, Christopher M., a civil assistant who joined the Irish Ordnance Survey on 29 May 1839 and worked with Tyndall in the 5th Division, C District, principally as a plotter. Although he was transferred from Leighlin Bridge to his native Youghal, he resigned from the Survey on 20 October 1840.

Roberts, John George ('Sam Weller'), a civil assistant who joined the Irish Ordnance Survey on 4 September 1839 and worked in the 5th Division, C District, principally in the District Office. His duties included writing boundaries and registering, plotting, drawing, and examining ground work. He was dismissed from the Survey on 9 January 1841 for 'idleness' and was married to Esther Armstrong of Corkbeg, County Cork in the autumn of that year. Although he worked closely with Tyndall, Roberts was not a congenial colleague and it also appears that he soon abandoned his wife.

Ryan, Phil, not identified.

Ryan, Mr, a school friend of Tyndall's father who emigrated to America, gained significant wealth, became a magistrate and supported the Repeal Association. He is wrongly described as 'a member of Congress for the district of New York' (letter 0007), but may have been the John W. Ryan who was elected to the House of Representatives in Pennsylvania, *c.* 1838.

Shean/Sheean/Sheehan, Mr (possibly H. J.), succeeded Mr Butler as inspector of schools for Carlow and first assessed the pupils at John Conwill's National School at Ballinabranna in February 1841(letter 0048). Conwill challenged his views on language.

Shine, Charlotte, daughter of the Postmaster William Hunter Shine and his wife Eliza, of Bridge Street, Bandon, a love interest of Martin Cuddy.

Sinnett, James (1807–44), enlisted in the 13th Company of the Corps of Royal Sappers and Miners on 21 April 1826 and was promoted to the rank of Corporal on 1 March 1839. He worked in the 4th Division, C District of the Irish Ordnance Survey, under the command of Lieutenant Robert Fenwick, and was respected as a skilled draughtsman, Major Marcus Antonius Waters describing him as 'attentive and useful' and of 'very good' conduct (NAI OS/1/15). Before being transferred to England on 5 August 1842 he worked closely with Tyndall in Cork and Kinsale. He died in Liverpool in August 1844.

Smith, John (1792–1861), served as editor of the *Liverpool Mercury* from 1842, when he succeeded the founding editor Egerton Smith, until 1849. Smith, according to an obituary in the *Liverpool Mercury*, was a 'warm and earnest advocate of the principles of civil and religious liberty, and he was one of the first to denounce the tax on the food of the people [i.e. the Corn Laws]' ('Death of Mr. John Smith', 7 January 1861, p. 2). After the apparent refusal by a number of other newspaper editors, it was Smith who published Tyndall's pseudonymous letters to Robert Peel in the *Liverpool Mercury*.

Steuart/Stewart, William Richard (1798–1852), descended from an established family of Scottish ancestry that settled in Ireland *c.* 1719, having obtained estates in Counties Carlow and Meath. The family's wealth was further increased in the mid-eighteenth century by a marriage settlement that included 2,000 acres of land abutting the south of Leighlin Bridge, where William Steuart built Steuart's Lodge. His grandson, William Richard, later inherited the estate and in 1820 married Elizabeth Dawson Duckett, from another very prosperous Protestant land-owning family. He held the rank of Captain, served as a local magistrate and was appointed High Sherriff of Carlow in 1821. He owned many of the properties in Leighlin Bridge, including the house where the Tyndalls lived and both John Tyndall, Snr and Caleb Tyndall were at various times in the employment of Steuart. Tyndall and his father looked to the Steuarts for patronage and approached them for support when Tyndall was considering various forms of employment. Tyndall also designed a porch for Steuart's Lodge.

Steuart/Stewart (Mrs) Elizabeth Dawson (née Duckett) (1802–93), born into a very wealthy family with an estate of 12,000 acres and a gothic mansion (Duckett's Grove) situated several miles north-east of Carlow. She married William Richard Steuart on 18 August 1820 and they lived at Steuart's Lodge, Leighlin Bridge. She took a keen interest in Tyndall's career and was probably his most consistent and long-lived patron, her husband having died in 1852.

Styles, Edward, married Tyndall's aunt Mary (née Tyndall) in 1821. He served as clerk of the Leighlin-Bridge Charitable Loan Fund, but owing to financial irregularities he was sacked in 1841 (letter 0056).

Taylor, Alexander, a civil assistant in the 2nd Division, C District of the Irish Ordnance Survey, under the command of Lieutenant Charles Edmund Wilkinson. He joined the Survey in July 1838, where his main duties were surveying and plotting. He was transferred to the English Survey on 8 April 1842, where he worked in the 5th Division and was stationed in Liverpool.

Tidmarsh, Christina (Xina), sister of John Tidmarsh, and a potential love interest of Tyndall.

Tidmarsh, John Francis (Jack; Tid) (1824–1906), born in Cork and educated at Dr. Radcliffe's School before becoming an assistant draughtsman at an architectural office in Dublin. He joined the Irish Ordnance Survey on 11 May 1840, working in the 5th Division, C District under the command of Lieutenant George Wynne, where his principal duty was plotting, although he also undertook drawing and tracing for examination on the ground. He worked closely with Tyndall in Youghal, where he lodged with him and William Ginty, and again in Kinsale and Cork, where Tyndall enjoyed the hospitality of Tidmarsh's family. He was separated from Tyndall after being transferred to the English Survey on 20 May 1842, where he worked in the 5th Division and undertook field work in Lancashire, Yorkshire and Westmoreland. In the summer of 1843 he contemplated emigrating to the Cape of Good Hope to take up a job with relatives, but instead remained in England where, by 1845, he was employed, alongside Tyndall, making surveys for the West Riding Union Railway. At this time, he and Tyndall were once more lodging together in Halifax. In March 1849 Tidmarsh emigrated to Australia, settling in Adelaide, where he initially worked as a surveyor and valuator for the city's first building society and later prospected during the Australian gold rush in 1852. He established a successful candle and soap manufacturing company, although overwork impaired his health. In November 1912 Tidmarsh ended his own life by firing a revolver into his head ('Death of Mr. John Tidmarsh', *Register* (Adelaide), 12 November 1912, p. 6).

Tidmarsh, Mary, of Merchant's Quay, Cork, mother of John Tidmarsh.

Tidmarsh, Mr, of Merchant's Quay, Cork, father of John Tidmarsh.

Tidmarsh, May, sister of John Tidmarsh.

Treacy, James (b. 1801), a Leighlin Bridge maltster who was declared insolvent in 1841.

Tucker, Henry, possibly born in Kent, where his father Stephen Tucker was vicar of Borden from 1797 to his death in 1838. He enlisted in the Royal Engineers *c.* 1815, and was made 2nd Lieutenant on 1 March 1816, and then 1st Lieutenant on 9 April 1825. He joined the Ordnance Survey in 1824, and was in Ireland by at least 1832, when he was stationed in County Antrim. He was in Ardee, County Louth in early 1836, where he was in charge of the E District, and was promoted to 2nd Captain on 25 December 1837. By January 1838 Tucker was stationed in Dublin, and was in command of the A District, which was operating in Counties Kildare, Wicklow, parts of Carlow, Queen's County and Dublin. In January 1840 he was stationed at Waterford, where he remained throughout the year and into 1841. He was then transferred to the English Survey, where he superintended much of the surveying of the six northern counties, commanding the 2nd and 3rd Divisions, which he ran from the Divisional Office at Wakefield. Tucker was subsequently stationed at the Ordnance Survey headquarters in Southampton, where he supervised the preparation of special plans of Windsor for Prince Albert. He became Brevet Major on 11 November 1851, and Lieutenant-Colonel on 29 March 1853, and in the same year returned to Ireland where he was put in charge of the Survey headquarters at Mountjoy House, Phoenix Park, Dublin. In 1854 the superintendency of the entire Ordnance Survey of Great Britain fell vacant, and Tucker was widely regarded as the most appropriate candidate. However, his profound deafness, alluded to by William Ginty in letter 0185, debarred him from the post, which went instead to the less experienced but well-connected Major Henry James. Outraged at being passed over, Tucker unsuccessfully tried to have the Irish Survey made independent of the rest, and then 'left the Survey in disgust to die a few years later in the West Indies' (J. H. Andrews, *A Paper Landscape: The Ordnance Survey in Nineteenth-Century Ireland* (Oxford: Clarendon Press, 1975), p. 246). He presumably died between 1856, when he received his final promotion to Colonel, and 1858, after which he no longer appears in the annual *Army List.*

Turner, John, a Private in the 13th Company of the Royal Sappers and Miners. He worked in the 2nd Division, C District of the Irish Ordnance Survey, having joined in *c.* 1834. A carpenter by trade, his duties on the Survey included laying down points, drawing and writing boundary remarks. He was transferred to the

English Survey on 5 August 1842, where he worked in the 5th Division and was stationed in Liverpool and Hull.

Tyndall, Caleb (*c.* 1798–1879), one of Tyndall's uncles who lived in Leighlin Bridge. He appears to have had various jobs including maltster and farmer (1855), while from 1819 he was employed by William Steuart as the Weighmaster of the Butter Crane in Leighlin Bridge. During the 1841 election he achieved notoriety by shooting Mary McAssey while facing a pro-Repeal mob. At his trial he was cleared of malicious wounding. He subsequently served as High Constable for the Barony of Idrone West and Overseer of the Poor. For reasons which are far from clear Tyndall's father had an intense dislike of Caleb.

Tyndall, Emma (1817–1904), Tyndall's sister whose education he repeatedly encouraged. She visited her brother in Preston in August and September 1843. She does not appear to have married and later in life suffered from a mental illness which was manifested as an extreme form of religiosity.

Tyndall, Emma (Tyndall's cousin), see Young, Emma.

Tyndall, John, Snr (1792–1847), grew up in Leighlin Bridge and married Sarah McAssey in 1813. He probably served in the Royal Irish Constabulary at the police station in Nurney from 1828 to 1833. The family then moved to Castlebellingham, County Louth for the next three years before returning to Leighlin Bridge, where he worked as a boot and shoe maker. He may also have served as a land agent for William Steuart, one of the principal landlords in Leighlin Bridge and also the main patron of the Tyndall family. Although self-educated he was a keen reader and sought to expand his own education and encourage his children's. A staunch Orangeman, he was a member of the Brunswick Club which sought to advance the Protestant cause and oppose the social and political advance of Roman Catholics. He is described by Louisa Tyndall as 'a vigorous man, [with a] tall and powerful frame, [who] suffered much from sleeplessness'. Plagued by a 'delicate stomach', he was 'fastidious in various ways ... and was very particular about the habits of his children' (LT, 'Biography', vol. 1, pp. 4–5). His character is apparent from the many letters of his included in this volume.

Tyndall, Robinson (d. 1842), an uncle of Tyndall's who emigrated to America.

Tyndall, Sarah (née Macassey/Macasey/McAssey) (1793–1867), descended from a prosperous farming family with lands in Ballybrommell, County Carlow, that included in earlier generations Quakers by the name Malone. According to Louisa Tyndall, she 'must have been a brave, loving, devoted soul ... Tyndall used to say that he had rarely seen such gentleness combined with so much courage. She was of a type occasionally met with in the south of Ireland, tender poetical, naturally refined, and well educated for her class'. (LT, 'Biography', vol. 1, p. 5)

She was also remembered for having encouraged the education of her children and for reciting poetry to them.

Tyndall, William (b. 1804), an uncle of Tyndall's who emigrated to Philadelphia.

Walsh, James (Jem or Jim), a young Roman Catholic from Leighlin Bridge who served as Tyndall's father's apprentice, but was dismissed for having helped lead the pro-O'Connell agitation during which Caleb Tyndall's house was attacked (see letter 0073). He was later listed in *Slater's* under 'Boot and Shoe Makers'.

Waters, Marcus Antonius (1793–1868), born in Dublin he enlisted in the Royal Engineers in 1809, being made 2nd Lieutenant on 30 September of that year, and 1st Lieutenant on 1 May 1811. He served with distinction in the Peninsular War against Napoleon and fought at the battles of Quatre Bras and Waterloo, becoming 2nd Captain on 2 November 1816. He joined the Irish Ordnance Survey on 10 April 1825, taking command of the C District, and being made 1st Captain on 19 June 1830, and Major Brevet on 10 January 1837. By late 1841 the work of the C District was winding down, and Waters personally took command of the 1st, 3rd and 4th Divisions in November when the officers that had previously done so left the Survey. He also took charge, in April 1842, of the 2nd Division, and remained stationed at Cork with only one remaining officer, Lieutenant Charles Edmund Wilkinson, under him. Waters stayed in Cork until the C District had completely finished its work in August 1842, after which he went to the West Indies (1843–49) and Mauritius (1852–4). He retired on full pay of £600 in May 1855, having reached the rank of Major General Brevet.

Whittingham, Paul Bernard (1819–94), born in the French Pyrenees he joined the Royal Engineers in early 1837. Less than a year later he was made 2nd Lieutenant and achieved the rank of 1st Lieutenant on 25 July 1840. After joining the Ordnance Survey of Ireland in February 1839, he initially worked in B District under Captain Richard John Stotherd, and from February 1839 to July 1841 assumed command of the 5th Division, C District. It would appear that he was not an efficient commander since in 1841, following complaints about his work in C District, Stotherd requested a report on his earlier work in B District. Whittingham subsequently served in Nova Scotia, Canada, Gibraltar, Hong Kong and Jamaica and was retired in 1871 on full pay of £600, having reached the rank of Major General (Brevet Honorary).

Wright, Catherine, of Leighlin Bridge, the wife of Robert Wright and mother of William Wright.

Wright, Gustavus Hamilton (*c.* 1812–72), a civil assistant in the 5th Division, C District of the Irish Ordnance Survey, where his principal duties were surveying and examining ground work. He left the Survey sometime between August and October 1841.

Wright, Robert, of Leighlin Bridge, the husband of Catherine Wright and father of William Wright.

Wright, Sally Maria (née Neale/Neill, 1815–88), from a prosperous family in Roscrea, County Tipperary, she was the girlfriend of William Wright. They married on 7 February 1841 and emigrated to America in early March 1841.

Wright, William (1818/9–92), from Leighlin Bridge, a close childhood friend of Tyndall's, with whom he subsequently maintained a long-term correspondence. In early 1841 he was working in Roscrea, County Tipperary, and courting Sarah Maria Neale, whom he married on 7 February 1841 at St Nicholas Parish Within, Dublin (Irish Church Records). Possibly due to opposition from her family, the newly-wed couple rapidly emigrated to America, sailing from Liverpool in early March 1841. They settled in Cincinnati, Ohio, where William worked in the dry goods and grocery trades. He frequently urged Tyndall to join him in America.

Wynne, George (1804–90), born in Dublin he enlisted in the Royal Engineers in *c.* 1825, becoming 2nd Lieutenant on 26 September 1826, and 1st Lieutenant on 12 March 1831. He joined the Irish Ordnance Survey in September 1835, working in County Galway and at Tyrellspass in County Westmeath, before being stationed at Leighlin Bridge from April 1838. By this time, he was in charge of the 5th Division, C District, under the command of Major Marcus Antonius Waters. In April 1840 he was transferred to Youghal, before handing over command of the 5th Division to Lieutenant Paul Whittingham on 18 September of the same year. He requested to be taken off the Survey altogether, apparently because of his wife's illness. Wynne later served with distinction in Greece, China and the West Indies, rising to General by the time he retired in 1877. In a letter to *The Times* in 1890, Tyndall recorded: 'In 1839, I being fresh from school, Lieutenant Wynne became my chief. A few years afterwards he was sufficiently my friend to offer me the use of his purse for the prosecution of my studies in Germany. Our friendship, afterwards, was of the most intimate and cordial character; and notwithstanding my heterodoxy, and his Christian piety, it was never for a moment shaken. Let me briefly say that a spirit more erect and pure I have never known ... The earth's noblest ones are sometimes found among its noiseless ones. A conspicuous illustration of this truth is furnished by the life and character of General George Wynne' ('The Late General Wynne', 12 July 1890, p. 11).

Young, George (1813–60), a tenant farmer at Kilgraney and the son of James Young, the jailer at Carlow jail, and his wife Elizabeth. He married Tyndall's cousin Emma in November 1841.

Young, Emma (née Tyndall) (b. *c.* 1825), Tyndall's cousin and the daughter of his uncle Caleb. She met George Young when visiting her father in prison (see letter 0111) and married him on 22 November 1841 at Dunleckney (Irish Church Records).

APPENDIX: JOHN TYNDALL'S ROLE IN THE ORDNANCE SURVEY OF IRELAND

Compiled by Mark Tynan

This appendix provides a chronological account of the activities of John Tyndall during his employment by the Ordnance Survey of Ireland between April 1839 and August 1842, when he was transferred to the English Survey. As stated in the Introduction to this volume, the Irish Survey had been established under the direction of Thomas Frederick Colby of the Royal Engineers in 1824 with the aim of mapping the entire island to a scale of 6 inches to 1 mile, a level of detail never before attempted. For organizational and administrative purposes the Survey divided Ireland into five separate Districts, lettered A to E, each District being under the command of a senior Royal Engineer. The Districts were divided into a number of Divisions consisting of a Royal Engineer, members of the Corps of Royal Sappers and Miners and a number of civil assistants. The work of the Survey, which had begun in the north of the country, steadily progressed southwards, and by the late 1830s the northern and eastern areas allotted to D and E Districts had been completed. With the surveyors of the A District operating in south-eastern areas of Ireland, and those of the B District based in the west, the divisions of the C District were making their way through the Midlands, traversing the counties of Westmeath, King's County and Queens' County, before moving south into Kilkenny and Carlow.[1]

The records of C District of the Ordnance Survey of Ireland, held at the National Archives in Dublin, provide a valuable insight into Tyndall's work on the Survey. The commander of each District was required to send monthly progress reports to the Ordnance headquarters in Dublin, and these monthly returns provide information on each employee, including the nature of his duties and rate of pay. Many of the progress reports also contain the dates when an employee was hired by the Survey, information about the locations where he was working at any given time, together with limited commentary on the manner in which the work was progressing, and the date when he left the Survey's employment. The Ordnance Survey records relating to Tyndall are contained in the progress reports with shelfmarks NAI OS/1/16–19. Various correspondence registers, indicated by the prefix OS/2, provide additional information on individuals who worked with Tyndall and also events that are mentioned in his personal correspondence. These records portray Tyndall's time on the Survey from the perspective of the Survey's administration and thus complement the contemporary Tyndall letters that provide a more vivid and rounded insight into his activities and routine. As such, cross-references will be given to relevant letters in this volume.

Tyndall joined the Ordnance Survey of Ireland on 1 April 1839 at about the age of eighteen. He was assigned to the 5th Division of C District commanded by Marcus Antonius Waters, a Major with the Royal Engineers and a veteran of campaigns on the Iberian Peninsula and in France.[2] The 5th Division had been stationed at Tyndall's home town of Leighlin Bridge, County Carlow since the previous April. The chain of command in each Division was hierarchically structured (see Figure A.1 below). At its head was a commissioned officer from the Royal Engineers who delegated authority to various members of the Corps of Royal Sappers and Miners under his charge. These non-commissioned officers in turn were responsible for an increasing number of civil assistants who worked both in the office and in the field. While the number of civil assistants employed on the Survey had initially been small, their numbers and their importance to the project had gradually increased during the 1830s, and by the end of the decade the progress of the Survey was dependent on the civilian workers. When Tyndall joined the Survey as a civil assistant he was initially employed at a rate of 1*s.* 6*d.* per day, which was an unusually high figure as the majority of new assistants received only 1*s.* or less. The Survey records fail to explain why he was paid such a high rate, and it is likely that Tyndall's educational background, and the grounding in mathematics he had received under the tutelage of John Conwill, provided him with the relevant skills that were perhaps lacking in other new recruits.[3]

When Tyndall commenced his work at Leighlin Bridge, the 5th Division was under the command of George Wynne, a Dublin-born Lieutenant of the Royal Engineers.[4] Wynne's subordinates consisted of two distinct groups: a surveying party led by Corporal James Mulligan, which contained twenty-four civil assistants, and a stationary office party that interpreted the data provided by these surveyors and produced plans and maps. This office party comprised eighteen civil assistants and eight members of the Royal Sappers and Miners, with the most senior of these non-commissioned officers, Corporal James Davy, serving as office superintendent. Although the office was based in County Carlow, progress reports indicate that the work undertaken by the Division during the spring of 1839 was centred on parishes in the north Kilkenny area, and a surviving personal logbook, which documents Tyndall's daily activities during his first year on the Survey, shows that he was initially working on plans of the Castlecomer area.[5]

The Survey as a whole was in a state of transition during 1839. Ironically, this instability was probably a factor leading to Tyndall's employment since new employment opportunities had arisen because significant numbers of trained assistants resigned their positions on the Survey in order to take up better-paid work on the English Tithe Survey.[6] In May 1839 Lieutenant Wynne expressed his frustration at the tendency of Ordnance Survey employees to leave for more favourable terms, claiming that his Division was being used as a 'training ground' for the Tithe Survey, while he also lamented that 'not a civilian will remain with him from the moment he feels capable of drawing a plan, or laying out his work'.[7] While the departure of trained assistants was obviously a hindrance to the progress of the Survey, it undoubtedly presented an attractive employment opportunity for numerous young men, including Tyndall.

In the absence of personal correspondence during the time that he spent working with the Survey in County Carlow, the Ordnance Survey records and the aforementioned logbook provide the only sources of information on Tyndall's activities during this period. The first duties that Tyndall was required to undertake in April 1839 were 'printing and writing fair plans', although by the following month he was working as a draughtsman in a role that would become familiar to him during his time on the Survey.[8] He continued to draw plans throughout the summer of 1839 as the work carried out by the Division progressed south-west through the parishes of County Kilkenny and into County Tipperary. In August 1839 Tyndall received a substantial pay increase; his daily rate rose to 2*s.* 3*d.* It typically took a civil assistant over a year to gain a similar pay increase, and the speed with which Tyndall achieved this suggests that he was perceived to have performed his duties particularly well. However, the leaders of the Survey were becoming frustrated by continuing boundary disputes that hampered progress in County Tipperary, while an unseasonably wet summer had also caused significant problems for surveyors. Both Colby and Waters remarked that the progress made during 1839 had been significantly less than that which had been achieved during 1838.[9]

Despite these difficulties, by late 1839 Lieutenant Wynne's 5th Division was surveying as far south as Carrick-on-Suir near the border of counties Tipperary and Waterford. The month of December saw Tyndall rewarded with a further pay increase, taking his daily rate to 2*s.* 6*d.*, while his duty was now listed as correcting plans. Tyndall returned to drawing during the early months of 1840, and by the end of March his party had completed plans of the two southernmost Kilkenny baronies of Ida and Iverk, as well as part of the plans of the barony of Lower Ormond in north Tipperary. During the early months of 1840 preparations to relocate C District stations to County Cork were well under way. In March,

Major Waters moved his District headquarters from the town of Portarlington in Queen's County, while in the following month Lieutenant Wynne also relocated his 5th Division to Cork, settling in the port town of Youghal in the east of the county. Tyndall first arrived at Youghal on 3 May 1840, and three days later he composed the first of many letters to his father, in which he describes his early impressions of the town and comments on the different appearance of coastal settlements in Cork compared with those of the inland Carlow region.[10]

Now stationed at Youghal, the 5th Division surveyed the parishes around Clonmel in south Tipperary during the early summer of 1840 and Tyndall continued to work as a draughtsman with the office party. He also spent some time in the field and described to his father a brief assignment that he had undertaken towards the end of May.[11] Although the Survey records make no mention of his work in the field, the letter reveals that Tyndall was enthused by the prospect of gaining experience in field work while stationed at Youghal, while his father certainly approved of the work that his son was involved in, considering it ideal preparation for a future career as a civil engineer.[12] Tyndall subsequently received a further pay increase during the summer of 1840, and was now earning 2*s.* 8*d.* per day. As the monthly returns for June are missing the exact date that this pay increase was granted is unclear, although by July Tyndall and his friend Phillip Evans, whose pay had simultaneously increased to 2*s.* 9*d.* per day, were the highest paid civil assistants in the 5th Division.

The number of civil assistants working in the 5th Division increased significantly during the summer of 1840. By the beginning of August there were two separate surveying parties consisting of a total of forty-two surveyors, while thirty-six men were now working in the office at Youghal. The Division was surveying the eastern parishes of County Cork, including Youghal itself, Aghada, Inch, Ightermurragh and Clonpriest, while parishes such as Ardagh in the north of the county were also being surveyed. From July 1840 Tyndall's principal duty was listed as examining ground work, and in a letter to his father on 18 August he mentions that he and Evans were operating in what Tyndall describes as 'wild' rural regions along Cork's coastline.[13] The autumn of 1840, though, was a period of upheaval within the 5th Division. On 18 September Lieutenant Wynne left the Survey and was replaced as the Division's commanding officer by Lieutenant Paul Bernard Whittingham, who had previously been stationed with a Division in B District in County Galway. Continuing health problems also meant that Corporal Davy was transferred to England in October, and he was replaced as office superintendent by Corporal James Mulligan.

Tyndall was engaged in laying down work, a duty which involved the transfer of measurements taken in the field to paper, for a daily rate of 2*s.* 8*d.* during the latter months of 1840. The beginning of 1841 brought further upheaval, as Major Waters gave an order for a number of civil assistants to be discharged from the 5th Division. In a letter to his father, Tyndall reported that Waters had warned that only the hardest workers would be retained and sent to the Survey of England after the Irish Survey was completed.[14] This suggests that the Division had not been functioning as effectively as the Major had expected. The letters that Tyndall composed during his time in Youghal indicate that he was generally happy in his work, although a further letter to his father of 23 January 1841 reveals a level of disappointment and disillusionment with the decision to replace Corporal Mulligan as office superintendent.[15] Tyndall had evidently grown close to Mulligan since Corporal Davy's departure, and he proceeded to organize a petition among his fellow civil assistants with a view to having Mulligan retained in his office role. On this occasion he had overstepped the mark and Lieutenant Whittingham curtly informed the office party that it was he who was in charge and that such insolence would not be tolerated. Mulligan was duly returned to the Division's surveying party, and Corporal Alexander Calder was drafted in from the 3rd Division as its work neared completion, taking over from Mulligan as office superintendent.

As the work progressed and its manpower was reduced, a restructuring of the C District became necessary. The 4th Division was incorporated into the 3rd Division under the command of Lieutenant Robert Fenwick during February 1841, which meant that the 5th Division to which Tyndall belonged now became the District's 4th Division. The work undertaken by the new 4th Division progressed westwards through County Cork during the spring of 1841, and took in the parishes of Cloyne, Kilcredan, Bohillane, Killeagh, Trabolgan and Midleton. The Survey records list Tyndall's principal duty from February 1841 as once again drawing, although he reveals in a letter to his sister Emma that among Calder's first instructions was a request for him to leave the office and instead to engage in fieldwork.[16] Tyndall was dispatched to trace and examine ground work in the parish of Killeagh, which lies around five miles to the west of Youghal.

On Corporal Mulligan's recommendation, Tyndall's daily rate of pay was increased to 2*s.* 10*d.* from the beginning of March 1841, and although he expressed his disappointment that he had received a rise of just 2*d.*, he was by now the highest paid civil assistant in his Division.[17] Although the Ordnance returns for March have not survived, Tyndall had returned to the office and was drawing plans, a duty he had not performed during the previous year.[18] He was now working in a room with Private Joseph Collins, with whom he appears to have enjoyed a close working relationship. The Division had been significantly reduced in size during the early months of 1841, and by April it included just eight Sappers and Miners and forty civil assistants. Many of the men that Tyndall had worked with during the previous two years had by now left the Survey, and in a letter dated 23 May, he lamented that his current Division was unrecognisable from the one he had joined in Leighlin Bridge.[19] This letter also indicates that the Division was required to survey a substantial extra acreage before the completion of work in Cork, and this delayed the departure of Tyndall and his colleagues to the Survey of England.

Tyndall continued to work as a draughtsman during the summer of 1841 as his Division proceeded further west to parishes around Cork city, and on 9 June he was informed by Corporal Calder that the office would be relocated to Kinsale within two weeks. The party was not long in Kinsale when Lieutenant Whittingham departed the Survey on 11 July 1841, the Division subsequently coming under the command of Lieutenant Fenwick. Despite the change of station, the Division continued to work on the same parishes as it had done in Youghal, most of which lay to the south of Cork city. Tyndall's duties during the month of July are listed in the Survey records as writing and drawing. While in Kinsale he enlisted in a local drawing school with the aim of improving his skills and thus his future career prospects.[20]

Of the last five months of 1841 only November's returns survive. Tyndall's letters show that he remained busy in his role as draughtsman in the office in Kinsale, despite the fact that the surveyors attached to his Division were in the process of completing their work. Writing to his father on 15 August, he suggested that the work undertaken in the office was behind schedule and that he himself was having to work up to fifteen hours per day.[21] By mid-September the surveyors had been sent to England to begin work on the Survey there. Corporal Calder also departed (much to the disappointment of Tyndall who considered him a 'fair' man) and was replaced by Sergeant James Carey.[22] The departure of Lieutenant Fenwick also necessitated the appointment of a new divisional commander, and Major Waters personally took charge of the 4th Division during the autumn of 1841. However, Waters was an absentee commander, remaining at the District headquarters in Cork while the Division was stationed at Kinsale.[23] The Survey records for November 1841 show Tyndall in the 3rd Division, and it is clear that a redistribution of the existing manpower had occurred within the District after the autumn departures. Very little had changed in terms of Tyndall's work and he continued to draw in Kinsale at a daily rate of 2*s*. 10*d*. There appears to have been a prevailing sense of uncertainty among Survey employees about how long they would remain in Ireland. He nevertheless remained in Kinsale throughout December. As Major Waters had set a deadline for the end of March, by which time the remaining civil assistants would be available for transfer to England, Tyndall's request for time off over the Christmas period was refused.[24]

Tyndall was re-stationed to Cork during the second week of 1842.[25] Although his duty for January 1842 is listed as drawing in the Survey records, a contemporary letter indicates that Tyndall was engaged in colouring plans of Cork city during January, February and March.[26] Waters's deadline of the end of March came and went, and during April a substantial number of civil assistants departed Cork for the Survey of England. The Survey records indicate that approximately half of the assistants that remained in the C District were transferred during this period, while a letter dated 8 April 1842 verifies this as Tyndall estimated that approximately 150 of his colleagues had departed.[27] That same month Waters took command of the remaining four Divisions within C District, with Lieutenant Charles Edmund Wilkinson, his last remaining officer, having left his post with the 2nd Division in Bandon. Tyndall's duty was listed as 'completing fair plans' in April, and during the same month he was finally granted leave to return home.[28] He left the Survey at the beginning of May and did not return until 10 June. While he was away a number of his close friends, including Evans and William Ginty, left for England. However, Tyndall appears to have approached the uncertainty surrounding life on the Survey rather pragmatically, as he informed his father that he had 'learned to treat such things with the most sublime indifference'.[29]

On his return to Cork, Tyndall was among a group of just ten assistants and two non-commissioned officers, including the latest superintendent, Corporal James Sinnett, remaining in the 3rd Division. The men worked on the parishes of Cork city itself during the summer of 1842, and Tyndall's duty involved 'Drawing the 50 feet Plans of the Forts and Barrack Grounds along the Coast and ... completing the Drawing of the 6 Inch and 5 feet Plans of Cork City and Town of Kinsale'.[30] A deadline of 28 July was fixed for the departure of the remaining assistants from C District, although Tyndall informed his father that this had been postponed due to the 'unfinished state of the work'.[31] This postponement was only brief, as Tyndall was finally transferred to the Survey of England on 5 August 1842. Having spent three years and four months on the Irish Survey, he left Ireland for a new station in the north of England with a number of other civil assistants. After their departure, only a handful of employees remained with Major Waters in Cork. Waters intended to pass all of the outstanding plans to the Ordnance head office in Dublin by the end of September, except for the plans of Cork city, which he proposed keeping until a later date for his own personal inspection. The Survey of Ireland was finally completed in 1846.

Although the archive of the Ordnance Survey of Ireland provides information for a detailed narrative of Tyndall's career and activities while working on the Survey, albeit from an administrative perspective, it is not possible to construct a similar account of his activities during his employment by the English Ordnance Survey, where he worked from early August 1842 until his dismissal on 7 November 1843. The monthly progress reports for the 1840s, together with many other records, were destroyed when the English Survey's headquarters in Southampton was bombed during the Second World War.

Notes

1. Introductory Notes for Ordnance Survey Parish Lists (NAI OS/105).
2. NA WO25/3913/21.
3. N. McMillan and M. Nevin, 'Tyndall of Leighlin', in W. H. Brock, N. D. McMillan and R. C. Mollan (eds), *John Tyndall: Essays on a Natural Philosopher* (Dublin: Royal Dublin Society, 1981), pp. 15–48, on p. 24.
4. NA WO25/3913/21.
5. RI MS JT/3/36.
6. Colonel T. Colby to Inspector General of Fortifications, January 1840 (NAI OS/1/17).
7. Correspondence Registers and Indexes, 1824–52, vol. 3, no. 8556, 28 May 1839, p. 121 (NAI OS/2/16).
8. Monthly Report, C District, April 1839 (NAI OS/1/16).
9. Return of Parishes to be Commenced, C District, 1 October 1839 (NAI OS/1/16); Colby to Inspector General of Fortifications, January 1840 (NAI OS/1/17).
10. Letter 0001.
11. Letter 0004.
12. Letter 0007.
13. Letter 0008.
14. Letter 0032.
15. Letter 0036.
16. Letter 0043.
17. Letter 0047.
18. Letter 0053.
19. Letter 0062.
20. Letter 0093.
21. Letter 0085.
22. Ibid.; Letter 0112.
23. Letter 0113.
24. Letter 0116.
25. Letter 0118.
26. Ibid.
27. Letter 0138.
28. Monthly Report, C District, April 1842 (NAI OS/1/19).
29. Letter 0148.
30. Monthly Report, C District, June 1842 (NAI OS/1/19).
31. Letter 0156.

Figure A.1: John Tyndall's hierarchy of command in C District of the Ordnance Survey of Ireland, April 1839 – August 1842.

Colonel Thomas Frederick COLBY, R.E., Director of the Ordnance Survey of Ireland
C District – Commanded by **Major Marcus Antonius WATERS**, R.E.
5th Division* – Commanded by **Lieutenant George WYNNE**, R.E. (– Sept. 1840)
Lieutenant Paul Bernard WHITTINGHAM, R.E (September 1840 – July 1841)
Lieutenant Robert FENWICK, R.E. (July 1841 – September 1841)
Major Marcus Antonius WATERS, R.E. (September 1841 – August 1842) †
Office Party – Led by **Corporal James DAVY**, R.S. & M. (- October 1840)
Corporal James MULLIGAN, R.S. & M. (October 1840 – January 1841)
Corporal Alexander CALDER, R.S & M. (January 1841 – September 1841)
Sergeant James CAREY, R.S & M. (September 1841 – May 1842)
Corporal James SINNETT, R.S & M. (May 1842 – August 1842)
Members include – Civil Assistants – **John TYNDALL** (April 1839 – August 1842)
Phillip EVANS (July 1838 – May 1842)
William GINTY (February 1836 – c. April 1839, October 1840 –May 1842)
John TIDMARSH (May 1840 – May 1842)
R.S. & M. – **Private Joseph COLLINS** (– May 1842)
Private Thomas M. FOY (February 1834 – June 1842)

* John Tyndall is listed as being in the 5th Division from April 1839. This became the 4th Division from February 1841, and Tyndall is listed as being in the 3rd Division from November 1841.

† A published source suggests that Lieutenant Andrew Beatty was to take command of the Division after the departure of Lieutenant Fenwick in September 1841 (*Cork Examiner*, 13 October 1841, p. [1]). As the Survey records are incomplete for 1841 it is unclear whether this actually occurred. Major Waters had taken command of the Division by November 1841.

Abbreviations: R.E. – Royal Engineers, R.S. & M. – Royal Sappers and Miners.

INDEX

The index refers to the letters using their serial numbers. The letter 'n' denotes a footnote reference and appears if an item is not otherwise mentioned in the letter to which the footnote pertains. A number in bold signifies a letter to or from a correspondent.